URBAN PUBLIC TRANSPORTATION SYSTEMS

Implementing Efficient Urban Transit Systems and Enhancing Transit Usage

PROCEEDINGS OF THE FIRST INTERNATIONAL CONFERENCE

March 21-25, 1999
Miami, Florida, USA

SPONSORED BY
Committee on Public Transport, Urban Transportation Division, ASCE

IN COOPERATION WITH
Advanced Transit Association
American Public Transit Association
Center for Urban Transportation Research, University of South Florida
Conference on Minority Transportation Officials
Export Import Bank of the U.S.
Federal Transit Administration, U.S. Department of Transportation
Florida Department of Transportation
Florida Transit Association
Miami-Dade Branch of ASCE
Miami-Dade Transit Agency
South Florida Section of ASCE
South Florida Women's Transportation Seminar
Transportation Research Board
The World Bank

EDITED BY
Murthy V.A. Bondada, Ph.D, F.ASCE

1801 ALEXANDER BELL DRIVE
RESTON, VIRGINIA 20191–4400

Abstract: Papers presented at the ASCE's First International Conference on Urban Public Transportation Systems held in Miami, Florida, during March 21-25, 1999, are included in this proceedings. Transportation professionals from 25 large international cities presented papers on 1) socioeconomic characteristics, land use, and travel patterns; 2) existing urban transportation system, its operation and use by the people; and 3) a topic of their choice pertaining to their city's public transit system. In addition to formal paper presentations on each city, proceedings also include the summaries of urban transportation experts who participated in three panel discussions: future of public transportation, innovative financing in urban transportation, and innovative techniques and transit technologies.

Library of Congress Cataloging-in-Publication Data

Urban public transportation systems : implementing efficient urban transit systems and enhancing transit usage : proceedings of the First International Conference : March 21-25, 1999, Miami, Florida, USA / sponsored by Committee on Public Transport, Urban Transportation Division, ASCE, in cooperation with Advanced Transit Association ...[et al.] ; edited by Murthy V.A. Bondada.

p. cm.

Includes bibliographical references and index.

ISBN 0-7844-0498-4

1. Urban transportation—Congresses. 2. Local transit—Congresses. I. Bondada, Murthy V.A. II. American Society of Civil Engineers. Public Transport Committee. III. Advanced Transit Association (Washington, D.C.) IV. International Conference on Urban Public Transportation Systems (1st : 1999 : Miami, Fla.)

HE305 .U684 2000

388.4--dc21 00-038953

Library of Congress Catalog Card No: 00-038953
ISBN 0-7844-0498-4
Manufactured in the United States of America.

Foreword

All countries are facing problems associated with increased automobile travel. Developing countries are finding that their infrastructure is outdated and insufficient due to an increase in automobile travel and that their foreign exchange reserves are being depleted due to importing oil at exorbitant prices. Developed countries are facing increased congestion, pollution, and unchecked auto travel created due to inefficient urban sprawl. Engineers, planners and policy makers in all countries are increasingly trying to propagate the advantages of public transportation as a solution to the problems caused by the increased auto travel. However, when it comes to implementing public transportation facilities, all countries face obstacles.

In order to provide a platform for an international exchange of information in the area of urban public transportation, the American Society of Civil Engineers (ASCE) has organized the ***First International Conference on Urban Public Transportation Systems*** in Miami, Florida, U.S.A. during March 21-25, 1999. The theme of the conference was *Implementing Efficient Urban Transit Systems and Enhancing Transit Usage.* This is the Proceedings of the Conference.

At the conference, transportation professionals from 25 large international cities presented their cities' urban public transportation systems. Each city was provided with one session consisting of three paper presentations covering: 1) socio-economic characteristics, land use and travel patterns of the city; 2) existing urban transportation system in the city and its operations and use by the people; and 3) a topic pertaining to the city's public transit system, chosen by the speaker. In addition to the presentations covering international cities, renowned members in urban transportation field spoke in three panels, encompassing the future of public transportation, innovative financing in urban transportation, and innovative techniques and transit technologies.

The conference provided an excellent forum for the exchange of ideas and innovations in the area of urban public transportation systems. This book includes ----- papers presented in the 25 city sessions and summaries of speeches made at the three panel discussions. The papers in the book are grouped by city and the cities are included in alphabetical order followed by summaries of presentations at panel discussions.

The success of this unique conference was due to the contributions of several people and cooperating organizations listed on the second cover. Special thanks are owed to the session coordinators of the 25 city sessions listed under "Acknowledgements." These session coordinators have put in unbelievable time and effort in assembling right speakers for the three topics in their sessions, obtaining papers and following the paper review process to include the final papers herein, and chairing their sessions at the conference. Without their cooperation and the power of electronic mail (email), it would have been impossible to coordinate 25 cities scattered in sixteen countries and four continents (nine from Asia, seven from Europe, seven from North America, and two from South America).

We must also remember several organizations, which contributed to the conference. Daniel Mann Johnson & Mendenhall, Kittleson & Associates, and Parsons Brinckerhoff Quade and Douglas have contributed cash for the conference. Mehta & Associates, Inc. has provided printed conference portfolios as part of registration material. Miami's Florida International University and Miami Chapter of the Conference of Minority Transportation Officials have jointly sponsored a luncheon at the conference. The Federal Transit Administration has organized a workshop on its International Mass Transit Program. Special thanks are extended to Miami-Dade Transit Agency for organizing three field trips: Intermodal Tour, Rail Maintenance Facility Tour, and Metrobus Central Tour.

The editor would like to express his gratitude to Mehta & Associates, Inc. for the facilities and secretarial help in editing papers of this conference proceedings, which took almost one year after the conference. Finally, the editor would like to acknowledge all the authors and their secretaries and other contributors who labored to produce these papers. This Conference Proceedings volume is the fruit of their effort and if there is any bitterness in the fruit, it is my fault.

Murthy V.A. Bondada, Ph.D., F.ASCE
Winter Park, Florida
USA

March 10, 2000

Acknowledgements

Conference Steering Committee

Dr. Murthy Bondada	Conference Chairman
Walter Kulyk	Program Chairman
Dr. Ramakrishna Tadi	Publicity and Exhibits
David R. Fialkoff	Local Arrangements
Jim Nadaska	ASCE Section Liaison

ASCE Public Transport Committee

Dr. Murthy Bondada	Chairman
Walter Kulyk	Vice Chairman
Dr. Ramakrishna Tadi	Secretary
Dr. Fred Coleman	
Raimundo Dovalina	
Dr. Foad Farid	
David R. Fialkoff	
David Leverenz	
Robert Stout	
Melanie Tilgner	
William Wilkerson	
Freank Zeinali	

ASCE Urban Transportation Executive Committee

Dr. A.E. Radwan	Chairman
Jerrold Kaplan	Vice-Chairman
Dr. W. Jeffrey Davis	Secretary
Dr. John Stone	Past-Chairman
Thomas Wholley	

Coordinators for City Sessions

Amsterdam -	Melanie Tilgner, USA	Beijing -	Eva Lerner Lam, USA
Budapest -	Andrew Bata, USA	Chennai -	Drs. S.B. Pattnaik
Chicago -	Walter Kulyk, USA	(Madras)	and R. Sivanandan, India
Curitiba -	Kenneth Kruckemeyer, USA	Helsinki -	Dr. Matti Pursula, Finland
Hong Kong -	Eva Lerner Lam, USA	London -	Dr. Paul Truelove, UK
Los Angeles -	Dr. Narasimha Murthy, USA	Mexico City -	Raimundo Dovalina, USA
Miami -	David Fialkoff, USA	New York -	Atefeh Riazi, USA
Osaka -	Dr. Yasunori Iida, Japan	Oslo -	Thor Haatveit, Norway
Paris -	Dr. Jean-Claude Ziv, France	Pusan -	Dr. Yong Eun Shin, S. Korea
San Juan -	Walter Kulyk, USA	Sao Paulo -	Mario Eduardo Garcia, Brazil
Seoul -	Dr. Jongho Rhee, S. Korea	Shanghai -	Eva Lerner Lam, USA
Singapore -	Mohinder Singh, Singapore	Tokyo -	Dr. Shigeru Morichi, Japan
Washington, DC-	Walter Kulyk, USA		

Conference Attendees

Botond Aba
Jose Abreu
Marcel Acosta
Mohamed Ahmed
Lorenzo Alexander
John Allen
Danny Alvarez
Nancy Anderson
Shereen Andrasek
Thomas Andrews
Audra Andrews-Wilson
Stephen Andrle
Pedro Andueza
Guillermo Anido
Thamizh Arasan
Celso Azevedo
Sadiq Baksh
Andrew Bata
Maria Batista
Peter Benjamin
Carlos Berlowitz
Joseph Berthet
Ladi Biezus
Richard Bishop
Felix Blanco
Mike Bolton
Murthy Bondada
Carlos Bonzon
Julio Boucle
Thomas Bradshaw
Lorraine Brown
Robert Brownstein
John Burrie
Susan Carlson
Robert Casey
Robert Cashin
Jorge Castillo
Carlos Ceneviva
Doris Chang
Wayne Chewing
Chi Gook Choi
Chik Cheong Choi
Francisco Christovam
John Claffey
Stuart Cole
John Collura
Benjamin Colucci
Edward Coven
Marty Covert
Rita Daguillard
Alan Danaher
James Davis
Jose DeAlmagro
James De La Loza
Steven Diaz
Maurice Dickerson
Ray Dovalina, Jr.
Thomas Dubail
Tracy Dunleavy
Armand Durrieu
Robert Dyck
Patricia Emard
Elyrosa Estevez
Wilson Fernandez
David Fialkoff
Daniel Fils-Aime, Sr.
Daniel Fils-Amie, Jr.
Eugenie Fils-Aime
Geoffrey Fosbrook
Claudio Frederico
Jack Furney
Jeff Gallagher
Mario Garcia
Gilbert Gardner
Kevin Gardner
King Gee
Othan Gilbert
Lucille Gomez
Antonio Gonzalez
Benny Gonzalez
Jose Gonzalez
Oscar Gonzalez
Alan Gray
Robert Gregg
Daniel Griffin
Thor Haaveit
Peter Haliburton
Yasumasa Hayashi
Michael Hester
Michael Hewitt
Garth Hinckle
Marvin Hinton
Hideo Hirasawa
Eric Ho
Mayer Horn
William Howell
Huy Huynh
Keeyeon Hwang
Hitoshi Ieda
Michael James
Zhao Jie
Angela Johnson
Sylvian Jolibois
Joel Jolinksi
Kenneth Jones
Ferenc Joo
Tore Kaass
Camille Kamga
Jack Kanarek
Hitoshi Kawata
Shinya Kikuchi
Ronald Kirby
Juergen Klemann
Gregory Knowles
Anthony Kouneski
Thomas Krieger
Kenneth Kruckemeyer
Walter Kulyk
Leong Kwok Weng
Jennifer Ladler
Peter Lai
Stephen Lam
Tore Langaard
Suzanne LaPlant
Jim Larkins
Richard Lear
Eva Lerner-Lam
Xiaosen Li
Lina Lim

Lucia Lindsey
Gordon Linton
Shit-Cham Lo
Diana Lopez
Jaime Lopez
Jian Lu
Kenneth Luebeck
Felipe Luyanda
Pedro Luiz Machado
Schone Malliet
Francis Mannion
James Martin
Kristin Martin
Mark Massman
Akira Matsushita
Christopher McCarthy
Fitz McLymont
Vipin Mehta
Lydia Mercado
Jose-Luis Mesa
William Millar
Kouroche Mohandes
Ja'nos Monigl
Woodrow Moore
Minnie Moreno
Shigeru Morichi
Philbert Morris
Theo Muller
Narasimha Murthy
Kyaw Myint
Anthony Nadaskay
Alassane-Balle Ndiaye
Lucas Nieri
Joachim Niklas
Kirsti Nost
Mario Nuevo
Robert Olmstead
William Orr
Robert Pearsall
Joaquin Perez
Karen Pinell
Stephanie Pinson
Richard Podolske
Richard Polo
Karma Prieto
Allen Proper
Jorge Pubillones
Matti Pursula
Harry Rackard
Benjamin Redd
Harry Reed
Theodore Renrick
Jongho Rhee
Atefeh Riazi
Tony Ridley
Antonio Rocha
Rene Rodriguez
Melissa Rolle-Scott
Alan Rumsey
Eugene Ryan
Paul Ryus
Micahel Sabatini
Richard Sanford
Jaime Santamaria-Serrano
Kengo Sato
Donald Schneck
David Schonbrunn
Fong Seck Kong
Mireille Segretain-Maurel
Sanjeev Shah
Peter Shaw
Fred Shields
Yong Shin
Victor Simuoli
Mohinder Singh
Craig Sivley
Christine Sizemore
Harriet Smith
Justin Smith
Gary Spivack
William Sproule
Jeffrey Squires
Charles Stark
Larry Stueck
Ramakrishna Tadi
Christopher Taylor
Terrence Taylor
Andre Testa
Edward Thomas
Margaret Thomas
Wayne Thompson
Melanie Tilgner
Bonnie Todd
Ole Torpp
Paul Trelove
Arthur Truman III
Vinod Tuli
Ike Ubaka
Toshimitsu Uebayashi
Eduardo Vasconcellos
Theresa Vassell
Seppo Vepsalainen
Alejandro Villegas-Lopez
Henk Waling
Jun Wang
Joel Washington
Claes Westberg
Arlene Willis
Jon Willis
Shelia Winitzer
Kok Wai Wong
Alan Wulkan
Carolyn Wylder
George Wynne
Jie Xiu
Takashi Yajima
Hongwei Yang
Hang Mook Yoon
Seongsoon Yun
John Zegeer
William X. Zhang
Wenshi Zhao
Samuel Zimmerman
Jean-Claude Ziv

Contents

Chicago

Curitiba

Helsinki

Hong Kong

London

SOCIO-ECONOMIC LAND USE AND TRAVEL PATTERNS IN AMSTERDAM

Author: H. Waling, Gemeentevervoerbedrijf Amsterdam

1. Introduction

Amsterdam, the capital of Holland has gone through major changes during the seven hundred years the city exists. This paper will deal with the socio-economic characteristics of the city and their structure.

2. Socio-economic elements

The number of inhabitants in Amsterdam is influenced by various developments such as the move to the suburbs, foreign immigration on the growth of prosperity. Important factors for the economic and social structure of the city are the average number of occupants in residence and the age of the inhabitants.

Figure 1
The development of the number of city inhabitants (inwoners), the number of households (woningen) and the average number of occupants per household (woningbezetting) between 1950 en 2005.

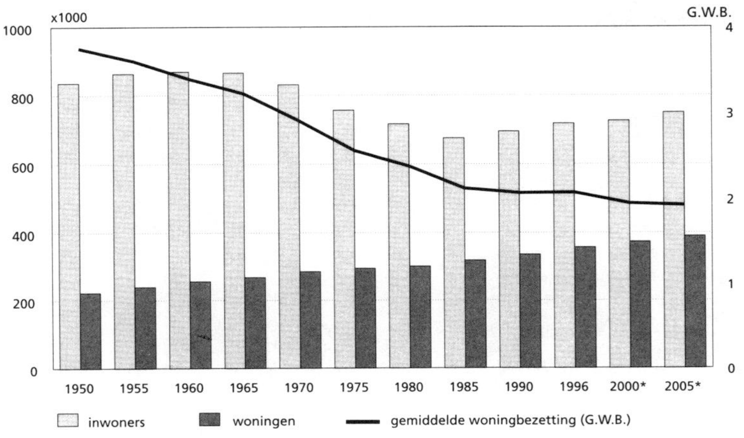

The strong decline in the number of people occupying a dwelling has helped determine that, despite a serious increase in residential building, the population of the city will hardly grow. Another effect of this is the shift in the concentration of inhabitants from the older existing, inner-city areas to the newly built neighborhoods on the edge of, and outside, the city.

An important characteristic of those who locate in Amsterdam is age. This is clearly dominated by young adults; about half are between 18 and 30. Migration out of the city is quite stabile. Before 1985, it was mostly families who left the city. Since then, a growing proportion have been single people.

Births and deaths, but more particularly immigration and social developments, have strongly altered the size of certain age groups.

Twenty-four to thirty-four year olds are a sizable population group in Amsterdam. Contrary to the aging population in the rest of the Netherlands, there is no such phenomena taking place in Amsterdam. In the last sixteen years, the oversixties have diminished by 20%.

This development, in which one comes to live in the city to study and work and leaves it for a number of years for a family-life, concluded with returning after this period has much to do with the quality and the size of housing and the quality of the public space (playing area for children).

Types of households in Amsterdam

household type	1988 abs.	1988 %	1992 abs.	1992 %	2000* abs.	2000* %	2005* abs.	2005* %	difference 1988 -1992 %	difference 1992 -2005 %
married, no children	55.158	15.0	53.768	13.9	53.096	12.9	54.751	12.8	-2.5	1.8
cohabiting, no children	32.068	8.7	35.233	9.1	39.719	9.7	41.515	9.7	9.9	17.8
married, with children	54.289	14.8	51.616	13.4	51.277	12.5	50.066	11.7	-4.9	-3.0
cohabiting, with children	5.840	1.6	7.509	1.9	10.645	2.6	11.857	2.8	28.6	57.9
single parent family	29.454	8.0	30.959	8.0	34.768	8.5	35.835	8.4	5.1	15.7
single person household	187.658	51.0	203.766	52.7	217.902	53.0	228.627	53.6	8.6	12.2
group	3.150	0.9	3.593	0.9	3.868	0.9	3.937	0.9	14.1	9.6
total	**367.617**	**100.0**	**386.444**	**100.0**	**411.275**	**100.0**	**426.588**	**100.0**	**5.1**	**10.4**

* Based on draft population projection of households

A large amount of the available housing consist of small units. Almost half are less than fifty square meters. Less than 10% is more than eighty square meters. About half of all the residences in Amsterdam are privately owned; and this number is growing.

Like any other city Amsterdam too is a destination for many immigrants from other parts of the country and from abroad. Amsterdam offers them a number of advantages such as cheap housing and varied employment.

Amsterdam gives much attention to new housing in and close to the city in qualities that better meet demand. This has its influence on travel patterns.

Economy:
The economic position of Amsterdam in the European economy is satisfying although employment figures show that job growth is slower than the rest of the region.

Table
Job opportunities in Amsterdam (1995)
(Economic Affairs, Amsterdam)

agriculture and industry	33.000
construction industry	14.000
trade	184.000
services	113.000
total	344.000

The sectors with high added value for the Amsterdam economy are oriented toward the regional, national and international markets. Commercial services, trade and distribution, tourism and industry are the most important reservoirs of affluence for the city. Each of these has a good chance to grow.

Policy of the city council is concentrating as much employment as possible in and close to the city. That reduces the travel distance of the inhabitants. Nevertheless it is more lucrative for companies to start their businesses in the region. That effect on the travel pattern is negative. The number of trips tot destinations outside the city grows. And, because the quality of public transport in the region is less attractive, the car is used more often.

The working population of Amsterdam did increase in the past years. On the one hand, caused by a growth in employment, and on the other, due to an increase in the total population.

The average income in Amsterdam grows at equal speed as the national increase in welfare. The effect on travel patterns is two-fold. The ownership of cars and car-use increase and the quality demands on public transport increase. Public transport fails often on speed and reliability.

Table
Automobile ownership in Dutch households (%)

	1985	1986
1 automobile per household	59.8	60.4
2 automobiles p. household	9.9	14.3
3 and more	1.1	1.1

In the center of Amsterdam, there is somewhat less automobile ownership than on the edge of the city.

2. Land Use Elements

Location policy
The growth of mobility had led to burdening the environment and traffic congestion. For these reasons, the objectives of urban planning are to control this growth, especially commuter traffic. Measures have been taken so that sites of high density employment and high frequency attendance activity such as offices, hospitals, schools are established at locations with good public transport.

Offices
The market for offices in Amsterdam has been tremendous growth whose end is not yet in sight. Most of the demand for space is focused on the south side of the city. The city council would like to see this growth spread more evenly throughout the city.

Companies are taking up more and more space in the city. Even in periods of relatively high unemployment, there still seems to be an increased demand for commercial/industrial space. There is the intention to stimulate the small-scale craft and industrial sector as a source for new jobs in the city center and the older surrounding neighborhoods.

Map

Concentrations of housing (red), employment (green) and recreation (light-green).

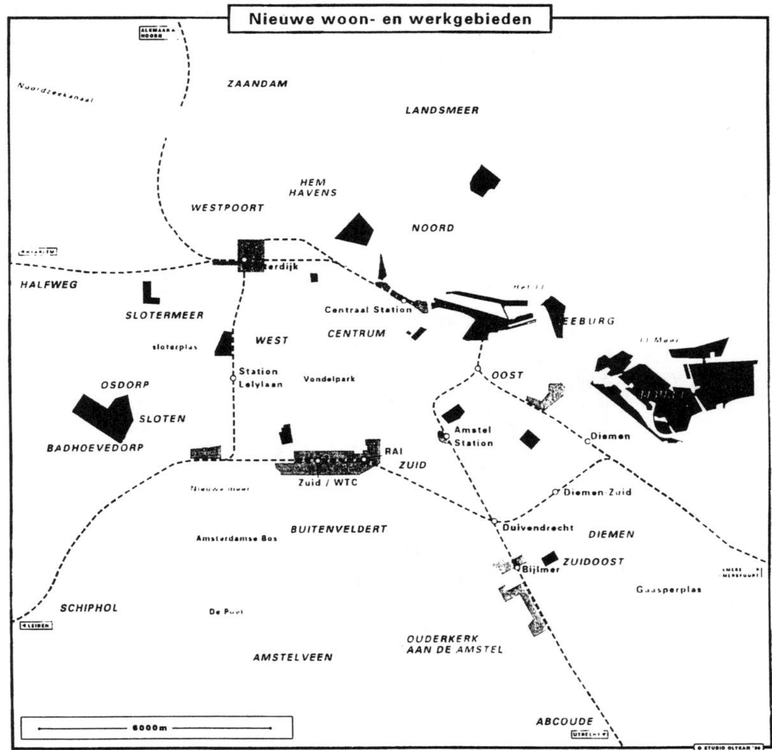

3. Travel Patterns

Travel patterns in 1999 are mainly the result of socio-economic developments and the effect of land use policy. Trip length, number of trips per day, trip motive and the market share of car, bicycle and public transport show a similar development as the changing role of the city:

· from a local to a regional scale and
· less of a radial structure

The role of Amsterdam as an self-supporting economic and social entity has ended. This is recognized by the city administration. They chose for cooperation with surrounding communities in the field of housing, employment, traffic and transport.

The traffic patterns in Amsterdam originally had a definite radial structure. This changed with the intense urbanization on the region. A growing section of the trips go around the edge of Amsterdam; an area with traditionally poor public transport.

Map
Intensity of evening rush-hour traffic (expected in 2005)

Scale -------- 30.000 AUTO'S

Map
Travel flow on public transport (expected in 2005)

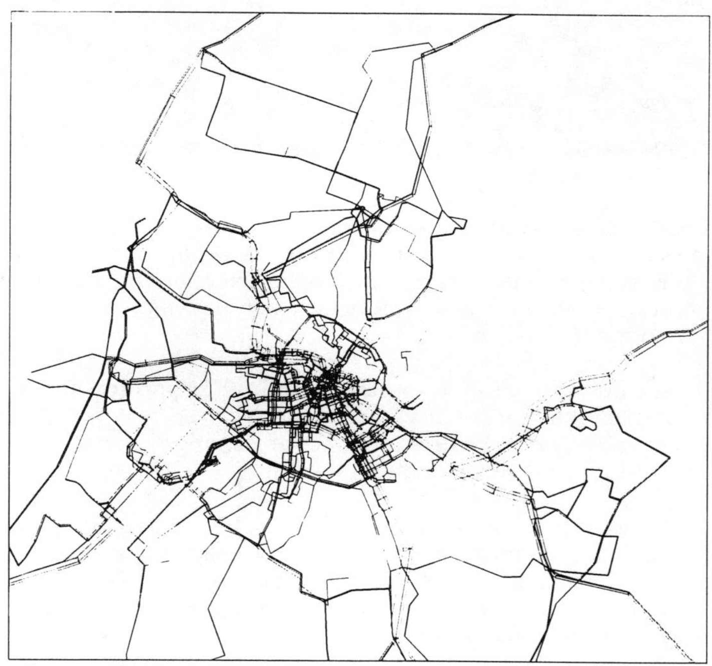

Scale --------- 30.000 Passengers

There is a strong variable picture of the roles of the automobile, bicycle and public transport. The differences can be explained by the inconsistent quality of public transportation and the short distances in which the bicycle is operational.

4. Conclusion

Public transportation is active on two fronts. One is maintaining its high market share in the center of Amsterdam by offering a product that is adapted to the changing demands. The other is to have new products which can increase market share outside the city.

For, in the outlying areas, there is still an important share of the market that can be achieved. What is needed is a regional public transport network that couples high quality with competitive travel times and real reliability.

The differences in travel quality with public transport in and around Amsterdam are indicated on the following maps. They show the number of inhabitants that can be reached with public transport within 45 minutes. The first map covers 1993; the second the situation in 2010.

< 800.000

800.000 - 1.200.000

> 1.200.000

Number of inhabitants reached within 45 minutes

The maps show a considerable increase in the quality of public transport in Amsterdam. This is the result of investments in new tracks for the tram, a ring metroline, the tramline to IJburg and the north/south metroline.

We expect a growth in the use of public transport by 30 - 40 % over a period of approximately ten years. The effect of the ring metro line (1997) is positive. The demand for this product is according to our expectations.

Public transport in the Amsterdam region is stimulated through investments in the extension of existing and in new raillines. Dutch Rail, local and regional suppliers develop together a new regional oriented public transport network: Regiorail.

Existing Urban Public Transportation system, its operations and use by people

Andre Testa
General Manager GVB, Public Transport Authority, Amsterdam

1. Introduction

Amsterdam, the capital of Holland, in more than 700 years, has developed into a city housing 720,000 people and providing 330,000 jobs. It has grown through a number of phases from a seafaring, trade oriented city with water in its most important arteries to a metropolitan community with specialisations in the area of harbor, airport and services.
The emphasis in transportation is now on the automobile, the freight truck, and bicycles for short distances. Public transport now reaches far beyond city limits.
Public transportation is getting ready to service that fast growing residential and employment region with projected two and a half million residents and one and a half employment opportunities.

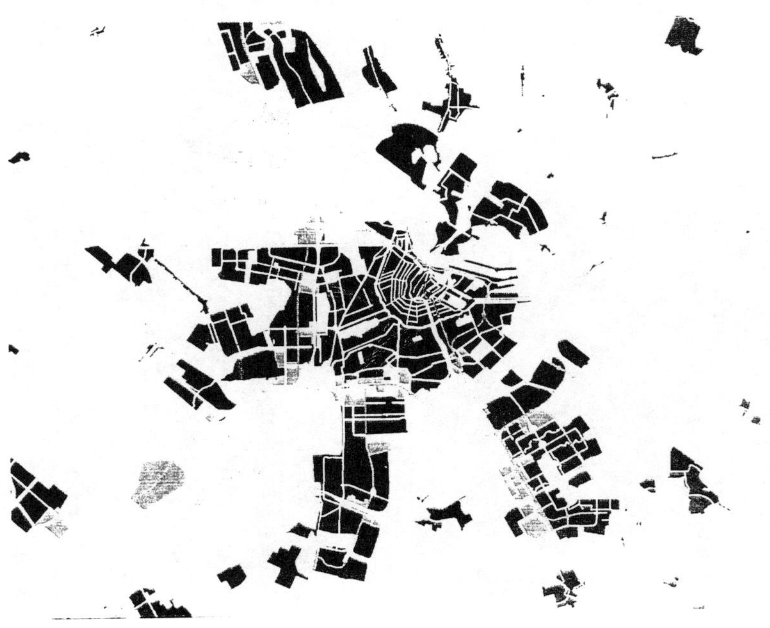

Characteristics

Public Transport's market share

Public transport in Amsterdam commands a good position with more than 20% market share during the rush-hours. There are differences between neighborhoods in the city. Public transportation's position can be sketched from the central inner-city to the outer areas as follows:

* A survey of shoppers in the downtown area indicates that about 60% have come by public transportation. In the entire central area of Amsterdam, public transport's market share is around 35%.
* In the area directly outside the center, public transportation is excellent, without making the districts inaccessible to the automobile.
* That is not yet the case in the ring of nineteenth century neighborhoods surrounding the center. However, things are moving in the right direction there as well. Travel by car is no longer, or only hardly, increasing. Both the bicycle and public transport are getting their chances, without drastically affecting accessibility by automobile.
* In the post-war neighborhoods on the edge of the city, the situation is less positive. Here, the position of public transport is worse. This is caused by the fact that many of the travel destinations of those who live and work here are not in the center, but are situated in a fan-like zone around the area.
* The situation is worse regarding transit to and from the surrounding region. Travel by car is definitely growing, and public transportation has a subordinate role. The impact of the bicycle is negligible.

3. Trends

* Many of the groups of transit passengers are only temporary customers. They vary from the tourist to the student, who leaves the city after a few years of study, to those who switch to the auto once they reach the age of eighteen.
* Demographic and societal developments such as extra free-time, part-time employment and an aging population result in more intensive use of public transport outside the rush-hours, as well as during them.
* The concentration of people residing and working in Amsterdam is diminishing. Public transport will have to be oriented to a larger area to move the same number of passengers.
* Employment concentrations are moving to areas on the edge of the city and outside it.
* Individualization creates diverse motives for travel and an increasing need for tailor- made travel. The already growing demand for more public transportation outside the rush-hours that has less and less connection to commuting is a result of this trend.

Task division within public transportation

Travel to destinations outside the region is handled by the national and international railroad services of Dutch Rail.
There are regional busses and train connections between Amsterdam and the other metropolitan centers in the region.

In the conurbation of Amsterdam, the majority of public transportation consists of a combination of tube, streetcar and bus. The maps are included to offer an image of the structure of the network.
The chart indicates of the supply and demand of public transportation in 1997.

	Kilometers per year	occupancy on workdays (%)	Number of workday trips per year
Bus	16.2 mln	22.9	51.9 mln
Tram	10.7 mln	29.8	128.7 mln
Tube	0.3 mln	27.6	56.4 mln

5. Public Transport system

Public transport in and around Amsterdam operates at three levels. First the city is linked to the national and international network of Dutch Rail. On regional scale a combination of metro, intercity buslines and regional train services operates. Operators have the intention to intensify the development of a special regional public transport product: Regiorail.

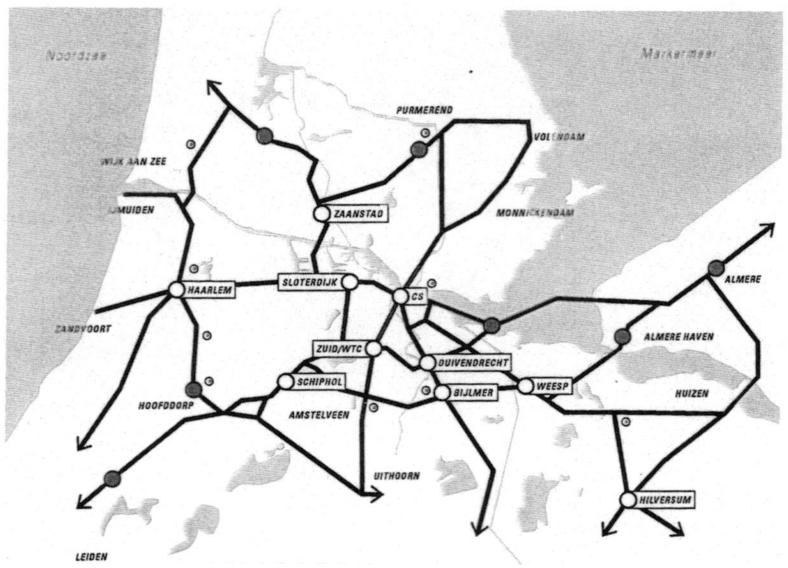

Netstructure for regional public transport

Ferry-services

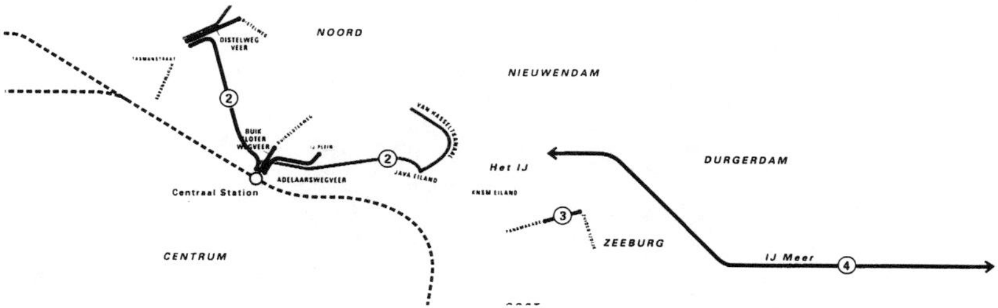

The third, local level of public transport consists of a combination of tram, bus and ferry-services

Metronet

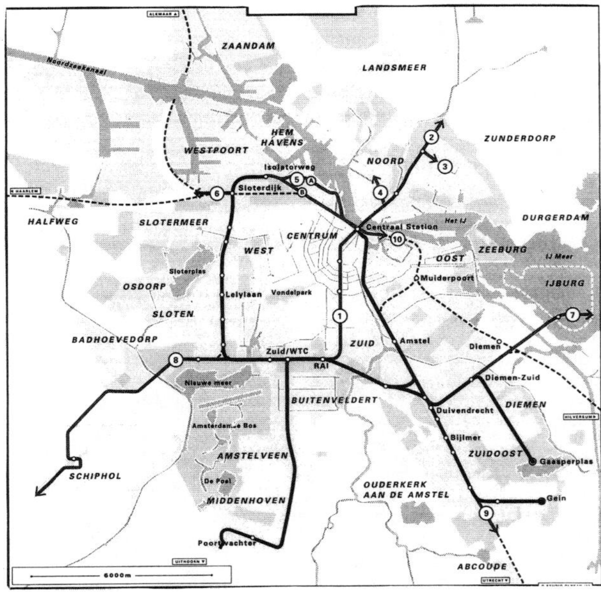

Busnet

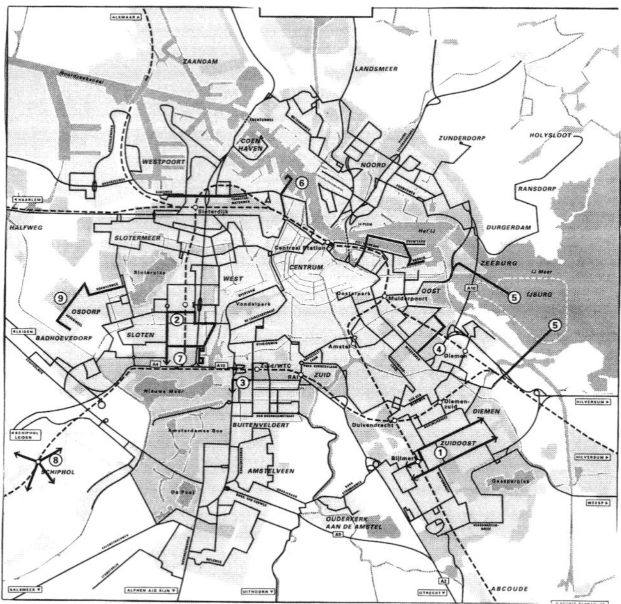

Tramnet

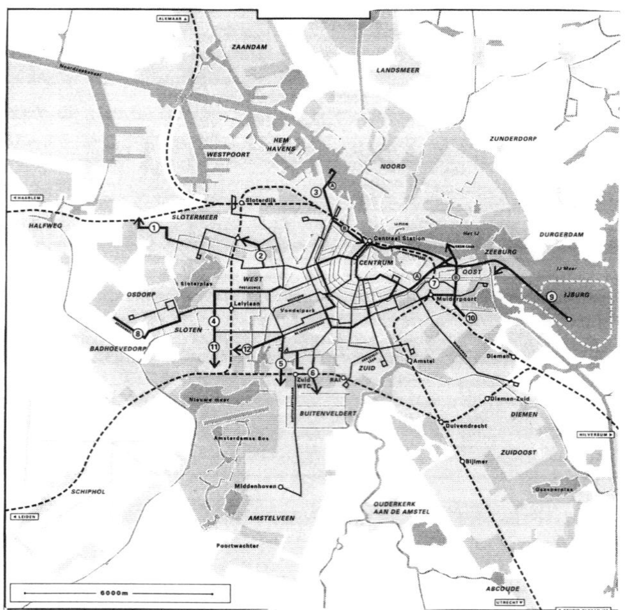

6. Public transport policy

More opportunities for public transport

Public transportation is most advantageous to Amsterdam if it is brought into a position in which it can perform better. The most long lasting effects are from a concentration of urban development around the stops and connection point.

a. Concentration of residence and employment

b. Good links in public transport:

More and more often travelers on public transportation use a combination of its services, for example, bus, train, streetcar. This is caused by the increased distances traveled.

c. The position of public transport in traffic:

Speed and regularity are more important to quality in public transport. The design of the road has big influence on that. Available space for public transport offers something extra: lower running cost. If vehicles operate faster, fewer divers and vehicles are necessary.

d. In combination with the bicycle and the car:

The city becomes more accessible if public transportation, in linking bicycle, car and public transport trips: for example more bicycles sheds available at tube stations and connection points for busses and streetcars.
Coupling car and public transportation journeys can also be improved. There is a trial project being prepared for the center so that people who park their cars can, with an extra charge, use their parking card for public transport.
Attention is also being paid to combining public transport and the care use. In this case, there must be parking for autos on the edge of the city, near rapid public transportation, and the improvement of public transport to these spots.

e. Clear-cut price policy:

Price also influences the use of public transport. Low prices can help it compete with the automobile. Therefore, Amsterdam does not want to raise public transportation prices at a rate higher than that at which the costs of auto use increase.

f. Involving costs of using a car:

In a few years, a pay as you drive system will replace some of the road taxes now paid. Then the motorist will be forced to make comparisons that more often favor the costs of public transportation contrasted with the amount of the new price system

and the additional petrol costs.
g. Investing in regional tube and bus routes:
Both longer distances and the locations of increasing destinations make altered demands on public transport. At the moment, it is particularly to the large local requirements. However, provincial transport requires bridging larger distances in reasonable time though a higher average speed. The tube network is an attractive product to this end.

h. Price policy:

The Ministers sets out price policy for public transport (p.t.) in the Netherlands by bus and streetcar. Every p.t.-operator in Holland accepts the tickets.
Only Dutch Rail has its own price- and ticketing system.

Gradually there is more space for experimenting with special prices for special products. This stimulates product differentiation and competition between operation.

i. Facilities for the disadvantaged

The design of streetcar, bus and metrocar takes the needs of the disadvantaged into account. New tramcars have low floors like bus and metrocars. Metro stations have escalators and elevators. Bus- and tramstops will in a few years in majority be adapted to the wishes of the user of a wheelchair.
But handicapped have also special bus services at their disposal. On request small busses with full facilities for the transport of wheelchairs bring their passengers from their home-address to any destination in Amsterdam.

j. Price subsidies

Public transport in the Netherlands is generally offered below cost-price. In several cities as high as sixty percent of the price is being subsidized. The Minister has changed his policy. In the future the percentage is lowered to fifty percent in flavor of raising the Governmental subsidy for new infrastructure of public transport.

7. Operations of the public transport company of Amsterdam (GVB)

• Head-office	staff	261
• Bus	staff	1,264
• Tram	staff	1,893
• Metro	staff	609
• Ferry	staff	64

Final results (x US $)

	1996	1997
Revenue:		
• price subsidy	231,095	219,635
• tickets	89,996	84,120
• other	35,310	21,297
Total	356,402	324,881
Costs	317,891	280,957
Result	38,510	43,924
Passengers	237 mln	236 mln
Passenger kilometers	855 mln	832 mln
Occupation rate	24.3 %	25.2 %
Bus	270 cars	276 cars
Tram	237 cars	240 cars
Metro	106 cars	69 cars

Conditional Priority
at Controlled Intersections
to Improve Public Transport Punctuality

Theo H.J. Muller[1]

Abstract:

The market share of public transport depends highly on the travel time ratio to private transport. The public transport travel time includes among other elements waiting times at stops and transfer points, both of which depend on schedule adherence. A method has been developed to reduce punctuality deviations by driver action (delay if early) and by giving conditional priority for public transport vehicles at controlled intersections.

Introduction

In order to cope with environmental and other problems caused by the enormous growth of car mobility, the market share of public transport (PT) should increase, especially during peak hours. Public transport has to function in a buyer's market in strong competition with the car system. An increase in the number of passengers can be achieved only by satisfying the service quality requirements of potential customers. However, motorists are very demanding, especially as far as travel time is concerned. A continuous effort is needed to improve both 'quality of transit system planning' and 'quality of transit operation'.

Quality of planning comprises the way public transport services are made available for potential users. It is the result of temporal and spatial system planning (connections, routes, stops, transfers and frequencies) and logistics planning (scheduling of personnel and available vehicles). This is not an easy task. Developments in land use, such as the shift of activities from city center to the periphery, have resulted in increasingly dispersed origin-destination patterns. This hampers the possibilities to provide adequate, direct, and high frequency public transport links. Transfers at interchanges become more frequent, and without specific measures they reduce the attractiveness of public transportation.

Quality of transit operation comprises the actual daily performance of the services on public transport routes in terms of speed, punctuality, regularity, and the possibility to transfer without delay. Operational quality should ensure a competitive travel time ratio. This paper addresses the possibility of increasing the market share of public transport by decreasing passengers waiting times by improving punctuality of operations.

---

[1] Head of the Transportation Research Laboratory of the Delft University of Technology, Stevinweg 1, PO-Box 5048, 2600 GA Delft, The Netherlands

The effect of travel time on the market share of public transport

Based on the regular Dutch National survey (OVG), Goeverden [1] found a strong relationship between the travel time ratio (R) and the market share (MS) of public transport in the Netherlands. The travel time ratio for a trip is the ratio between door to door travel times by public transport (Tpt) and by private car (Tpc).

$$R = Tpt/Tpc \tag{1}$$

Goeverden found the following relationship:

$$MS = e^{(-0.36 \cdot R^2 - 0.17 \cdot NT - 1.35/F + 0.23)} + 0.03 \tag{2}$$

In which:
NT: Number of Transfers in public transport trip
F: Frequency [veh/hr]

Figure 1 shows this relationship for trips with no transfer and with one transfer respectively. The frequency is two departures per hour.

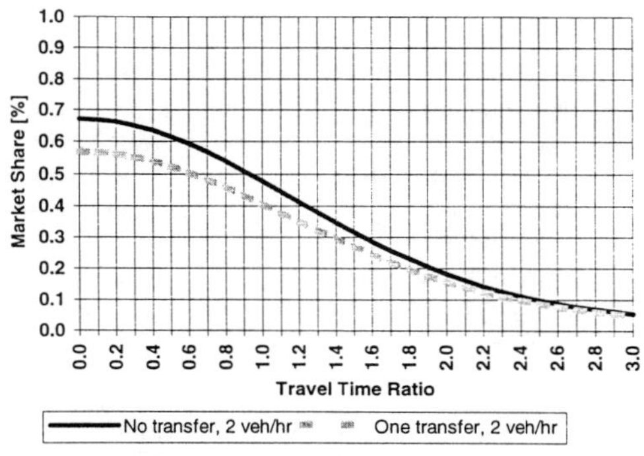

Figure 1. Relationship between travel time ratio and public transport market share

Figure 1 shows a considerable increase of the market share if the travel time ratio decreases, especially in the range of 1 to 2. To reduce travel time, public transport companies should pay special attention to operational speed by increasing punctuality and by synchronizing the arrivals and departures of vehicles at connecting lines at transfer stops.

Process management and operational control

A system for Process Management and Operational Control (PMOC) aims to:

- improve quality of operation in order to make public transport more attractive and to increase its market share.
- support personnel with both on-line and off-line information.
- inform passengers about the planned services and about current deviations.

Quality improvement is not only the drivers' task, it is a goal for everyone working at the transportation company.

PROCESS MANAGEMENT involves the guidance and support of schedule planning, service execution and company management. The basis for process management is the exchange of information between all personnel involved. It should improve understanding of each other's activities, problems, and mistakes and heighten the personnel motivation and cooperation to raise the quality of services. Positive motivation can only be achieved if all levels of the organization keep supporting the idea of quality management. Process management should include regular consultation and progress discussion based on unbiased information about the services performance.
The main conditions for effective process management are:
- Procedures to inform the personnel
- Procedures to discuss, to decide and to plan actions
- Motivation to perform the planned actions.

To determine whether both planning and operations meet the formulated goals, information about the execution of trips should be available. On a number of bus and tramlines in the Netherlands this information is gathered automatically by on board processors.

Primarily, the improvement of quality and efficiency of service performance will be a case of process management rather than operational control. The starting point is: activate the sense of ***responsibility and the motivation of the personnel***. Quality circles (audits and consultations), work structuring methods (the right man on the right place), zero defect programs (motivation to prevent errors) and contract management (personal responsibility) are tools that can be used in process management.
Before and after introduction of a PMOC-system, frequent consultation with personnel is a must. Using their experiences and responding to their remarks prevents repetition of errors and improves quality of operation.

OPERATIONAL CONTROL is applying technical tools to improve service performance. It involves providing real time information to drivers and to traffic controllers at intersections about current deviations from the schedule. An on board computer informs the driver about the schedule deviations using a simple display. It also communicates with the intersection controllers which, based on the status of the transit vehicle (early, on time, or late) are programmed to request priority or not.
For the ***drivers***, this implies an extra task. They should try to minimize the schedule deviation by adjusting the operational speed. It appeals to drivers' skills to reduce punctuality deviations to reasonable proportions. In this case operational control is a human behavior problem. Drivers can adjust the progress of a vehicle mainly in one direction. It is relatively easy to decrease velocity or to postpone departures. For safety reasons: speeding up can rarely be done by drivers.
Traffic controllers nowadays are able to provide priority to public transport vehicles. After being detected, a bus or tram can get an extra realization in the cycle, its stream and parallel streams can be extended, and conflicting streams can be terminated. Many cities in the Netherlands apply these *absolute priority* control tactics. Absolute priority control helps to reduce travel time, but punctuality deviations persist due to stochastic processes such as passenger arrivals at stops resulting in varying stopping times and drivers characteristics (slow or fast). *Conditional priority* helps to keep the transit vehicles on schedule.

Absolute priority for public transport can disrupt other modes of traffic. The negative effects are justified only when priority is provided to vehicles behind schedule. Compared with absolute priority, then conditional priority also reduces interference with other traffic.

The first condition for successful operational control is an accurate timetable. It should take into account the actual traffic circumstances during the period of the day, the day of the week and the season. Extensive data about route times and section times is required to construct such a timetable. On board computers can automatically gather the needed data. Performance can then be analyzed and compared with objectives for operational quality which should be defined in terms of speed, punctuality, and synchronism.

To achieve these objectives, the responsibilities and skills of the personnel must first be mobilized. Much attention should be paid to convincing drivers of the importance, necessity, and potential benefits of such a system. Drivers should also be involved in defining the system's requirements.

Operational quality

In the Netherlands, **speed** of public transport has been increased by the application of dedicated bus and tram lanes and by active priority control at intersections. To allow detection while approaching the intersection, almost all buses and trams have an electromagnetic vehicle identification system (VETAG™ or VECOM™).

Regularity deviation is the difference between the scheduled intervals and the actual intervals between public transport vehicles. Theoretical research by Welding[9], Heap and Thomas[3], and Osuna and Newell[7] resulted in an expression for the expected waiting time E(W) for a high frequency operation. The expected waiting time depends on the average headway E(H) and the variation in the headway Var(H)

$$E(W) = \tfrac{1}{2} E(H) \cdot (1 + Var(H) / E^2(H)) \qquad (3)$$

To calculate passengers' mean waiting time at boarding stops on lines with a high frequency, Hakkesteegt and Muller[2] use the relative regularity deviation RH, defined as the ratio of the headway deviation of an individual vehicle H with respect to its predecessor and the planned headway HT in the timetable:

$$RH = (H - HT) / HT \qquad (4)$$

The expected waiting time is calculated by:

$$E(W) = \tfrac{1}{2} E(H) \cdot (1 + E^2(RH)) \qquad (5)$$

Theoretically equation (5) is not quite sound. It confuses the mean of squares with the square of the means. In practice however, the error is minor. The advantage of equation (5) over equation (3) is that it can be used for individual vehicles and in the presence of variations in the scheduled headway.

Punctuality deviation is the difference between the scheduled moment and the actual moment that a public transport vehicle leaves a stop. On low frequency routes, passengers will arrive at the boarding stop based on the departure time in the published timetable. Experienced passengers arrive some time before the expected departure time E(D). The arrival moment (and the waiting time) depends on the departure punctuality deviations and the desired probability not to miss the vehicle. The expected time of arrival E(A) of experienced passengers is

$$E(A) = E(D) - \alpha * \sigma_p \tag{6}$$

Where is:

σp: the punctuality standard deviation

α: factor depending on the desired certainty and the shape of the probability distribution.

The expected passenger waiting time depends on the difference between the expected arrival moment of the passenger and the expected departure moment of the vehicle, assuming that is sufficiently large that the probability of missing the vehicle is small:

$$E(W) = E(D)-E(A) = \alpha \cdot \sigma_p \tag{7}$$

Schedule deviations result in undesired long waiting times at bus stops and increase passenger trip times substantially.

Schedule deviations are almost inevitable. They are caused by variations in stopping times and delay times, different characteristics of vehicles and drivers, and interaction with other modes of transport. Tram and bus drivers have to be observant for dangerous situations. They have to anticipate and react properly.

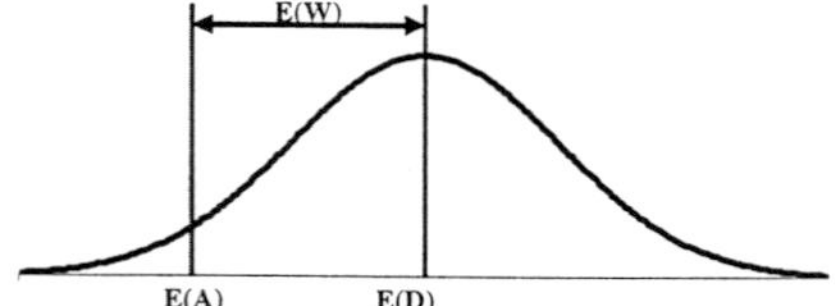

Figure 2. Passenger waiting time E(W) at boarding stop

According to Jolliffe and Hutchinson[4], in practice, passenger arrivals become less random as headway increases. Passengers arrive practically at random if the headway is below some threshold. Seddon and Day[8] found a value of 10 minutes in Manchester; O'Flaherty and Mangan[6] 12 minutes in Leeds.

However, the best strategy for a passenger (whether to arrive at random or based on the timetable) depends not only on the planned headway but also on the regularity and punctuality deviations. As operation becomes more punctual, experienced passengers will change their strategy accordingly.

A combination of equation (5) and (7) provides an estimate of the waiting time. It can be applied for both long and short headway operations.

$$E(W) = Min\{ ½ * E(H) * (1+E^2(RH)) , \alpha * \sigma_p \} \tag{8}$$

The expected transfer waiting time depends on the scheduled transfer time and the probability of missing the connection accepted in the planning. A short scheduled transfer time increases the probability of missing the connection and can result in longer waiting times. Knoppers and Muller[5] found an optimum solution by minimizing total (weighed) waiting time. Minimization of transfer disutility can be achieved by applying static and dynamic measures.

Static measures try to optimize the planned transfer time. For interchange terminals, the scheduling system gives high priority to attune arrival and departure moments. The arrival and departure moments for single line terminals are less important.

Dynamic measures continuously adjust the departure moments based on information about schedule deviations communicated by on board processors of other vehicles.

Scheduling

We present a new approach for scheduling and controlling the departure moments at stops en route that will reduce the punctuality standard deviation. The scheduled route time (Tx%) is based on a x-percentile completion time. This means that x% of the vehicles needs less time than scheduled. Most Dutch companies apply a 85% feasibility condition. This way, most drivers arrive in time. They have sufficient recovery time at the terminal stop and can start the next trip in time.

Assuming a normal distribution of route times, the scheduled route time is:

$$T_{x\%} = \mu_{i,t} + \alpha_{x\%} \cdot \sigma_{i,t} \tag{9}$$

In which:

$T_{x\%}$:	Scheduled route time (with feasibility condition x%)
$\mu_{i,t}$:	Mean route time from initial stop i to terminal stop t
$\sigma_{i,t}$	Route time standard deviation
$\alpha_{x\%}$:	Constant depends on the probability density distribution (for normal distribution: α_{80}=0.84, α_{85}=1.04, α_{90}=1.28, α_{95}=1.64)

The scheduled arrival moment at the terminal stop is:

$$M_t = M_i + T_{x\%} \tag{10}$$

In which:

Mt:	Schedule arrival moment at terminal stop t
Mi	Schedule departure moment at initial stop i

This condition guarantees that most trips need less time than provided. Figure 3 shows an example of the route time scheduling for a bus route with $\mu_{i,t}$=50, $\sigma_{i,t}$=3, and x%=85%.

This approach of scheduling increases the route time compared to using the mean value. The increase is proportional to the route time standard deviation. Therefore, management information should focus on actions that will reduce route time standard deviations by decreasing the influence of disturbing points.

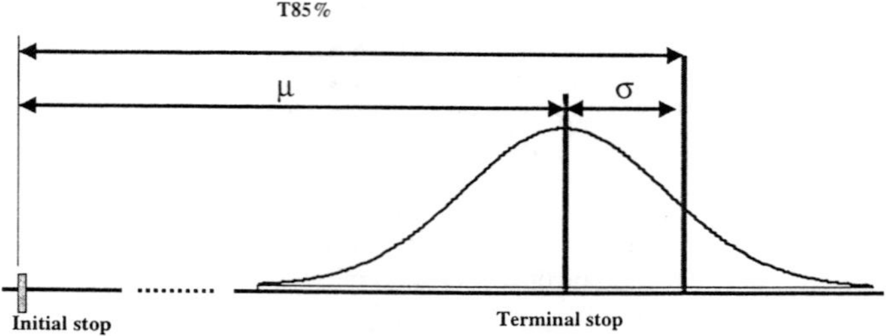

Figure 3. Calculation of the scheduled route time $T_{x\%}$.

Using this timetable, without operational control most vehicles will arrive too early at the terminal stop and at intermediate stops as well. Operational control is therefore needed to delay vehicles that are earlier than scheduled. Operational control needs a so called passing moment (Mp,x%) for each stop. That makes each stop serve as a control point in that vehicles that arrive at that point earlier than the passing moment should wait until the passing moment. Vehicles arriving after the passing moment should go on as soon as possible. The passing moment should be chosen so as to make it possibility to run the remainder of the route within the x-percentile completion time (x% feasibility condition) and to arrive in time at the terminal stop. The passing moment for a given point on the route is:

$$M_{p,x\%} = M_t - \mu_{p,t} - \alpha_{x\%} \cdot \sigma_{p,t} \tag{11}$$

In which:

Mt	Scheduled arrival moment at terminal stop
Mp,x%	Passing moment at point p.
$\mu_{p,t}$:	Mean time to go from any stop p to terminal stop t (mean section time)
$\sigma_{p,t}$	Section time standard deviation from any stop p to terminal stop t

The larger the feasibility margin the more opportunity to control, the less the punctuality standard deviation and the less passenger waiting time at stops.

Figure 4 illustrates the effect of operational control with a passing moment upon the arrival moment of the passenger and the departure moments of the vehicles.

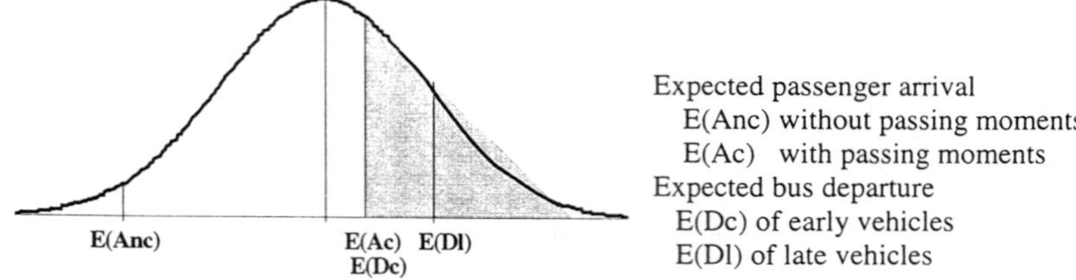

Figure 4 Arrivals of passengers with and without operational control

Suppose ß is the proportion of vehicles that arrive earlier than scheduled at a given stop. With operational control, these vehicles wait till the calculated passing moment. So, passengers do not need to arrive (much) earlier than this moment. Passengers only wait if a vehicle is late. The expected waiting time (E(W)) is:

$$E(W) = (1-ß)*(E(Dl) - E(Ac)) \qquad (12)$$

By definition, the value of ß is greater than 50%. This control approach therefore decreases the waiting times at boarding stops en route by at least 75%. If the waiting time according to (7) was six minutes (p=3 and =2), a reduction of waiting time of more than 4.5 minutes is possible.

Also the arrival moment reliability at transfer stops increases. As a result, the individual passenger's trip time by public transport decreases still further. The effect on the travel time ratio may be substantial, depending on the initial travel time.

Step by step introduction

Only the combination of practical experience of drivers with unbiased information about the services can lead to constructive discussions that result in drivers involvement, motivation and cooperation.

Step 0. Document present situation. In the present situation the schedules of public transport services have often been based on rough trip time information. Passengers experience excessive punctuality deviations resulting in long waiting times at boarding stops.

Step 1. *Collect data and quantify quality and processes.* On board processors gather data automatically. Analysis of data results in information about the quality of planning and operation of services. Analysis of the data collected (see figure 5) often shows bizarre patterns in individual (thin lines) and average (thick line) punctuality deviations en route as a result of an improper route schedule and/or distribution of the route time over the sections).

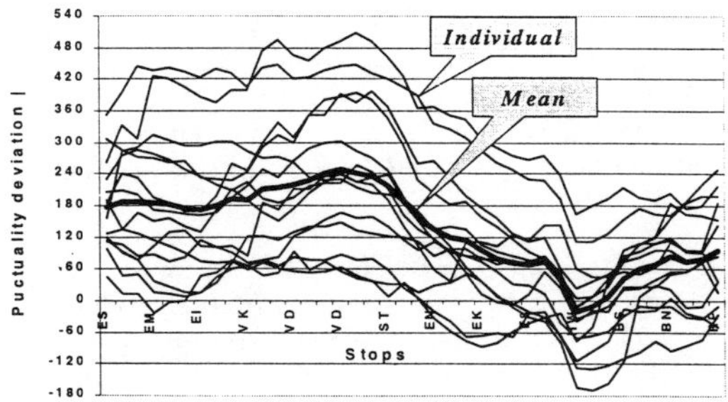

Figure 5. Mean and individual punctuality deviations per stop en route in seconds

Mean punctuality deviations above the x-axes indicate late; blow the x-axes early arrivals. If the average punctuality deviation varies consistently along the route or over the day, it is necessary to adjust the timetable. Before asking drivers to improve the operational quality, planners should correct the timetables based on trip time information.

Step 2. *Improve the schedules.* The timetable is the first subject of discussion in process management. In addition to the objective information about trip times, this step of process management also needs subjective information on drivers' experiences. If the scheduled route time is too short, drivers tend to run faster than comfortable. On the other hand if the route time is too long they run slower. Production of an adequate timetable requires both objective data and personal experiences of the drivers.

Step 3. *Aim for schedule adherence* After each trip, drivers are given information about their schedule adherence (similar to figure 6). The idea is to stimulate drivers to use their skills to reduce punctuality deviations as much as possible. Their experiences (problems, insights, proposals) in this step will be used in the cyclic processes of the former steps.

Step 4. *Produce management information.* While the average punctuality deviation is a quality indicator for the service planned, the variation of punctuality deviations is a quality indicator for the service performed (waiting time of passengers at boarding stop). For example, if a passenger accepts a risk of 2% to miss a bus or tram, he should be at the stop twice the standard deviation before the average departure moment of the vehicle. When the timetable is more or less correct, drivers are asked to identify the disturbances en route causing punctuality deviations. The data collection system provides the driver with the possibility of logging different types of delay.

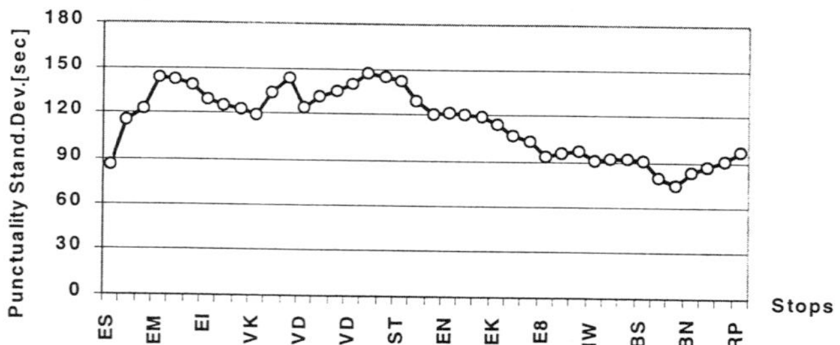

Figure 6. Punctuality standard deviation per stop en route in seconds

A list of quantified disturbances makes it possible to inform authorities and to ask for the proper infrastructure improvements to minimize (variation in) delay causing punctuality standard deviations and passenger waiting time.

Step 5. *Punctuality control.* The next step is to develop a system to respond to schedule deviation due to the usual stochastic processes in operations (traffic conditions, delays at intersections, passenger arrival rates, small dispatch delays). An incident is a marginal deviation from the schedule. Because it is much easier to control an early vehicle (by delaying) than a late one, effective control can only be exercised when there is a significant probability that busses arrive early. For this reason, the schedule route time should be based on a feasibility condition of at least 80% (see section 5).

Step 6. *Guarantee quality.* If the quality of public transport is high, it can be advertised as a competitive alternative for private transportation. Punctuality, speed, and reliability should be guaranteed. In order to evaluate customers' complaints, the actual performance should be measured and stored in a database.

Step 7. *Incident control.* An incident is an extreme schedule deviation mostly due to failures of vehicles, accidents, temporary infrastructure closures, or unwell drivers. Incident control needs exchange of information between the vehicles involved and a central control station. This concerns information by exception. Radio equipment (voice or data) is necessary and an operator support system can be helpful to find the proper remedies.

Step 8. *Quality certification.* In the near future, only certified public transport companies will be allowed to run public transport services. A working process management and operational control system will be necessary for quality certification.

Application in Eindhoven

Bus line 1 of the transportation company Hermes in Eindhoven has been rescheduled according to the methodology described above. For each homogenous period of the day, passing moments have been calculated. On board computers are loaded with these schedules and inform the bus driver about their schedule deviations en route. They can then slow down operation when early and try to speed up when late.

Traffic controllers detect busses arriving at intersections and receive information about the schedule deviation. Early busses wait for their regular green phase; late busses get absolute priority and will pass the intersection with minimum delay. The effect of this conditional priority system is shown in figure 7 (schedule deviations of trips on may 26, 1998 without priority) and figure 7.2 (trips of may 29, 1998 with conditional priority).

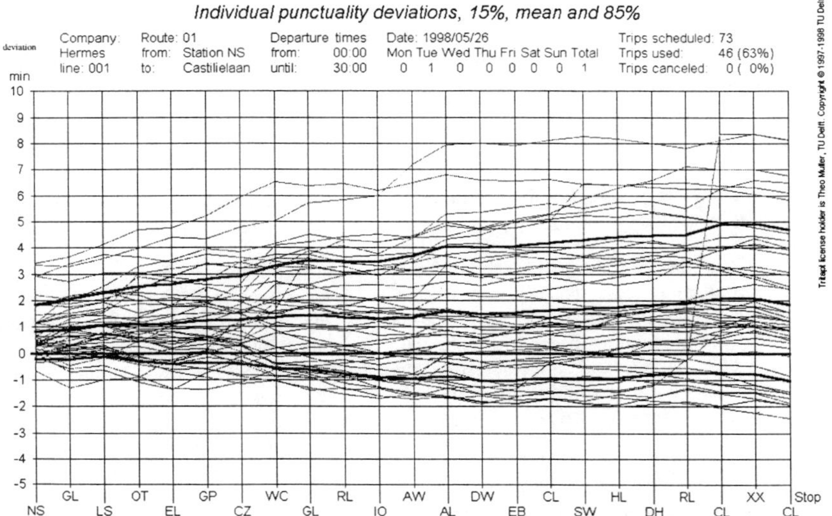

Figure 7.Punctuality deviations during individual trips without priority at intersections (the mean, 15% value and 85% values are indicated by bold lines).

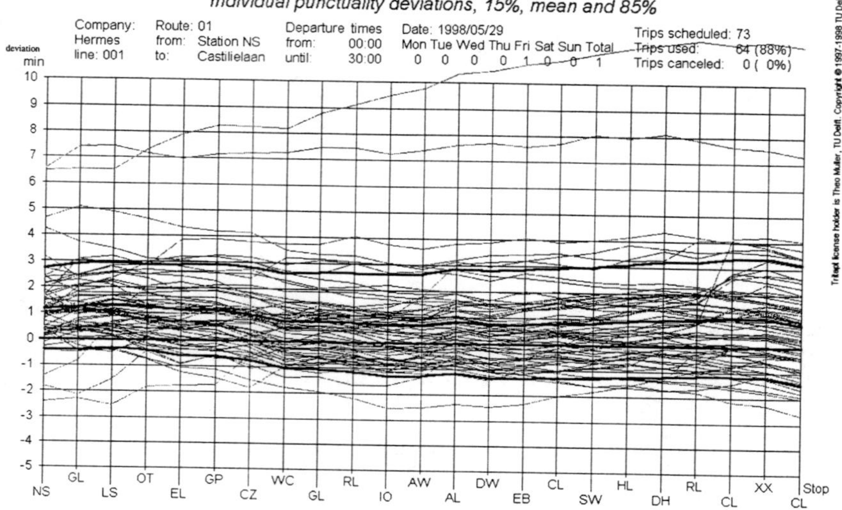

Figure 8 Punctualty deviations with conditional priority

Vehicles are not allowed to leave early. So, the average punctuality deviation should be positive (to late) but as less as possible. The results show that with conditional priority most vehicles operate with a punctuality standard deviation of less then 2 minutes. Without priority this standard deviation is 3 minutes. Also the average trip time may 29 is 1 minute less then on may 26.

Discussion

1. A process management and operational control system requires both non-technical expertise and technical tools.

2. Adequate factual information about the actual performance of services forms the backbone of process management. Exchange of ideas and willingness to solve the problems encountered are the starting points for improvement of operational quality. The primary functions needed for automatic data collection and information provision have been defined and implemented in practice. At present, the Delft University of Technology has developed modules to analyze vehicle trips data, produce management information reports, calculate passing moments, and simulate public transport operations.

3. Conditional priority (no priority if early, absolute priority if late) reduces the standard deviation in schedule deviations and decreases long trip times (higher speed), reducing passenger travel time.

4. If the punctuality standard deviation exceeds three minutes, operational control can decrease passenger waiting time at the boarding stops by more than five minutes and decrease the travel time ratio considerably.

References:

1. Goeverden C.D. and M.G. van den Heuvel, "Relationship between travel time ratio and mode choice" LVV rapport VK 5304.301 ISSN 0920-0592 (1993) (in Dutch)

2. Hakkesteegt P. and Th.H.J. Muller, "Relation between passenger waiting time and regularity in the service executed of public mass transport in towns", Proceedings of Verkeerskundige werkdagen 1981, 415-436 (1981) (In Dutch)

3. Heap R.C. and T.H. Thomas, "The modelling of platooning tendencies in public transport", Traffic Engineering and Control, Vol 8(9), 360-362 (1976)

4. Jolliffe J.K. and T.P. Hutchinson, "A behavioral explanation of the association between bus and passenger arrivals at a bus stop", Transportation Science, Vol 9, 248-282 (1975)

5. Knoppers P. and Th.H.J. Muller, "Optimized transfer opportunities in public transport", Transportation Science Vol 29, no 1, 101-105.

6. O'Flaherty C.A. and D.O. Mangan, "Bus passenger waiting time in central areas", Traffic Engineering and Control, Vol 11, 419-421 (1990)

7. Osuna E.E. and G.F. Newell, "Control strategies for an idealized public transport system", Transportation Science, Vol. 6(1), 52-72 (1972)

8. Seddon P.A. and M.P. Day, "Bus passenger waiting times in greater Manchester", Traffic Engineering and Control, Vol 15, 422-445 (1974)

9. Welding P.I. "The instability of a close interval service", Operations Research Quarterly, Vol. 8(3), 133-148 (1957)

Socio-Economic Characteristics, Land Use and Travel Patterns of the Municipality of Beijing, China

Gladys D M Frame[1]

Introduction

Beijing is a superb and unique example of the principles of Ming and Qing Dynasty urban planning. The city is at risk of losing this special character due to the rapid development of real estate, in the form of high rise commercial and residential buildings in the historic city centre, and of transport facilities focused on providing for the private motor vehicle. A key issue today is how the character of Beijing can be maintained, or at least not further eroded, while at the same time providing the transport infrastructure that a modern capital city requires. The key elements in any transport strategy in order to achieve this for the city are: (1) making the best use of the existing road network by utilising traffic management techniques; and (2) developing public transport facilities. The Beijing authorities have tackled the inherited problems of access and mobility, and have aimed to cater for increased transport demands and higher expectations of level of service by residents, by an extensive road building program focusing on Ring Roads and widening of existing roads. At the same time, the development of public transport has lagged behind and the trend has been towards individual private modes of transport, firstly the bicycle, then the private car. These policies are having a severe impact on the quality of life, the human scale of the city and its historic legacy. There is a need to develop an integrated and coordinated package of policies with traffic management and public transport at its heart and focusing on how a balance can be achieved between conserving the old and catering for the new. In this context, this paper discusses the key historical factors affecting the development of the public transport network.

---

[1] Gladys D M Frame, Independent Consultant Traffic Engineer, 106 Spottiswoode Street, Edinburgh EH9 1BY, United Kingdom

The Development of Beijing

The inheritance of the past: Since the establishment of Kublai Khan's Dadu on the site in 1272 during the Yuan Dynasty, Beijing has been the capital for more than 700 years, except for brief periods in the early Ming Dynasty and the late Republic. Beijing is a city that never really evolved as other centres did, but was created and planned. When the present government came to power in 1949, they inherited an historic city that was a relic of Ming and Qing Dynasty urban planning: a grid network of streets enclosed within crumbling city walls, a city centred on a symbolic north/south axis, with the Forbidden City in the centre. The former 'Imperial' and 'Tartar' cities to the north of present-day Changan Jie had a reasonably well laid out street system characterised by the unique Beijing lanes called *hutong*, while the former 'Chinese' city to the south was a jumble of winding lanes. There was a lack of east/west routes and the Forbidden City in the centre posed problems for through traffic and the general mobility of the population; even going north/south through the city was difficult. However, the important point was that there was a sound urban plan to build on and to form a basis for upgrading to meet the needs of a modern capital city.

New China builds: The new government responded by embarking on a programme of reconstruction to improve access, mobility and amenity for the population. Tiananmen Square was enlarged to its present-day size, a major east/west link called Changan Jie was created, other east/west and north/south links were widened, the old wooden triumphal arches (*pailou*) across roads were removed to create wider carriageways, and the city walls torn down to build the Second Ring Road (2RR). Two metro lines were constructed. Line 1 was an east/west line from Pingguoyuan to Fuxingmen and was 17km long with 12 stations. It started trial operation in 1971 but became fully operational in 1981. Line 2 was a circular line built under the 2RR, 23km long with 18 stations and it became fully operational in 1988. The building of the metro was seen to be a symbol of the national capital rather than a serious attempt to tackle transport problems by focusing on public transport. Developing the bus network would have been a more viable option. Like the recent construction of other infrastructure, the subway was built with the aim of symbolising power through grandiose schemes. The main railway station was consolidated southeast of Jianguomen and rail links were upgraded. The former military airport, 18km to the northwest, was consolidated as the capital's airport. There was considerable investment in infrastructure and network development, often at the expense of other cities, such as Shanghai.

The post 1980s symbolic city: Although today's Beijing may seem to some observers to be a city of omnipresent grey tower blocks and sterile suburbs, a city with no vibrant hub, a city of walls and vast expanses of road, it is its symbolism, both past and present, rather than the actuality, that characterises Beijing. Its urban plan has always symbolised the responsibilities and aspirations of those in power. Today, Beijing symbolizes the modernization and the open door policy of China.

The rapid pace of reform: In recent years, Beijing has relied on generating its own revenue through an increased role in business and trade. From the 1980s onwards, the economic reform policies and the steps taken towards the creation of a socialist market economy have accelerated the pace of change, not only in Beijing, but also throughout the country. Beijing has expanded rapidly in area and population. As controls on movement were relaxed, people flocked to the city from the countryside in search of better lives. The dismantling of the 'iron rice bowl' meant that people had to become more responsible for more aspects of lives; if they had money, they began to have more freedom of choice on where to live and what modes of transport to use.

In summary, the physical essence of Beijing comprises of three elements:

- ***walls - defining the city boundary and boundaries within the city, delimiting the hutong and the courtyard***
- ***symmetry - the north/south axis, the checkerboard layout***
- ***horizontality - the low and quiet city profile***

These elements were maintained but transformed by socialism. The city walls were lost but walls continue to play an important role - defining the danwei and the da yuan in addition to the hutong and the courtyard. The imperial space at Tiananmen was transformed into peoples' space and a new east/west counteraxis developed. The first new buildings to be built in the 1950s were not high-rises, they were public buildings that kept the monumentality of horizontal form.

Land Use

Beijing is located in the north of China on a plain with mountains behind. It was chosen because of the excellent *feng shui* of the site. Beijing Municipality consists of four urban districts, four suburban districts and several rural counties. The total land area is 16,807 km^2, 1,282 km^2 of which is the built up area - defined as being within the Fifth Ring Road (5RR) - and 87.1 sq. km. which comprises the four urban districts in the central area, the East District, West District, Xuanwu District and Chongwenmen District within the 2RR.

Pre-1980s Land Use

The centre of government and culture: As the capital city of China, Beijing has traditionally been a centre of government and culture rather than trade. Government and cultural buildings occupy the most prestigious sites. Commercial areas were centred around four junctions which formed the corners of a rectangle around, but sufficiently distant from, the Forbidden City - Dongsi, Dongdan, Xisi and Xidan. The Qianmen area was another thriving shopping and business area and Wangfujing Dajie became one of the premier shopping streets in the city. Heavy industry was mainly located to the southeast and to the far western suburbs in Mentougou and Shijingshan. The eastern suburbs of Jianwai and Sanlitun started to be developed as the new diplomatic and trade areas, and embassies were relocated from the former Legation Quarter east of Tiananmen Square. Universities were located in the northwest in Haidian District.

Housing: However, the key land use issue facing the new government of 1949 was the lack of housing and a system was implemented whereby the work unit or *danwei* became responsible for housing. *Danwei* often occupied huge tracts of land (the *da yuan*) and their legacy has left Beijing with large block lengths with no through roads and a lack of intermediate roads in the road hierarchy. Where there was space, residential apartments were built for its workers in the *danwei* compounds; where there was no space, areas were allocated in the suburbs. Soon there was insufficient room in *danwei* compounds and more and more housing was required in the suburbs. Travel patterns changed. Instead of walking to work within a compound, more and more people had longer and longer journeys to work and they went by bicycle or bus.

The mixture of land uses: Although the city walls were not saved, sites of national historic importance were put under preservation orders. In Beijing there are more than 150 such sites. However, the environment around these sites was generally considered suitable for redevelopment and tended to be encroached upon to a considerable degree. Countless other sites of relatively lesser historic importance have completely vanished. Many historic buildings were transformed into *danwei.* Temples became factories, courtyard houses became schools. The repercussions of these changes of land use are now being felt in the central city area as access to public transport for workers and access for servicing by motor vehicles on unsuitable lanes is now required. This incompatible mixture of land uses and the difficulties of relocating established work units meant that transport planners traditionally looked towards building new road infrastructure rather than making the best use of the existing network.

No space constraints: Unlike the commercial cities of Shanghai or Guangzhou which were constrained by lack of space, narrow streets and physical features such as rivers, Beijing had space to expand and it did.

Post-1980s Land Use

Commerce flourishes: More and more commercial, business and shopping areas were developed and small private businesses flourished. Traditional shopping areas were upgraded and the eastern suburbs were developed as a major business and commercial area. Firstly, the area around Jianguomenwai was developed with joint venture hotels and enterprises stretching eastward to the China World Trade Centre. Then areas further north in Sanlitun were consolidated with more hotels, office blocks and multi-function developments. Suburban centres began to take trade away from traditional areas and became hubs in their own right. In the university area, Zhongguancun became the centre of computer industry. More environmental areas began to lose out to new infrastructure construction as the low density hutong in prime areas become targets for redevelopment and historic sites lost their historic context.

Sport becomes a symbol: The accelerated pace of development and infrastructure construction was only partly due to the loosening up of controls. Beijing's Olympic bid formed an impetus to upgrade transport links citywide. In 1990, Beijing had held the Asian Games and the Asian Games village was constructed in the north around the Fourth Ring Road (4RR).

Current and future plans: Existing county towns such as Changping in the north and Tongxian in the east are being developed. Plans to concentrate major urban construction away from the city centre proper along radial routes such as the Airport Expressway are being implemented. Satellite towns are being developed along the Beijing-Tianjin-Tanggu expressway. Tourist sites in Beijing Municipality such as the Ming Tombs in Changping and the reservoirs at Miyun and Huairou to the north are increasingly catering to local tourism since the introduction of the five-day working week a few years ago. In the city centre, the following key elements will greatly impact the transport network: (1) the consolidation of the East District as the new CBD and the redevelopment of the Chaowai/Hujialou and the Liangmahe areas on the Third Ring Road (3RR); (2) within the 2RR, the upgrading of the north/south route linking Dongsi and Dongdan; (3) the upgrading of the west side of the Xidan area into a multi-purpose complex for shopping, entertainment and commerce; (4) the redevelopment of the Wangfujing shopping street; (5) the construction of a 'Financial Street' on the west 2RR; (6) the construction of a new main railway station at the West Station and downgrading of the existing main railway station.

Current land use development has not taken sufficient notice of public transport needs. By regulation, all new developments have to provide large underground car parking areas, thus encouraging the use of the private car. Links to the public transport network are often not well developed. In new areas, public transport development has lagged behind.

Population and Employment

Permanent Population: The built-up area of Beijing - defined as being within the 5RR - currently contains a permanent population of about 6 million, but with neighbouring counties and rural areas making up about a further 4.5 million people. About 2.3 million (36%) of the population live in the four central area sectors within the 2RR. About 70% of the population is classified as non-agricultural. Average annual income for residents almost doubled during 1993-1996 - the third highest in China.

Employment: Of those employed, almost half are in manufacturing, although this is expected to change. In the Beijing 8th 5-Year Plan, (1996 - 2000) the tertiary sector is earmarked to grow faster than the agricultural and manufacturing sectors, and the service sector will be expanded to include telecommunications, consultancy and real estate. Total employment is estimated at 3,861,700, with 1,672,000 (43%) located in the central area, but with about 50% of the office jobs and 57% of the commercial and service employment being centrally located. Underemployment is prevalent, and the hope among planners is for a large increase in jobs to come from the service industries. The efforts to increase productivity in the state sector are already increasing redundancy.

Future Population: Population is expected to grow 9% between 1992 and year 2000, and by an overall 15%, reaching 7,231,000 by 2010. The inner sectors within the 2RR are expected to decrease in population while the outer sectors will increase. By 2010, it is estimated that about 458,000 people will have left the area within the 2RR. This is due to the forced relocation of residents of old one story houses (*pingfang*) to make way for new road building and widening. Most of the relocated residents are being relocated beyond the 4RR and in the county towns. However, an increase in suburban population, especially in the NE Chaoyang District of the city, is also expected.

Floating Population: Migration of people from the hinterland and rural areas to the coastal cities and urban areas has been occurring as a result of increased prosperity in the urban areas. In 1990, the size of the 'floating population' of migrant rural labourers was estimated at 50 million nationwide. This was a result of more relaxed control of internal migration, or the increased difficulty in controlling them. In Beijing, the current floating population is estimated at about 4 million and they have a significant impact on public transport usage:

The distribution of employment determines the patterns for the journey to work. Therefore the key item for public transport planning is the distribution of employment, rather than the total numbers. Much of the commuting problems of the journey to work arise due to the imbalances between where people live and work.

The Vehicle Industry

The automobile industry in China is increasing production at a rapid rate. The Japanese are at the forefront of this drive and initially flooded the market initially with Japanese-made vehicles but latterly with small cars, minivans, minibuses and trucks built through joint-ventures with the Chinese. The small minivan taxi made by Daihatsu and others, and the small hatchbacks Tianjin Toyota (xiali) are particular examples. Figures for minivan output show that in 1988, 20,000 vehicles were produced in China; by 1992, this had increased to over 54,000 vehicles per year. However, there are also French, American and other joint ventures. The rise in demand for the luxury car is also evident.

The rapid growth of China's own automobile industry reflects an increasing consumer purchasing power and freedom to buy small cars and minibuses. The authorised private use of *danwei* vehicles is on the increase. Local leaders and high level cadres, having got their cars, now want to pass on the advantages to their staff. The strict controls on motorcycles in Beijing have resulted in few people being able to utilise this mode; those who aspire to motorised transport now see a small car within their reach. The small Daihatsu minivan is increasingly used as a taxi, costs between RMB40,000 - 50,000 (US $4,938 - 6173). It is predicted that soon even cheaper cars, costing around RMB15,000 (US $1852) will be available and within 10 years, a car will be on everyone's 'wish list'.

In contrast the bus industry has been slower to develop. In Beijing until recently, the bus fleet consisted of old Eastern European and domestically produced vehicles including trolleybuses and articulated buses. However, in recent years, the influx of new and second-hand double-decker buses, minibuses and new air-conditioned buses is evident.

The plans to produce an affordable family car will lead to increased congestion that cannot ultimately be solved by new road building. There is a need to review policies and to invest more resources in the domestic bus building industry.

Vehicle Trends

1950s - early 1980s: bicycles encouraged: In the 1970s, bicycles were encouraged as a suitable means of personal transport and to relieve pressures on public transport. As household incomes rose, bicycle usage developed rapidly. Bicycles were easily catered for as road space was not a problem. Private cars were almost non-existent, *danwei*-owned vehicles were few and taxis a rarity. An extensive public transport system of bus routes to serve the new suburban residential areas was developed but it was still not enough to cater for demand. Buses tended to be overloaded and underpowered.

1980s - 1990s: the rise of the taxi: From the late 1980s onwards, it has been the rise of the taxi that is the most significant change on the streets of Beijing. In 1988 there were some 140 taxi companies and a total fleet size of 14,000 cabs; by the 1990s this had grown to 600 companies. There is now a glut of taxis on Beijing streets and the new low priced minivan taxi or *miandi* is within the price range of many residents. Supply exceeds demand and the Beijing authorities have recently decided to phase out the *miandi* taxi.

The increase in *danwei* owned vehicles and the private usage of these is evident. Private ownership by individuals has increased significantly. In 1980, just 20 private vehicles were registered; by the 1990s this had grown to 5,000 vehicles. The growth of small passenger vehicles (cars, taxis, *miandi*) has been significant, from around 110,000 vehicles in 1989 to 425,000 vehicles in 1996. In 1996, vehicle ownership was 82 veh/1000 population.

Public transport has not been developed and remains severely overloaded. The rise of the minibus, both public and private has alleviated some routes. Bicycles remain a significant mode and although they are considered a traffic problem, are not yet seen as the major cause of congestion as in Shanghai and Guangzhou.

From 1980 to 1993, motor vehicle (MV) registrations in Beijing Municipality increased fourfold. Total MV registrations were 109,552 in 1980 and 466,846 in 1990. By the end of 1996, MV registrations totalled 982,000. Bicycle registrations are about 7 million, but it should be noted that this is a cumulative figure since records began - there is no system for "deregistering" scrapped, stolen or disused bicycles.

The unrestricted growth of private cars is having a serious impact on the road network, which cannot be solved merely by new road building. There is a need for fiscal and ownership control policies to complement the development of public transport.

The Road Network

1950s to Pre-1980s: In 1950s, the principles for the development of the road system were established under the influence of Soviet road planners. These principles remain the same today. Primary roads were considered to be few and *hutong* numerous. The central axis was considered to be the city's most important traffic artery, apart from the fact that it ran through the Forbidden City. The three lakes, the Forbidden City and the Temple of Heaven were considered to be hindrances to east/west traffic movements and the small *hutong* were unable to serve traffic needs. Five new east/west and one new north/south routes were widened or opened up. The railway around the city was considered a hindrance to road traffic and was demolished. The grid system was to be maintained for the Inner City while a ring and radial system was to be developed for the outer areas. The First Ring Road (1RR) was defined as the roads linking the four key commercial areas of Dongdan, Dongsi, Xisi and Xidan but these routes were never developed as such nor do they function as such today. However, plans in the 1950s show proposals for the 2RR, 3RR, 4RR and 5RR.

There were few intermediate level roads; the north/south roads and the new east/west ones, which were opened up, were essentially primary distributors, which linked directly onto *hutong* or access roads. There were few district or local distributors which should have had a natural intermediate place in any road hierarchy and which would form the most appropriate routes for developing a comprehensive bus route network. The categorisation of roads by their physical characteristics, in particular, by width and cross-section, was favoured over a functional classification.

In 1953 the 'Red Line', *hong xian*, right-of-way planning width was developed. New buildings along road frontages were required to be set back beyond the future planned road width. Most 'Red Line' demarcations were extremely wide - typical Red

Lines for *hutong* were 30m. These Red Lines were practically 'cast in stone' - once set, they would be extremely difficult to change (except to widen even further) to adapt to new circumstances or new policies. The fundamental thinking was that new road building was the solution to increasing traffic growth. Building a way out of the traffic problem was established and the basis for comprehensive wide road policies was set.

The road network was not conducive to the development of public transport because of the lack of intermediate roads in the hierarchy. Buses were too large to access the lanes or hutong. Passengers were not well served by the bus route network, which was developed on the major streets as they still had considerable distances to walk into the danwei compounds or to the interior lanes.

Post 1980s and Current Situation: Beijing rapidly upgraded its transport network to cater for the increased growth in vehicles. Bicycles were still designed for most of the new and planned roads incorporated physically segregated bicycle lanes. The concept of Ring Roads, inherited from the era of Soviet style planning (in contrast to the existing grid network), was consolidated and the 3RR was built, construction on the 4RR commenced and plans for a 5RR considered. The 2RR and the 3RR were upgraded to form urban expressways with no at-grade junctions or crossings. More road widening of roads within the 2RR commenced with the recent widening of Chongwenmenwai, Chaoyangmennei and Pingan Dajie. A new east/west metro line under Changan Jie is currently nearing completion. There are plans for a light rail line to serve the northern suburbs.

The total length of Beijing’s road system is 14,500 km of which 3,500km comprise urban roads and 11,000 km suburban and rural roads. In the built up area, main arteries account for 160.73km of road. Including the 2RR and the 3RR there are 9 main east/west arteries and 7 main north/south arteries. The large block lengths mean that the average interval between cross-streets is about 1km. Beijing has more than 10,000 junctions of various types of which 434 are signalised. The majority of Beijing's 3,400 *hutong* are unsuitable for through traffic or buses. The average width is about 6.0m but many are much narrower. However, the congestion on the main arteries has meant that the *hutong* are often used as short cuts or 'rat runs' for motor traffic. There are 74 intersections on the main arteries with over 4000 veh/h and 27 intersections with more than 10,000 veh/h at peak hours.

The changing function of the Ring Roads: The impact of the changing function of the Ring Roads is important. The function of a road depends on its adjoining land use and the number of accesses. When the 2RR and 3RR were upgraded, there were no changes in the number of accesses and junctions but the improved roads tended to encourage more commercial development alongside. Examples are seen on the 2RR at the planned Financial Street alongside the western sector, the huge Fangzhuang residential area in the southeastern sector, and the developments at Chaoyangmen in the east (as part of the Chaowai redevelopment); on the 3RR examples are the developments at Yansha Qiao and Liangmahe (Lufthansa

Centre, Landmark Offices, hotels) in the east and at district shopping centres along the northern sector. So the function of these Ring Roads is now to cater for both through traffic avoiding the city core and traffic requiring access to district developments, in other words both primary and district distributor functions.

Bus routes: Bus stop spacing is too great. National standards set the distances between urban bus stops to be between 500 - 800m. But even this is too long, and Beijing's bus stops are even more widely spaced. There is also no good provision for passengers at stops although this is improving with the introduction of more hi-tech bus shelters. Interchange facilities are very poor. Passengers usually cannot interchange routes at the same stop and typically have to walk up to 0.5km and cross roads to interchange. Design of grade-separated interchanges did not consider bus routes - bus routes have had to make long detours to turn left and this is very evident at Xizhimen. The nearside lanes of the wide bicycle lanes alongside the 2RR and 3RR have been converted for bus and taxi use. Bus stops and taxi ranks are also sited along here. Recent introduction of bus lanes on Changan Jie, Chongwenmenwai Dajie and in Haidian District demonstrates belated efforts to give priority to public transport.

Road network development from the 1980s onwards has not favoured public transport. The aim has been to provide for free-flowing private motor car trips at the expense of public transport. The road widening and new road building programmes did not result in a greater spread of public transport. However, in recent years, the belated introduction of bus lanes has begun to redress some of these issues.

Future Network Development

The 1992 Beijing Master Plan Revision: Transport Strategy: The transport strategy is to have high speed and free flowing traffic as its mainstay with emphasis on grade-separation and segregation of traffic modes. The strategy focuses on providing for the motor car with the provision of public transport, bicycle and pedestrian facilities straggling behind the unbridled promotion of motorization. It anticipates a doubling of person trips over the next 15 years together with a doubling of freight tonnage.

The urban road system is to have the four Ring Roads envisaged in the 1950s starting with the 2RR and culminating with the 5RR. There are to be ten primary radial routes and 15 secondary radials starting at the 3RR and linking to the national and provincial highway network. Within the 2RR, the Inner City should have six east/west primary roads and three north/south roads. A four level road hierarchy is envisaged for the urban area consisting of urban expressways, primary distributors, secondary distributors and access roads with the focus on constructing standardised grade-separated interchanges, *lijiaoqiao*. The old roads and the *hutong* are seen by many people as symbols of old Beijing whereas the *lijiaoqiao* and the new wide roads symbolize a modern capital for the 21st Century. The *lijiaoqiao* have become the new 'scenic spots' of Beijing.

The planned road network density is high with on average 2.44km of road per square kilometre of land area and road space to be 11.1% of land area. Inside the 3RR, road density is to be even higher at 4.64km/sqkm and 21.18%. Within the 2RR road space is planned to be 25.85% of land area. Road widths and Red Lines set in the 1950s have not been changed.

For Old City renewal and new district development, the transportation focus is to be on providing car parks, not only private ones for new residential and office developments but also public ones for commercial areas which are large traffic generators. By 2000, it is planned to have 23,000 car park spaces within the 3RR. According to Beijing's master plan, it is estimated that by the year 2000, Beijing's road system will comprise 20,000 km, of which 10,000 will be urban roads and 12,000 will be suburban and rural roads.

Future Metro and Light Rail Network: The construction of a new east/west metro line under Changan Jie to serve the city centre is nearing completion. Consideration is being given to a new north/south metro line in the East District extending between the north and south sectors of the 4RR. Current light rail plans are focusing on the north of the city.

Planners are finally recognising that the development of public transport in conjunction with traffic management of the existing road network is the key to catering for trip demands. The extension of the existing metro and the consolidation of a light rail line are steps in this direction. However, the mainstay of any public transport system is typically a bus, feeding into main metro and light rail lines and Beijing is no different. There is still a need for the level of service of the bus network to be improved in terms of spread of services, bus stop spacing and bus priority measures.

Bibliography

(1) Beijing Municipal Institute of City Planning & Design, *Beijing Chengshi Zongti Guihua 1991-2010 (Beijing Master Plan 1991-2010)*, (1992)

(2) Frame, Gladys D M, *The Preservation of Historic Beijing and the Threat from Motorization*, Qinghua University, Beijing (1994)

(3) Hou Ren Zhi (Ed), *Beijing Lishi Ditu Ji (A Collection of Historical Maps of Beijing),* Beijing Chubanshe, (1985)

(4) MVA Asia, *Beijing Urban Transport Study Final Report*, prepared for the Asian Development Bank, (1996)

(5) The MVA Consultancy, *Beijing Traffic Management Study Final Report*, prepared for the UK Overseas Development Administration, (1992)

(6) The MVA Consultancy, *Beijing Transport Planning Study Final Report*, prepared for the UK Overseas Development Administration, (1994)

Urban Public Transportation System of the Municipality of Beijing And its Operations and Use by People

Eva Lerner-Lam[1]
Zhao Wenzhi[2]
Zhao Boping[3]
And Zhuang Jianhua[4]

The objective of this paper is to describe the total existing urban public transportation system in the city in terms of its size, operations, and use.

Beijing's Urban Public Transportation System

The Beijing Municipal Government operates a subway rail system and a surface system of trolleybuses, buses, jitneys and taxis. The Beijing Subway Corporation runs the subway system, and the Beijing Public Transport Corporation (BPTC) operates the surface public transport modes. These two municipal organizations serve a general population of nearly 12 million. The basic organizational structure of the administration of subway and bus transport is shown in Figure 1.

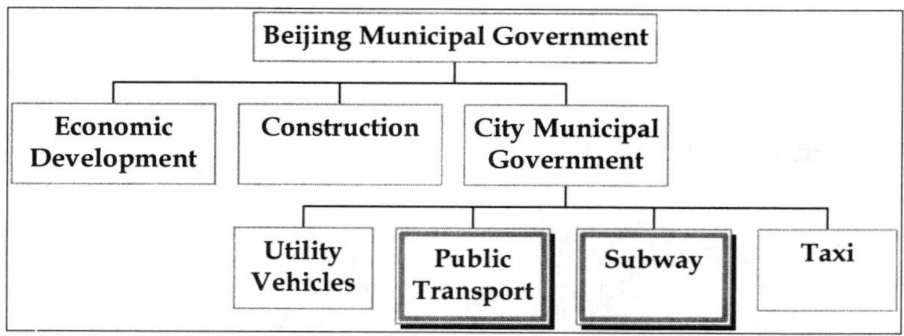

Figure 1. Organizational Structure of the Beijing Public Transport and Subway Administrative Units of the Beijing Municipal Government

[1] President, Palisades Group International, New York, New York
[2] Chairwoman of the Board of Directors, Beijing Public Transport (Group), Ltd., Beijing, China
[3] Director, Urban Transportation Institute, China Academy of Urban Planning and Design, Beijing, China
[4] Engineer, Beijing Municipal Administration Committee, Beijing, China

Beijing Subway Corporation

The BSC operates two subway lines. The East-West Line is a 24-km line with 17 stops that runs from the main station at Beijing Railway Station to Pingguoyuan, a western suburb of Beijing. The Circle Line is a 16-km line that presently has 13 stations; the line begins at the Beijing Railway Station, circles the northern area of Beijing, and ends at Fuxingmen, on the western edge of the central Beijing.

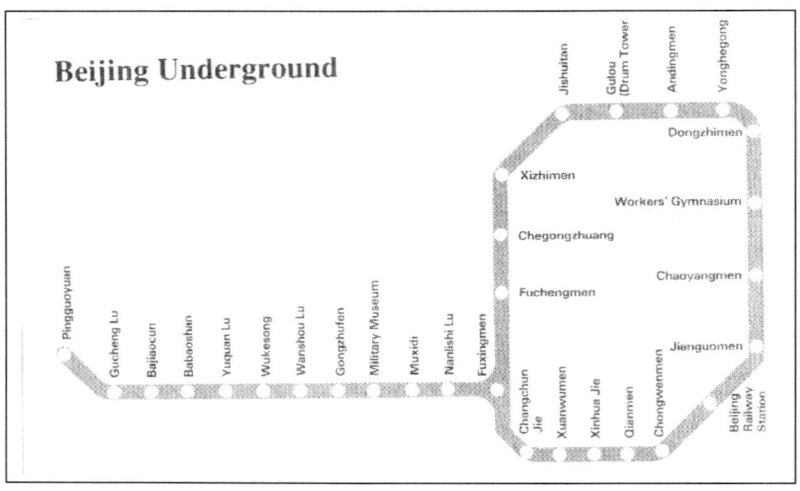

Figure 2. Map of Beijing Subway System

In 1997, the BSC carried a total of 447 million passengers. Fares are 2.00 RMB (about US 25 cents). Operating costs for the "core" business of running the two subway lines totaled 58,700,000 RMB: 17,600,000 million RMB for salaries and benefits, 13,200,000 million RMB for repairs, 6,900,000 million RMB for electric power and 4,300,000 million RMB for other expenses. The BSC employs 14,300 workers.

Ridership trends are shown in Table 1.

Year	Ridership (in millions)	% Change
1971	8	NA
1972	15	88%
1973	11	-27%
1974	11	0%
1975	19	73%
1976	26	37%
1977	28	8%
1978	31	11%
1979	49	58%
1980	55	12%
1981	65	18%
1982	73	12%
1983	82	12%
1984	103	26%
1985	140	36%
1986	158	13%
1987	192	22%
1988	307	60%
1989	311	1%
1990	382	23%
1991	371	-3%
1992	428	15%
1993	491	15%
1994	533	9%
1995	551	3%
1996	444	-19%
1997	445	0%

Table 1
Beijing Subway Corporation (BSC)
Ridership Trends, 1971-1997

The BSC has seven operating divisions:

1. Passenger Service
2. Old City
3. Tai Ping Lake
4. Electric Supply
5. Signal Systems
6. Equipment and Electrical Maintenance
7. Construction and Line Maintenance

There are also five branch companies:

1. Beijing Subway Coach Factory
2. Subway Building Company
3. Subway Security Company
4. Subway Construction Assembly and Engineering Company
5. "Eight Lines" Subway Company, Inc.

And five human resources bureaus:

1. Personnel Training Center
2. Party Education Center
3. Public Security Bureau
4. Technical Training College
5. Design and Research Center

Beijing Public Transport Company

More than 85% of all public transport passengers in the municipality are carried by the Beijing Public Transport Corporation (BPTC) on buses and trolleybuses, taxis, long-distance buses and minibuses. The BPTC operates 645 bus lines and 308 electric trolleybus lines. In 1997, BPTC provided 73,380 route-km of bus service and 7,533 route-km of electric trolleybus service, and carried more than 3.3 billion passengers.

The bus fleet consists of more than 12,500 vehicles, as shown in Table 2.

Electric Trolleybus	5,072
Tour Bus	396
Articulated Bus	121
Minibus	1,527
Taxi	4,343
Long Distance Bus	1,005
Others	43
TOTAL VEHICLES	**12,507**

Table 2
Beijing Public Transport Corporation (BPTC)
Vehicle Fleet Inventory
As of June 1998

Service Trends

The BPTC has experienced significant growth in ridership over the past fifty years since the establishment of the People's Republic of China.

1949	372.8
1966	12,232.3
1978	19,141.0
1990	36,748.8
1995	38,914.0
1996	50,036.2
1997	73,380.0

Table 3
Beijing Public Transport Corporation

Vehicle Service Miles (buses, minibuses and trolleybuses)
Unit: 10,000 kilometers

Organizational Structure

The BPTC operates fifteen units:
Beijing No. 1 Bus Company
Beijing No. 2 Bus Company
Beijing Trolley-bus Company
Beijing Long-distance Bus Company
Beijing Taxi Company
Beijing Bus Assembly Plant
BPTC Cadre School
BPTC Training Center
BPTC Driver School
BPTC Technician School
BTPC Advertising Company
BPTC Materials Supply Comopany
BTPC Real-estate Development Company
BPTC Transportation Research Institute
BPTC Enterprise Culture Institute

In all, the BPTC employs 69,076 workers, and carries 3.3 billion passengers annually.

Organizational Reform

Beginning in 1996, the Beijing Public Transport Co. has undergone major organizational reform. In particular, it has set up new business divisions that can target specific markets groups. Resources are then deployed in the manner of private enterprises, with marketing and operating plans that are geared towards attracting new riders and providing them with a desired level of service at a competitive price. Such market-oriented reforms have resulted in the creation of new, long-distance tour operations for the fast-growing leisure travel market as well as new, popular "commuter" services for laborers in the outlying agricultural areas that provide access to jobs in the urbanized area. The new organizational structure is shown in Figure 3.

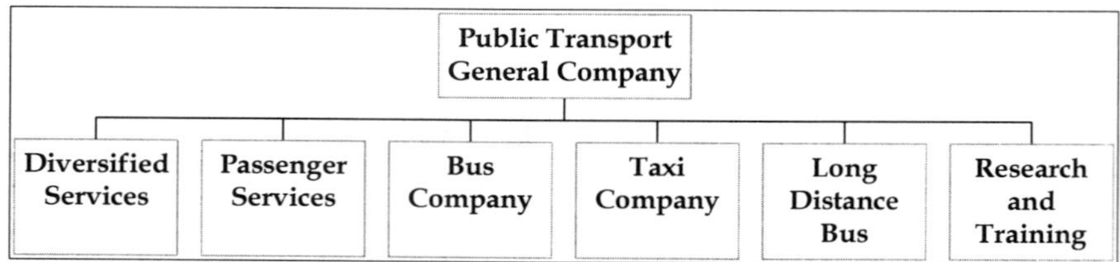

Figure 3. Organizational Structure of the Beijing Municipal Public Transport General Company

Future Challenges and Activities

Continued Market-Driven Provision of Services

The public transport organizations within the Beijing Municipal Government must generate revenues and increase the amount of capital available to improve service and invest in more advanced systems and equipment. To this end, the BPTC is aggressively privatizing profitable market services such as long distance bus tours. Profits generated by these activities are used to offset operating deficits and to invest in new buses, equipment and training facilities.

Deployment of ITS Technologies

The BPTC is in the process of procuring automated technologies (known as Intelligent Transportation Systems (ITS), or Advanced Public Transportation Systems (APTS)). Such technologies can significantly improve the operating efficiencies of the BPTC by automating the vehicle location, fare processing and other functions that are currently performed manually or not at all. One of the difficulties encountered is the lack of international equipment, hardware and software standards, thus preventing the BPTC from being able to pick and choose among various vendors for different functional components. The Research Institute of the BPTC is seeking ways to incorporate standards into the procurement of new technologies in order to make the deployment of such systems more economically efficient.

References

1. Beijing Public Transport Company, Ltd., Annual Report, 1997.
2. Beijing Subway map, Odyssey Illustrated Guide to Beijing, edited by Don J. Cohn, 1992, p. 40.

Beijing Public Transport Company: Facing the 21st Century

Zhao Wenzhi[1]

Abstract

As the major operator of Beijing conventional surface public transportation, the Beijing Public Transport Group Company, Ltd. (BPTC) provides services not only to the urban and suburban area but also some long distance transport to nearby provinces.

Rapid growth in population and private automobile use has worsened the traffic condition in the Beijing region. The general public realizes that public transportation can improve traffic congestion and improve efficiency. The BPTC recognizes the important responsibility it bears for resolving key public transport issues and has strategically prioritized its future activities to meet the challenges of the 21st Century.

Introduction: Beijing Municipality's Transportation Conditions and Characteristics

Beijing is the capital of China. It is a famous cultural city with a 3,000-year history. It covers an area of 16,800 square kilometers. The total population is more than 11,000,000, including 7,220,000 urban population. As a famous tourist city and the political and cultural center of the country, Beijing attracts enormous numbers of international and domestic tourists, in addition to business people. The average transient population reaches 3,500,000 yearly.

Beijing City consists of 8 city and suburb districts. This comprises an area of 1,000 square kilometers, including 500 square kilometers of developed area. The

[1] President of the Board of Directors, Beijing Public Transport (Group), Ltd., Beijing, China

center of the city has kept the historical grid network of streets. A ring road and the connector road system connect the suburban newly developed areas. Starting from the center of the city toward the suburb, there are 2nd Ring Road, the 3rd Ring Road, the 4th Ring Road and the 5th Ring Road (also named as Ring Highway). Within the 2nd Ring Road is the "old city" that covers 62 square kilometers. Within the 3rd Ring Road is the developed area that covers 150 square kilometers. The 4th Ring Road surrounds a partially developed area and the 5th Ring Road encompasses largely undeveloped land. The primary transportation mode in the Municipality of Beijing is the bicycle. The number of the bicycles in Beijing is approximately 9,000,000. More than 35% of the population uses bicycles. The next most popular transportation mode is public transport, which accounts for less than 40% of all trips made. Other transportation modes are company vehicles and private vehicles, which account for about 24 % of all travel.

Beijing public transportation services include subway, trolley bus, bus, minibus and taxi. Among these services, trolley bus and diesel bus service carry the greatest proportion of passengers. Together, they carry 74% of all passengers, while taxis carry 15% and the subway carries 10% (1997 China Statistical Year Book).

Although private automobile use is not predominant, it has increased rapidly in recent years. The average growth rate exceeds 20%. At present, private automobiles comprise 20% of the total vehicles.

Beijing's Urban Transport Characteristics

The characteristics of Beijing's urban transport can be summarized as follows:

<u>Population density is not evenly distributed.</u>

70% of the permanent residents and almost the entire transient population lives in 10% of the land area. This high population density causes the city center, especially within the 3rd Ring Road, to become overcrowded with residents and too many vehicles. The road network is insufficient, and transportation demand exceeds supply.

<u>Because of historical and economic reasons, Beijing's road system has always been under capacity and poorly structured.</u>

The rights of way for streets in the city center is only 10%; this percentage will only increase to 20% with the road improvements specified in the city long-term plan. This is only about half of the street network density of other, similar large cities such as Washington, DC (45%) and London (35%). There are other problems as well, such as the distant spacing of major arterials and the imbalance between the density of land use and the layout of tertiary roads. These problems are very difficult to solve at the municipal planning level.

Too many bicycles, and no separation of motor vehicles and non-motor vehicles.

The traffic density of bicycles and the lack of separation of bicycle and motor vehicle flows not only cause frequent traffic accidents, but also create traffic jams that greatly reduce the capacity of the overall street network.

A Lack of a large capacity transit system and will not change in the near future.

The conventional passenger transport modes such as trolley bus, minibus and taxi will continue to dominate public transport.

Therefore, the only solution for Beijing's transportation problems is the adoption of a strategy of transit priority. More efforts should be devoted to develop high capacity railway and bus systems in order to accommodate more passenger demands on limited road networks.

Major Conflicts and Problems in Public Transportation

Beijing Public Transportation (Group) Company, Ltd. (BPTC) is the largest public transport provider in Beijing. The Group consists of 11 passenger and maintenance branches as core enterprises, 7 sole ownership companies, 4 shared ownership companies (BPTC is the major partner and controls these companies), and several related enterprises, including training and research institutes. BPTC has 70,000 employees and its total assets are RMB 4.98 billion.

As the major operator of Beijing's conventional surface public transportation, BPTC provides services not only to the urban and suburban areas, but it also provides some long distance transport to nearby provinces. BPTC has 12,525 various vehicles, and 645 operating routes that cover 71,438 operation kilometers. The vehicles include 5,072 trolley buses/buses, 353 tourist buses, 115 double-deck buses, 1,535 minibuses, 4,395 taxis and 1055 buses for long distance transportation to other cities or provinces. The core of the transport system is trolley buses/buses, which number more than 5,000 vehicles. There are 308 routes that provide more than 5733 kilometers of service. The annual ridership reached 3.3 billion in 1998, comprising 70-80% of total system ridership (subway and surface transport).

Rapid growth of population and private automobiles has worsened traffic conditions in the Municipality. The general public realizes that public transportation can reduce traffic congestion and improve efficiency. Political leaders have established the strategy of prioritizing public transportation. As the largest public transport provider in the capital, BPTC bears the very important responsibility of resolving key public transport issues.

Here are some aspects of the problems:

1. Insufficient supply of vehicles cannot keep up with the need for transit

During the last few decades, the population of Beijing has expanded rapidly. Right now, the total number of residents has surpassed 11 million, which is an increase of 1 million residents since 1990. The total number of transients is 3.5 million, or five times the number in 1990. Meanwhile, the urban area has been extending at a rate of 11 sq. km annually. All of these factors lead to multiplying trip demands.

In contrast to the above situation, during the same period, the amount of buses reached 5,072 or 4.23% of the total number of motor vehicles. The increase in the rate growth in the number of buses is less than 10%. According to surveys, many lines are over capacity. There are 60 lines on which the amount of passenger trips per line exceeds 1,500 in the peak hours. Among the 209 newly built residential areas, 131 (or 62.7%) have good transport service; however, 57 (27.3%) of these areas have insufficient service and 21 (10%) are without transit service due to deficiency of roads and transit facilities. Consequently, there is an urgent need to raise the capacity of transit.

2. The performance and structure of vehicle fleet cannot satisfy various passenger needs

The active buses, manufactured mainly in the 1950's and 60's, have many shortcomings, such as high floors, loud and inferior noise and temperature insulation. In addition, a significant number of buses do not meet national regulations, and 1,779 active vehicles will be removed from the fleet by the year 2000.

In selecting new vehicles for procurement, there are three factors that should be taken into account. First, the average income level of ordinary citizens is increasing steadily; secondly, passenger trip needs are diversifying; thirdly, more and more families have access to private cars. Therefore, it is difficult to attract more people to use public transit modes with current vehicle performance and service pattern.

3. Deteriorating conditions of roads and decreasing speed of vehicles

Surveys indicate that the average commercial speed of buses has dropped by 15.6% since 1990. It is calculated that the loss of capacity amounts to 1,000 buses; the average travel time by transit passengers has increased by nearly 9 minutes. In other words, the number of lost hours suffered by society exceeds more than 0.75 million hours per day (assuming an average of 5 million daily passenger trips). In the central area of Beijing, the commercial speed of buses drops from 16.7km/h to 9.2km/h, even slower than that of bicycles.

4. Out-of-date administrative measures and patterns

Due to inadequate utilization of modern technologies and equipment, especially in the area of communications, the dispatchers of transit companies do not receive timely and relevant information regarding the actual running buses. It is therefore impossible to take swift actions in the light of field situations. This generates another problem: the low efficiency of the dispatcher means that each dispatcher is only able to manage a limited number of vehicles; therefore, the organization of passenger service has to be based on the level of lines. As a result, BPT is obliged to employ more workers for management and dispatching. This significantly increases the cost of human resources dedicated to operations.

5. The management system and operating mechanism needs to be improved

A public transport company ought to take into account the interest of the public as well as itself. On one hand, it should meet the needs of the society, sometimes even sacrificing its own benefits. On the other hand, it needs to care for its own interest and obtain payment for the transport service. Under the conditions of the market economy, how to adequately coordinate the relation between the two aspects is very important.

From 1985 to 1996, the deficit of BPTC has been increasing with the sum of the deficit in 1996 reaching RMB 900 million-yuan (US $). The main cause is that fares charged on passengers are far below the costs. For example, right now the consumer cost of a monthly ticket is RMB 70.18 yuan (US $) per ticket, but the average selling price of a monthly ticket is only RMB 19.53 yuan (US$), or 27.8% of its cost. This is because the government controls the price of the monthly tickets; it has only made four adjustments during the last four decades, and the rate of increase in the cost far exceeded that of the price of the tickets.

With respect to the financial aspects of operations, there are several issues that need to be resolved, such as the fact that there is only a single investor (namely, the government), limited financing channels and high operating costs. It is a difficult task for BPTC to attract more funds from various resources and enhance its share of passenger transport market in light of these constraints.

In the field of management, there are also some challenges in order to meet the requirements of modern enterprises, such as simplifying the managerial structure, enhancing the efficiency of management and reforming the mechanism of subsidy, labor and revenue distribution.

Accordingly, there are three issues to be resolved:

a) How to establish and regulate the price of tickets.
b) How to distinguish the loss between mismanagement and the loss of suffering from government reasons.
c) How to acquire enough subsides and investments for further development.

Solutions and Measures

Obviously, there are many causes of the above-referenced problems such as outdated management procedures, slow application of science and technology, challenging economic conditions, policy constraints, and other external and internal factors. Hence, it is impossible to use a simple measure to solve all the problems. Instead, a set of systematic and comprehensive measures should be taken, such as adopting the strategy of transit priority, establishing specific development goals, utilizing technology to improve efficiency and effectiveness, etc.

Development Goals by 2010

The objectives of transit priority are to ensure transit has the leading status in urban transport and attract more people to use transit instead of bicycles and private cars. As the primary provider of transit service, BPT has set the following goals:

a) Increase the number of annual passenger trips from the current 3.3 billion to above 5 billion and maintain the market share in passenger transport.
b) Expand the bus fleet from approximately 5,000 to 10,000, or 15 buses per every 10,000 inhabitants;
c) Increase the number of bus routes from 300 to more than 430 and increase the total route length to 7,500 km.
d) Increase the length of the "busway" network to more than 100km, and in this way, form a speed transit system;
e) Raise the average commercial speed to 20km/h on urban main roads and 25km/h on ring roads.

After the accomplishment of these goals, there will be significant improvements in capacity, quality and scope of passenger transport service in Beijing.

Relevant Measures

In order to efficiently solve the exiting problems concerning transit industry and realize the goals of BPT, the following measures should be taken:

1. <u>Improve the operating conditions of transit companies and give them priority over traffic management</u>

The Beijing municipal government increasingly has paid more attention to transit development all the time, and recently adopted the traffic management strategy of giving priority to public transport while limiting the total number of motor vehicles. Since 1997, municipal government has opened 5 bus lanes on main streets like Chang'an Avenue. It is estimated that subsequently, the operating efficiency of buses has been raised by about 15%.

In the opinion of the author, there are three issues that should be emphasized in the development of urban transit. They are as follows:

a) Strengthen the management of planning with consideration of land use and population distribution in order to generate a unified and effective traffic management system.
b) Control the total amount of urban transport at a reasonable level and adjust its structure with emphasis on transit.
c) Reinforce the coordinated management of the transit market, work out relevant transit laws and polices, and introduce a system of franchise on transit routes. For the sake of creating better operation conditions for the passenger transport market, both legal and administrative means are needed.

2. <u>Accelerate the process of transforming the transit industry to be market-oriented and establish an open and secure mechanism of investment and finance</u>

It is an urgent and difficult task for transit companies to transform themselves to be market-oriented. In this process, there exist two basic contradictions: transit demand vs. supply and developments vs. available funds. The key to solve them both is funding, but can reliable sources be found? The answer to this question is to find more financing channels and pay more attention to the following three areas:

a) **Reinforce assets management**;
BPT now owns quite large sum fixed assets and the ratio of assets to liabilities is 28%. So the key is, through asset reorganization, form a sizable and beneficial capital industry. In the construction of large hubs of communications and clean-fuel filling stations, we will follow the way of co-operation with other partners to draw more funds.

b) **Improve current subsidy policy;**
According to an old practice, two sides on an annual basis determined the sum of subsidy that transit operators received from government. Now the municipal government has introduced a new subsidy method, which freezes the total amount of subsidy for three years and then reduces it progressively. Such a

policy also requires that transit operators streamline services and justify their requests for subsidies.

c) **Establish a rational price system;**
With consideration of affordability of ordinary passengers, a rational price system will allow the fares to float regularly and the price level to be determined in light of per capita income and the general price index. At the same time, it is also important to reform the structure of tickets and use advanced techniques like IC farecards, to prepare transit operators for future challenges.

3. <u>Reform current management system and operation mechanism to conform to the requirements of market economy</u>

During the last several years, reforms were focused on the following three aspects: management systems, operations and fare structure. Since 1996, BPT has been modifying the composition of passenger transportation, vehicle fleet, business range and labor force.

With regard to the needs of the passenger market, we have increased the ratio of special lines, touring coaches and taxis, and established a new structure of passenger transport with buses taking the leading role.

In the first half of 1998, BPT streamlined its management system. Approximately thirty-one work units were dissolved and 1,500 administrative personnel and 2,000 workers were relocated. At the same time, BPT divided bus services into 11 specialized subsidiary companies and 5 new corporations in terms of business range and geographical area. These reforming measures have brought about positive effects on the performance of transit.

Summary

In the coming years, we will continue to complete the reforms on the structure of business and operation mechanism with emphasis on the following tasks:

a) Construct a modern enterprise system which is characterized by clear property rights, definite rights and duties, and scientific management;

b) Reorganize the business structure in order to set up a new organizational structure for the transit group, with bus service as its main business;

c) Spread the use of clean fuels like VNG (compressed natural gas) and develop new vehicles with low floors and lower levels of pollution;

d) Make full use of current assets, including vehicles, maintenance facilities, and real estate to develop diversified business areas;

e) Establish a rationalized subsidy mechanism and cost management system;

f) Set up a reasonable distribution system on the basis of efficiency and benefits;

g) Work out an efficient and flexible employment system and reduce the rate of nonproductive staff from 24% to 17.5%

4. Utilize new scientific and technological achievements

The development of urban transit should leverage the advancements of science and technology. By October 1,1999, BPT will complete the construction of a transit dispatching center with the use of Intelligent Transportation System (ITS). This project has applied the use of many modern technologies like synthetic voice communication, vehicle positioning, computer network, big screen monitor and non-contact IC farecards. Through the application of these new technologies, the management level of transit will be greatly enhanced.

Prospects and Characteristics of Urban Public Transport in China

Zhao Boping[1], Zhao Jie[2], and Jun Wang[3]

Abstract

This paper describes the policies of urban transport and public transport, the characteristics of land use, and the operation management of state-owned bus companies in Chinese cities. It also presents current status of public transport and discusses the development trend of urban public transport in China. To better understand an overall situation of urban public transport in China, the paper presents statistics and detailed information obtained from sample cities around the country.

Introduction

It is clearly said in a published national policy of science and technology that development of public transport would be a priority in developing China's urban transportation system. The public transport system should be well developed in more than 600 cities all over the country. In addition, mass transit systems need to be developed in large cities with over a million populations.[1] In 1997, a policy of national sustainable development, again, pointed out that the development of public transport system should be an important strategy for urban transportation development as well as urban environment improvement.

Because of above-mentioned national policies, urban public transport has made a great improvement in the nation's urban infrastructure and transportation system. Some of the highlights about development of urban public transport systems in Chinese cities are listed in follows:

---

[1] Senior Engineer, Director, Urban Transport Center, Ministry of Construction, China

[2] Senior Engineer and Deputy Director, Urban Transport Center, Ministry of Construction, China

[3] Transportation Engineer, Turner Fairbank Highway Research Center, McLean, VA, USA

- During the 18 years from 1978 to 1996, China's bus vehicles and length of bus route increased 4.7 and 2.8 times, respectively. The bus ratio reached 0.7 vehicles per 1,000 persons (Figure 2). Meanwhile, the quality and standard of public buses had improved greatly. During the period of "the Eighth-Five Plan" years, China's minibus industry had developed from zero to about 100,000 minibuses all over the country by the end of 1995. 1994 marked a peak demand of minibuses in all Chinese cities.

- The most rapid growth during "the Eighth-Five Plan" period was taxicabs. The number of commercially operated cabs increased from 1,628 in 1978 to 585,000 by the end of 1996, about 359 times increase. Among the increase, there were 5,300 in Beijing, 39,000 in Shanghai, and 15,000 in Guangzhou. As of August 1995, the number of taxicabs in Beijing had increased to more than 80,000. It is anticipated that, in "the Ninth-Five Plan" period, the taxi service will enter a stabilized development period, with room for growth.

- The cities where subway systems exist are Beijing (43.5 km), Tianjing (7.4 km) and Shanghai (16.1 km). It covers about 67 km of subway routes. Currently, there is a 18.45-km subway under construction in Guangzhou, and several subway systems are under consideration including a 14.8-km line in Shenzhen, a No. 3 Line in Shanghai and a Fu-Ba Line in Beijing. The mass transit systems are expected to be developed significantly during the period of "the Ninth-Five Plan."

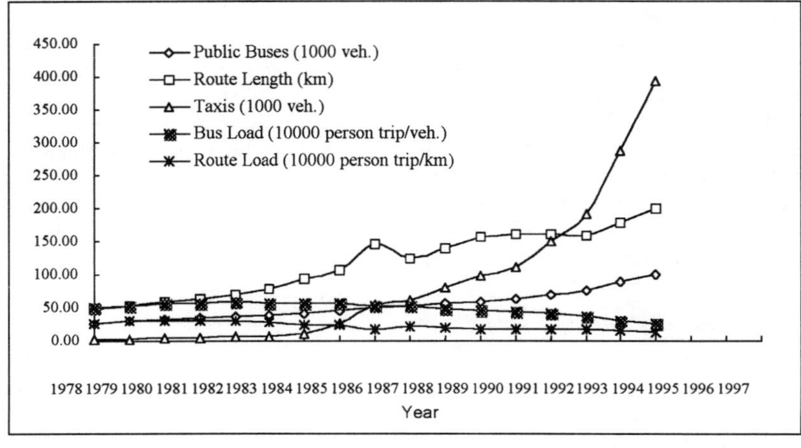

Figure. 1 Capacity and volume of public transport from 1978 to 1997

At the same time, with China's economic development, private motor vehicles use had grown more rapidly, compared to the public transport. From 1978 to 1995 when motor vehicles had grown most rapidly, the annual increase rate of China's motor vehicles was 21.97% and that of the cars or other passenger vehicles (excluding motorcycles) was 13.19%. Among this, more than 70% of motor vehicles were in the 600 cities mostly in big cities and metropolises. According to statistics, motor vehicles in 34 metropolises such as Beijing, Shanghai, Tianjing, Guangzhou and Shengyang account for 50% of total ownership in the entire country. The increase rate of motor vehicles in large cities and metropolises is often higher than that of the country's average rate. The trend of private car ownership has been more concentrated in urban China. For example, from 1994 to 1996, private car ownership in Beijing had increased 130%, mounting up to 145,000.

A competition between private vehicles and public transport has arisen. Although city's public transport facilities and services have improved significantly, the public transport equipment and facilities still fall into a low level. Most public buses in China are still out-of-date single vehicles or articulated vehicles. The cost-effectiveness of public transport system is still low. As a result, the ridership of public transport has been decreasing. For example, the ridership in 1994 was 6.5% lower than that in 1992 for the entire country. Lack of cooperation and coordination between public transport and whole transportation system becomes an obvious problem. There is still a long way to go for the government to carry out "public transport priority" policy pertaining to allocation of urban street resources. Decreasing transportation efficiency resulting from traffic jams undermines the transportation efficiency brought up by the development of public transport facilities. The speed of buses falls from 12~14 km/h to 5~10 km/h. The increased capacity in public transport is eroded by such a decrease in transport efficiency. Survey in many cities indicated that the ratio of passenger transportation taken by public transport vs. bicycles has been declining. For instance, the ratio in Shanghai has fallen from 6:4 in 1980s to 4:6 in 1990s. In both Tianjing and Xiamen, the ratio has fallen from 2:8 to 1:9. It is more difficult to coordinate the development between urban mass transport system and growth of using motor vehicles. Furthermore, the existing urban rail transport systems have poor coordination to regular public transport systems, which make the efficiency of entire public transport system decrease substantially and far behind that of other transportation modes.

Challenges and Problems Faced

With increase of economic standard and people's income coupled with intensifying the pace of modern life and dramatic growth of automobile industry, the philosophy of travel behavior among all urban inhabitants has changed greatly. People demand more comfortable, convenient, accessible transportation and shorter travel times. They are no longer satisfied with the traditional travel modes such as walking, bicycle, and substandard public transport service. They expect to have their trips more comfortable, flexible, and faster. Trips using automobiles have increased rapidly and they have even extended to commuter travels - a main travel purpose in urban areas. When private cars are unavailable to normal families due to various limitations, taxicabs become a better choice to the inhabitants in large cities. Demand has resulted in a dramatic increase in taxicabs. The number of taxicabs becomes much higher than that in most other countries. This may reflect a potential of private cars entering into urban families. In the meantime, it also challenges the services provided by public transport systems, and the development of sound urban transport systems.

The speed of urbanization has been accelerated along with motorization in many Chinese cities. Due to lack of suitable land, densely populated urban structure has to be developed in Chinese cities. Because of the development of economy and auto industry, Chinese cities are imposed to accommodate more population to support growing market of national industry. This leads to an ultimate conflict between the land use of urban areas and motorization development. It is expected that within the next 15 to 20 years, China will face a great pressure of growing population, land use and motorization development.

Currently, the land development pattern in most Chinese cities is the so-called centralized "pie-shaped" pattern. The city's radius is relatively small. A survey conducted in 1995 indicated that the average radius of 26 cities, including duchy cities, province capital cities and sub-capital cities, which hold a million population or more, is about 7.1 km. Among these cities, the inhabitants' average travel time is 20~30 minutes.

The average travel time of public buses, including the time spent waiting for buses, walking or riding bikes from/to each destination, is about 50 minutes. The average travel distance of urban residents is less than city's radius. The long-distance travel by public transport is no more than 10% of total trips. Because of such high density of urban land-use, travel density is highly concentrated. Therefore, a high-quality public transport system would be the best choice for a sound urban transport system.

The suitable travel range of a car is generally longer than 20 km, which falls into the range of suburbs of most Chinese metropolises. In recent years, many metropolises have established higher-class highways and radiant roads that stimulate the suburbanization to many Chinese cities. With the rowing number of automobiles in cities, urban China has evolved from centralized-layout pattern to distributed-layout pattern, and this accelerate suburbanization in to urban areas. However, with highly concentrated urban population and unavailability of suitable urban land, it is questionable that China can supply the land demand for this kind of change in the urban pattern. For example, a national regulation on land use policy (GB 137-90) being carried out now requires that the planned average areas per capita is 60 to 120 square meters. According to development experiences from foreign countries, it is difficult to meet the demand of urban motorization from this land-planning index

(Figure 2). A recently issued "Land Administration Law of P.R. China" attests that urban areas cannot supply the required land use for motorization needs.

Table 1 Land Use Radius for 26 Cities in China

Cities	City Land Area (km^2)	City Radius (km)	Cities	City Land Area (km^2)	City Radius (km)
Beijing	476.8	12.3	Wuhan	200.0	8.0
Shanghai	390.2	11.1	Changsha	101.0	5.7
Tianjing	359.3	10.7	Guangzhou	259.1	9.1
Shijiazhuan	97.0	5.6	Chengdu	129.0	6.4
Taiyuan	170.0	7.4	Guiyang	85.7	5.2
Shenyang	185.7	7.7	Gunming	99.5	5.6
Changchun	124.0	6.3	Xian	148.0	6.9
Harbin	156.	7.0	Lanzhou	162.5	7.2
Nanjing	151.0	6.9	Urmqi	82.5	5.1
Hangzhou	102.2	5.7	Dalian	217.7	8.3
Nanchang	68.0	4.7	Qingdao	106.0	5.8
Jinan	113.8	6.0	Chongqing	184.0	7.7
Zhengzhou	108.3	5.9	Shenzhen	88.0	5.3
			Average		7.1

Note: The cities listed above are those of duchy cities, capital cities of provinces and sub-capital cities, which have a million population or more. Data sited from "Annual Statistics Book of Urban Construction", Ministry of Construction, P. R. China, 1995.

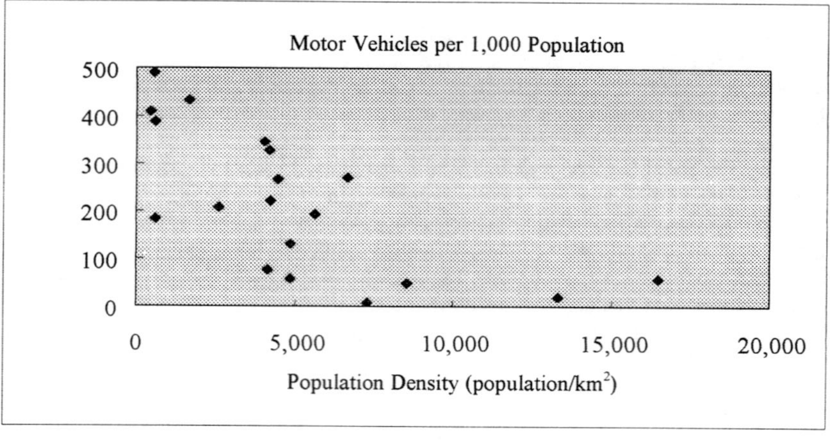

Figure 2. Motor Vehicle Ownership vs. Population Density (1980 data).

It is believed that, in the next fifteen years, urban transport development will bring a profound and long-term influence to the development of urban China and its public transport systems. Unfortunately, the urban policymakers both in central and city governments have not yet recognized the importance, urgency and priority of developing sound public transport systems. Thus, in reality, the policies and implementations of the public transport development appear inadequate. They clearly reflect in following aspects:

- **Financial policy**: From central to local governments, there are no stable financing sources for public transport development. As to the investment of urban infrastructure, the main focus from the governments is still on building expressway system and grade separation interchanges. In most cities, the investment in bus transport has only reached to 5% of that in road development. It is even harder for developing the urban mass transportation system. Most of such investments depend on foreign loans and land concessions. The investment in public transport technology is even less. Moreover, there has been no definite and stable national finance policy and regulation to support public transport. Therefore, the survival and development of local bus companies lies in the extent to which the local government pays attention to it.

- **Institutional management**: The administration of national public transport is generally only on the management of bus companies nationwide. As for the development of public transport system and the coordination with other transportation modes, there is no specific government agency that is in charge. Although passenger transportation market has introduced strong competitions among transit companies, the self-interest and non-regulated market competition affect those transit companies greatly. It sometimes makes the passenger transportation market chaotic with low operational efficiency for the operations. The overall administration of urban transport system is still in a low-level period.

- **Planning and construction**: Most cities' public transport development is still planned by individual public transport companies concentrated on purchasing new buses and opening new routes. It is not coordinated with programs of entire urban transport development, land-use strategy, and the urban inhabitants' expectations. In addition, the speed of public transport development is not comparable with that of urban development.

- **Urban transport policy**: In many Chinese cities, the current policies on use of limited urban road resources and policies on individual subsidy given by the local governments are apparently favored to private transport modes, including those with low passenger capacity modes.

- **Staff training and technical standard**: The lack of technical training for transportation professionals and out-of-date technical standards in public transport system also adversely affect development of public transport. Research and technology development in public transport area is apparently inadequate. Less attention has been paid to so-called "soft technology," such as the management technique for urban transport development.

In the past forty years, like other state-owned enterprises, bus companies have operated with a comprehensive management style which covered both urban and suburban, bus operation and repair, existing and new bus routes attempting to wait for, depend on and request the government's help. With economic reforms, state-owned enterprises have experienced changes and innovations that led old ways of operation to be reformed. However, they are faced with many challenges of mechanism, system and fare regulation transforming from planning economy to market economy. In addition, they are facing new challenges of information, new technology, and better use of human resources.

Future Policies and Actions

Although the public transport has been developed slowly and many problems still exist, with the urban transportation becoming increasingly congested and causing environmental deterioration, in recent years, both central and local governments have began to enforce their power to develop sound public transport systems. There is still a distance between the efforts and overall urban development; however, it is obvious that more attention is being paid to urban public transport than any time before. This is reflected from following sections:

- **Planning and construction**: Many cities enforce urban transport planning with developing public transport, especially when developing a comprehensive urban transport plan and carrying out that plan. Many municipal governments emphasize optimization of bus routes and give buses a priority of using urban streets. Many cities, such as Guangzhou, Shantou, Luo yang, Beijing, Nanjing, Xiamen, Chongqing, Shenzhen, Shanghai and Baoji, indicate clearly in their recent urban transport plans that exclusive bus lanes and bus transfer hubs be implemented and built. For example, Shanghai proposes to establish an exclusive bus network system with "five horizontal, five vertical, and one big ring routes." It will build 7 to 8 large-scale bus hubs within its inner ring area. It also sets up a goal that the city's periphery residents need only transfer once with the public transport system going to downtown area. In June, 1997, Beijing established a exclusive bus lane on its well known Changan Street (Before its opening, the bus speed was 16 km/h. After the opening, the speed increased to 20 km/h. The bus on-time arrival rate improved 43.6%) consequently, a plan of building an express bus network with the routes on two rings, three vertical, five horizontal and eight radiation roads was proposed. A so-called "green channel" of public transport corridor in Shenzhen was also developed. Besides these developments, the state began to ease a two-year ban on building urban rail transport system set up two years before. In 1998, with a policy of increasing investment to urban infrastructure, the urban rail transport system has listed as a priority area for the investment. Public transport development began to enter a new era.

- **Regulation and administration of urban passenger transportation market**: According to local development stages, many cities established a single office to administer passenger transportation, and issued appropriate regulations on managing the city's public passenger transportation, city taxis, mini buses and private bus companies. This makes the passenger transportation market more orderly for a high quality service and fair competition. Many cities also implemented regulated bus operation routes, such as those in Shenzhen. These routes have been in service for the past five years. Currently, about 46 such bus routes are authorized by the municipal government to Shenzhen public Transport Company, Ltd. for operation. This kind of operation changes the traditional operating system, which mostly depends on government subsidy. Other cities in China learned from the Shenzhen's experience and combined it with their own situation to carry out appropriate bus operation reforms. For example, in 1997 specialized bus lines in Yiyang City began to operate. The operation mileage, passenger volume and company profit increased by 38.6%, 66.2% and 66.8%, respectively, compared to the year before.

- **Transportation policies:** In most cities, transport policies have experienced notable changes. For example, Guangzhou has ceased to license motorcycles; Shanghai no longer licenses bicycles assisted by diesel power in urban areas and limits private cars, and motorcycles by way of auction. Due to policy reasons, Dalian's number of bicycles has decreased. Shenzhen removed 1000 minibuses from its urban routes and Nanjing pulled out 400 minibuses as a means to regulate the minibus operation. In many cities, taxi operations are better regulated and controlled. The transport policies of many Chinese cities have begun to favor bus operations and public transport development.

- **Enterprise innovation**: With the changes of government policy to limited subsidy, object management, unified planning and price regulation coupled with competition pressure brought by more public and private companies entering the market, public transport companies have had to speed up their innovations. This is reflected with the following facts:

  Establish new operating mechanisms: Public bus companies are undertaking reforms to their old operating system, which had been used for the past forty years. They establish a new system focused on an economic operating scale, specialization, regionalization, and market driven. For many years, the traditional system in most cities is a comprehensive one, ranging from urban to suburban, transport operation to vehicle repair and maintenance. Now they generally evolve into companies specialized in either transport operation or vehicle repair and maintenance. Many companies put in effect a system where individual bus operators are contracting with the company. The driver is responsible for the operation of his or her bus on a given route. Some companies adopt a system of shareholders. New operation systems are still under development in different cities and different companies. Shanghai, Beijing and Shenzhen are examining different models for their public bus operations. The overall trend of such innovation is shown to be very helpful for expanding the market and changing to more efficient management system.

Reform employment system: Similar to other state enterprises, many Chinese bus companies began to abolish a so-called iron-bowl employment system and change to a contracting system. The enterprises are free to choose their employees and employees are also free to choose their jobs. This reform makes the employment and competition for the job open to the public and bases pay on job performance.

Increase capital investment: Aside from the state capital investment, enterprises issue stocks internally and publicly, and collect capital investment from multiple channels for the development.

Improve bus conditions: Although most buses in urban China are traditional single vehicles or articulated bus vehicles, to meet the market needs, luxury buses, mid-size buses, and other types of buses are used. They include buses using natural gas instead of petrol, having low chassis, automatic speed shift, and air-conditioning or double decks. In December 1997, Guangzhou's last articulated bus, which served for 36 years, was retired. Buses using natural gas run in Urmqi and Zigong cities. Utilizing buses that cost less and are more environmentally friendly is a win-win situation.

Reform fare system: With a help from the government, many cities have begun to reform their monthly-pass system, which used to be much cheaper than buying ordinary tickets due to a planned economic system. The reform is basically to limit the use of the monthly-pass system and instead, use IC cards, automatic tickets and a single ticket system. Many cities have raised the prices of ordinary bus tickets simply because the long time low-fare system did not reflect the level of society development.

Enhance overall staff quality: By combining education and on-the-job-training in both the preliminary and advanced levels, many bus companies have begun to provide technical training to their staff. This also includes professional ethical education and behavior training. This should enhance the customer service of the staff and the enterprises in public transport services.

Strengthen management: According to the concept of enterprise management in modern society, the public transport enterprises need to implement a series of management techniques on contract, audit, operation, inspection, and quality control in the field. For example, a cost management system can be used to improve fiscal management of a company. The system should focus on enhancing the object management of income and expense, while reducing operation losses, breaking down accounting units, and separating the losses caused by public policy from those of operations. All of these not only can be a benefit to a company's operation but also reduce the burden of the government.

Increase level of service: The willingness of high quality services should be enhanced in a public transport company. The high quality services include improving appearance of buses, stations and facilities. In addition, bus operation structure and service aptitude need to be improved as well.

In general, state-owned bus companies are undertaking reforms in all aspects. They are now moving towards establishing modern enterprises. At the same time, several purely private bus companies are emerging in some coastal cities. It clearly shows that China's public transport is entering a new era.

Socioeconomic Characteristics and Transportation in Berlin

Jürgen Klemann[1]

Introduction

When the wall fell in Berlin in November 1989, this city moved back from its decade-long marginal and insular position into the center of a Germany which was reunified shortly afterwards and - from a geographical point of view - into the center of a Europe which was moving closer together. The new capital city has been under construction in Berlin since 1991. The total annual volume of construction is now around $15 billion (30 billion DM). The transfer of the Federal Government and Parliament into the city, the establishment of a new center, the reconstruction of streets, squares and highways - this is a task, which normally spans many decades. However, this task must be completed in the briefest possible space of time. Construction specialists, architects, engineers and politicians face a true challenge.

In the final analysis, Berlin's future as a site of commerce and services, as an interface between East and West, North and South, as a significant goods handling center and as an important focus for commercial relationships, including those with Eastern Europe, is dependent upon an intact transportation infrastructure which will ensure mobility in the future, too.

The primary goal of the city's traffic policy is therefore to ensure mobility in an intact environment. For this reason, Berlin is also actively participating in the development of new forms of mobility for conurbations on the basis of state-of-the-art information and communication technology.

Population, Land Use, and Travel Patterns

Berlin has 3.45 million inhabitants in some 1.8 million households, distributed throughout a polycentrally-structured major city with an area of approximately 889 square kilometers and encircled by thinly populated surrounding countryside. Every

[1] Minister for Construction, Housing and Transport, Berlin, Germany

day, each of Berlin's inhabitants undertakes an average of some three journeys. This adds up to 10.3 million journeys per day within the public domain of the city. If the journeys made by the inhabitants of the surrounding countryside are added to this, because these partially occur in Berlin as leisure and shopping journeys and partially as professional journeys, then approximately 12.5 million personal transportation journeys are undertaken within 24 hours. These journeys are made 26% on foot, 5% with bicycles, 23% with public means of transportation and 45% with private cars. The last two figures form the center of our attention.

The ratio of local public transportation (LPT) to motorized individual transportation (MIT) in Greater Berlin including the closer interconnected areas is currently 34:66. But this will not remain the case. The two halves of the city continue to grow together. This is borne out by the construction cranes on the Potsdamer Platz, in the Spreebogen, in the water-front districts of Oberhavel and Rummelsburger Bucht and in the areas of development of Adlershof and Johannisthal. However, the population of Berlin does not necessarily have to grow along with this process. More recent prognoses are rather careful. In the year 2010, approximately 3.6 million inhabitants are expected to live on the Spree.

Since the reunification, the focus has been on repairing the ailing transportation infrastructure in the eastern sector of the city, on rehabilitating and modernizing the train systems (city railroad and long-distance railroad) in West Berlin which have been neglected under the reign of the GDR, and on recreating the many interrupted transportation connections between the East and West. But due to the immense amount of work involved, the recreation of the railroad connections with the surrounding countryside ("closing the gaps"), in particular, has not yet been completed.

In order to estimate the transportation situation, it is necessary to examine certain geographical and demographic data. The city zone of Berlin stretches 38 km from north to west, and 45 km from west to east. Berlin has more than 3.4 million inhabitants, approximately 870,000 people live in the nearby surrounding countryside. Berlin has 1.46 million gainfully employed persons, the nearby surrounding countryside a further 300,000. 980,000 motor vehicles are registered within Berlin, with a further 230,000 in the nearby surrounding countryside.

The settlement structure comprises two inner-city centers and various sub-centers. In addition to this, settlement is concentrated predominantly along the city railroad routes. The number, structure and distribution of the inhabitants and employees (workplaces) form the basis for calculating current and future passenger transportation within the Berlin region. The face of Berlin's inner city will change radically within a few years due to the higher-density development in the inner-city areas, e.g. in the government district and around the area of Potsdamer Platz. This means a significant increase in transportation and particularly applies to professional commuters and commercial transportation.

In the inner-city areas, only a few large sites have been planned for the new construction of dwellings. In addition to the planned main service centers which have a desired dwelling percentage of up to 30%, a multitude of smaller construction projects have been planned as gap-filling and attic conversion measures.

With regard to the development of workplaces, we currently assume an increase within the city from 1.6 million to approximately 1.8 million. The majority of this growth will predominantly be the result of a significant increase in the number of persons employed in the service sector, who should partly become settled in the city center (e.g. at Potsdamer Platz) and partly outward in the so-called "ring-area" along the city railroad.

Outside of the city center, on the outskirts of Berlin, larger housing development areas are defined by the land use plan (LP 94). Large grounds, such as the banks of the eastern Spree, are meant to be used by service-providing companies in the future. The subsequent addition of workplaces is planned for the new housing estates in Hohenschönhausen, Marzahn and Hellersdorf in order to compensate for the current deficit. The main areas of future housing construction (northeastern area and southeastern area) should also be provided with an appropriate number of workplaces.

The requirements of transportation planning are based on balancing structural "catching up" and the connection of Berlin to supraregional traffic routes. Further requirements include the city's integration into and its orientation towards Central and Eastern Europe, something which has become necessary as a result of historical developments, and the expansion of local public transportation.

In concrete terms, this means that the avoidance and shifting of traffic as well as an environmentally-compatible and efficient transportation management must be realized. In the city center, in which more than 1 million people now live in an area of 100 squarekilometers, transportation planning is also responsible for maintaining the quality of life and for solving the problems of noise and air-borne pollution.

Concepts predominantly have to be developed and realized for three main focal points:

- The connection of Berlin to the international and national transportation network,
- The connection of Berlin with the surrounding countryside in terms of transportation, particularly with regard to ordered and traffic-avoiding structural development and
- The management of inner-city transportation against the background of increasing ecological strain on the city.

For this reason, transportation-political and transportation planning considerations regarding passenger services must be oriented towards the following:

- Avoiding traffic in the city and the surrounding countryside via ordered structural development, i.e. locating accommodation and employment in the city as close together as possible, in order to make large-volume commuter traffic superfluous,
- Shifting the unavoidable percentage to modes of transportation which are as city-friendly and environmentally-compatible as possible and
- Managing the remaining necessary motor vehicle traffic in an optimum manner together with local public transportation surface traffic.

The ratio of distribution between passenger transportation by motorized individual traffic and local passenger services reveals significant differences for the various areas within the city, and is dependent on the relevant traffic situation. Determining factors in this regard include the volumes of traffic and the available capacity of local passenger services or the road network, the reachability (journey time) of the city areas for local passenger services and motorized individual traffic as well as available parking space. It is revealed that in the future, particularly in the densely developed areas of Berlin's inner city, the percentage of motorized individual transportation journeys can only be approximately 20% to 30% of the total passenger transportation. In the inner-city areas (within the city railroad ring) which are farther away from the city center, the percentages of motorized individual traffic are between 40% and 50%.

In the outlying areas of the city which are less well serviced by local pubic transportation and which have significantly higher road capacity, the percentage of journeys undertaken with individual passenger cars will continue to predominate in the future. Mean motorized individual traffic percentages in this case are between 60% and 80%.

Public Transportation: Now and Future

Local public transportation in Berlin is based on a hierarchically structured local public transportation network for the creation of large-scale connections and for the connection and development of partial areas. To this end, all local public transportation carriers are integrated in accordance with their range of tasks, whereby rail-based transportation takes precedence. Regional railroads, city railroads, subways and streetcars, supplemented by busses operate transportation.

Using the Berlin-Brandenburg interconnecting transportation system, which commenced operation in 1997, a coherent local public transportation system is to be created in Berlin and Brandenburg. It should coordinate timetables and therefore guaranteed connections, adopt a universal tariff system and tickets that are valid in the entire area, and develop an identical image as well as coordinated marketing and public relations work.

In the future, characteristics such as the disabled-friendly designed installations and vehicles, connection of the transportation systems with one another and with other modes of transportation (bike-and-ride, park-and-ride), punctuality, safety, cleanliness and user-friendly operation will be assured for new infrastructure measures through the quality standards of the interconnecting system. Significant characteristics of this new quality include the good reachability of the city as a whole and its parts, and the unimpeded use of the diversity of this city as seat of government, service center and European commercial and cultural metropolis, for which a high-performance and internationally recognized transportation system represents an important prerequisite.

Priority is being given to the extensive recreation of the city railroad network as it was in operation up to 1961. Its routes are being operated again and thoroughly renovated. The completion of the inner-city railroad ring, the connections from Tegel to Hennigsdorf (Velten), from Lichterfelde Ost to Teltow, and via Spandau to Falkensee must be realized by Deutsche Bahn AG, the German Railroad corporation, which is responsible for the planning and execution of the project.

In the long-term, the direct north-south city railroad connection of the future Lehrter railroad station represents an important project. Routes are being kept free for a city railroad connection via the eastern outer ring with a network completion at Karower Kreuz, for a city railroad link from Jungfernheide to Gartenfeld with a possible extension in the direction of Spandau (Hakenfelde), and for the city railroad from Wannsee to Stahnsdorf.

The subway opens up Berlin's urban area and connects the districts and important parts of the city to one another. In order to realize the Berlin government's transportation-political goals and to achieve a modal split of 80% in favor of local public transportation in the central area, precedence is to be given to expanding the subway network there. In the medium and long-term, further subway plans are particularly to be realized in the eastern districts.

A new connection to the city railroad network will be created by shifting the subway U2 to the north from the Vinetastrasse subway station to the Pankow station. In a later stage, the U2 is to be extended to Pankow Kirche, linking Pankow's district center, Breite Strasse, to the subway network. Further subway expansion plans concentrate on extending the U5 from Alexanderplatz via Unter den Linden, in a first stage to Lehrter station. In the south, the extension of the U7 line is planned, from Rudow subway station to Schönefeld's Berlin-Brandenburg International Airport, which is to be expanded, too.

Berlin's streetcar network is to be developed into a modern and attractive mode of transportation. In doing so, the planning goal is not to compensate historically related imbalances in the network in the east and west of the city. However, the streetcar network should, in particular, be expanded in tangential relations into the western districts again, and should also be sensibly linked to the city railroad network and the long-distance and regional railroad stations.

The expansion of the streetcar network is being carried out in stages - starting at the eastern edge of the city - in a westerly direction. In doing so, Alexanderplatz and Berlin's old center, Mitte, will be connected to the streetcar network again. In addition to the expansion of the rail network in the inner-city districts, the extension of the streetcar network is planned in order to connect and open up new residential and commercial areas, particularly in the north and south-east.

The attractiveness of regional transportation should especially be increased by providing services, which are coordinated and connected with the remaining local public transportation network, and by improving changing possibilities. The task of rail-based regional transportation is to link the centers in the state of Brandenburg with each other and with Berlin. From Berlin's point of view, it is most important to transport passengers directly to their destinations within the city, so that they do not choose to make use of their private car.

In the future, transportation routes in Berlin will become longer, but also more numerous, the volume of traffic will increase. This is the inevitable consequence of urban development in and around Berlin, and does not necessarily have to be evaluated in a negative manner. However, the fact, this development is nevertheless portrayed as an apocalypse by many people. It can be attributed to the fear that these increases would favor individual passenger car transportation with its known consequences of pollution. This fear is not unjustified. In a society, in which the individual occupies the uppermost position of the value scale, (motorized) individual transportation, combined with comfort and convenience, is given preference over and above "collective" transportation with mass modes of transportation, i.e. local public transportation: This is not a conclusion drawn by transportation planning, by the way, but by behavioral psychology.

Therefore, complaining about the increase in motor vehicles in the eastern section of the city - which was, after the fall of the wall, liberated from the bonds of a socialist economy -, but also in West Berlin - which continues to open out into the surrounding countryside - is at best hypocritical. In 1990, a total of 1.16 million motor vehicles ran on Berlin's roads. In 1997, the figure was 1.27 million motor vehicles, or 367 motor vehicles per 1,000 inhabitants. In comparison with other conurbations in the Federal Republic (e.g. Hamburg with 475 motor vehicles and Munich with 515 motor vehicles per 1,000 inhabitants), the number is still low and reveals that a further increase is to be assumed here in the course of "normalization".

However, this must not - and this also applies to practically all other conurbations – find expression in a corresponding increase in transportation volume. Even if one should not realistically hold any great expectations with regard to the effects of traffic avoidance strategies: Limits have been placed on the growth of motor vehicle traffic within the city, due to the space required for both flowing and stationary traffic. "Space requirements" for motor vehicle traffic therefore represent a limiting situation similar to the "financial constraints" for local public transportation: In the strictly limited and not significantly expandable urban traffic space within Berlin, the future realization of mobility requirements via motor vehicle transportation will be subject to severe restrictions. And this is what the Berlin government's transportation policy is based on within the framework of a "classic" transportation displacement strategy: Part of the current, and future, motor vehicle passenger transportation should be shifted from the road to local public transportation.

The aim of this is to create space on the road for commercial transportation, which occurs almost exclusively as motor vehicle traffic in urban areas, the order of magnitude of which is frequently underestimated. At present, commercial transportation already represents 35% of road traffic in Berlin; to put it in simple terms, every third motor vehicle on the road is involved in commercial transportation. If passenger road transportation were to increase further, this would inevitably lead to an unsolvable conflict.

For this reason, the Berlin government has announced its strategic goal for the central region as being the division of passenger transportation in the ratio 80:20 in favor of local public transportation, in full realization of the fact that this may be an order of magnitude which may not be achieved within the short-term. The ratio of local public transportation to motorized individual traffic in the inner city is currently approximately 60:40. The route to "80:20" leads via the management of parking space in the inner city and also the continued expansion of local public transportation.

Extensive parking space management has the goal – whilst granting privileges to residents - of driving the remaining long-stay and continuous parkers, who generally represent avoidable journeys between the place of residence and the place of work, from public roads. Experience in pilot areas in the east and west city and in the old town of Spandau, where extensive parking space management has been in operation since 1995, is positive. For this reason, the Berlin government has resolved to continue its policy of parking space management in the two inner-city areas and, in addition, to encourage the districts to introduce similar parking space management, for which they themselves will be responsible, within the relevant district centers, after the problem of monitoring has finally been clarified.

The goal of predominantly shifting passenger transportation to local public transportation unavoidably requires an attractive local public transportation system. In addition to operational attributes (safety, speed, cleanliness, punctuality, etc.), this also involves the availability of a corresponding high capacity, hierarchically-structured network with correspondingly interlinked interfaces. These prerequisites are extremely favorable in Berlin: The city railroad and the subway are available as high-performance high-speed railroad systems. These look back at different lines of development in the past: Whilst the subway was extensively developed as competitor for the unpopular (because administered by the GDR) city railroad in the post-war period in West Berlin, and as a substitute for the discontinued streetcar network, the city railroad was the number one mode of local public transportation in East Berlin, followed by the streetcar. The eradication of the invisible walls, which were consequently erected in the local public transportation network, is one of the current prerequisites of the fusion of the two halves of the city.

In the case of the city railroad, the discontinued routes in West Berlin and in the surrounding countryside are being brought into operation again, and those which are in operation are being rehabilitated and brought into shape. Almost no new construction is planned. Basically it is the recreation of the network as it was in 1961 - before the construction of the wall - with a total length of 341 km, in comparison with a network of 247 km prior to reunification. In view of the required expenditure, some 5 billion DM, which must be financed by the Federation. This is an extremely large-scale goal. This is also revealed by the fact that the period of time required for realizing this is longer than originally assumed: The most important inner-city measure, the completion of the city railroad ring in the north, will take place at the end of 1999.

In the case of the subway, two projects are at the forefront of additions to the existing 153 km-long network: Extending line U5 via Alexanderplatz to the future Lehrter station (3.5 km) and linking U2 in the north with the city railroad network at Pankow city railroad station, and the possible extension of U7 to Schönefeld airport, must be examined. The U5 extension is of disproportionately greater significance: Management of traffic arriving at Berlin's future largest long-distance and regional station, the Lehrter Bahnhof, which is still undergoing construction within the framework of the mushroom concept, is dependent on the realization of U5. This project is costing approximately $0.65 billion (1.3 billion DM), part of which is being financed directly by the Federation. The section between Lehrter station and Pariser Platz is already under construction. The considerable cost of subway U5 is hardly surprising if one considers how sensitive the affected area of Berlin's old city is to construction. There, any reduction in groundwater is strictly taboo.

The bus, which has been much criticized in public discussion regarding the streetcar system and which has a route network covering 1,230 km in Berlin, is better than its reputation, with regard to emissions, too. Whilst the natural gas-driven bus trial, which was commenced in 1994 in Berlin, could not be continued on a large-scale, technical developments in conventional busses also indicate that considerable progress is to be expected as regards energy consumption and pollutant emissions.

As surface modes of transportation, both the bus and the streetcar are dependent on road traffic: It is therefore understandable that efforts are being made to accelerate these systems within the framework of a continuous process. With approximately 90 km of realized bus lanes, Berlin's effort is very positive even in comparison with other major cities in the Federal Republic. It is understandable, however, that what has been achieved is only one stage towards the goal, and is therefore not yet sufficient. Efforts to optimize surface transportation are being continued, and will increasingly integrate light signal systems in the future. However, every effort must be made to maintain a sense of proportion and to consider the problems within the whole context, particularly with regard to commercial traffic.

The investment need in local public transportation infrastructure is immense. In the immediate future, the Berlin government will invest approximately $108 million (315 million DM) per year in the subway and streetcars. The annual Federal investment in the city railroad should be even greater. The operation of local public transportation also requires considerable expenditure. The S-Bahn Limited, which is operating the city railroad network and the Berlin Transporation Services, BVG, are provided with approximately $0.75 billion (1.5 billion DM) subvention per annum by the Berlin government. The value, which the Berlin government places in local public transportation in Berlin, is therefore made perfectly clear. Nevertheless, the goal is to reduce this deficit and to increase the degree of cost coverage in Berlin's local public transportation. The recently founded Berlin-Brandenburg interconnecting transportation system should make a considerable contribution towards this as a result of the expected synergistic effects.

Further increase in the attractiveness of local public transportation can be achieved by networking this with the other partial transportation systems within the framework of comprehensive urban transportation management. The forthcoming years will continue to be characterized by completion deadlines and operational start-ups. These include two outstanding events, which may also serve as milestones in the realization of Berlin's transportation system. Firstly: The initial operation of the Lehrter long-distance and regional railroad station in 2003/4 will, in all probability, represent the preliminary completion of the transportation measures within the inner-city area, which will by then have been home to the government and parliament for some time. Secondly: he completion of the Berlin-Brandenburg International airport in Schönefeld in 2006/7 should simultaneously mark the final cycle of the realization of significant transportation measures in the outer region of the city.

The basic prerequisite of this is the introduction of transportation telematics amongst all transportation carriers, including commercial traffic. At present, the necessary technology is not yet fully ripe in the sense of marketability, as recent experience has shown in Berlin. Nevertheless, urban transportation management remains the great goal, which must be the subject of all efforts in the near future in order to ensure the city-friendliness of transportation in Berlin.

As a result of this multitude of measures, Berlin will succeed in doing justice to the mobility requirements of a Capital City and global metropolis. The Berlin government and the public transportation carriers are prepared for this development and are doing their best to avert "gridlock".

The Berlin Local Transport Company (BVG) – A Part of Berlin

Joachim Niklas[1]

Introduction

Berlin is a fast city. This characteristic is attributed to the capital of Germany not only on account of the breakneck pace of its development but also because of its dense and well-functioning public local transport network. Public transport has been in existence in Berlin for more than 150 years. And for almost 70 years, buses, trams, and underground trains have been combined in a single company, whose initials BVG are part of Berlin, just as the Brandenburger Tor is. And like the Brandenburger Tor, the Berliner Verkehrsbetriebe, or BVG for short, has also had a chequered past. Founded in the 1920s, the BVG turned in a remarkable performance in its early years. With more than 5000 vehicles and 25,000 employees it operated 89 tramlines along a length of 1600 km, 35 buses along 338 kilometres of bus routes and 8 underground connections. The Second World War left behind a heap of ruins. 76.5 percent of trams, 98 percent of buses and 45 percent of underground carriages were destroyed. Laboriously, work was commenced on reconstruction. The division of Berlin into West Berlin and East Berlin not only cut the traffic network into two, but also separated the BVG Company into two separate operations in West Berlin and East Berlin.

Following the fall of the Berlin Wall in November 1989, both operations quickly became a single entity once more. Lines, which had been truncated or closed down between both sectors of the city, are joined up and new links are created. Amongst these rank such spectacular connections as the reopening of the U1 over the Oberbaumbrücke, the link between "Zoo and Alex" on the U2 and the extension of tramline 23 via Bornholmer Straße in the western sector of the city. There has been a combined BVG company in existence again since 1992.

[1] Member of the Board of Management, Berliner Verkehrsbetriebe (BVG), Berlin, Germany

I Range of Services Provided by BVG

The combined company with about 28,000 employees in 1992 had to deal with massive rationalisation at a single stroke while offering the same level of performance. In six years, staffing levels have been reduced by more than 11,000 by offering generous compensation packages. Today, about 16,000 employees at BVG transport about 780 million passengers a year on 9 underground lines, 28 tramlines and 161 bus lines. The route network traverses the entire city area of Berlin with 150 km of underground lines, 359 km of tram lines, and 1888 km bus routes.

BVG offers Berliners and their guests a travel service with local coverage every few minutes at peak travel times. At night too, the transit services in Berlin are second to none in Germany. 58 buses and 4 trams run round the clock every half an hour during the nights. During the nights of weekends, two underground night lines running every 15 minutes supplement the night-time service. The culture vultures and art lovers in the city of 3.5 million inhabitants value this service.

II BVG as a Business Enterprise

BVG in Turmoil

At present BVG is undergoing a process of upheaval. It was formerly an independent company owned by the Berlin region, with the result that the region paid up for its losses. Now, BVG is on the way to becoming an independent business and service provider. In order to smoothen the transition, the Berlin region managed to conclude an inter-company agreement for the years 1996-1999, which guarantees the BVG a grant totalling $2.15 billion (4.3 billion Deutsche Marks). The grant is to be paid annually on a declining basis. Since savings will also have to be made in the regional budget, the Berlin region reduced this sum by $24.25 million (48.5 million Deutsche Marks).

The current situation in the market is unfavourable. The population of Berlin is falling. Many Berliners are moving out into the surrounding area, and since the local transport network is not as dense there, they use the suburban fast train, regional railway or their own car to travel to work. It is precisely in the area surrounding Berlin and in the Eastern sector of Berlin that the number of car owners has risen in recent years. Added to this the number of jobs in Berlin has fallen. With the level of unemployment in the city at 17%, this also affects people's travelling patterns, so that BVG has recorded a drop of the number of its passengers.

In spite of this, the BVG has already passed significant milestones on its way to becoming a business enterprise. This means that for the first time it has passed the 50 percent mark covering its costs i.e. revenues exceed 50 percent of the running costs of the company. Initially this figure was 32 percent. This improvement has been made possible as a result of the various measures, which have been introduced.

The financial accounting has been completely reorganised and aligned on computer to SAP R/3. An internal market has been created with the introduction of cost centres with their own profit and loss accounts, in which market prices are charged just as they are by the competition. BVG is divided into 214 cost centres, in which heads of cost centres become "businessmen within the BVG company". The successes are apparent in an increase in profit-awareness, in terms of a critical review of expenses, and in the systematic implementation of measures to increase revenue.

The whole company has been reorganised. The turning of company sectors into divisions led to levels of the hierarchy being dismantled. As a result the number of departmental heads has been reduced from 170 to about 70.

The conclusion of lease-in lease-out contracts with US investors on rolling stock has had a beneficial effect on the earnings position of BVG. This gave rise to BVG managing to earn $30.6 million (61.2 million DM) in revenue in 1997. Moreover, the manifold successes in rationalisation had a favourable effect on the trading profit. Consequently BVG managed to increase productivity by a total of 30 percent. Expressed in actual terms, this means that in spite of a reduction in staffing levels of 11,000 employees, BVG has managed to increase its transportation performance by 5,016 million kilometres up to 97,264 million kilometres.

At the same time, the BVG opened up new spheres of business. As a result, proper shopping centres or shops have been established on mezzanine floors at important underground railway stations. BVG rents out this space to a BVG subsidiary company as part of a Public-Private-Partnership (PPP) concept. In turn, the subsidiary acts as an investor. At present, more than 40 such sales outlets already exist at 28 locations. This figure will increase to 58 by the end of 1999. In addition to the additional revenue generated, this new sphere of business has other beneficial aspects. The favourable opening hours of the shops from 6 o'clock in the morning to 9 o'clock in the evening, make shopping very straightforward for passengers, either on their way to work or on their way home. They can purchase the goods they want even after other shops are closed. Moreover these shops sell books of BVG tickets, so that passengers do not have to resort to buying their tickets or books of tickets from automatic vending machines. At the same time such shopping arcades transform a dreary railway station into a lively, friendly, and bright place.

Stations became cleaner as the construction of shops eliminates obscure nooks and crannies in which rubbish and dirt collect.

The conversion in BVG's own workshops of double-decker buses which have come to the end of their useful service life into open-top, double-decker buses for trips around Berlin is another sphere of business which has been started up. Following the success in using these "Topless buses" for this function in Berlin.They also have a telescopic folding roof to cater for inclement weather. Now, these buses are also used for sight-seeing purposes in Hamburg and Copenhagen. Even Brussels has registered an interest in them.

A newly formed BVG's Charter and Tourism Section offers bus-trips within Germany and Europe.

Furthermore, various BVG-owned bus workshops have acquired contract workshop status for Volvo and Neoplan vehicles. That means that vehicles owned by other companies can also be serviced and looked after here in addition to BVG's own buses.

III Customer Orientation

Offer

At the same time as BVG is developing into a business, it is being organised to become a customer-oriented enterprise. A raft of measures has been initiated for this purpose. One of which is the replacement of the vehicle fleet. BVG has managed to procure 164 new buses, which, in addition to cheerful fittings, have air-conditioning. They not only ensure a more pleasant temperature on hot summer days but also prevent the windows from becoming misted up in the autumn and winter, and reduce humidity.

There is a "kneeling" equipment fitted to buses to make them easy for passengers to board and alight. Passengers can board and alight the low floor-level tram carriages with ease. About 90 vehicles of this type travel on the tram routes. Above all they are indispensable for wheelchair drivers and passengers with prams.

Older trams have been put through a modernisation programme, so that altogether 447 Tatra trams have been fitted with upholstered seats, heat and sound insulation, visual and sound information systems for passengers, new doors, a better design of restyling, etc.

The first of a new generation of vehicles are already in service on the underground. Designated as the H-train, it is particularly passenger-friendly. What is completely new about this train is that you can walk through its entire length. This increases the feeling of safety for the passengers. In addition to this, it has more space for taking bicycles. This vehicle model has already been fitted for driverless operation.

Speed plays a significant role above all in competition with the other carriers in Berlin. About 90 kilometres of bus lanes have been set up so that the buses do not get stuck in a traffic jam,. The trams are also to have their travelling time cut now. With systems fitted to the light signal equipment and on the newly introduced computer-assisted operational management system, the signal switches to the "go" signal if there is a tram approaching so that the tram can cross a crossing without having to stop. BVG financed this equipment on the light signal equipment in anticipation of funds coming from other sources.

BVG is also taking completely new roads in fitting its bus and tram stops with appropriate protection from the elements for its passengers. To date, it had a contract with a company, which financed the construction of waiting rooms with advertising. Since this company only advertises products it was only interested in waiting rooms in exposed locations. Being a customer friendly organisation, BVG set itself the objective of fitting all its bus and tram stops with shelters by the year 2000. There are about 3000 of them. BVG's own advertising department developed a completely new concept for this. Indeed, this concept too is to be financed by advertising, but most advertisements will be for events within Berlin and shopping centres and shops in the locality. The waiting rooms are to be manufactured on a conveyor belt in a workshop owned by BVG. More than 1000 of them have already been erected.

The fact that BVG vehicles and equipment are immediately recognisable by everyone is due to its good Corporate Design. But it is not just nice design, but above all function which counts. It is clear and makes it easy for the passengers to find their way around from the beginning of their journey to the end. In 1994, BVG was awarded the coveted "Beacon Award" for its Corporate Design by the American Centre for Design.

Fares and Sales

The customer-friendliness of BVG is also expressed in a simple, clear, and fair fare scale. BVG went from a standard ticket to a 3-zone fare system even before the integrated transport system was set up. The A-B-C fare system, which is in operation since March 1997, met with the approval of the passengers, because it is easy to understand. This fare structure and the ease of use of the automatic vending ticket machines have put Berlin in top place in a comparison of six German cities, which was conducted by the business journal of the television channel ZDF.

Buying tickets is made as easy as possible for customers. About 80 percent of customers have a season ticket. The appropriate stamps can be acquired at about 3500 ticket machines mounted in the vehicles or at the railway stations. In addition to the 28 points of sale owned by BVG, which are opened to meet customer needs during seven days of the week from 6:00am to 8:30pm, privately owned sales outlets also sell the tickets. The number of shops has been increased from 350 to 800.

Passengers can receive information on bus, tram, and underground train times at their convenience via "Call and Ride". This new service has been in operation since June 1997. With this service, a customer can order his annual or monthly ticket and it will be delivered free of charge to his address within a few hours. It is also possible to purchase individual tickets in this way, however, delivery is only free for quantities of more than 50 tickets.

The combination-ticket is also convenient for the passenger and profitable for BVG. More and more interest is being shown in this type of ticket. It all began with tickets for major sporting or cultural events also being valid as return tickets for public transport to the event. For this, BVG negotiates with the individual event organisers about the receipts, which are due to it per ticket sold, and concludes the appropriate contracts. In addition to the receipts, this type of ticket prevents BVG from making a loss. If a football match is staged in the Olympic Stadium, 60 to 70 thousand spectators have to be transported. When such large numbers of people are travelling on BVG transport, it is hardly possible to check every passenger's ticket. A combination ticket therefore prevents people from travelling without having a purchased ticket.

In addition to organisers for major sporting events, all three operas in the city and some theatres, amongst them also ten childrens' theatres have also signed contracts with BVG for this type of combination ticket. Increasingly, the organisers of congresses in the city are also recognising the advantages of a combination ticket for their participants. Given the limited amount of parking space in the city, most congress participants, who do not know their way around Berlin, can in this way travel to their congress venue without having to trouble themselves with the conditions of using their tickets and to bother with the instructions for using automatic ticket vending machines.

Recently, travel agents have also become partners with BVG in the combination ticket scheme. A Berlin hotel also offers its guests the free use of public transport if they show a hotel pass. The service of a car repair workshop, which gives customers mobility within the city with a sticker on their workshop card while their car is being repaired, is an entirely new innovation.

In 1997 BVG managed to achieve a turnover amounting to $1.4 million (2.8 million DM) as a result of combination ticket agreements.

Tickets in the form of company tickets have been in existence for several years in Berlin. $0.6 million (1.2 million DM) was earned by BVG with this type of ticket in 1997. But this year, and in years to come, the company ticket will experience an enormous upturn, since BVG has managed to conclude this type of agreement with the Deutsche Bundestag (Lower House of Parliament). In 1998 and 1999, the bulk of its staff including those of ministries and political parties will be moving to Berlin. The debis headquarters on Potsdamer Platz also rank amongst those companies, which have reached agreements with BVG on company tickets.

The company ticket is an officially approved fare in which a reduction of between 5 and 15 percent can be granted. At least 100 employees in a company have to buy a season ticket for this type of ticket to be profitable for the carrier. Smaller companies can get together to form a user-association. Once purchased, the company ticket is not transferable.

Although about 80 percent of BVG's passengers are regular customers and have a monthly or annual season ticket, some users do try to use the public transport system without buying a ticket. However with a cleverly devised system for checking BVG is increasingly succeeding in catching out the "black sheep". The number of ticket inspectors has increased from 150 to 500. Most of these are staff, who are no longer fit to drive, or staff whose jobs have disappeared as a result of rationalisation. They go out mostly in civilian clothing around the clock from five bases, to all part of the city. The routes to which they are assigned are changed from day to day as required. Added to this, there is a full-scale duty check once a week. This full-scale check is conducted by BVG in close collaboration with the police and other security staff. Such checks involve a railway station or line being cordoned off without prior notice for about two hours, so that no passenger can evade having his ticket checked. If the average ratio of people travelling without a ticket is 3 percent, such full-scale checks mostly achieve a ratio of 10 percent.

BVG Guarantee

The BVG is convinced that it offers a good service. Just as manufacturers of brand goods give customers a guarantee with their products, BVG has given a performance guarantee since November 1997. To date no other German company in the local transport industry has done this. The customer guarantee makes it clear that the BVG takes its claim to be a customer-oriented service company seriously. The image of BVG benefited as a result. The customer guarantee refers to punctuality and standards of cleanliness. Consequently every passenger is sent a free ticket if he reaches his destination more than 20 minutes later than the time shown in the current timetable. Should this delay occur from 11 o'clock at night to 5 o'clock in the morning, and the customer is forced to take a taxi, the BVG will, upon production of the receipt for the taxi fare, pay the taxi fare up to a maximum of $25 (50 DM). If a passenger's clothes become dirty as a result of dirty BVG vehicles or equipment, BVG will reimburse the costs of having them cleaned.

Prepared guarantee certificates are on display in customer centres and are kept by driving staff on their person. Based on the number of daily passengers, in the early months the number of complaints was 0.002 percent.

BVG – Club

In order to tie the regular customers to BVG and look after them individually and to gain new customers, there has been a BVG club since the end of 1997. Membership to this club is free and it is open to everyone. Club membership does not entail any obligations, however a club member can take advantage of so many savings. There are diverse and attractive offers for ways in which you can spend your spare time in a club magazine, which appears on a regular basis. This magazine is delivered to every club member, who can book the offers via his personal club card and over a special club hotline. Highlights are a trip through the underground tunnels in a customised underground train with the roof removed, during the night when there are no other trains running, or a trip to the bus manufacturer's factory.

IV Outlook

Electronic tickets

BVG is currently preparing itself for the introduction of an electronic ticketing system. It provides for the non-contact debiting of kilometres travelled by means of a chipcard, which the passenger keeps on his person. The first field trials are to be conducted shortly. Then 25,000 participants in the test will try out the check-in/check-out on two underground trains, two bus-lines, a tramline as well as a section of the fast suburban line. BVG has been working together with other partners on this project since 1995.

In 1999, BVG will put a computer assisted management operation system (RBL) into operation for trams and buses. The quality of the service will be increased enormously as a result. With the assistance of a satellite location system the location of every vehicle can be shown and it shows whether vehicles are running on schedule. In the event of a breakdown, the best possible detour for other vehicles will be worked out.

In parallel to RBL, a dynamic information system (DAISY) will also be introduced progressively. This will enable passengers at all underground stations and at the most important tram and bus stops to see on a display how long it will be before the next service is due.

Fare Integration Ambitions for Public Transport in the Budapest Region

János MONIGL[1]

Abstract

The paper gives an overview on the fare integration ambitions in the Budapest region including 170 settlements and approximately 3.3 million inhabitants and serving more than 85% of the commuters to Budapest.

The institutional and legal framework of establishing the Budapest Transport Association is discussed in detail. The basic contract of the authorities responsible for public transport provision – the Municipality of Budapest in case of urban and the Central Government in case of interurban transport – would be the fundamental document to express the political and economic intentions.

The main principles of a uniform fare system are described, allowing for the universal use of the services of the three operators (Budapest Transport Company (BKV), Hungarian State Railway (MÁV) and the state bus company, Volánbusz), wich operate now independently from each other.

Two alternatives are considered for the ticketing system: a conventional with paper media and an electronic one with contactless smart cards, magnetic stripe cards and individual paper tickets.

As a possible first step towards the association the extension of the local transport pass system is also mentioned.

Recent Changes in the Transport of Budapest

The changes of the 90's have brought several changes in the urban transport sector in Hungary. These changes and the associated problem areas can be summarised as the following:

- Due to the liberalisation of access to cars, car ownership has grown significantly. In specific passenger groups car use was encouraged by the more intense external relationship system of small companies replacing the already closed great state enterprises. All this lead to an increase in traffic congestion.
- The functional changes caused by new activities (banks, insurance companies, offices) appearing in city centres further increased parking demand. Because of this and the exaggerated car use, parking conditions have become chaotic, disturbing the living environment.

---

[1] Managing director, MSc eng. PhD MSc ec., TRANSMAN Consulting, Hercegprímás utca 10. H 1051 Budapest, Hungary

- The situation is worsened by the fact that financing public transport (PT) services is only possible with a continuously growing deficit in spite of the fast increasing public transport fares. As a result PT service frequency and the general quality level of the services is decreasing.
- All these phenomena contribute to the further increasing air and noise pollution, which is accompanied by an increasing public sensitivity towards these circumstances.

These changes are well illustrated by the data contained by Table 1.

Table 1: Some basic figures of the transport sector of Budapest

Indicators	1984	1991	1994	1997
Inhabitants (1000)		2018	1996	1870
Inhabitants (%)		100	99	**93**
Car ownership (car/1000 capita)		252	279	310
Car ownership (%)		100	111	**123**
Traffic on Danube bridges (%)		100	122	**118**
BKV passengers (%)		100	94	**83**
BKV passenger-km %		100	94	**80**
Modal split (public/car %)	~80:20		~62:38	**<60:40**
BKV place-km (%)		100	93	**82**
BKV operational costs (real %)		100	79	**72**
BKV fare level/tickets (real %)		100	104	**154**
BKV fare level/passes (real %)		100	96	**126**
Car fuel price (real %)		100	72	**80**
BKV revenues (real %)		100	78	**64**
BKV degree of financing (%)		100	83	**77**

Transport Association: a Solution

Similarly to many big cities of the world, the establishment of the **Budapest Transport Association (BTA)** can be an efficient tool to stop the further decrease in the share of collective (public) transport and the deterioration of transport and environmental conditions in Budapest and its vicinity.

There are a number of possible forms and phases of transport partnerships, all aimed at more effective utilisation of different passenger transport capacities on economic level and making it easier for passengers to use the services of public transport providers (BKV Rt, MÁV Rt, Volánbusz) currently having separated networks, fare, concession and ticketing systems:

- **tariff association:** based on a uniform fare system and the distribution of common revenues
- **transport association:** in addition to applying a uniform fare system it harmonises supply while retaining the independence of the operators.

A study Examination the implementation conditions of the association was prepared by TRANSMAN with respect to the earliest possible introduction, the Investigation for the Foundation of the Budapest Transport Association (BTA) – Final Report existing financial constrains and the legal framework.

Institutional and Financing Scheme of Public Transport in Hungary

Control and operation of public transport in Hungary is settled on two levels - corresponding to the structure of the administration system:

- **urban public transport** is the responsibility of local governments (municipalities), defined by the Law of Local Governments
- **interurban public transport** is the responsibility of the state government, derived from the Law of Railways, Law of Road Transport and the Law of Concessions

In spite of the fact that county level administration does exist, these bodies have only a formal role, without own budget and any powers delegated to them regarding public transport.

Responsibilities of authorities in relationship with the provision of PT cover the following main issues (see Figure 1):

- **network and service definition,**
- **fare and discount determination,**
- **operational licence issuing,**
- **subsidising the operation.**

The above responsibilities should not change even after the establishment of the association.

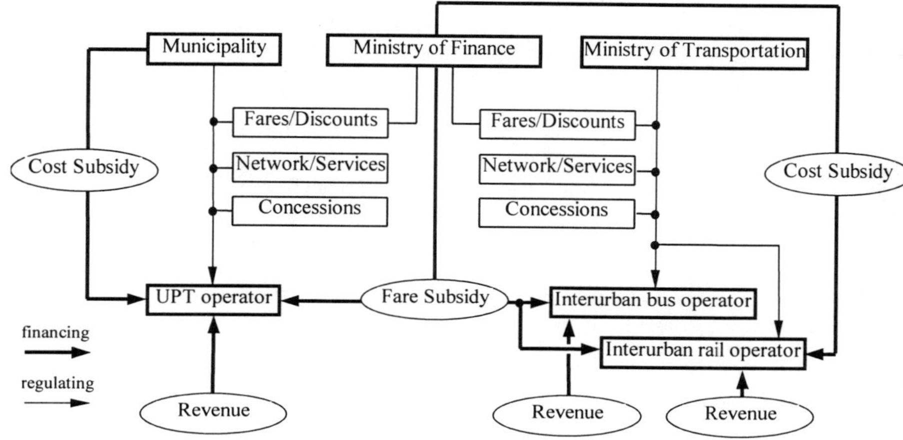

Figure 1: Institutional and financial framework for urban and interurban PT in Hungary

The Present Fare and Financing System of the Operators

In the Budapest region the most important public transport operators are:

- the **Budapest Transport Company** (BKV), a 100% municipality owned shareholder company, which operates more then 250 lines within the boundaries of Budapest exclusively (tram, bus, trolley, subway) and some suburban rail (HÉV) and bus lines;
- the **Hungarian State Railway** (MÁV), a state owned company, which operates 11 lines and runs suburban, long distance and international trains in all radial directions to/from Budapest;
- **Volánbusz**, a state owned regional bus company, which operates in the suburban area of Budapest.

The services of MÁV and Volánbusz can not be used with local fare within the boundary of Budapest only by national interurban fares for journeys beyond the boundary.

The main elements of the fare and ticketing system are the following:

- **in urban public transport (in Budapest)**
  - flat fare system for tickets (only single tickets for one boarding);
  - passes for adults (monthly, 30-day, annual + daily, weekly)
  - passes for students/pensioners (monthly, annual with 67,5% discount);

- **in interurban public transport:**
  - single tickets by distance (10-km rail and 5-km bus zones)
    - full price for adults (50% for public servants)
    - for students (67,5% discount between home-school, and 50% otherwise)
    - 16 trips with 50% or 8 trips with 90% discount for pensioners
  - passes for commuting employees
    - 86% of rail pass price is to reimburse by the employers
    - 80% of bus pass price is to reimburse by the employers
  - passes for commuting students (90% discount).

For small children and people over 65 years the public transport is free of charge in urban as well in interurban relations.

The fare levels for interurban buses are 20-50% higher then rail fares, which is one of the main problems at the unification the fare system.

The **main financing recourses** for the operation are:

- the **fare-box revenues** from the passengers;
- **fare subsidies** from the state compensating the discounted tickets/passes;
- **cost subsidy** from the responsible authority
  - the Municipality of Budapest in case of BKV;
  - the state in case of MÁV;
  - for Volánbusz there is no cost subsidy (because of the higher fare levels and no infrastructure costs)

The cost recovery ratios are different for all operators, as well as the cost subsidy paid by the owner.

Main Tasks Regarding the Establishment of the Association

The Government and the municipality of Budapest — as bodies responsible for the provision of public transport services in the area — shall express a joint political and economic will in a basic contract on the formation of the association and on securing the conditions thereof.

The association's company (as an independent legal entity) would be based on the basic agreement taking over some tasks from the municipal authority (e.g.: demand planning, fare proposals) and from the operators (e.g.: supply requirements planning).

According to the findings of the preparatory study the **most important preconditions for setting up the tariff association** in the first phase are as follows (see Fig 2.):

- the introduction of the association fare system valid for the services of the operators (BKV, MÁV, and Volánbusz), the elaboration of a uniform system of tickets and passes and a ticket validation system,
- harmonising regional fares in compliance with the legal regulations on setting prices, as well as the smaller standardisation of the system of discounts,
- agreement on the apportionment of the revenues of the association,
- the elaboration of the organisation of the association.

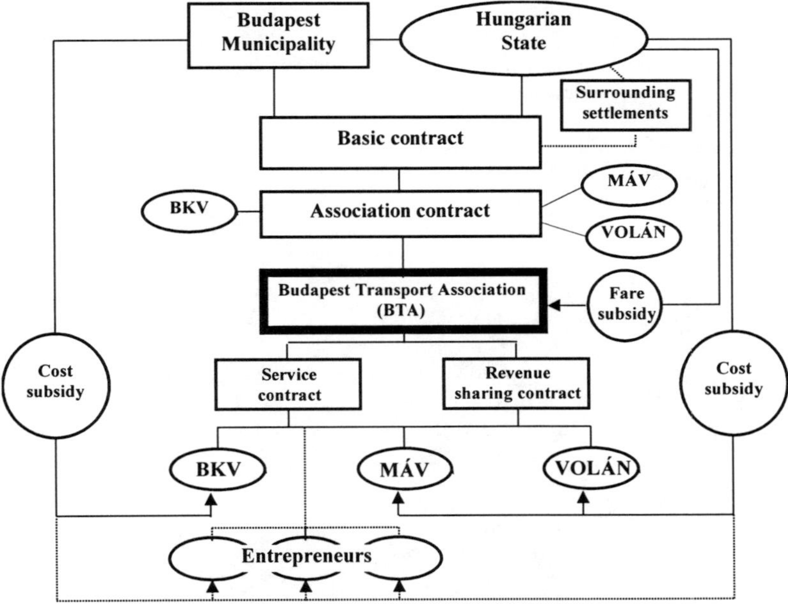

Figure 2: The Proposed Organisation Model of the BTA

Definition of the Association Area

In the politically sensitive definition of the association the main arguments were:

- to facilitate the satisfaction of travel demand by making uniform services directly available in Budapest and in those neighbouring settlements from which many people commute to Budapest on a daily basis,
- to ensure that the use of the association services after the trip into the association area is advantageous for those regular passengers travelling

to Budapest, who find no association services at the beginning of their journey.

Therefore, in the first place the transport association can be defined as a **pool of services** facilitating transport into Budapest together with the feeder services. It can also be defined as a certain area with its settlements but only in the second place. As a result of the territorial investigation of the number of daily commuters a boundary has been defined (see Fig.3.). This is located approximately 20-25 miles from Budapest, and contains 105 settlements apart from Budapest in the inner circle. In addition, there are settlements and stations/stops located along the rail and bus lines crossing the boundary of the ring. Altogether, the association shall cover 170 (106+64) settlements and 3.3 million residents providing services to nearly 86% of the people who commute to Budapest.

About 3.8 million trips are expected in the BTA using public transport: 3.35 million in Budapest, 0.33 to/from Budapest and 0.12 outside of Budapest (1996 data).

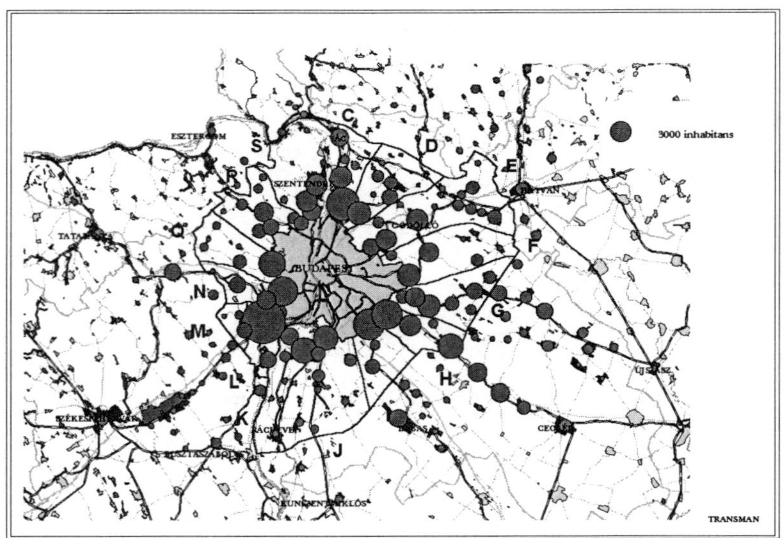

Figure 3: Daily Commuters to Budapest from the Settlements of the BTA

The Principles of the Association Fare System

The application of a uniform **fare system** and uniform fare levels would be desirable for the creation of the association.

The significant differences (up to 50%) between the fares of different companies — especially between MÁV and Volán — will probably be impossible to eliminate in one step when setting up the association. Therefore the problem might be resolved in the following ways:

- **uniform fare levels** could be introduced after a gradual, several-year long harmonisation of fares,
- **mixed fare levels** could be applied outside Budapest, allowing for differences between the rail and bus services.

The following **travel documents** are considered for regular and occasional users:

- **for the central zone** within the boundary of Budapest:
  - single and transfer (for maximum 60 minutes) tickets;
  - passes for unlimited use within the time of validity;

- **for the outer zones**
  - tickets for inter-zone trips for given number of zones;
  - passes for inter-zone trips valid for travelling in specified zones;

- **combined passes** for one outer sector + Budapest inner territory

Each ticket/pass type is considered for adults, students and pensioners.

In case of student and pensioner tickets a common discount rate should be applied for the whole area, which needs legal harmonisation.

Ticketing System Alternatives

Two technologies have been considered from the aspect of the value carrier:

a) **conventional** pre-printed passes and multi-stripe paper tickets with stamping validators;
b) **electronic** media and validation using contactless chip-cards for the regular users and magnetic stripe paper cards and individual printed paper tickets for occasional users.

In case a) (see Fig 4) the multi-stripe paper ticket can be used universally for single journeys as a uniform means of payment within the association when using different association services. The system ensures that the validity of the coupons could be inspected when using the services.

Besides of the simplicity and lower installation costs it should be mentioned as a big disadvantage that the system does not provide sufficient data about the use of the services of the different operators for revenue sharing.

In case b) (see Fig 5 and 6) where the whole process of ticket selling/card recharging as well as the validation at boarding and the data collection should happen automatically.

The higher investment costs would be compensated by well-based data for bookkeeping and revenue sharing and by more efficient demand-supply management.

Changes in the Revenue Volumes and Allocation

The unification of the fare system and fare levels causes through tarification revenue losses and needs more fare subsidy resources because of the extension of the discount system. This loss must be covered by the Municipality and the Government.

At the moment the basis for the revenue calculations is provided by the TRANSURS model system, which contains an integrated multi-mode passenger transport model with a revenue calculation module. The effect of the fact that the traditional distance categories (MÁV, Volánbusz) must be converted into ring-zones on revenues is shown by Figure 7.

The revenue sharing is one of the most sensitive issues, which has to be handled very carefully. Patronage (passenger km) should be taken a basis of the revenue sharing process. However, it is well shown by Table 2 that other indicators determine different proportions because of the initial conditions (e.g.: operation costs and fare levels) of the operators. The rules of the apportionment for a given time period should be agreed and included in an other contract.

Table 2: Indicators of the Companies Co-operating in the Association

Operator	**Passenger km %**	**Place km %**	**Costs %**	**Revenues %**
BKV	75.3	74.7	78.7	73.9
MÁV	16.1	17.7	14.8	13.6
Volánbusz	8.6	7.6	6.5	12.5
Total	100.0	100.0	100.0	100.0

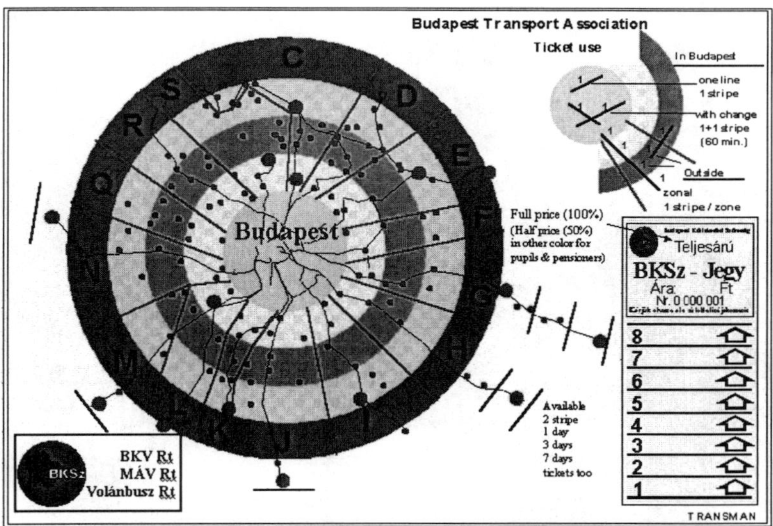

Figure 4: BTA Fare System with Conventional Ticketing

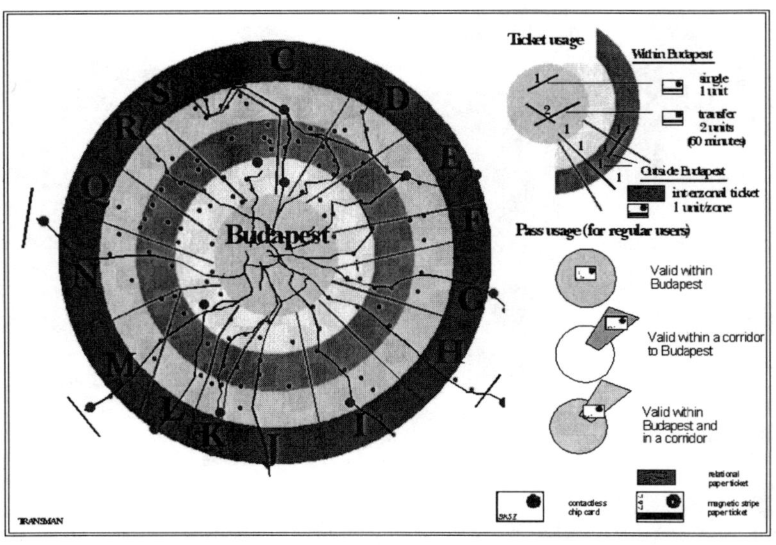

Figure 5: BTA Fare System with Electronic Ticketing

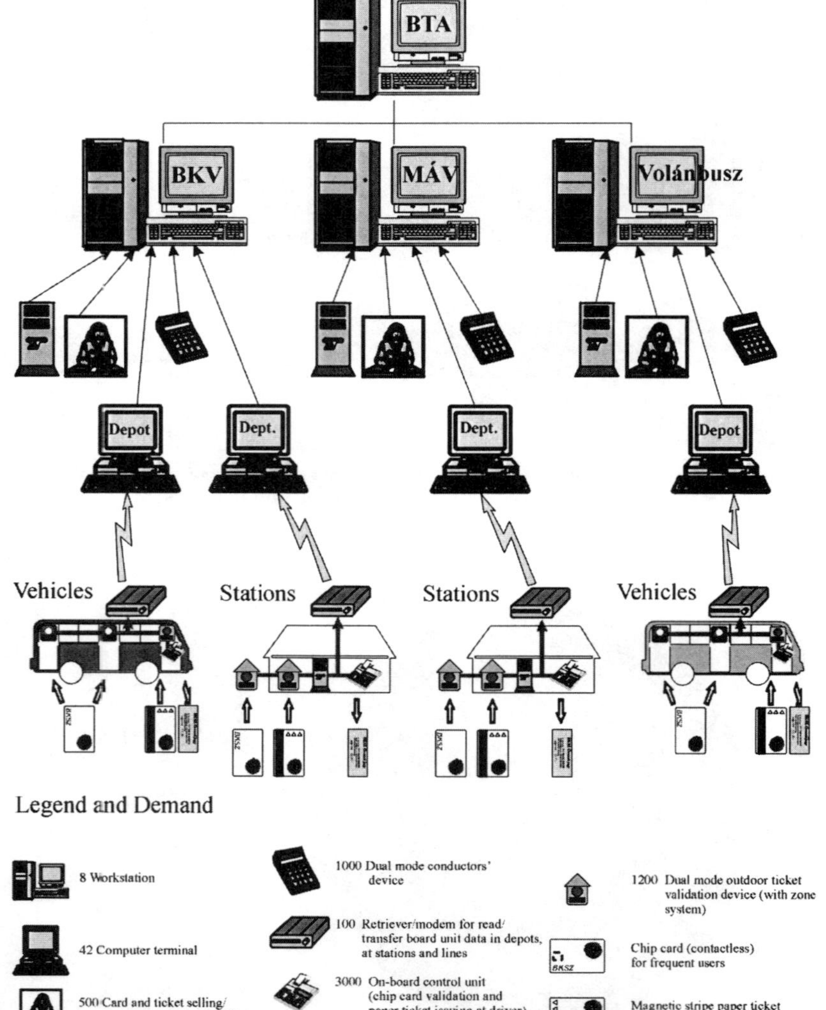

Figure 6: The Structure of the Electronic Ticketing System of the BTA

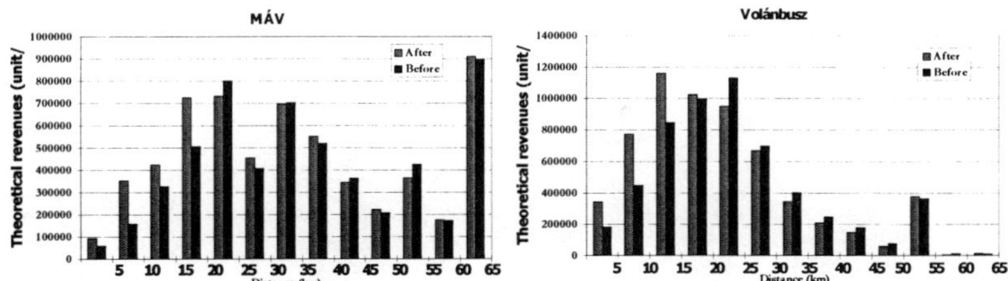

Figure 7: Impact of Zone System Change on Revenue Distribution

A Possible First Step Towards the Realisation of the Association

As a first step towards the tariff association the **extension of the pass-system** could be introduced in the region, which means:

- the introduction of a new integrated inner pass for Budapest which would allow the use of the MÁV train and Volánbusz coaches within the boundary of Budapest for 10% higher price;
- realisation of a combined outer+inner pass for commuters giving some discount which can be bought in one place and which allows a better access to the services.

This step would not need substantial legal changes and subsidy, so there is a chance to introduce it by the year 2000.

This intermediate solution would give the experts an opportunity for further investigations into a more complete transport association with a new fare and ticketing system. Additionally it would allow the politicians to choose the best solution and make the necessary political and economic decisions.

References

Monigl, J. – Ujhelyi, Z. – Koren, T. – Berki, Zs.: Investigation for the Foundation of the Budapest Transport Association (BTA) – Final Report; TRANSMAN, February 1997; client: BTA Preparatory Office

Monigl, J. – Berki, Zs. – Molnár, B.: Proposal for an Electronic Ticketing System for the Budapest Transport Association (BTA) – Report; TRANSMAN, November 1998; client: BTA Preparatory Office

Monigl, J. – Ujhelyi, Z. – Koren, T. – Nagy, E. – Berki, Zs.: Space-Time-Cost Approach in Modelling and Impact Evaluation of the Transport in Budapest and Its Region (TRANSURS Model System) – Report; TRANSMAN 1998

SOCIO-ECONOMIC CHARACTERISTICS, LAND USE, AND TRAVEL PATTERNS OF CHENNAI (MADRAS)

V. R. Rengaraju[1] and R. Sivanandan[2]

Introduction

Chennai, formerly known as Madras, is the gateway to southern India. It is also the capital of the state of Tamil Nadu and the fourth-largest city in the country. It lies on the Coromandel Coast of the Bay of Bengal, about 650-km (400 mi.) northeast from the tip of the subcontinent. Its population of 3,841,000 (1991) is composed primarily of Tamil-speaking Hindus. Industries manufacture cotton textiles, aluminum utensils, matches, hand-woven cloth, and bidi (handmade cigarettes). The main exports are leather, iron ore, textiles, and coffee. A large film industry is centered in the Kodambakkam section.

Chennai is a port city, with railway and road networks extending in all three land directions available for expansion. Chennai's history goes back to 1693, when a nucleus of 67-sq. km. called Chennapatna was established. This nucleus expanded to include more settlements around it and the city grew. Today, the city limits comprise of 172-sq. km. and the metropolitan area (known as Chennai Metropolitan Area (CMA)) covers an area of 1172 sq. km. The development pattern of the city was generally controlled by the City Corporation before Chennai Metropolitan Development Authority (CMDA) took over.

This paper is intended as an introduction to the city of Chennai, and to serve as a stage-setter for the other two papers of the session. This paper is organized into three broad areas covering socioeconomic characteristics, land use characteristics, and travel patterns. Following this introductory section, section two on

[1] Professor, Dept. of Civil Eng., Indian Institute of Technology Madras, Chennai, India
[2] Associate Professor, Dept. of Civil Eng., Indian Institute of Technology Madras, Chennai, India

socioeconomic characteristics covers the topics of population, employment, and household characteristics. Land use characteristics including the geographic distribution of population and employment, and travel corridors are included as part of section three. Section four delves on the travel characteristics of the city. This is followed by conclusions.

Socioeconomic Characteristics

Population

As per the 1991 census, the population of Chennai City was 3.841 million and that of the metropolitan area (CMA) 5.846 million. Table 1 below shows the growth of population in the city as well as the metropolitan area since the year 1900. It is seen from this table that the population growth has been particularly rapid since the 1960s.

Table 1: Population Growth in Chennai
(In millions)

Year	Chennai City	Chennai Metropolitan Area (CMA)
1900	0.541	0.800
1911	0.556	0.830
1921	0.579	0.876
1931	0.713	1.056
1941	0.855	1.450
1951	1.416	1.762
1961	1.729	2.324
1971	2.467	3.475
1981	3.266	4.386
1991	3.841	5.846

(Source: Census Data)

Presently, the population of CMA is believed to be over 7 million. The CMDA has predicted that the population of the city and the metropolitan area will reach figures of six and nine million, respectively, by the year 2011.

Employment by Type

The employment data for Chennai has been obtained by RITES (1994) from secondary sources considering 1991 as the base year. The employment data for the CMA under the three categories of basic employment, educational employment and other employment, and their percentages have been shown in Table 2. The zone-wise distribution of different types of employment and the area consumed by these activities are shown in a later section.

Table 2: Types and Number of Employment in the CMA (1991)

Basic Employment	Educational Employment	Other Employment	Total Employment
789,856	60,515	907,212	1,757,583
44.9%	3.5%	51.6%	100%

(Source: RITES, 1994)

Age and Sex Characteristics of the Population

The population characteristics excluding children below five years of age was analyzed as part of the Household Interview (HHI) survey conducted by Pallavan Transport Consultancy Services Ltd. in 1992 (PTCS, 1993). The male and female population was found to account for 53.2% and 46.8%, respectively, in the CMA. The population distribution by age and sex, based on the above survey, is shown as percentages in Figure 1.

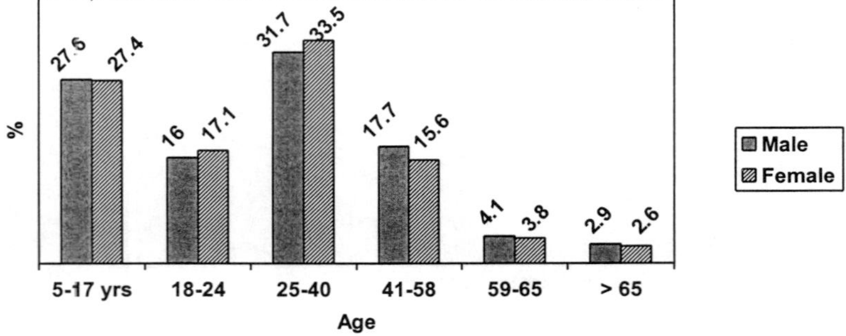

Figure 1: Percentage Distribution by Age Group and Sex (1992)
(Source: Based on data presented in PTCS (1993))

Household Size

One of the interests in the HHI survey was to determine the distribution of households by family size. The survey revealed the following distribution for the city, rest of the metro, and the CMA (Tables 3 and 4).

Table 3: Average Household Size in Chennai (1992)

CMA	City	Rest of Metro
4.55	4.67	4.35

(Source: PTCS, 1993)

As can be seen from Table 4, family sizes of 4 and 5 members are predominant. This observation holds good for the city as well as for the rest of metro area. As per the census data, it is found that the average household size has dropped from 5.21 and 5.12 for the city and rest of the metro in 1981, to 4.81 and 4.63, respectively, in 1992.

Table 4: Distribution of Households by Family Size in Chennai (1992)

Persons in Household	Number of Households in				
	City	%	Rest of Metro	%	CMA
1	9320	1.14	5598	1.22	14918
2	62508	7.70	48111	10.50	110619
3	124425	15.28	90192	19.70	214617
4	207661	25.50	120797	26.35	328458
5	189171	23.23	95926	20.93	285097
6	114714	14.10	53344	11.64	168058
7	53382	6.55	24557	5.36	77939
8	32410	3.98	14019	3.11	46429
9	12782	1.60	3995	0.87	16777
10	3451	0.42	932	0.20	4383
> 10	4428	0.54	883	0.19	5311

Family Income

The family income data revealed by the HHI survey is presented in Table 5. It shows that the earning of the city households are higher compared to metro area households. The average income per month per household (for 1992) in the city, rest of metro, and the CMA were Indian Rupees (Rs.) 2210 ($73.67), 1584 ($52.8), and 1985 ($66.2), respectively.

Table 5: Distribution of Households by Family Income Level in Chennai (1992)

1992 Monthly Income Level (Rs.)	Households in					
	City (No.)	%	Rest of Metro (No.)	%	CMA (No.)	%
Up to 500	62117	7.63	59940	13.1	122057	9.59
501-1100	215246	26.43	153643	33.52	368889	28.98
1101-1800	167144	20.53	102883	22.45	270027	21.20
1801-2300	99757	12.25	55824	12.18	155581	12.20
2301-3600	144096	17.70	58805	12.83	202901	15.90
3601-5700	80752	9.92	20650	4.50	101402	8.00
> 5700	45139	2.54	6602	1.44	51741	4.10

(Source: PTCS, 1993)
(An exchange rate of approximately US $ 1 = Indian Rupees (Rs.) 30 may be assumed for the period concerned)

Auto Ownership

As per the survey estimates, the availability of vehicles are shown in the table below (Table 6).

Table 6: Vehicle Availability in Chennai (1992)

	Car/Van/Jeep	Motorcycle/ Scooter	Pedal Cycle
Vehicles Per Person of Over 5 Years Age	0.01	0.05	0.13
Vehicles Per Household	0.04	0.20	0.55

(Source: PTCS, 1993)

Ownership of cars and fast two wheelers was higher in city while the cycle owning population was more in rest of CMA. Ownership of fast moving vehicles, particularly the two-wheelers, has greatly increased since the survey. As of 1998, approximately 0.7 million of these vehicles are running on the roads of Chennai.

Industrial Growth

The industrial growth of the city in the CMA began in the 50s, with very rapid growth in the first two decades. There is a large concentration of engineering industries in CMA, most of which are located in the north and the west. Small-scale industries are mostly located in the south and southwestern sides of Chennai. An Export Processing Zone (EPZ) has been developed in the southern end of the CMA. A major electronic complex has been built in Perungudi area in the south, generating considerable basic employment and resulting service employment. A number of handicraft centers, farms, fun parks, and holiday resorts have come up on the East Coast Road, south of Thiruvanmiyur. These are further sources of service employment.

Recently, several multinational car manufacturers, such as Ford, Hyundai, and Mitsubishi have set up manufacturing plants, jointly with Indian companies. Chennai prides itself as the major automobile center of India. Another recent development that the city is witnessing is the State Government's initiatives to make this City the "haven" and ideal destination for Information Technology (IT) companies. In this direction, several software technology parks are under development at Taramani, Sholinganallur, and Kelambakkam, forming a "golden triangle."

Land Use Characteristics

The land use characteristics of Chennai will be elaborated in this section. The total area consumed by different types of land uses in the CMA are as shown below in Table 7.

Table 7: Total Area in Different Types of Land Use in Chennai (1991)

	(sq. km)
Residential Use	290.5613
Basic Employment Location :...................	107.9731
Educational Employment Location	22.5333
Other Employment Location	35.6246
Area Unusable for Development	599.0547
Vacant Land	116.2533
Total Area Under Use	456.6924
Total Area of CMA	1172.0000

(Source: RITES, 1994)

Geographic Distribution of Population and Employment

The geographic distributions of population densities and employment are shown diagrammatically through Figures 2 and 3. From Figure 2, it can be seen that the highest density of population is in and around the Georgetown area, which represents the Central Business District (CBD). The density drops towards the suburbs as one moves away from the city. Figure 3 compares the population and total employment in different areas through bar plots. This figure indicates that the highest number of employment is offered by areas in and around the CBD.

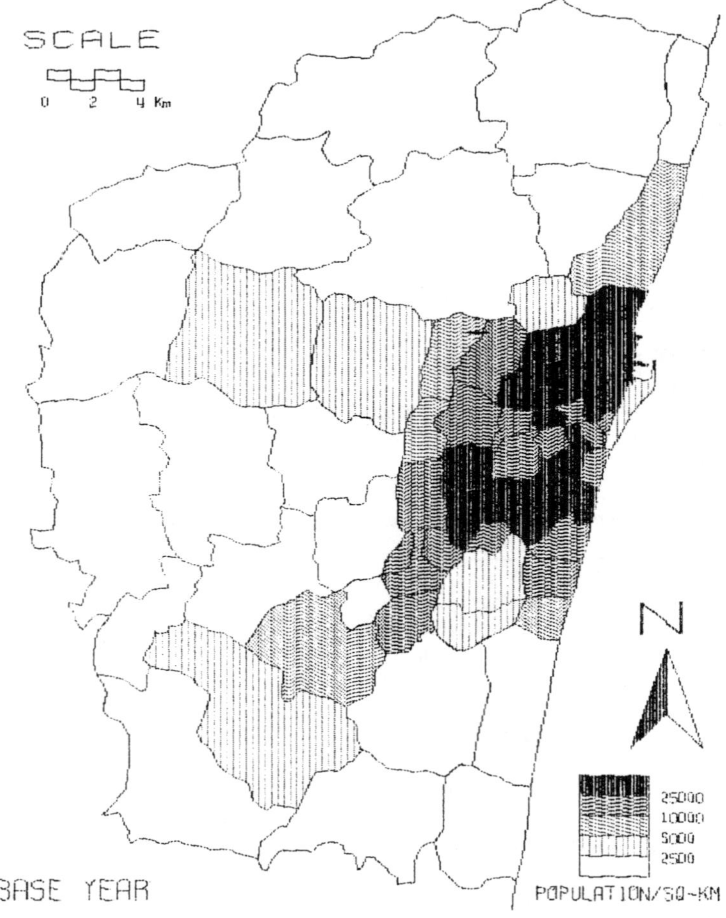

Figure 2: Population Density in Chennai (1991)
(Source: RITES, 1994)

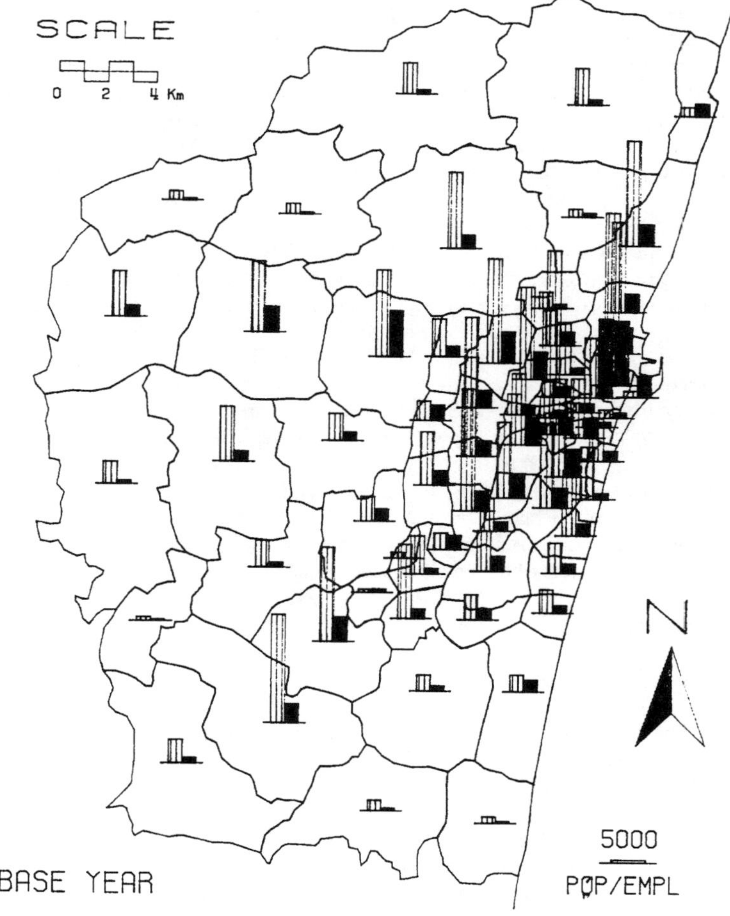

Figure 3: Geographic Distribution of Population and Employment in Chennai (1991)
(Source: RITES, 1994)

Current Transportation Systems in Chennai

The transportation systems in Chennai are currently supported by both road and rail networks. The road network in the CMA is a radial pattern with three principal radial arterials, with National Highway (NH) designation. These are NH 4 to the north, NH 5 to the west, and NH 45 to the south. In addition, there are two more radial arterials, one along the coast between NH 5 and NH 45 (Arcot Road), and the other parallel to NH 4 (Thiruvallur High Road). Recently, an Inner Ring Road (IRR) has been added to the road network on the periphery of the city in the western side to cater to the travel needs along this corridor. Apart from increasing number of private vehicles, the CMA's huge fleet of public buses also use the road network. The bus transit system, operated by the Metropolitan Transport Corporation (MTC), is one of the best in the country, and works under the direct control of the State Government of Tamil Nadu. This system operates over 2800 buses in over 550 routes. Paratransit mode, primarily in the form of motorized three wheelers (called auto rickshaws) is also a popular means of public transportation.

In terms of railway infrastructure in the metropolitan area, the following three suburban sections have been catering to the travel needs: Broad Gauge (BG) line towards Gummidipoondi (North), BG line towards Arakonam (West), and Metre Gauge (MG) line towards Chenglepet (South). All these lines are electrified, and they also cater to commuter traffic. In addition to this, a rail rapid transit, named Mass Rapid Transit System (MRTS) is currently under construction along a north-south eastern corridor. Chennai also boasts of a major port on the East Coast. Detailed presentation on the public transportation systems and their operations form one of the following papers of the session, and hence have not been elaborated here.

Travel Characteristics

Major Transportation Corridors

The Chennai metropolitan area has eight major transportation corridors, as identified by the Madras Area Transportation Study Unit (MATSU, 1974). These travel corridors (as shown in Figure 4) are:

I. North-South Eastern Corridor
II. North-South Central Corridor
III. East-West Northern Corridor
IV. East-West Central Corridor
V. East-West South Corridor
VI. The Inner Circular Corridor
VII. The Intermediate Circular Corridor
VIII. The Outer Circular Corridor

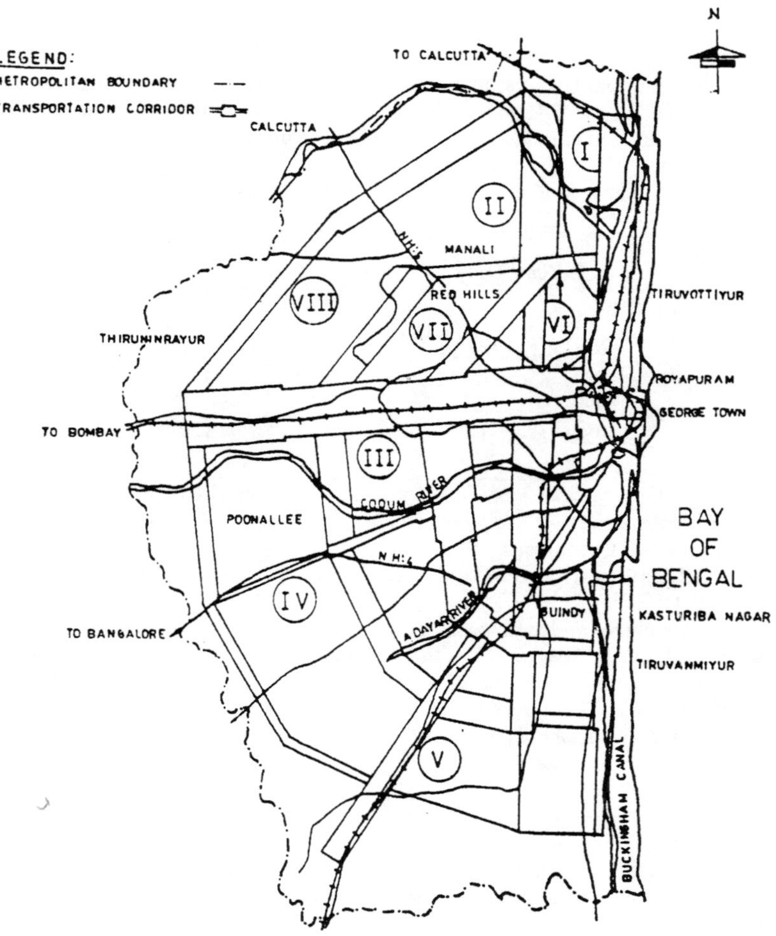

Figure 4: Major Transportation Corridors of Chennai
(Source: MATSU, 1974)

Of these, the Corridor I has the highest concentration of trips. Based on this, the existing transportation facilities on this corridor are being supplemented by the construction of a higher capacity Mass Rapid Transit System (MRTS).

Travel Characteristics

The total number of trips in the CMA has seen a three-fold increase from 2.45 million in 1970 to 7.45 million in 1991. Per capita trip rates have gone up from 0.866 in 1971 to 1.282 in 1992. Table 8 depicts the trend in growth of trip rates.

Table 8: Growth in Per Capita Trip Rate in CMA

Year/Study	Per Capita Mechanized Vehicular Trip	Per Capita Total Trips
1971 - MATSU	0.39	0.866
1984 - Kirloskar Study	0.70	1.146
1992 - HHI Survey	0.73	1.282
2011 – Projected	0.93	1.500

MATSU – Madras Area Transportation Study Unit (MATSU, 1974)
Kirloskar Study – Transportation Study by Kirloskar Consultants Ltd. (Kirloskar Consultants, 1986)
HHI Survey – Household Interview Survey (PTCS, 1993)

The average trip length has also increased from 7.2 km in 1970 and 9.5 km in 1984 to 10.1 km in 1992. About 70% of the trips are within 4 km in distance. Average trip lengths of home based work trips and home based education trips were found to be 7 and 3.5 km, respectively. Percentages of person trips by purpose and predominant mode are shown in Table 9.

Modes of Travel

Several modes of travel are used in this City. These include bus, train, car, two-wheeler, bicycle, walking, and others. Table 10 shows the share of various modes and also the trend in change of modal split.

Table 9: Person Trips (%) by Purpose and Mode (1992)

Trip Type	Walk	Perso-nalised Modes	Paratransit	Public Transport			Total	Total Trips (Mill.)
				STU Bus	Train	Others		
HBW	2.2	12.9	1.2	23.8	2.4	0.4	42.9	3.19
HBE	19.6	5.2	3.3	8.0	0.9	0.6	37.6	2.80
OHB	7.6	4.0	0.4	5.8	0.5	0.1	18.4	1.37
EB	0.2	0.6	-	0.3	-	-	1.1	0.09
Total	29.6	22.7	4.9	37.9	3.8	1.1	100.0	7.45

(HBW: Home Based Work; HBE: Home Based Education
OHB: Other Home Based; EB: Employer's Business)
(Source: PTCS, 1993)

Table 10: Trend in Change of Modal Split

Mode	% of Total Trips Shared by Various Modes		
	1970[a]	1984[b]	1992[c]
Bus	41.50	45.53	37.90
Train	11.50	9.03	4.10
Car, Van, Jeep	3.20	1.45	1.50
Two Wheelers	1.70	3.24	7.00
Bicycle	21.30	10.70	14.20
Walk	20.70	28.07	29.50
Others	0.10	1.98	1.00
IPT (Auto Rickshaw, Cycle Rickshaw)	NA	NA	4.80

Sources: a. MATSU, 1974
b. Kirloskar Consultants, 1986
c. PTCS, 1993

The figures in the above table indicate a trend of decline in public transport trips. Yet, over 40% of the total trips in 1992 were carried by the two predominant public transport modes of bus and train. An interesting point to note is the increasing trend in the two-wheeler share of trips. This trend seems to be continuing even today, with more and more two wheelers flooding the roads of the City.

Travel Problems in the City

A major transportation problem of the city currently is traffic congestion. Several arterial roads get choked with congestion, especially during peak periods. A steep rise in vehicle population, inadequate capacity of roads, poor geometric and surface conditions, roadside encroachments, and on-street parking are the primary causes of congestion. Clearly, the road system is inadequate. Insufficient and poor facilities for pedestrians and cyclists further compound the problem. Added to these, inadequate traffic management, laxity in traffic enforcement, and drivers non-compliant to road regulations aggravate the situation further.

Bus transit is the major mode that satisfies the travel demand in the city, while the rail transit system serves as a good supplement to the bus system. However, each of these systems function independently and there is lack of coordination among them. Inaccessibility of the rail network from a number of population centers and a high difference of fares for single journeys between bus and rail make the bus more popular. This is particularly more glaring in the case of recently opened Phase I of the MRTS, which is operating well below its capacity. Though the patronage of public transit systems is high, as compared to cities of western countries, their inadequacy, inaccessibility, and inconvenience are causing shifts to private vehicles, particularly, to two-wheelers, further adding to road congestion. This has led to increased travel times, fuel consumption and pollution, and accidents.

Conclusions

Mobility in Chennai is becoming a concern, not only to the travelers, but also to the providers of transportation systems and services. Primary causes of mobility problems arise due to increasing demands on transportation facilities and services, which are unable to expand at the pace of demand. This is resulting in the deterioration of quality of transportation – severe traffic congestion, air pollution, accidents, reduced speeds, higher travel times, and increased travel stress. With the improvement in economic status of the Chennai residents (with higher incomes and affordability) and inadequacies of public transportation, a shift to private transportation for travel is the current trend. This is conspicuous through the explosion in the number of two-wheelers. The fast growth in car ownership is also visible. While these may offer convenience and comfort to travelers, they have rendered the city's road network inadequate. Though Chennai can boast of having one of the best public transportation systems in India, the system is still not fully adequate to meet the growing demands of travelers.

The pace and nature of growth of the city, influenced by migration of the rural population to the city, haphazard development of settlements, and growth of slum areas are also serious concerns that threaten to further tax the city's transportation infrastructure. In this context, what the city's transportation systems call for is enhancement of network capacities, by adding road space within budget constraints and removing traffic bottlenecks; more effective traffic management and enforcement; better integration of the bus and the rail systems; and cautious land use control.

Acknowledgement

The authors would like to thank Rail India Technical and Economic Services Ltd. (RITES), Chennai branch for providing us with much of the data that has been used in this paper. This data has been gathered from reports of the study entitled, "Comprehensive Traffic and Transportation Study for Madras Metropolitan Area," conducted for Madras Metropolitan Development Authority. Thanks are also due to Pallavan Transport Consultancy Services Ltd. and Kirloskar Consultants Ltd., who were associated with RITES in conducting the above study, and to Chennai (Madras) Metropolitan Development Authority, which sponsored the study.

References

Kirloskar Consultants Ltd. (1986). "Madras Metropolitan Area Traffic and Transportation Study," Chennai, India.

Madras Area Transportation Study Unit (MATSU) (1974). "Comprehensive Traffic and Transportation Study," sponsored by Department of Town and Country Planning, Chennai, India.

Pallavan Transport Consultancy Services Ltd. (PTCS) (1993). "Household Interview Survey – Final Report," Comprehensive Traffic and Transportation Study for MMA, Chennai, India.

RITES (1994). "Comprehensive Traffic and Transportation Study for Madras Metropolitan Area – Landuse Transport Model for MMA," prepared for Madras Metropolitan Development Authority, by Rail India Technical and Economic Services (RITES), Pallavan Transport Consultancy Services and Kirloskar Consultants Ltd., Chennai, India.

RITES (1995). "Comprehensive Traffic and Transportation Study for Madras Metropolitan Area – Final Report," prepared for Madras Metropolitan Development Authority, by Rail India Technical and Economic Services (RITES), Pallavan Transport Consultancy Services and Kirloskar Consultants Ltd., Chennai, India.

URBAN PUBLIC TRANSPORTATION IN CHENNAI AND ITS OPERATIONS AND USE BY PEOPLE

M. Imtiyaz Ahmed[1] and S. Ponnuswamy,[2] M. ASCE

Introduction

Presently, the Chennai (formerly known as Madras) Metropolitan Area (CMDA) covers an area of 1,172 km^2 of which 172 km^2 forms the city proper. It housed a population of 5.8 million in 1991 and is possibly over 7 million now. The population performed on average 1.35 trips per capita in 1991 and at the rate at which this has been growing, it would be reaching a rate of 1.50 trips per capita by 2001. Over 42 % of these trips are made by the public transit system, while about 44% depend on walking and bicycles. Chennai is one of the three metropolises in India that has a rail-based transit system in the form of suburban train services (a combination of commuter rail and Metro) provided and operated by the main line railways (Figure 1). Chennai, like most of the Indian cities, has a well-established bus transit system, which in fact can be considered one of the best-run systems in India.

Transport Network

The Chennai Metropolitan area has three old radial rail corridors and a short leg of an orbital rail corridor under development. It has five radial road corridors of which three principal ones are parallel to the rail corridors for their major lengths. It has a number of orbital road layouts, but except for the recently developed Inner Ring Road (at the periphery of the city), none of them is continuous for the full length. The total road network covers 105 km of arterial roads and 450 km of collectors and distributors. The public bus transport operates over the entire area consisting of 500 links within the city and 200 links in the outer area.

Of the three rail corridors, one is with one-meter wide Metre Gauge (MG) track and the other two are run on 1.676 meters wide Broad Gauge (BG) track. The

[1] Chief Project Manager, RITES, 56, Veerabhadran Street, Valluvar Kottam, Nungambakkam, Chennai, India

[2] Formerly Additional General Manager, Southern Railway and Guest Faculty, IIT, Chennai, India

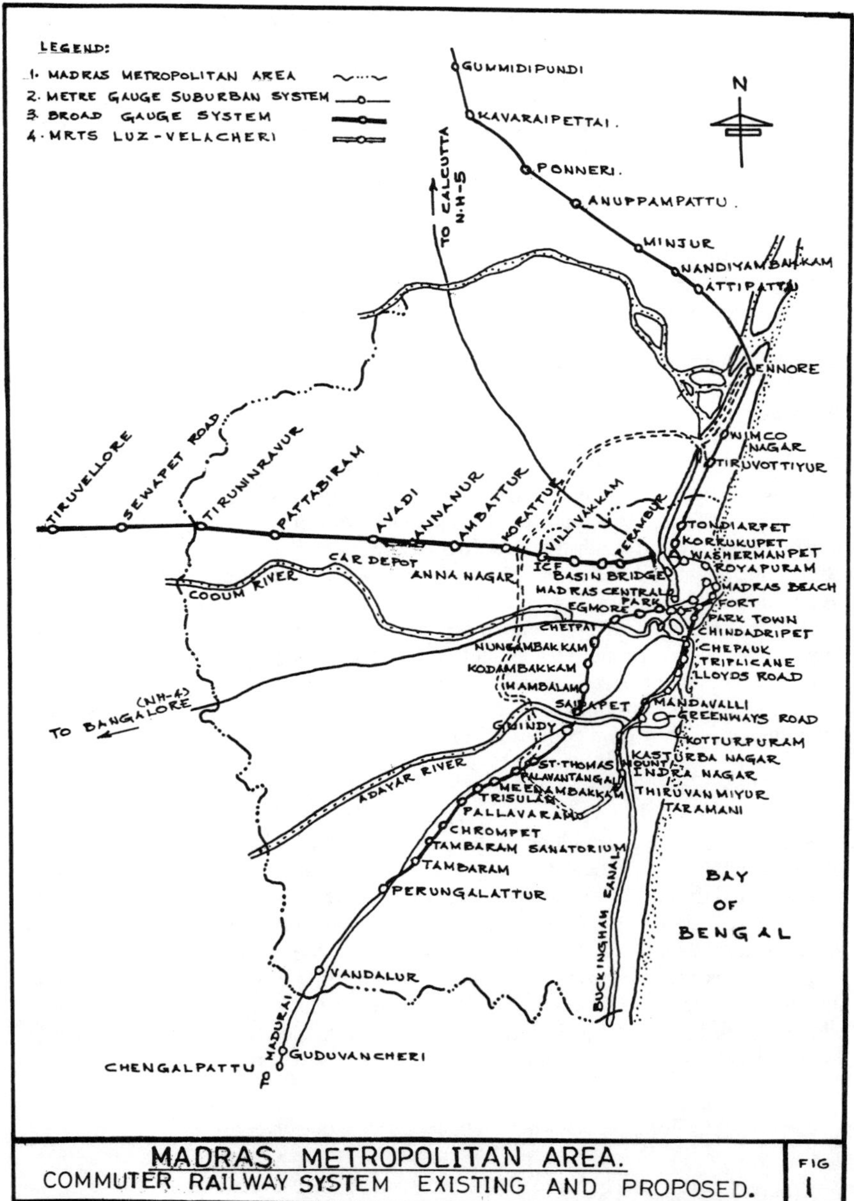

MADRAS METROPOLITAN AREA.
COMMUTER RAILWAY SYSTEM EXISTING AND PROPOSED. FIG 1

Metre Gauge corridor towards the south from Madras Beach (i.e., Port) and Tambaram, a distance of 29 km has a pair of dedicated lines for suburban trains and one line for inter-city trains. Recently one BG line has been added, on which eight suburban services run daily. The BG corridor towards the west has four lines up to Pattabhiram, a distance of 14 km and three lines beyond up to Tiruvellore (41.6 km from Chennai). The corridor towards the north to Gummudipundi has two lines for a distance of 46.8 km and it is treated as suburban section. Rail and bus transport systems are operated by the State. The two operations function independent of each other. Apart from the rail and bus systems, the public depends on over 25,000 auto rickshaws (three wheeled scooters) operating as taxis which are permitted to carry up to three passengers each. There are about 2,500 cycle rickshaws permitted to carry two passengers each. The latter are not licensed and operate on short distance mostly in local areas, predominantly for transporting school children.

Historical Development of Public Transport Systems

Railways

In 1857, four years after the first railway line was laid in India, the first line in Broad Gauge was laid in Chennai from the city westward up to Wallajah, a distance of 112 km. This line was later extended west towards Bangalore, northwest towards Bombay and south towards Cochin Port. The line was soon doubled up to Arakonam, a distance of 68 km. The main workshops of the broad gauge rail system in this area are located on this line. The line was electrified at the end of 1979, when they started running EMU's operated on 25 kV AC system. The Broad Gauge line towards the north was laid as a single track and subsequently it was doubled. The electrification for this was carried out in 1979. A few worker specials and local trains using steam traction were run earlier on this section during morning and evening hours. They were converted to EMU commuter trains in 1979.

The main suburban section in Chennai is the one towards Tambaram in the south, which is a MG section. This system was principally passenger-oriented. It passes through some populous residential areas like Chetpet, Mambalam, and Saidapet within the city and thus attracts some intra-urban travel demand. It was doubled in early part of the century and was electrified as early as 1931. Then, it carried only 11,000 passengers per day. The introduction of EMUs on this section served as a catalyst for development of this corridor. A third line, added later on this section, serves solely inter-city trains and a few fast EMU trains in peak hours, while the suburban lines are reserved mostly for intra-urban and suburban traffic. An addition of further length of 33 km from Tambaram to Chengalpattu was doubled in 1984 and EMU services were extended.

New Corridor: Madras Beach-Luz (Thirumayilai)

A comprehensive traffic and transportation study conducted in early seventies by the Chennai Metropolitan Development Authority for the metropolitan area recommended high priority to be accorded to provide vast transit on the length from the city core to Adyar (Kasturbha Nagar) in the south. This was first proposed as a grade separated modern Mass Rapid Transit System (MRTS), following the alignment of an old unused canal, in order to avoid the expense of acquiring private land. This was to be on Broad gauge with metro coaches to be operated on 750-volt DC system. Due to lack of funds, the scheme could not be taken up. After considering a number of alternatives, utilization of the existing BG corridors by upgrading them and constructing a new line as an extension from Madras Beach where all the three rail corridors meet was approved. It was decided to implement the first length of about 8.6 kilometers from Madras Beach to Luz, where connecting bus services could carry the commuters bound for areas beyond. It was proposed to use the same type of EMUs as were being used on the BG corridors (i.e., 3.6-m wide cars operated on 25 kV AC). Proposal was approved in 1983 and taken up in 1985. It was completed in October 1997. Approximate cost for this length was $60 million (Rs. 2400 million). The next phase of 10 kilometers, estimated to cost about $ 158 million (Rs. 6300 million) up to Velacheri, a major new residential complex, with a major bus terminal was taken up in continuation of the first phase work. The cost for this phase is being shared in the proportion of 2:1 by the State government and Central Government respectively. This type of implementing rail-based urban transport schemes was first tried in Mumbai in 1987-93 and is being extended to other Indian cities wherever the State Government comes forward with state share.

Bus Transport

Chennai is one of the five Indian cities that have used trams for intra-city commuter movement since the early days of electrically operated streetcars. The tram system in Chennai was closed due to uneconomic operations and gave way to the now popular automobiles and buses. In the last year of operation, it carried as many as 170,000 passengers. Chennai has had a city bus transport system for well over five decades. Before 1947, private entrepreneurs operated them. In 1947, after the Independence of the country, there was a move to nationalize the bus transport to avoid unhealthy competition where more than one operator operated and to avoid monopoly where single or oligopolies groups operated. Also, the private operators tended to neglect unremunerative routes. The State Government went in for nationalization of bus transport by first entering the city bus system in Chennai in 1947 introducing thirty buses side by side with the private buses. They started taking over the private buses soon after and took over complete city bus system in the city by July 1948. It was administered directly until 1971, when they converted the departmental setup as a company under the Indian Companies Act and formed the Pallavan Transport Corporation. The bus transport undertaking in the city is an autonomous corporation registered with 100% equity provided by the state government, in the form of initial capital. Pallavan Transport Corporation then was managed as a single entity until four years back, when it was split into two units on an operational territorial basis. The company is expected to pay all taxes just like other operators, and also provide concessional fares for senior citizens, handicapped persons, and also carry students of classes up to twelfth standard level free of charge. For these special transportation functions they are given some subsidy. This corporation has been consistently performing well, except that they have not been able to provide enough services to meet the growing demand, mainly for want of funds to buy enough buses. Since 1972, the fleet strength has increased from 1,030 to 2,849 in 1999, while the total passenger trips have gone up from 550 million to over 1,500 million per annum (about 4.61 million per day). The growth rate was 6.42% up to 1990 and 2.48% since then. This growth is due to its spread of operations, which is within an accessible distance of less than half a kilometer for the commuters within the city and the fare levels having been kept low so as to be affordable to the lowest income groups.

Operations

Bus Transit

Organization and Management

The Pallavan Transport Corporation, which has now been renamed the Metropolitan Transport Corporation (MTC), is comprised of two units. A Board administers each unit with a Chairman, who is an Administrator. The Chief Executive is a full time Managing Director, who is generally an engineer. There are six other Directors on the Board, who are ex-officio, representing different departments of the government or the Managing Director of adjacent similar Transport Corporation. The Managing Director, who is assisted by one or two General Managers, a Financial Officer, some Managers and Deputy managers, monitors the day to day work. Each of these divisions has a central workshop of its own for major repairs. Daily garaging, cleaning, washing, fueling and tire replacements and minor repairs are performed in 22 depots. Each depot is designed to hold 50 to 100 buses, depending on location. The two units together have a total of about 21,800 staff members, of whom 13,400 are on operational duties, and 4,000 work in the workshops and depots. Staff per bus works out to 8.96 as of now. This is the lowest bus-staff ratio for city transit systems in India.

Operating Characteristics and Performance

There are 558 routes operated, with a daily total number of services of 2,545, excluding 53 night services. Two hundred forty of these routes and 1,258 of the services are limited-stop or express services, a majority of which are operated in peak hours. The headway varies from 5 minutes on very busy routes to 20 minutes on others during peak hours and about double the same during off-peak hours. There are a number of overlapping routes due to a number of services converging and or diverging to a number of population centers and places of work or shopping centers located in different parts of the city. The express services reach an average speed of 25 to 30 km/h, while the ordinary services rarely exceed 15 km/h. This is mainly due to passenger loading /alighting delays and time taken for ticketing on board in peak hours. Some of the latest important performance indicators for bus system are given below:

Number of buses in the fleet	2,849
Fleet utilization	93%
Average number of passengers/bus /day	1,709
Operating kilometers /bus/day	255
Operating kilometers (in 1997)	207 million
Percent of trip losses due to technical	5.91
Due to operations	6.68
Fuel consumption (liters/bus km)	3.65
Average occupancy ratio	78%
Bus Capacity	48 seated +24 standees excluding crew

The bus chassis used for the city bus is the same as that of a regional bus and rather heavy. The seating arrangements have been altered to provide for more standees. The floor height is 1.1m with three steps. There is no special provision for physically handicapped persons. Two pairs of seats are provided for handicapped and elderly persons inside. About 45 % of seats (i.e., seats on one side) are reserved for ladies. The heavy but sturdy construction of buses has facilitated their being used on rough roads and overcrowding (carrying over 150 passengers per bus in peak hours over some links) without major breakdowns. The MTC is planning to introduce low floor, lighter and more efficient urban buses.

Fare Structure and Financial Results

Table 1 shows the comparative fares on rail and bus as at present. The state government gives an annual subsidy to the operators towards losses incurred on account of concessional passes to student's etc. During 1997-98, a total of 1,203.7 million traveled on individual tickets, 75 million on monthly season tickets, and 156 million students (of whom 116 million were carried free) on the two systems. The cost per bus kilometer of bus operation for the larger of the two units in 1997 was \$0.34 (Rs. 14.65).

Table 1: Fare structure of Suburban Rail and Bus Transit in Chennai As of August 1998

Distance Slab (in km)	Suburban Rail II Class				Bus Transit			
	Single Journey		Monthly Pass		Single Journey		Monthly Pass	
	Rs	US$	Rs	US$	Rs	US$	Rs	US$
0-2	3.00	0.08	45.00	1.13	1.00	0.03	40 times	
2-4	3.00	0.08	45.00	1.13	1.00	0.03	Single	
4-5	3.00	0.08	45.00	1.13	1.25	0.03	Journey	
5-8	3.00	0.08	55.00	1.38	1.50	0.04	fare	
8-10	3.00	0.08	55.00	1.38	1.75	0.04		
10-15	5.00	0.13	75.00	1.88	2.25	0.06		
15-20	5.00	0.13	90.00	2.25	2.75	0.07		
20-25	5.00	0.13	95.00	2.38	3.00	0.08		
25-30	8.00	0.20	105.00	2.63	3.25	0.08		
30-35	8.00	0.20	110.00	2.75	3.25	0.08		
35-40	8.00	0.20	120.00	3.00	3.50	0.09		
40-45	10.00	0.25	130.00	3.25	4.00	0.10		
45-50	10.00	0.25	135.00	3.38	-	-		
50-60	11.00	0.28	150.00	3.75	-	-		

Personnel

Each bus has a crew of two, a driver at the wheel who operates each bus, and a conductor with a seat in rear (i.e., near the entrance). On average, there are 6.81 passengers per bus and the buses are on road from 0500 hours to 2100 hours, involving two shifts. The conductor issues tickets to the passengers while the bus is on the move in order not to delay at stops. As an incentive to the crew not to skip stops, they are given an incentive payment based on sale of tickets, which presently is 2.764% in ordinary services and 2.412 % in express services, to be shared by the two. Former drivers with experience were appointed after tests. Some years back, the State Government started a Driver Training Institute at Gummudipundi, an industrial satellite town being developed at 46-km north. Initially, this Institute was used to give intensive refresher course training to the serving drivers of State Transport Corporations. More driver training institutes have been started in other cities. The Institute now takes people with a basic knowledge of driving and who hold Light Commercial Vehicle drivers licenses and offers a longer course to train them as bus/ truck drivers. Recruitment of drivers for the city service is made from those who possess this training qualification. Drivers who are involved in accidents are sent to the Institute for a refresher course. Periodical special courses are conducted at the State-run Institute of Road Transport in Chennai.

Safety and Security

The accident rates in Indian cities are generally high, due to the mixed type of vehicles on the roads. Traffic includes vehicles such as buses, trucks, vans, cars, three-wheeled auto rickshaws for passengers and similar light goods vehicles, motorcycles/scooters, cycles and on some roads hand carts and cycle rickshaws. During December 1997, there were 474 accidents in the city involving buses. The accident rate is 0.88 per 100,000 effective kilometers operated per bus. While severe punishments are awarded to the erring crew, 'safe driver' awards are given to those drivers who have a minimum of 100 steering duty days, and are not responsible for any accidents in the half year of the award period. Those who have not been involved in any accidents for five continuous years are given gold plated medals and a certificate at a special function, generally during Road Safety Week.

Rail Transit

Organization and Management

The suburban rail system in Chennai, as is the case in the other three Indian metropolises where similar systems are functioning, forms part of the Indian Railway system. A separate organization under the Railways does the implementation of the new urban rail projects linked to their system, but once the project is operational, it is taken over for maintenance and operation by the zonal railway. Thus the Southern Railway operates the entire suburban rail network including the recently completed MRTS length. The Indian Railways are organized into a number of Zones (until recently nine in number, now being reorganized into fourteen) and the Southern zone is a major zone. Each zone is divided into a number of divisions for purpose of day-to-day operation and maintenance. The suburban rail system is in the Chennai division with its headquarters at Chennai, the same as that of the zone. The BG rail sections work on an automatic block system. The rail has three aspect color light signals, and stations at intervals are provided with control panels manned by the station managers. The Metre Gauge section however, is operated from a central control on CTC (centralized traffic control system). Local control panels are provided in some stations en-route for taking over control in case of emergencies and when some switching work has to be done for any of the trains, but not normally on EMU trains. The power collection is with overhead pantograph. There are three different combinations of trains being operated on the older corridors viz., 2MC (motor coach) +6 TC (trailer coach); 3 MC+ 6 TC; and 2 MC+4 TC. The busier lengths are provided with track circuiting and axle counters at ends of block sections to protect the less busy ones in the outer lines. The average speed of the trains varies around 30 km/h, irrespective of the gauge of section. The maximum speed capability is 80 km/h. The train dwell time is restricted to 30 seconds. In all cases, services start at 0400 hours and close at midnight. Track maintenance is done using off-track portable machines on the meter gauge and with track tampers on the broad gauge. For traction equipment maintenance, special tower wagons hauled by diesel shunters are used. Some of the physical characteristics of the three earlier systems are given below:

<table>
<tr><th></th><th></th><th colspan="2">Broad Gauge</th><th>Meter Gauge</th></tr>
<tr><th></th><th></th><th>East-West Corridor</th><th>Northern Corridor</th><th></th></tr>
<tr><td>Route length (major suburban) km</td><td></td><td>41.6</td><td>46.8</td><td>29.0</td></tr>
<tr><td></td><td>(extended)</td><td>42.0</td><td>33.0</td><td>30.0</td></tr>
<tr><td>Number of stations in major suburban areas</td><td></td><td>17.0</td><td>13.0</td><td>18.0</td></tr>
<tr><td>Number of trains per day (up and down total)</td><td></td><td></td><td></td><td></td></tr>
<tr><td></td><td>Suburban</td><td>182.0</td><td>68.0</td><td>31.0</td></tr>
<tr><td></td><td>Inter-city</td><td>52.0</td><td>35.0</td><td>33.0</td></tr>
<tr><td>Rolling stock</td><td>Motor coaches</td><td colspan="2">88.0</td><td>69.0</td></tr>
<tr><td></td><td>Trailer coaches</td><td colspan="2">161.0</td><td>243.0</td></tr>
<tr><td>Headway-Number of trains in peak hour</td><td></td><td>4.0 to 5.0</td><td>2.0</td><td>9.0 to 11.0</td></tr>
<tr><td></td><td>Off peak</td><td>2.0</td><td>1.0</td><td>5.0 to 6.0</td></tr>
</table>

On the newly opened MRTS, which is grade separated for major length, the peak hour headway is kept at 20 minutes, as demand is yet to pick up. Doubling the frequency with shorter trains can induce more demand but the unit formation, which on this section is at present six coaches (2MC + 4 TC), cannot be reduced due to cab arrangement.

Operation

A comparative study of the number of commuters who have used the rail for the past 36 years has shown that the annual rail trips increased from 114 million in 1971-72 to 189 million in 1979-80 but decreased to 138 million in 1983-84. It increased to 162 million in 1986-87 and gradually increased to 208 million last year. The reduction was mainly in the years when there was a steep increase in railway fares. Since the suburban rail forms part of the main line railways, the fare system adopted for inter city traffic is applied to the former also. Until the late seventies, the fares on railways were comparable with those of buses. But since 1979, the railways adopted a policy of hiking the minimum fare, and for the short distance, the fare stages were reduced in number. The result was that the fare on railways for short commuting distances was much higher in comparison with those on city buses. However, they offer a very high concession for regular travelers in the form of monthly season tickets. Such a fare structure is so disadvantageous to non-regular traveler that whenever there is a revision of basic fare on railway; the total patronage drops and starts picking up gradually in conformity with general growth until the next fare revision stage.

Various operational parameters of the BG and MG systems for the last five years for which data is available are tabulated below in Tables 2 and 3. These have been compiled from data collected from Southern Railway.

Table 2. Operational Statistics of Suburban Rail in Chennai

Year	Broad		Gauge		Metre		Gauge	
	PKM	EPKM-(Ps.)	TKM	CKM	PKM	EPKM-(Ps.)	TKM	CKM
1992-93	1338	8.23	3.12	24.96	2131	8.91	3.14	25.77
1993-94	1406	9.46	3.21	25.74	2206	9.61	3.15	25.95
1994-95	1518	9.66	3.41	25.16	2381	10.16	3.27	27.03
1995-96	1590	10.88	3.61	26.80	2454	11.00	3.32	27.53
1996-97	1803	11.10	3.93	30.42	2473	11.39	3.33	27.40

PKM-Passenger kilometers; TKM-Train kilometers; CKM- Car kilometers (all in million)
Ps.-Paise 1 (US $ would equal to about 4000 paise now but about 2500 paise at the beginning of this period).
EPKM-Earning per passenger km.

Table 3. Performance Indicators During 1996-97

INDICATORS	MG	BG
Train km / Route km	153	96
Passenger km /Route km	110813	49309
Passengers per day	384850	245454
Passenger /Route km	6207	2215
Vehicle km / Car	331	299
Vehicle km/ Route km	1227	763
Fleet utilization	86.6 %	89.2 %

During the five-year period, the average length of journey for commuters has increased only marginally from 20.3 km to 20.97 km on BG and from 17.3 to 17.81 km on MG respectively.

Stations are spaced at 1.11 km to 2.12 km on the MG corridor, 1.00 to 4.00 km on E-W and Northern BG lines, except for the last two stations on Northern corridor which are farther apart. Major stations, which are spaced at about 4 to 5 km are provided with switching arrangements and loop lines. The single journey or return tickets are card tickets pre-printed and stocked or punched and taken from manually operated ticketing machines by the booking assistants and issued to passengers. Monthly and quarterly season tickets are in the form of cards with photo identity cards issued at the beginning of each period for which they are required. These season tickets are not transferable, but they are for unlimited number of journeys for the period of validity. Ticket checking on trains is done at random and is also done randomly at destination stations.

Trains are not vestibuled and doors are not closed during run. They are operated with a two-man crew viz., a motorman (a skilled operator) in front and a guard (a traffic staff) in the rear. Communication is provided between the two. There are two depot-cum-workshop types of car sheds to maintain and repair the cars, one each for BG and MG. Additional stabling yards also are provided in the proximity of the sheds. Rail operation is highly labor intensive. A study in 1991 showed that the numbers of staff for the 29-km MG section were: station/operation staff-590; Workshop and Depot staff - 892; track maintenance staff-318. These numbers exclude supervisory staff, executives, accountants and administration staff in the division and headquarters. The rolling stock is maintained in two depots and generally major repairs are carried out on a unit exchange basis. In 1991, the BG car shed-cum-workshop had a staff of 815, excluding supervisors and executives. Track maintenance is partially mechanized, and despite this, they have about 1,050 men on the E-W line and 970 on the northern line.

Economics of Operation

The Indian Railways nationally incur a loss of over $50 million (Rs. Two billion) every year, out of which about 10% is in operating suburban rail systems of four major cities in the country, including Chennai. The losses are cross subsidized by the gains in freight traffic. There is no separate financial accounting maintained for suburban working, as it is difficult and complicated to separate incidences of expenditure on day-to-day basis, since lot of assets are common. Cost of replacement of assets is met from a Depreciation Reserve fund to which each zone contributes an amount in proportion to the assets on their books. The working results of the suburban working for a few years are indicated in Table 4.

The cost includes depreciation, but does not include cost of capital. An analysis made gauge-wise, showed that the operations on the Metre Gauge section, which has been functioning for a long time, were better even though it does not break even. The disparity can be attributed to higher patronage (about 0.33 million trips per day on MG Line to 0.184 million on the two BG lines). With fewer vehicles, more round trips could be run on MG lines. The MG system stations are closer to and more easily accessible to commuters. The catchment area is also more densely populated on the MG corridor than on the other two BG corridors, particularly the northern corridor. According to the latest data available for 1996-97, the fare-box ratio for BG system without depreciation was 0.56 and 0.43 with depreciation. On the MG system, it was 0.80 without depreciation and 0.70 with depreciation.

Table 4. Working Results of Suburban Rail in Chennai

(Conversion US$=Rs40.00)

Year	Earnings		Costs		Earning/Cost ratio
	Rs. Million	$ Million	Rs. Million	$ Million	
1984-85	109.0	2.73	282.9	7.07	0.39
1987-88	165.7	4.14	442.5	11.06	0.37
1989-90	173.3	4.33	469.2	11.73	0.37
1990-91	217.8	5.45	618.8	15.44	0.35
1991-92	257.2	6.43	676.1	16.90	0.38
1992-93	320.1	8.00	842.7	21.07	0.38
1993-94	359.3	8.98	929.9	23.25	0.39
1994-95	402.3	10.06	908.7	22.72	0.44
1995-96	478.5	11.96	1134.6	28.36	0.42
1996-97	594.1	14.85	1095.7	27.39	0.54

Source: Based on data collected from Southern Railway

Safety and Security

The rail mode is naturally a safer mode. With a fail-safe signaling system and well-regulated and monitored maintenance of the track, rolling stock as well as traction and other infrastructure, accidents on suburban systems are hardly heard of except for fire in coaches. The only fatal accidents have been run-over cases of trespassers and cattle. The lines pass very close to built up areas and in some areas, there are unauthorized squatter settlements close to tracks.

Intermodal Centers and Transfers

Intermodal integration of the public transport system in Chennai has not been seriously attempted in the past. The only transfer facilities available are the parking areas provided at Rail stations for cycles, motor cycles/ scooters and auto rickshaws (which serve as taxis). No such facilities exist at any bus terminal or station. No bus transfer facility exists in any station except at Chennai Central (BG) for transfer of passengers to and from Chennai Egmore (MG), the two main terminals. As part of a future strategy, intermodal transfer facilities are being developed at all-important stations on the new MRTS corridor.

Use by People

Modal Split

The proportion of commuters traveling by public transport (bus and rail) has been going down since 1984 (from 54.56% to 42%), while those by private vehicles has almost doubled from 4.69% to 8.5%. Other public transit trips (vans, school and company buses, auto rickshaws and cycle rickshaws) have shown an increase from 2% to 5.8 %. Out of the latter, the share of auto and cycle rickshaws was 2.2% and 2.8%, respectively. During 1984 - 1992, walk and cycle trips have shown a slight increase from 38.77% to 43.70%. The share of rail within bus and rail trips has reduced from 17% to 10%. The main reasons are the poor accessibility of rail stations to the newly developing residential areas, high fare differential for non-regular trip makers and inadequate frequency of services on two of the rail corridors, where there is considerable unsatisfied demand.

Ridership by Major Systems

On average, about 0.62 million commuters travel by rail and over 4.0 million by bus daily. The average trip length of all buses, including school and company buses, is 10.1 km while by city buses only, it is 8.4 km. Average trip length by rail is about 21 km.

Peak/Off-Peak Ridership

The trips, in the period from 0900 hours to 1030 hours and from 1700 to 1830 hours, are about 10% to 11% of the Average Daily Traffic. About 60% to 65% of the peak period trips are in the peak direction (depending on the corridor). During the peak period, trains carry dense crush loads, i.e. 2,000 passengers in an 8-car train of MG and 2,500 plus passengers in a BG train of 8 to 9 cars.

Cross-Section of Travelers

No long-term statistics are available regarding the socio-economic composition of ridership on transit systems. The details furnished below are for all trip makers in the city from two recent studies. About 69.6% of trip makers were male. About 35% of trip makers were in the 5 to 17 year age group and 31.5% in the 25 to 40 year age group. Only 5% of travelers are older than 58.

Conclusions

The main problem in the city is the decreasing share of mass transit trips by rail. The planners desire that the public transport trips (out of total trips) should be 60%. They also desire that in Chennai, 40% of transit trips should be by train. The long-term aim is to have four exclusive radial rail corridors and one orbital corridor as shown in Figure 1. Rail can be made more effective only if the bus and rail systems are integrated, not only in terms of service and intermodal transfer facilities but also in fare.

Public view of both systems is low due to public transit's inability to meet demand, in terms of number of buses and frequency of trains on two corridors, resulting in overcrowding. Due to this view (image), people are switching to personal modes. To some extent they take Intermediate Public Transit or IPT (Vans, auto rickshaws and Company buses) due to necessity than attraction. The cycle rickshaws and auto rickshaws are more accident-prone. The IPT taxi operators in other cities (Mumbai and Calcutta), provide shared service on a point to point basis in busy areas and thus are able to increase their share of commuters. There are no takers for this concept in Chennai.

Lack of funds has been the main reason for delayed implementation of many of the schemes identified for expansion of rail network and even augmentation of existing services in terms of train units. Since the Central Government has to meet demand for expansion and extensions of the system in other cities also, the annual fund flow for individual cities is low. Based on the success of joint sharing of costs of such projects in Mumbai, the State Government also has come forward to share the costs with the Central Government. The next phase of MRTS is now expected to be completed by 2002.

It is known that a good rapid transit facility accelerates development of the city. Any city, which has a population over a million people should start planning for a good arterial high capacity transit system, preferably rail. When the population reaches the two million mark, it ought to have a basic established system which would not only ease congestion on roads, but also accelerate and guide the development of the city in desired direction. The city should provide for such corridors with a long-term view and reserve the land required for it. It should control the development on the reserved portion in order to avoid delays and inconvenience later at the time of implementation of the rail transit. In India, the implementation should be a joint responsibility of Central and State Governments.

Acknowledgment

Authors are thankful to the Southern Railway and Metropolitan Transport Cooperation from whose reports basic data have been taken for preparation of this paper. Some data have been taken from the Journal of Transport Management published by Central Institute of Road Transport, Pune.

NOTABLE FEATURES OF THE PUBLIC TRANSIT SYSTEMS IN CHENNAI CITY

V. Thamizh Arasan[1] and S.B. Pattnaik[2]

Introduction

Urban transportation in India uses a variety of modes. On average, the foot and bicycle modes comprise about 50% of the trips made in Indian metropolitan cities (the share by these two modes is much more in the case of small and medium size cities in India (Thamizh Arasan et.al, 1998)). The other personal transport modes used for urban travel are the motorized two wheelers (motorcycle, scooter and moped) and cars. Bus is the predominant public transport mode used in most Indian cities except in Delhi, Mumbai (Bombay), Calcutta and Chennai (Madras) where well-established rail transit systems also exist. Apart from these, an Intermediate Public Transport (IPT) system serves as a para transit mode. The different types of vehicles that constitute the IPT system are car taxis, auto rickshaws (a 3-wheeled motorized vehicle to carry a maximum of three passengers, and cycle rickshaws (a non-motorized tricycle to carry a maximum of two passengers). In some cities, animal (horse/bullock) drawn vehicles also form a part of the IPT system. Of all these vehicles of the IPT system, the auto rickshaw and cycle rickshaw are more popular than the other types of vehicles.

The public transport system in Chennai City comprises both bus and rail transit systems, and it caters to about 42% of the daily trips made in the city; the shares of bus and rail are 38% and 4%, respectively (CMDA, 1995). The bus transit system, with the fleet size of 2,849, carries about 5 million passengers by providing 2,545 services on 558 routes, per day (MTC, 1998). This bus transit system is considered to be one of the most efficient systems in the country. The rail transit system operates along three traffic corridors (routes) in the city. The total length along the three routes is about 105 km, and the number of stations over this length is

[1,2] Associate Professor, Transportation Engineering Division, Department of Civil Engineering, Indian Institute of Technology, Madras, Chennai-600 036, India

48. The rail transit system carries about 0.6 million passengers per day. There is a Mass Rapid Transit System (MRTS) being newly developed along a circular traffic corridor in the city.

Certain unique features of the bus and rail transit system of Chennai City are presented in this paper. These features are related to the aspects concerning the development, operation and management of the systems. The first part of this paper deals with the notable management and operational aspects of the bus transit system, whereas the latter part of the paper deals with certain important aspects related to the development of the new MRTS in the city.

Bus Transit

Organizational Structure

Presently, the bus transit system of Chennai city is operated and managed by the Metropolitan Transport Corporation (MTC) Limited which is a registered company under Indian Companies Act. It is an autonomous body with 100% equity provided by the Tamilnadu State Government. The organisational structure of the MTC has been framed in such a way that there are independent units with the required amount of autonomy coupled with corresponding accountability at different hierarchical levels of administration. Considering the huge size of the fleet, the corporation has been primarily divided into two independent units with units I and II having their predominant operational jurisdictions in the southern and the northern regions of the city, respectively. A board of directors comprising ex-officio, representing different departments of the State Government administers the two units. The chairman of the board is the chief administrator, and the chief executive of the board is a full time Managing Director. The Managing Director is assisted by a few General Managers, and a Finance and Accounts Officer. A set of Managers and Deputy Managers monitors the day-to-day work of the units. There are two central workshops (common for both units) to attend to major repair works. There are 11 depots attached to each unit. Each depot is an independent unit administered by a Depot Manager. The Depot Manager is responsible for the operation, garaging, fuelling, cleaning, and day-to-day maintenance of a given subset of the fleet (usually 50 to 100 buses, depending upon the location of depot). The Depot Manager is also responsible for the management of the personnel (crew, and other supporting staff) attached to the depot. Thus, a depot is the unit in the lowest rung of the organisational structure, which directly bears the brunt of the day-to-day operation of the system. The Depot Managers are vested with requisite administrative authority to enforce the rules and regulations, within the general framework of the administrative jurisdiction, without waiting for the approval by the higher authorities for his/her actions. Thus, the breaking up of the organisation with independent units at different hierarchical levels of administration enables the bus company to perform better. The flow chart shown in Fig.1 illustrates the organisational structure of the transport company.

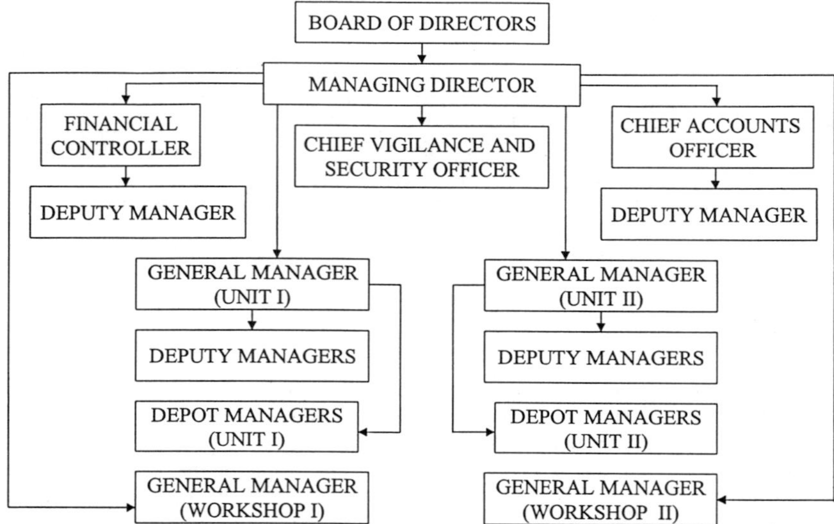

Fig. 1. Organisational Structure of MTC

Need-Based Operating Strategy

The metropolitan city of Chennai is now spread over an area of about 1200 km^2; and this has resulted in a significant increase in the trip length of the commuters. As the city grew in size, a continuous restructuring of the routes was to be done for the operation of the bus transit, the net effect being an increase in the length of the routes. Now (1998), the route length varies between about 10 and 40 km in the city. The average journey speed of the bus transit is very low (15 to 20 km/hr.), due to poor roadway and traffic conditions. Long–distance commuters therefore spend a considerable amount of time in the journey. To mitigate the problem of the long–haul trip makers, Limited Stop Services (LSS) have been introduced. The LSS will stop for approximately 50% of the total number of stops on a route. This has enabled an increase in the journey speed by about 30%. The fare for the LSS is about 30% more than the fare for the ordinary services. The recent steady increase in the patronage for the LSS has encouraged the bus transit operators to introduce more and more of the services. Now, the transport company has introduced two other upgraded versions of LSS. These are: (i) Express Service and (ii) Point-to-Point Service. The express service has very limited stops (about 25% of the total number of stops) on any route; and the point to point service does not stop between the origin and destination except for one or two important locations. The fare for the express and point-to-point services is about 90% and 120% more than the fare for the ordinary service respectively.

The said three types of fast bus transit services are found to be advantageous both by the affordable section of the long–distance commuters (due to significant reduction in travel time) and the operator (due to significant savings realised in the vehicle operating costs). In the context of the introduction of the LSS, to facilitate the bus transit operator to have a better understanding of the relative influence of the different factors that affect the Vehicle Operating Costs (VOC), a study was conducted by the Transportation Engineering Division, Department of Civil Engineering, IIT, Madras (Thamizh Arasan et.al, 1990).

Study on Fuel Consumption

The scope of work for this study was to estimate the relative influence of the important vehicle and traffic factors cn fuel consumption. The fuel cost in India is very high constituting a major part of the variable cost of operation of bus transit system as shown in Table 1. Hence, it was felt necessary to identify the significant factors, under the purview of the operators that influence this segment of the VOC, to enable introduction of appropriate measures to save fuel.

Table 1. Cost of Operation of Bus Transit

Item	Cost Per Effective km Run of a Bus					
	MTC unit I		MTC unit II		Average for MTC	
	Paise*	US$**	Paise	US$	Paise	US$
Personnel	674.4	0.161	775.0	0.185	724.7	0.173
Fuel & Lubricants	331.2	0.079	340.3	0.081	335.8	0.080
Tires & Tubes	54.5	0.013	45.2	0.011	49.9	0.012
Spare Parts	78.5	0.018	97.6	0.023	88.1	0.021
Interest on capital	104.4	0.025	92.4	0.022	98.4	0.024
Depreciation	151.5	0.036	142.6	0.034	147.1	0.035
Motor vehicle Tax	24.0	0.006	25.4	0.006	24.7	0.006
Others	46.3	0.011	54.3	0.013	50.3	0.012
Total Cost	1464.7	0.349	1572.8	0.374	1518.8	0.362

* 100 Paise is equal to 1 Indian Rupee; ** 42 Indian Rupees is equal to 1 US$ (in 1998).

The factors (independent variables) initially identified for this purpose were:

(i) Speed, (ii) Stops, (iii) Load, and (iv) Make of vehicle. The dependent variable, namely, fuel consumption, was obtained as Vehicle Kilometres Per Litre (VKMPL). The value was calculated as follows:

$$\text{VKMPL} = \frac{\text{Total run of the bus in km in four months}^{@}}{\text{Total diesel consumption in the period}}$$

@ Four months period was considered to cover the rainy season (during which period the roadway condition will adversely affect the vehicle operation) also.

The formulation of the independent variables was attempted as follows:

(I) Speed: two aspects of speed, namely, journey speed and running speed were considered.

(a) Journey Speed (km/hr): the value of this variable was obtained by dividing the route length (km) by the average journey time (hours) of the route.

(b) Running Speed (km/hr): the route length (km) was divided by the average running time (hours) of the bus to obtain the value for the variable.

(II) Stops: number of bus stops on up and down trips, and the number of controlled junctions on a route were considered to formulate the variable. The value of the variable for each route was arrived at as, $NS = (N_1+N_2)/2L$. Where,

NS = number of stops per km; N_1 = number of bus stops both on up and down trips of the route; N_2 = number of controlled junctions on the route on up and down trips; L = route length in km.

(III) Load: the city buses, most of the times, are overloaded, and during peak hours, the overloading is maximum. Hence, it was considered logical to take the excessive loading as an important factor affecting fuel consumption. An independent variable, as shown below, was introduced in the analysis.

$$\text{Excess Turn over} = \sum_{i=1}^{n} [(P_i - C)\, l_i] \div L \;\; ; \text{ if } P_i > C, \text{ and zero otherwise}$$

Where,

P_i = the number of passengers in the bus in link i; $L = \sum l_i$ = route length in km; l_i = Length of link i in km; n = number of links on the route; C = capacity of the bus, taken as 70 passengers for all the cases.

(IV) Make: two makes of buses, namely Tata and Ashok Leyland were used in the city-bus service. The makes being different, the fuel consumption rate of the two vehicles was found to be different. Hence, the dummy variable technique was applied to take this factor into account. The variable was assigned the value of One in the case of Ashok Leyland vehicles and Zero for Tata make vehicles.

The data of the independent and the dependent variables pertaining to fifteen routes were considered for the study. Multiple linear regression analysis was used to develop a relationship between the dependent variable and the set of independent variables. After an intercorrelation analysis, the independent variables that finally entered the regression analysis were: (i) Number of stops per km, (ii) Make of vehicle, (iii) Excess turn over, and (iv) Journey speed (km/h).

The calibrated model, using the regression analysis, is as follows:

$$Y = 4.02 - 0.13X_1 - 0.17X_2 - 0.02X_3$$

$R^2 = 0.87$, $t = 3.14$, 2.39 and 4.29 for X_1, X_2, X_3 respectively.

Where,

Y = fuel consumption in VKMPL; X_1 = number of stops per km; X_2 = make of the vehicle; X_3 = excess turn over (passenger km per km)

From the regression equation, based on the t values, it is found that the independent variables, excess turnover, number of stops per km, and the make of the vehicle, in that order, significantly influence the fuel consumption of the city bus transit. The variable journey speed was found to be insignificant in influencing fuel consumption, and hence, did not find place in the final regression equation. The result implies that apart from the make of the vehicle, the number of stops and overloading significantly influence fuel consumption. Thus, the study provided a very useful input to the MTC with regard to the extent of the benefit of limited stop services, and the drawbacks of overloading of buses; and this encouraged the MTC to introduce as may LSS as possible.

Fare Structure

By virtue of the spread of the city over a vast area, the land use pattern that has emerged over the past, the distance between residences and other activity centres, including work places, has gradually increased. It has become difficult to have the place of residence nearer to the work place due to economic and land use constraints. Thus, a considerable proportion of the commuters who belong to the middle and lower income group of the society are found to travel long distances every day. To mitigate the burden of travel expenditure on this section of the commuters, a 'telescopic' fare structure, as shown in Figure 2, has been adopted for the bus transit in the city. It can be seen that the fare is 2.4¢ (Rs. 1) for travel from 0 to 4 km, 4.8¢ (Rs. 2) for 10 to 15 km, and 9.5¢ (Rs. 4) for 40 to 45 km. Considering the upper limit of the distances in each case, it can be seen that the per-km fare is 0.6¢ (Rs. 0.25), 0.31¢ (Rs. 0.13, and 0.21¢ (Rs. 0.09) for 4 km, 15 km, and 45 km of travel respectively. Thus, the fare structure helps to reduce the burden on the long distance commuters, most of whom are captive to public transit, in the city. It can also be seen in the figure, that the fare structure for the rail transit system is also 'telescopic' though the per-km, fare is higher than that for bus transit over the whole range of the distance.

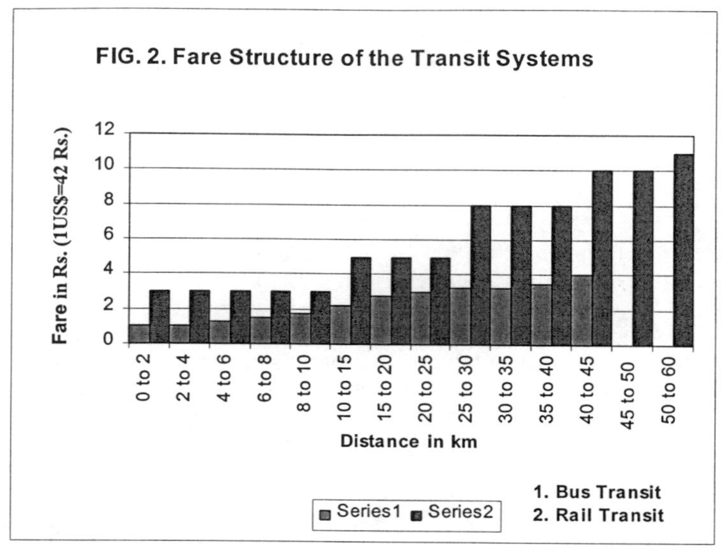

FIG. 2. Fare Structure of the Transit Systems

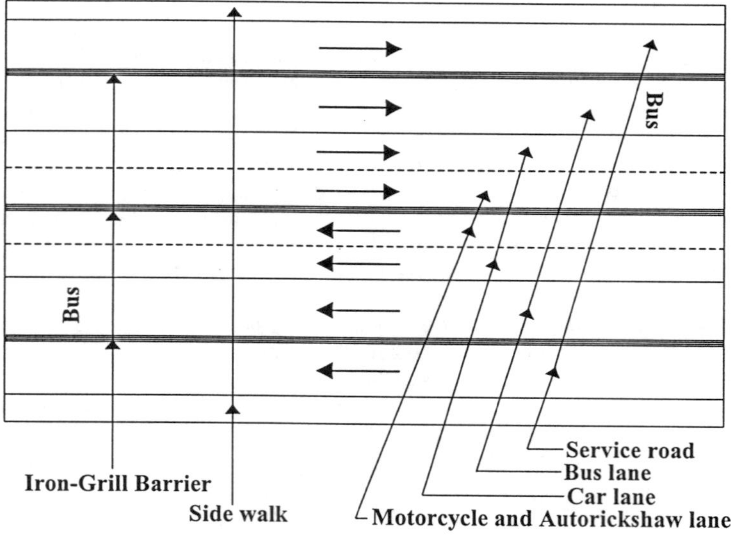

Fig. 3. Traffic Segmentation on Anna Salai

Bus Priority System

One of the arterial roads that cater to the flow of a major portion of the traffic to and from the Central Business District (CBD) of Chennai City, is Anna Salai. The land on both sides of this arterial is heavily built up, and the access to roadside land uses is unrestricted. This created a considerable amount of haphazard movements of both the vehicles and pedestrian traffic impeding the general traffic flow along the road. Moreover, the vehicular traffic in Chennai City is highly heterogeneous, comprising different kinds of vehicles such as heavy vehicles (buses and tracks), cars, vans, auto rickshaws (three wheeled motorised paratransit vehicles to carry a maximum of three passengers), cycle rickshaws (non- motorised tricycles to carry a maximum of two passengers), motorised two wheelers (motorcycles, scooters and mopeds), bicycles and animal drawn vehicles (relatively small in number), all sharing the same road space. This sharing of road space by different categories of vehicles results in a drastic reduction in the speed of traffic, particularly during peak hours. The city buses being relatively larger in size, find it difficult to negotiate in a complex traffic situation, and thus the speed of the buses along Anna Salai was found to be much less (about 17km/h) during peak hours. With the objective of improving the traffic speed in general, and the speed of the buses in particular, the different types of vehicles, depending on their static and dynamic characteristics, were divided into four subgroups, and specific traffic lanes were allocated for each of these subgroups on a stretch of about 2 km length on Anna Salai on an experimental basis. The available road space was divided into 8 traffic lanes (4 lanes for each direction) with a barrier type median provided at the centre. Among the four lanes available for each direction of traffic flow, the outermost lane was separated from the other three lanes by a barricade, and was allotted for access to the adjacent buildings, and for flow of all types of non-motorized vehicles, including bicycles. Each of the other three lanes, starting from the outer side, was allotted for heavy vehicles, cars and vans and, auto rickshaws and motorised two wheelers respectively. Figure 3 depicts the lane allocation arrangement in detail. This arrangement facilitated a streamlined flow of traffic with an increased speed on the stretch. Trucks being prohibited on this stretch during peak hours, the lane allocated for heavy vehicles is available, exclusively for buses during peak hours. It was found, after the introduction of the system, that the bus transit speed increased to 21 km/h on this stretch (Srinivasan 1995). The system has been found to be working well for the past five years; and there is a move now to introduce this system on selected stretches of other arterials of the city where sufficient width of road is available.

Transit-Route Network Design

A study on Urban Bus Transit Route Network Design using Genetic Algorithm was taken up by the Transportation Engineering Division of IIT, Madras (Pattnaik et al. 1998), and the work is in progress. The suggested method involves determining a route configuration with a set of transit routes and associated frequencies that achieve the objective of minimising the overall cost (both user and operator) incurred in materialising the travel. The study is aimed at overcoming the difficulties encountered by conventional approaches to the combinatorial nature of the problem. The final results of the study would, it is hoped, go a long way in solving the complex problems of the bus-transit-route network design of the Chennai Metropolis.

Public-Private Partnership

As elsewhere, in India, after the liberalisation of the economy, there were intense moves to encourage privatisation in almost every sphere of economic activity. Transport being a vital sector of economy, an urgent need was felt to explore the possibility of involving private entrepreneurs not only in the development of transport infrastructure but also in the operation of transport systems. In this regard, a study was conducted to explore the feasibility of private participation in bus transit operation in Chennai City (Pattnaik, and Rath 1997). Opinion of a number of experts involved in transit system-related issues were collected on several issues related to private participation in transit operation and recorded on a qualitative scale. The results obtained after the analysis of the data, using Fuzzy Logic, revealed that private participation with certain government controls to protect the economically weaker section of the travellers will be suitable for Indian conditions.

Mass Rapid Transit System (MRTS)

Chennai (Madras) is a port city, with the Bay of Bengal constituting its eastern boundary. The city has developed around a small nucleus of about 70 km^2 in the vicinity of the port (1700), by spreading towards the other three directions to reach the present (1998) state; covering an area of about 1,200 km^2. Figure 4 shows the layout of the major roads, railway lines, and the waterways of the city. It can be seen that, there are three major roads (NH45, NH4, and NH5) radiating in three directions from the CBD area (Chennai Beach, Chennai Fort, Park Town). It can also be seen that three railway lines also radiate to form the CBD in the three directions. The residential development of the city, expectedly, had been along these major roads and the railway lines. This has resulted in the development of three distinct major traffic corridors, diverging away from each other as the distances from the CBD increased. The interconnection between these three major corridors has been provided through a network of roads forming a series of concentric arcs (circular sub corridors), connecting the three major radial corridors. However, the thick residential growth along the coast towards south, and the recent increased intensity of the land uses in the peripheral areas of the city, made the concentric circular roads inadequate to meet the travel demand. Hence, a Mass Rapid Transit System (MRTS) was planned to meet the demand. Accordingly, it was proposed to have the layout of the MRTS as shown in Figure 4. The extreme points connected by

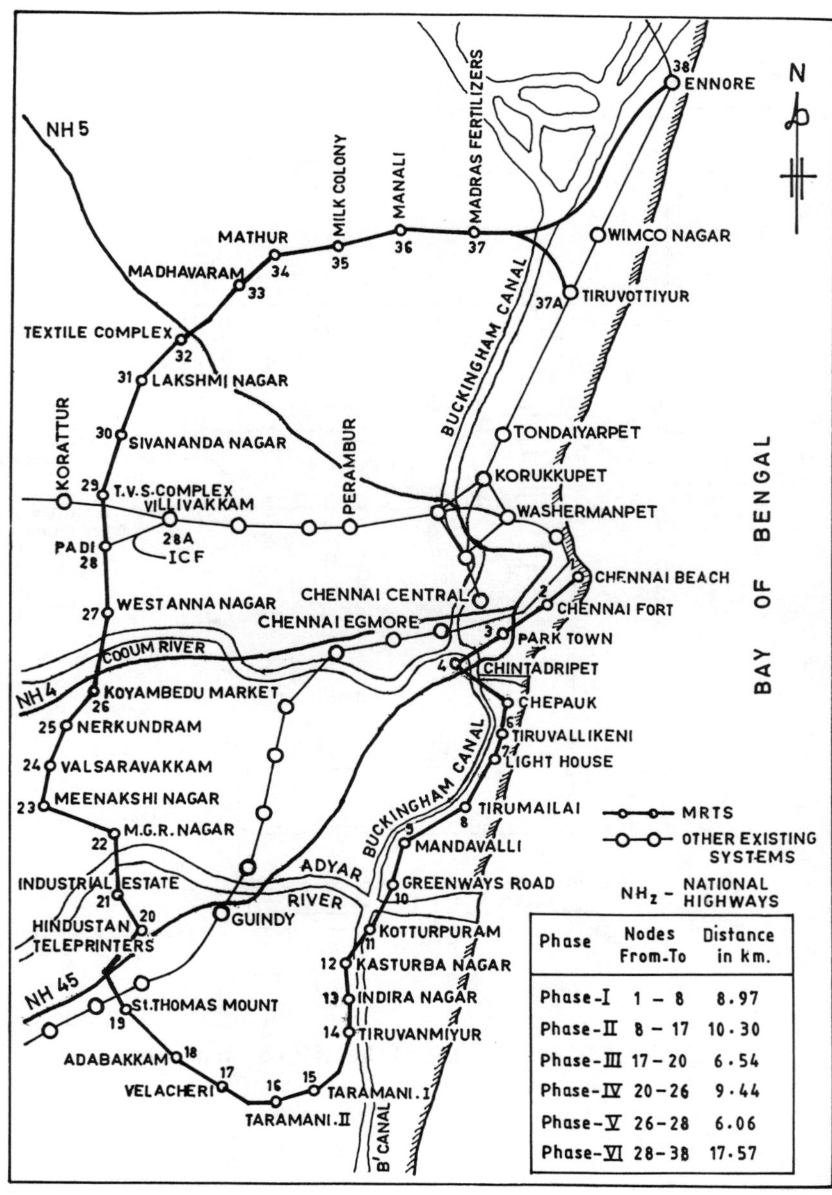

Phase	Nodes From-To	Distance in km.
Phase-I	1 – 8	8.97
Phase-II	8 – 17	10.30
Phase-III	17 – 20	6.54
Phase-IV	20 – 26	9.44
Phase-V	26 – 28	6.06
Phase-VI	28 – 38	17.57

Fig. 4. Layout of the Alignment of MRTS

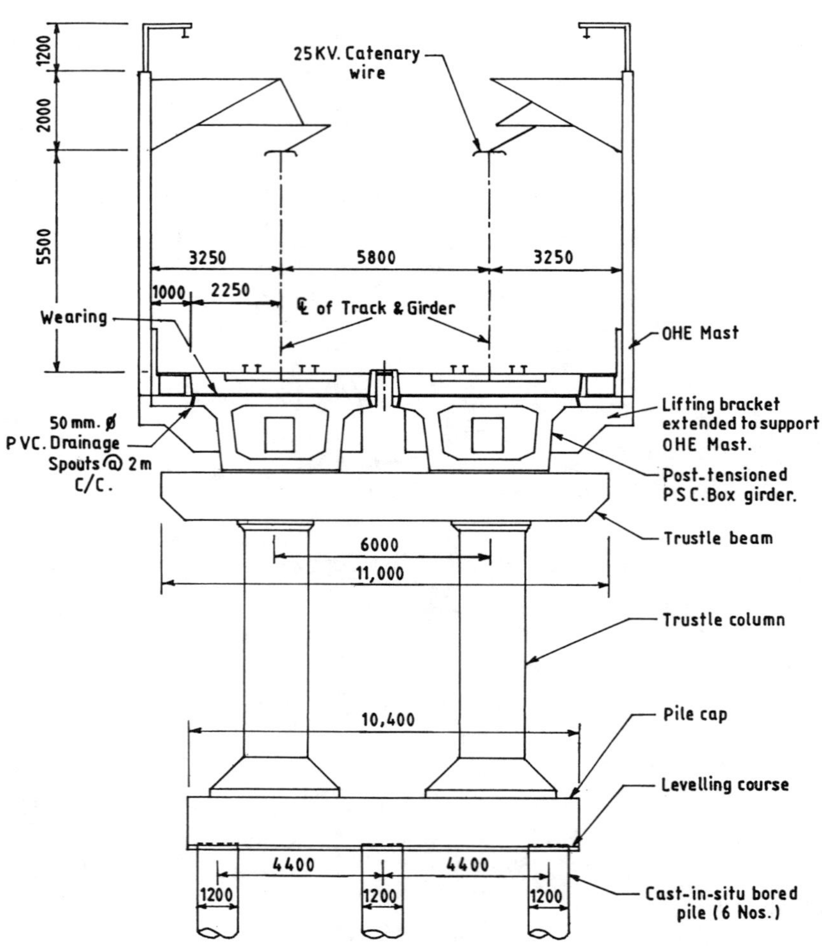

Fig. 5. Cross Profile of the MRTS Structure

the system on the east, south, west and north respectively are, Chennai Beach, Tharamani II, TVS complex and Ennore (Figure 4).

The construction of the MRTS has been planned to be carried out, in phases, as indicated in the figure. Phase I has already been completed and train service has been provided on the stretch. The construction of phase II is now in progress. About 70% of the length of the track, both in phase I, and phase II, is elevated, the rest being at the ground level. The details of the two phases of the MRTS are shown in Table 2. The cross profile of the structure supporting two broad gauge (1676 mm) railway lines, on the elevated portion of the track is as shown in Figure 5.

Table 2. Salient Features of MRTS

Details	Value	
	Phase I	Phase II
Total length of track (km)	8.97	10.8
Length of elevated track (km)	6.22	7.58
Total number of stations	8	9
Number of elevated stations	5	7
Estimated total cost (US$)	62 million[a]	175 million[b]
Project duration (years)	15	4
Extent of completion (%)	95	5

[a] cost of land alone met by state government; [b] 67% of the total cost met by the State Government, and the rest by the Central Government.

Partnership in the System Development

At present, in India, all the railways (both intercity and intracity lines) are owned and operated by the central government at Delhi. Usually, the land required for the railways will be acquired by the Central Government, free of cost, from the State Government (except in the case of private land) in whose jurisdiction the land falls. Thus, the development of new railway lines depended on the availability of resources with the Central Government (which had always been scarce) and the extent of help and co-operation provided by the State Government in acquisition of the land. Phase I of the MRTS was constructed on this basis. The total cost (1998) of the project is Rs. 2.6 billion (US$ 62 million). The project was approved in 1983 and has taken 15 years to complete, due to a resource crunch in the Central Government. The extraordinary time overrun has resulted in an enormous increase in the project cost (due to monetary inflation). Based on this experience, and the experience of development of a MRTS for the city of Calcutta, the Central Government insisted that the state governments should bear 67% of the project cost of intracity railway projects, as these railways are confined to a city within a state, and the accruing benefits are going to be enjoyed only by the people in that state. After prolonged negotiations, the stipulated condition was agreed to by the State Government, and Phase II of MRTS is now being constructed with central and State

Governments sharing the cost (Rs. 7.33 billion = US$ 174 million) in the ratio of 33:67. It has been planned to complete the project in 4 years; and the progress of the work so far is as per schedule (MTP-Railways 1998).

Minimisation of Private Land Acquisition

It can be seen in Fig. 4 that a watercourse, namely Buckingham Canal, passes in the North–South direction along the seacoast. The canal and the adjoining land are the properties of the State Government. The canal, on the stretch south of the CBD area passes through dense residential land uses. Thus, it was planned to fix the alignment (between stations 4 and 14) of the MRTS (phases I and II) along this canal. This has helped the State Government to avoid the problem of private land acquisition (the cost of which is very high) on this stretch. Thus, fixing of the alignment along the canal has resulted in a considerable saving in time and land cost of the project.

Conclusions

Certain notable features of the bus transit system related to organisational and operational aspects, and a few important aspects related to the mass rapid transit system development in Chennai City, India were presented. The gist of the features is as follows:

1. It is found that the administrative structure of the bus transit system has facilitated creation of independent administrative units accountable for both their successes and failures.

2. An optimal mix of the limited stop services and ordinary services provided in the bus transit satisfies, the requirements of the long-haul and short-haul trip makers respectively.

3. The 'telescopic' fare structure adopted for both the bus and rail transit systems of the city helps the long-distance commuters, most of whom are captive to transit, to meet the travel expenses without much financial strain.

4. A study on fuel consumption of city buses revealed that, excess turnover (over loading), number of stops per km, and make of the bus, in that order, influence the fuel consumption rate significantly.

5. The government of India has recently made a decision that the share of the cost for urban rail transit system development, for the central and state governments, will be 33% and 67% respectively for enabling faster development of the system. A new rail transit (MRTS) system is now being developed for Chennai City based on the said cost-share basis.

References

Central Institute of Road Transport. (1998). " Performance of STUs for the quarter ending December 1997." Indian Journal of Transportation Management, 22(4), 303– 333.

Chennai Metropolitan Development Authority (CMDA). (1995). "Report on comprehensive traffic and transportation study." Chennai, India.

Metropolitan Transport Corporation (MTC). (1998). "Annual report – (1997-98)."

Metropolitan Transport Project (MTP) (Railways) – Mass Rapid Transit System for Chennai. (1998). "Report for parliamentary standing committee on railways." Chennai, India.

Narayana Rao, T.P. and Govindarajan, P. (1997). "Mass rapid transit system – Phase II." Proceedings of the Workshop on Impact of MRTS Phase II, Chennai, 1 – 30.

Pattnaik, S.B., and Rath, T.C. (1997). "Feasibility of privatizing the bus operation in an Indian metropolis using fuzzy set theory." Proceedings of the symposium on Mathematical Methods and Applications, IIT, Madras.

Pattnaik, S.B., Mohan, S. and Tom, V.M. (1998). " Urban bus transit route network design using genetic algorithm." Journal of Transportation Engineering, ASCE, 124(4), 368 – 375.

Srinivasan, N.S. (1995)."Traffic lane system on Anna Salai in Madras." Journal of Institution Engineers, Chennai, India. Vol. 75, 268 - 272.

Thamizh Arasan, V., Rengaraju, V.R. and Govinda Rao, K. (1990). "Modelling city bus fuel consumption." Indian Highways, Indian Roads Congress, 18(7), 23 – 29.

Thamizh Arasan, V., Wermuth, M. and Srinivas, B. (1998). "Modelling of stratified urban trip distribution." Journal of Transportation Engineering, ASCE, 122(5), 342 – 349.

SOCIOECONOMIC AND TRAVEL CHARACTERISTICS OF CHICAGO

Eugene Ryan[1]

Abstract

This paper discusses the socioeconomic, land use and travel patterns of the Chicago area as they relate to the provision and use of public transportation. The purpose of the paper is to explain the type and extent of the success of public transportation service throughout the area in relation to the differing socioeconomic and land use conditions present.

Introduction

A mention of Chicago conjures images of towering office buildings in the historic Loop, bustling retail on North Michigan Avenue, and the beaches of Lake Michigan. Others may include O'Hare Airport, new suburban office parks and the farms in the outer southern and western areas. Neighborhoods vary from new residential loft districts adjunct to the Loop, the bungalow belt of Chicago's southwest side and mature inner-ring suburbs to new single-family developments in the outer suburbs. The region's employees are as diverse as traders at the Chicago Board of Trade, auto workers on the South Side, hotel managers in Rosemont, and farmers in Kane County. All of these images and lives are brought together under a regional transportation system that has and will continue to help shape the Chicago region. The challenge for the regional transportation system is to serve varied interests in a region that stretches more than 130 kilometers north to south, 80 kilometers east to west and encompasses almost a million hectares.

---

[1]Associate Executive Director, Chicago Area Transportation Study, 300 W. Adams St., Chicago, Illinois 60606

Historical Development

As early as 1673 Pere Marquette knew what the Pottawatomies already understood: the Chicago River, the Mississippi River and the Great Lakes watersheds are within a few kilometers of each other. The prairies to the west and the white pine forests to the north, refrigeration technology, the emergence of railroads, along with a team of impressive local boosters all played roles in the growth of Chicago as a major destination. Standing at the center of the Midwest, along the shores of Lake Michigan, Chicago's location has allowed its economy to thrive. From its early years as a center of shipping and railroads, to its recent growth spurred by the booming Midwestern economy, the region's comprehensive transportation system has allowed the region to meet changing transportation needs.

Historically, Chicago has developed in a traditional urban model, with a strong urban core and suburbs stretching along radial transportation corridors. While suburbanization is popularly viewed as a post World War II phenomenon, suburban growth in northeastern Illinois occurred as early as the 1860s. The growth of railroad lines radiating west, north and south from Chicago created access to new communities. The railroads and real estate developers marketed the peaceful country life that was waiting for families along these rail corridors. By the 1890s, the ideal situation for a middle class businessman was to work downtown but commute to his suburban, country home.

The emergence of the car and the demand for housing after World War II contributed to changing the urban landscape in Chicago and all across the country. An aggressive statewide road building program, the start of the tollway system and the creation of the federal interstate highway system supported the suburbanization drive.

As in many other cities in the United States, historic patterns of development have evolved into a more complex system where the central city is no longer the dominant destination for travel. Employment clusters have emerged in numerous suburban locations and residential development has filled the gaps between historic radial transportation corridors and now stretches far from the traditional core.

The Existing Landscape for Public Transit

The Chicago area is home to more than seven million people of whom slightly less than three million reside in the city of Chicago. Figure 1 is a map of the area which shows the major transportation facilities and the boundaries of the city of Chicago and the counties.

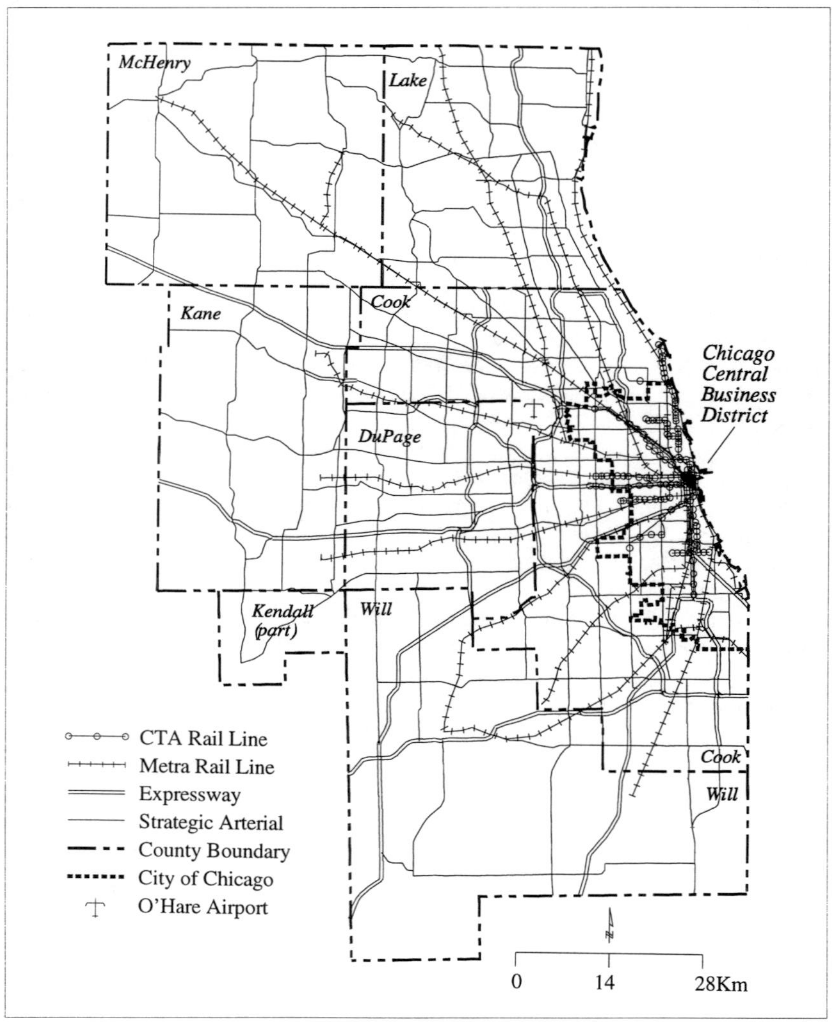

Figure 1. **Northeastern Illinios Transportation System**

Public transit is provided by three entities under the umbrella of the Northeastern Illinois Regional Transportation Authority. Combined, their operations constitute the second largest rail and third largest bus system in North America. The Chicago Transit Authority provides bus and rail service in the city of Chicago and into some adjacent suburbs. The METRA system provides commuter rail service from the suburbs (and limited city of Chicago stations) to the Chicago Central Business District. PACE is the provider of bus service in the suburbs. The operating characteristics are shown in Table 1.

	Route Kilometers	Routes	Stations	Vehicles	Annual Riders
CTA Bus	3,260	139	n/a	2,035	302 million
CTA Rail	530	n/a	140	1,192	124 million
Metra	880	n/a	240	1,071	71 million
Pace	n/a	233	n/a	948	36 million
Total				5,246	533 million

Source: *1996 RTA Annual Report* Bus data is for fixed route and paratransit services.

Table 1. **Transit Operating Characteristics**

On an average weekday, public transit carries about 8% of all trips and about 15% of work trips. The biggest markets for public transit are work trips destined to the Chicago Central Business District (which includes the Loop area) and trips within the city of Chicago.

The viability of the public transit systems across the region is largely shaped by the variable socioeconomic and land use characteristics of the area. Population and employment densities are shown in Figures 2 and 3.

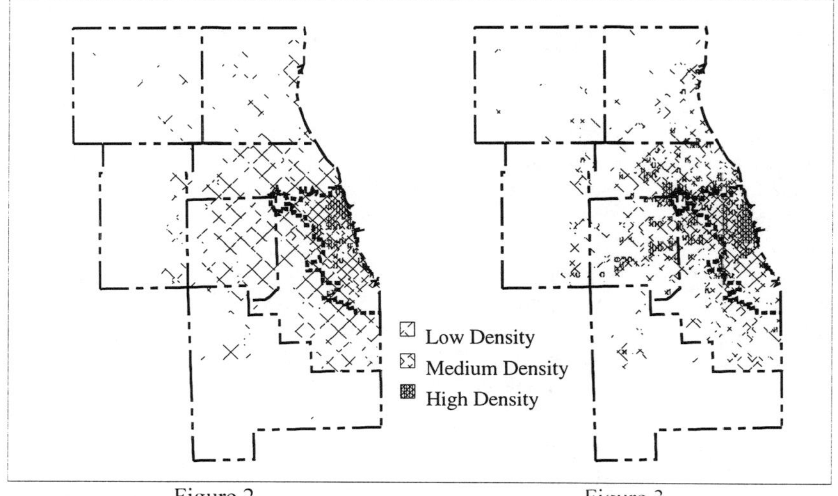

Figure 2 Figure 3

Northeastern Illinois Population and Employment Densities

The category marked high is over 50,000 people per square kilometer in Figure 2 and over 13,000 jobs per square kilometer in Figure 3. Generally, densities are only high enough in the city of Chicago to support an extensive grid oriented fixed route transit system. The high employment density in the Chicago Central Business District supports both the CTA rail system and the METRA commuter rail system. Although there are areas of high employment density outside the city, particularly around O'Hare airport, the relatively low suburban population densities present a challenge to providing good public transit service.

Auto ownership patterns as shown in Figures 4 and 5 reinforce the density situation.

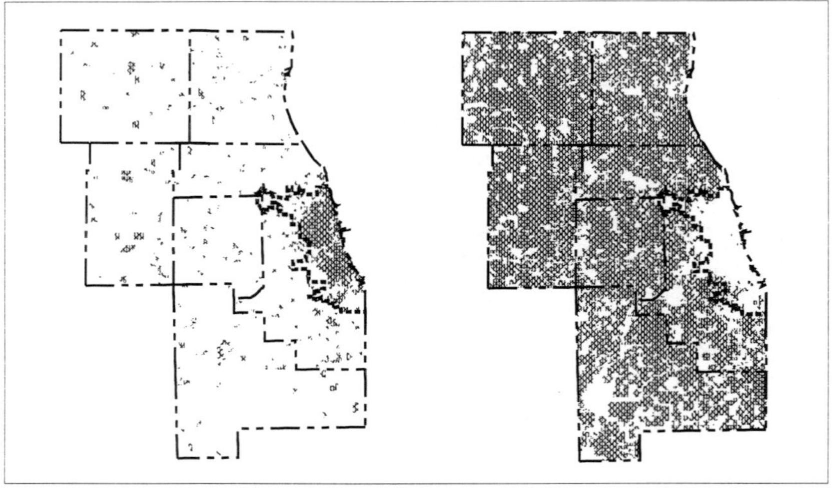

Figure 4. **Low Auto Ownership** Figure 5. **High Auto Ownership**

The only extensive areas with appreciable concentrations of zero auto households are within the city of Chicago. Almost all the suburban area is characterized by the majority of households owning two or more automobiles.

In 1995, private vehicles carried over 90 percent of the region's trips - approximately 16 million person trips per day. The system consists of 87,500 lane kilometers of freeways and expressways, arterials and collectors and local streets. Between 1985 and 1995 vehicle kilometers of travel in the region increased by almost 40 percent, but the number of highway lane kilometers increased by only 5 percent. This trend has led to ever-increasing traffic congestion on the region's roadway system but, in spite of this congestion, public transit usage has slightly declined. Of course, except for the rail system, public transit has no separate right-of-way and experiences the effects of congestion on travel times as do private vehicles. Work is underway to implement bus activated traffic signal preemption. Studies and field testing have demonstrated that signal preemption can significantly improve schedule adherence and reduce bus travel times. In the future, highway congestion may induce more transit usage and, at the same time, help alleviate the congestion. The pattern of traffic congestion across the region as measured by morning peak hour average roadway volume to capacity ratios is shown in Figure 6.

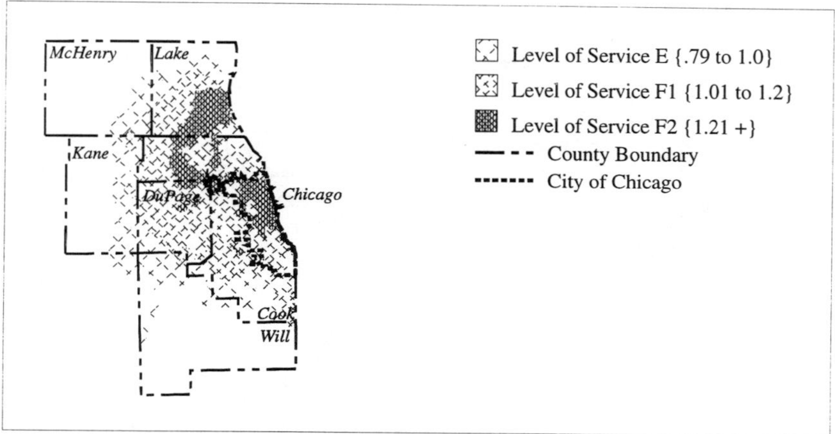

Figure 6. **Northeastern Illinois AM Peak Congestion**

As can be seen, much of the city of Chicago has the highest level of congestion. Here the public transit service is the best in the region but the density of development is high. High levels of congestion are also present in the north and northwest suburbs. Congestion has increased in these areas in recent years and continues to increase. The challenge is how to serve these relatively low density high auto ownership areas with public transit.

Socioeconomic and Land Use Trends

<u>*Population*</u> - Since 1950, most of the region's population growth has taken place in the suburbs. Population totals for 1950, 1970 and 1990, by county, are presented in Table 2.

	Population			**Households**			**Employment**	
	1950	1970	1990	1950	1970	1990	1970	1990
City of Chicago	3,621	3,369	2,784	1,088	1,138	1,025	1,864	1,482
Suburban Cook County	888	2,124	2,321	249	628	854	836	1,294
Du Page County	155	488	782	43	136	279	146	530
Kane County	150	251	317	43	75	107	103	145
Lake County	179	383	516	47	103	174	116	229
McHenry County	51	112	183	15	33	63	36	66
Will County	134	248	357	37	71	117	83	99
6-County Total	**5,178**	**6,975**	**7,260**	**1,522**	**2,184**	**2,619**	**3,184**	**3,845**

Numbers in thousands

Table 2. **Population, Household and Employment Trends**

Between 1950 and 1970, the population grew by 35 percent, with suburban Cook County accounting for two-thirds of the region's 1.8 million increase. The total population in the five "collar" counties more than doubled to 1.5 million.

The region's population grew much more slowly from 1970 to 1990; only 285,000 people were added, or 4 percent. The growth in suburban Cook County slowed dramatically, while the collar counties' growth continued at almost the same rate. Combined, their population increased by 673,000, slightly more than the decrease in the city of Chicago.

Households - The average household size in the region has decreased from 3.3 in 1950, to 3.2 in 1970 and to 2.7 in 1990. Because of this, households have grown faster than population. Table 2 shows county household totals for 1950, 1970 and 1990. Between 1950 and 1970, population grew by 35 percent, while households grew by 43 percent. The difference was more dramatic from 1970 to 1990, when population increased only 4 percent, but households jumped 20 percent.

Employment - Similar to population and household trends, suburban jurisdictions have led the region in employment growth since 1970. Table 2 includes employment totals by county for 1970 and 1990. The 21 percent employment growth during this period greatly outpaced the 4 percent population growth and nearly matched the 20 percent household growth.. Increases were shown in all suburban areas, but the city of Chicago posted a 25 percent decline.

Except for the continued high job concentration in the Chicago Central Business District, all the socioeconomic trends work against transit usage. Population and employment become more dispersed and harder to serve by public transit. The employment growth relative to population growth has resulted in a more affluent area with higher auto ownership and a lower transit dependent population.

Travel Patterns

Changes in regional development have contributed to a rapid growth in suburb to suburb trips. At the same time, the demand for the traditional suburb to downtown trip has remained strong. This combination of travel patterns has added a new dimension to the challenge of serving regional travel needs. Between 1980 and 1990, the number of work trips destined for the suburbs increased 23 percent, while the region's total work trips increased just 9 percent. Cook County's northern suburbs, DuPage County, and Lake County all experienced a significant increase in the number of trips related to employment, while the city of Chicago experienced a decline.

In addition to a shift in travel patterns, the individual choice of mode in the Chicago region has continued to change. The role of the automobile has grown dramatically, as transit ridership has slipped and the number of single-occupant automobile trips has increased. In just ten years, between 1980 and 1990, the percentage of individuals driving alone to work increased from 58 percent to 68 percent (see Figure 7).

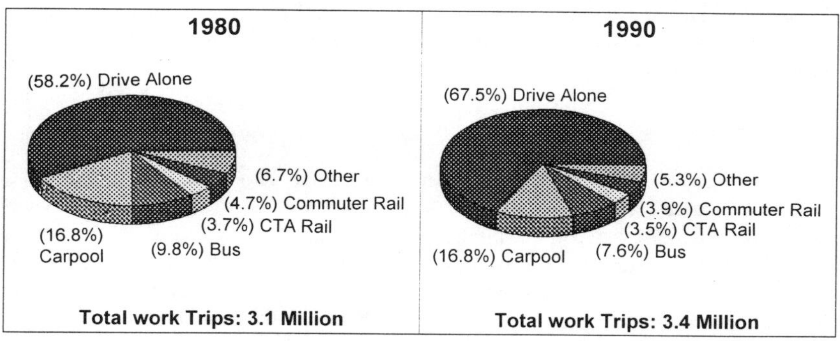

Figure 7. **Commuting by Mode**

Despite an increasing reliance on the automobile, the Chicago region still maintains the second lowest drive-alone to work rate among urban areas in the United States. Transit, carpooling and other modes continue to play a critical role in serving the region's transportation needs.

Since 1970, the growth in vehicle kilometers of travel (VKT) has outpaced regional population, household and employment growth, while overall, transit ridership has dropped. Figure 8 compares the relative change in VKT and transit ridership to the population change for the 1970-1995 period.

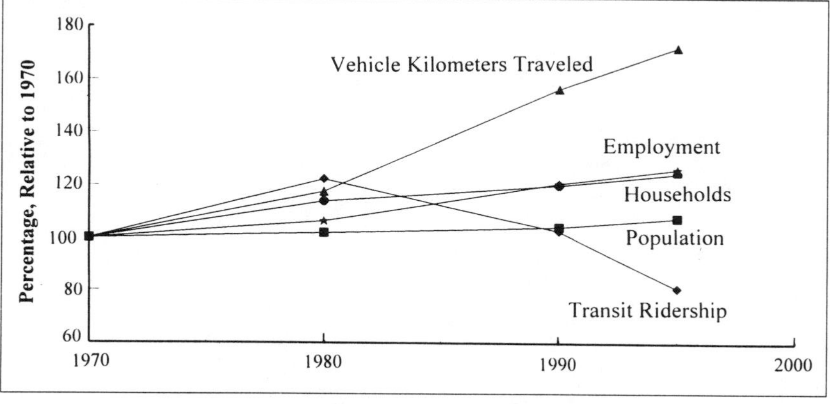

Figure 8. **Trends in Transit and Highway Use**

Both transit ridership and VKT grew by roughly 20 percent from 1970-1980. The healthy increase in transit ridership was due in part to automobile fuel price increases during this time. In the subsequent fifteen years, from 1980 to 1995, transit ridership decreased substantially. At the same time VKT was up 46 percent, while population increased only 6 percent, households increased 9 percent and employment increased 20 percent. The cost stability of auto fuel along with the increase in affluence of the population and the dispersion of land use as discussed earlier all contributed to this result.

The Future

Public transit serves a vital role in the Chicago area. The Chicago Central Business District, critical to region's economy and where one out of every six jobs are located, could not exist without public transit. Highway congestion on the major arterials in Chicago and radial freeways throughout the region during peak hours is a problem that would be much worse without the availability of public transit. The mobility that public transit provides to those without the financial means or otherwise inability to use auto transportation is very important. Even though public transportation serves a minority of the population and usage has declined in recent years, regional public policy is to support it and attempt to reverse the declines. The recent demographic and land use trends present a challenge to achieving that end.

Chicago's Urban Public Transportation System, Its Operations and Use by People

Marcel C. Acosta[1] and Paul Ulrich[2]

Introduction

The six counties that make up the Chicago area are served by three public transit systems: the Chicago Transit Authority (CTA), which operates urban bus and heavy rail, Metra commuter rail, and Pace suburban bus. Together they serve over 200 million rail riders a year and about 330 million bus passengers. Metra operates on 12 rail lines with 240 stations, while the CTA has 7 lines (with a total of 10 branches) and 140 stations. Pace operates on 234 bus routes and the CTA on 129. Overall, however, the CTA, on which this paper will focus, provides around 80% of all transit trips in the six county region and serves 38 suburbs as well as the city of Chicago. Highlights of the CTA are listed below:

- CTA is the second largest transit system in the US
- CTA serves the city of Chicago and 38 suburbs
- CTA covers nearly 375,000 miles a day
- CTA serves 424 million passengers a year
- CTA has 1876 buses, 129 bus routes, 12,200 bus stops, and 292 million bus passengers/year
- CTA has 1190 rail cars, 7 rail lines, 140 rail stations, and 132 million rail passengers/year
- CTA employs 11,300 people
- CTA's annual operating budget is $782 million
- CTA's recovery ratio is 52.36% (including non-fare revenue)

[1] Senior Vice President, Department of Planning and Development, Chicago Transit Authority, 120 North Racine Avenue, Chicago, IL 60607

[2] Transit Research Analyst II, Department of Planning and Development, Chicago Transit Authority, 120 North Racine Avenue, Chicago, IL 60607

The Role of the RTA in Transit in Northeastern Illinois

The three service boards are overseen by the Regional Transit Authority (RTA) which was created in December 1973 to ensure that public transportation in northeastern Illinois is both comprehensive and coordinated. Most broadly speaking, it is a policy and financial oversight body. Specifically, since 1983 the RTA has been involved in regional transit planning and has received sales tax, which provides funding to the three service boards; it has no day-to-day operating responsibilities. Half of the cost of RTA system operations must, by law, be covered by system-generated revenue. The RTA gets the other half through appropriate grants from federal and state sources, investments, and from the sales tax it levies in the six counties it serves. In Cook County, Chicago's county, the RTA receives a 1% tax on the purchase of food and drugs and a 0.75% tax on all other sales. The state of Illinois provides the RTA an amount equivalent to 0.25% of non- food and drug purchases. In Lake, Will, Kane, DuPage, and McHenry counties--the "collar counties"--the RTA collects a 0.25% tax on all purchases. Eighty-five percent of the tax the RTA receives is distributed according to the following formula: 100% of that collected in Chicago and 30% of that collected in suburban Cook County goes to the CTA. Fifty-five percent of the tax collected in suburban Cook goes to Metra and 15% to Pace. Seventy percent of the tax collected in the collar counties goes to Metra, and 30% to Pace. Fifteen percent of RTA's tax revenue--along with an amount from the State of Illinois equal to 25% of RTA sales tax receipts that makes up a Public Transportation Fund--covers the RTA's own expenses and bonds, and provide funds the RTA distributes to the service boards at its discretion.

With population and number of jobs growing much faster in the suburbs than in the city, revenue from the RTA tax is also growing at a much faster rate in the suburbs than in the city, and so the CTA finds itself less financially secure than Metra or Pace. While Metra is able to use mandated RTA tax revenue not just for its operations but even to fund capital improvements, the CTA relies more and more on uncertain discretionary funds from the RTA; over the next 10 years the CTA expects its formula funds to grow at just 0.3% annually. The CTA is thus forced to continue to find ways to cut costs as well as to increase revenue generated from its system.

CTA Ridership and Trip Characteristics

The CTA is the second largest transit system in the United States, and it is so large in part because it is used by a wide variety of people who have a wide variety of purposes for using it. When we look at how the system is used and by whom, we find that the CTA neither merely serves commuters to and from downtown, nor does it serve only those who do not or cannot drive or those who cannot afford to own a car.

Of the more than 1.4 million trips made per workday and 424 million trips made per year on the CTA (as of 1998), a little more than twice as many are made by bus as are made by train. Of all trips, about 39% are made for work or are work-related, 11% are for school, and half of all trips are made for shopping, socializing, or other personal business. Thirty-nine percent of all trips to downtown are on CTA, as are 42% of all CTA trips to downtown for work. But despite the heavy use by commuters to and from downtown, only 20% of all CTA trips originate downtown, and a full 2/3 end outside the Central Business District. Over half, 54%, of all CTA trips either stay within a neighborhood or go from one neighborhood to another; only 46% of trips go to, from, or within downtown. The CTA is truly used for a wide variety of purposes, all over the city.

A very wide cross-section of the people of the Chicago area use the system, or perhaps it is more accurate to say that CTA ridership is very representative of the population of the city as a whole. Nearly half of riders are Caucasian and nearly 40% are African-American. About 15% are Hispanic, Asian, or of another ethnicity. For reasons that may or may not be related to the system itself, over 60% of all riders are women. Two-thirds of riders are between 18 and 54 years of age and nearly the same percentage of riders are over 65 as are between 12 and 17. Just under 30% of riders earn $15,000 or less, but while the median income of riders is somewhat lower than that of non-riders, over 20% make over $60,000 a year. The percentage of riders with an income over $40,000--just over 1/3--is roughly the same as that of riders who earn under $15,000. Given the overall income of riders, it is not surprising that over 2/3 of all riders have at least one car in their household, though half of riders either don't have a car available to them or don't have a car at all. Of those who do have a car available but choose to use the system, it is worth noting that they count among the major factors in their choice not only the directness of service and the money saved they would spend on driving and parking, but also that they find riding CTA to be much less stressful than driving.

CTA Service Description

The CTA has 105 rail route miles on 7 lines with 10 branches and 140 stations. The Red Line and the O'Hare and Forest Park branches of the Blue Line operate at all times, every day. The Yellow Line and the Cermak branch of the Blue Line do not run on weekends, nor does the Purple Line Express, which is strictly an AM and PM rush service. The other lines, including the Purple Line Shuttle, are out of service for just a few hours a day, usually from about 1:00 or 2:00 in the morning until about 4:00 or 5:00 in the morning, though the Brown Line does not go south of Belmont on Sundays. (See Figure 1 for map of CTA Rail Service.) CTA's 120 bus routes have well over 12,000 stops and cover nearly 2,000 miles. 19 of those routes operate at all times; a few operate only during peak periods. CTA's trains travel over 170,000 miles per day and its buses over 200,000. It covers these nearly 375,000 combined daily miles with 1876 buses and 1190 train cars.

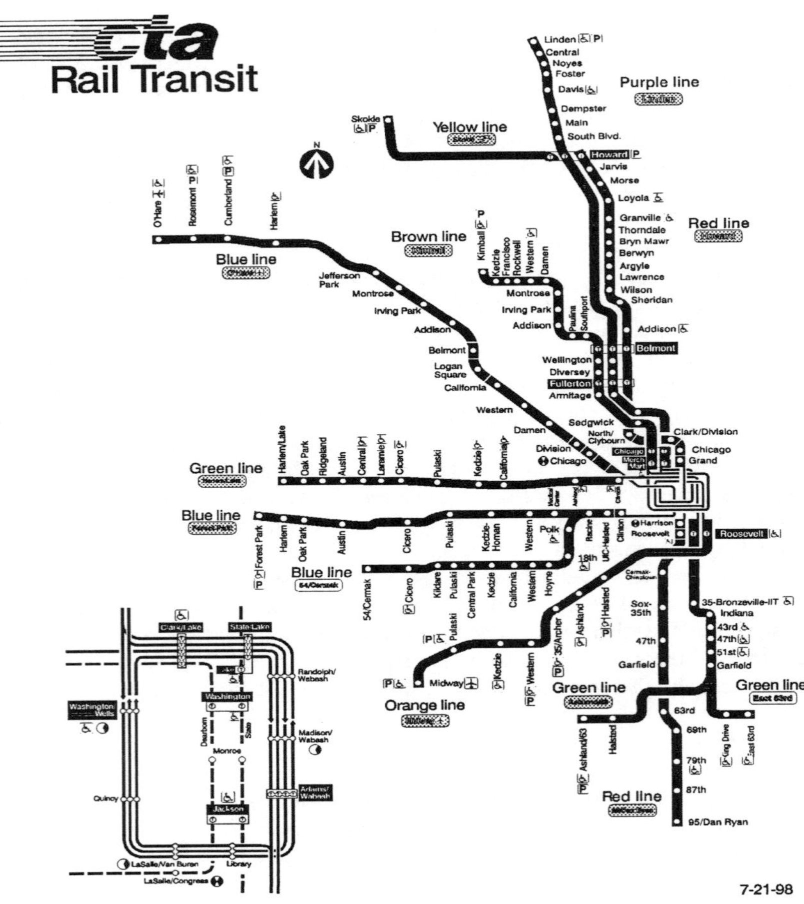
cta
Rail Transit
Purple line
Yellow line
Red line
Brown line
Blue line
Green line
Orange line
Linden
Central
Noyes
Foster
Davis
Dempster
Main
South Blvd.
Skokie
Howard
Jarvis
Morse
Loyola
Granville
Thorndale
Bryn Mawr
Berwyn
Argyle
Lawrence
Wilson
Sheridan
Addison
Belmont
Wellington
Diversey
Fullerton
Armitage
Sedgwick
North/Clybourn
Clark/Division
Chicago
Grand
O'Hare
Rosemont
Cumberland
Harlem
Jefferson Park
Montrose
Irving Park
Logan Square
California
Western
Damen
Division
Kimball
Kedzie
Francisco
Rockwell
Paulina
Southport
Harlem/Lake
Oak Park
Ridgeland
Austin
Central
Laramie
Cicero
Pulaski
Merch Mart
Forest Park
Kedzie-Homan
Polk
Racine
UIC-Halsted
Clinton
Harrison
Roosevelt
18th
54/Cermak
Kildare
Central Park
Hoyne
Cermak-Chinatown
35-Bronzeville-IIT
Indiana
43rd
47th
51st
Garfield
Sox-35th
35/Archer
Ashland
Halsted
Midway
63rd
69th
79th
87th
95/Dan Ryan
King Drive
East 63rd
Ashland/63
Clark/Lake
State/Lake
Randolph/Wabash
Washington
Washington/Wells
Madison/Wabash
Dearborn
State
Monroe
Quincy
Adams/Wabash
Jackson
LaSalle/Van Buren
Library
LaSalle/Congress
7-21-98

Figure 1

CTA's Workforce and Finances

The CTA employs about 11,300 people to provide this service, of whom almost 60% are directly involved in providing transit service and almost 33% are employed in maintenance. Fewer than 8% of all employees work in administration, and this administration-light structure is very favorable for CTA's finances. It gets nearly 145,000 vehicle miles per administrative employee, serves 545,000 passengers per administrative employee, and has revenue of $436,000 per administrative employee.

With a total operating budget of $782 million, CTA's fare revenue is about $366 million or 46.5% of operating cost, and the total of system-generated revenue is about $422 million, or just under 52% of operating cost. These percentages make CTA's recovery ratio compare favorably with that of other major transit systems, as among major cities only New York (NYCTA), San Francisco (BART), and Washington D.C. (WMATA) recover a greater percentage of their expenses from their systems. Though NYCTA is far ahead of everyone, BART recovers a modest 4.5% more of its operating expenses than CTA does and WMATA only about 2% more. The CTA does far better than New York's PATH, Philadelphia, Boston, Atlanta, San Francisco's Municipal Railway, and Los Angeles (see Table 1 for chart on Comparison of Farebox Recovery Ratios). And the CTA is finding ways to increase system-generated revenues that are not from fares. For example, revenue in 1998 from investments, advertising, concessions, and charters was 37% higher than originally budgeted.

Ridership Initiatives: Service Changes

Increasing ridership and market share was CTA's focus in 1998, and will continue to be its focus in 1999. The CTA is doing what it can to improve, adjust, and fine tune service so as to increase ridership, and it does so with an advertising and promotion budget of just about $2.5 million. Time and again in recent years it was forced to raise fares and cut service. Most recently, in 1997-98, it made $25 million in service reductions affecting 10% of its service. To give some examples of those reductions: ten of the lowest performing bus lines were eliminated (and plans are underway for replacing five others with flexible service); 14 bus routes lost "owl" or all night service, as did the Green and Purple rail lines as well as the Cermak branch of the Blue Line, which also had its weekend service eliminated; and in all 106 bus routes had service cut. All cuts were made with an eye to actual use and demand, and to minimizing impact on riders.

With these reductions, however, the CTA has become able to redirect its resources to where service most needs to be expanded. It extended service on the rapidly growing Brown Line as well as on a number of bus lines. It also created some useful new Neighborhood Express bus lines and actively plans to create several more. Those

Table 1: Comparison of Farebox Recovery Ratios (1996)

City (system)	Fare Revenue	Expense	Recovery Ratio (excluding non-fare revenue)
Chicago (CTA)	**$358,666,833**	**$770,678,527**	**46.54%**
New York City (NYCTA)	$2,039,814,397	$2,818,836,792	72.36%
San Francisco (BART)	$123,691,285	$242,925,683	50.92%
Washington D. C. (WMATA)	$314,417,501	$647,309,034	48.57%
New York (PATH)	$66,122,000	$157,231,000	42.05%
Philadelphia (SEPTA)	$268,696,632	$664,201,783	40.45%
Boston (MBTA)	$213,423,899	$546,276,892	39.07%
Atlanta (MARTA)	$84,664,730	$222,496,852	38.05%
San Francisco (Muni)	$94,745,896	$273,115,361	34.69%
Los Angeles (LACMTA)	$210,145,673	$719,379,822	29.21%
Cleveland (GCRTA)	$42,987,573	$182,024,160	23.62%

lines are on the 46 Key Bus Routes the CTA has identified that provide critical coverage in its service area, and on which 66% of all CTA bus trips are taken. Those routes all will receive consideration for increased service frequency and other improvements that will help passengers better make connections with rail service and other bus lines. All this is the result of CTA's continuous search for possibilities for modifying existing lines and adding new ones to attract new riders and better serve existing ones.

Ridership Initiatives: Fares

As to fares, CTA has held them constant for several years now and recently even lowered the price of some passes. The basic fare has been $1.50 since December 1991. A first transfer is $0.30, and a second transfer is free. Reduced fare passengers pay half those rates, and include children from 7 to 11--children under 7 ride free--grade school and high school students with a CTA Student Riding Permit, persons with disabilities, and those over 65. Passes can now be purchased that are good for 1, 2, 3, 5, or 30 days, and they are good for the set number of days not after the purchase of the pass but after its first use. These new passes, as well as the new inexpensive pass for university students (the U-Pass) may slightly decrease fare revenue in the short term. But they are designed with a view to the long term: to increase non-peak time ridership, to encourage those who might not otherwise use the system to ride it, and to build a ridership base for the future.

A fare can be paid with cash, but the system has been made easier to use with the introduction of Automatic Fare Cards. The fare cards may be purchased at train stations from vending machines for any amount up to $100, and in fixed amounts at some super markets and other locations around the city, as well as over the internet; the CTA is looking for ways to sell them at even more places, perhaps even from vending machines at bus stops. Riders receive an extra dollar of value on the card for every $10 they put on it at one time, and are thereby encouraged to add value to their cards quickly using 10 or 20 dollar bills. This moves passengers through stations more quickly and has led to more fares being collected, and collected more efficiently. Further, passengers appreciate that with the fare card they don't pay for a transfer until they actually make it. The transit card seems to have both encouraged use of the system and improved fare collection: in its first full year of use it made an $11 million difference to the CTA by reducing expenses and increasing revenues.

Ridership Initiatives: Safety, Security, and Maintenance

The CTA is also trying to take into account what makes riders and potential riders uncomfortable about the system, and it has tried to find out especially what would make riders feel safer while both riding and waiting for a bus or train. In response to riders' concerns, the CTA will add more uniformed police and security guards on vehicles and at stops in 1999. It has budgeted just over $25 million for security services in 1999--an increase of about 50% over 1998--to cover the new officers and to implement new programs. Graffiti and etched windows on buses and trains will not be tolerated; CTA aims to replace all etched windows within one day of finding them. And a pilot program will put security cameras into 25% of the bus fleet. Enabling riders both to feel and be safe will help CTA add new riders, keep its current ones, and in general encourage wider use of the system. Further, starting in 1999 operators and rail station customer assistants will receive new customer service training in order to make the system more friendly and pleasant.

In general, the initiatives to increase ridership are paying off. Even though service was reduced, 1998 ridership increased by almost 5 million over 1997 ridership, and any increase is significant on a system that has steadily lost riders for so long. The simplified fare structures, which have made the system more affordable and easier to use, are expected to lead to another 5 million more riders in 1999 than in 1998 and seem to have finally and for good reversed the trend that lost the CTA 47% of its riders over 36 years (see Figure 2 for CTA Ridership from 1980 to 1998).

CTA's Needs and Challenges

Looking to the near and distant future of the CTA, we see that a great deal depends on the agency's ability to secure public funding for rebuilding, replacement, and expansion. For example, to receive full federal funding for the badly needed replacement or rehabilitation of the Cermak branch of the Blue Line and for the expansion of capacity on the Brown Line, the CTA must secure 20% matching state funds. Overall, the CTA estimates that it needs $4.1 billion over the next five years for renewal of its assets simply in order to keep the system in a state of good repair. A similar amount of money will be needed over the following five years. The source of the problems with maintaining rolling stock and facilities could be said to go back at least to the time the CTA was formed--1947--and bought out the private rail companies in Chicago that could not afford to maintain their equipment. The problem was made worse in the 1960's when the CTA deferred maintenance and equipment replacement in order to comply with the law requiring it to fund all operating costs with farebox revenue. Today much of the needed renewal is the routine replacement of buses and rail cars that have gone beyond their expected life span and hence have become unreliable and excessively costly to maintain. Indeed, maintenance costs are growing in part because maintenance facilities themselves have grown too old. Some of the needed work would provide adequate service on lines through neighborhoods whose populations have grown enormously since the original stations and platforms were built; that is why platforms on the Brown Line must be extended so that eight-car trains can run where only six-car trains can run now. Most urgent, however, is work on rail structures and stations. Just one part of the work to be done is the $60 million job of bringing 41 rail stations, already wheel chair accessible, up

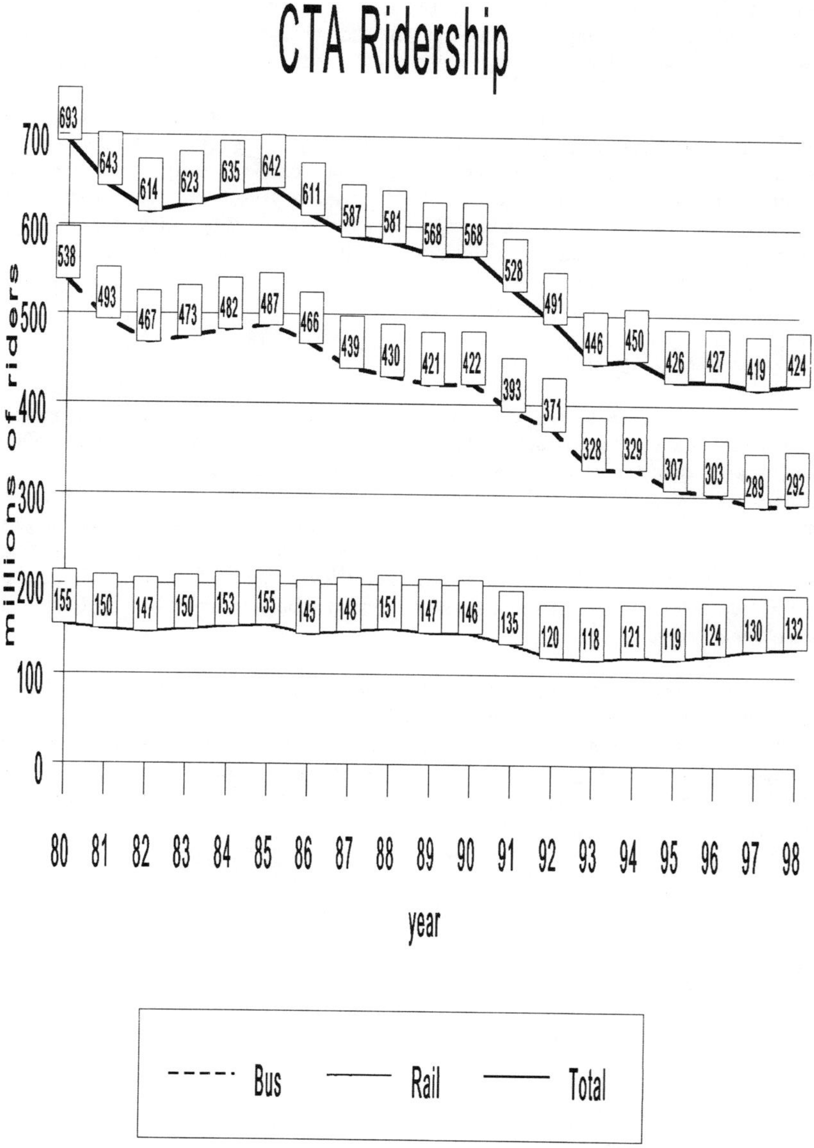

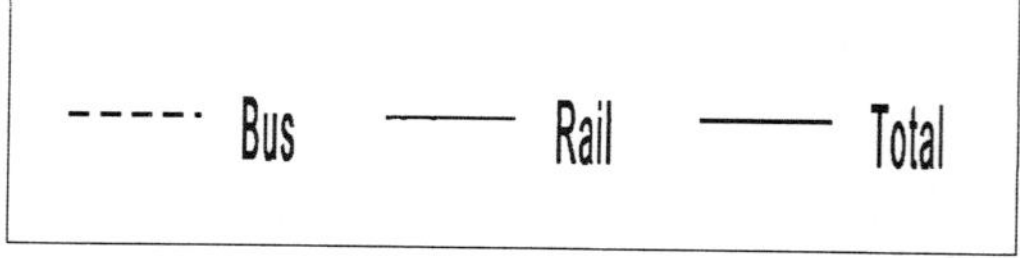
CTA Ridership
700
600
500
400
300
200
100
0
millions of riders
693 643 614 623 635 642 611 587 581 568 568 528 491 446 450 426 427 419 424
538 493 467 473 482 487 466 439 430 421 422 393 371 328 329 307 303 289 292
155 150 147 150 153 155 145 148 151 147 146 135 120 118 121 119 124 130 132
80 81 82 83 84 85 86 87 88 89 90 91 92 93 94 95 96 97 98
year
Bus
Rail
Total

Figure 2

to the standards of the Americans with Disabilities Act. By making station improvements such as adding or improving elevators and ramps, adding or improving gates and signs, the CTA will benefit all passengers and allow stations to function more smoothly.

A great deal of renewal work must be done simply to keep the system safe. The rail communication and signal system will be upgraded at a cost of $112 million, and an additional $74 million could be put to good use in that area. 15 miles of track are in poor enough condition to require reduced speed operation over them. At present it is anticipated that only 4 of those miles will be repaired soon, though many other miles of track will be given preventative maintenance so that they will not become slow zones or require replacement. In total, the CTA can expect to spend only $1.9 billion on renewal projects or $2.2 billion less than truly needed.

On-Time, Clean, Safe, and Friendly

The CTA faces a number of challenges of all kinds in its attempts to gain and keep riders. But most important is finding ways to serve and satisfy its customers. One new way it is trying to be more responsive to its riders' needs is by implementing a Bus Service Management System. This communication system will first of all provide more information to passengers, for instance, telling them at bus stops the expected arrival time of the next bus. And by allowing the Control Center to locate buses with great accuracy, the system will make service more consistent and reliable: it will eliminate bus bunching and protect connection points so that passengers will be able to make transfers even when one bus is late.

Lastly, and perhaps most importantly of all, the CTA is focusing attention on its customers' needs rather than just on moving buses and trains. The CTA aims to treat passengers in such a way that they will feel more comfortable with the system and any problems they experience on it. The CTA recognizes that it faces challenges that cannot be met with simply more money and better technology; some problems have to do with the way one person treats another.

Personal Rapid Transit (PRT): Solving Traditional Mobility Problems With New Technology - The Chicago RTA Experience

John DeLaurentiis[1] and Angela Johnson[2]

Abstract

The Regional Transportation Authority's (RTA) Personal Rapid Transit (PRT) development effort began as a vision. Guided by statutory mandate, staggering gridlock and the unique travel demands of emerging suburban activity centers-- all against a backdrop of declining ridership on conventional transit modes, the RTA turned to technology for solutions. Known as Personal Rapid Transit, PRT attracted the attention of the RTA because it promised to be an innovative addition to the existing family of mass transportation services. This paper covers the RTA's experience with a new transit technology, PRT, as a solution for the traditional mobility problems.

The Regional Transportation Authority

The RTA was created in 1974 upon the approval of a referendum in the northeastern Illinois counties of Cook, DuPage, Kane, Lake, McHenry and Will to ensure the development of a comprehensive and coordinated mass transportation system in the six-county region. Among its responsibilities, which at the time included the operation of some transit lines, the RTA was to provide financial support for the regional public transit system.

In 1981, growing financial difficulties resulted in the fiscal collapse of the RTA. As a result, the Illinois State Legislature reorganized the RTA in 1983 to protect the system from future financial crises. The amended *RTA Act* includes three key provisions.

[1] Deputy Executive Director, Planning

[2] Project Manager, Engineering & Technology Regional Transportation Authority, 181 West Madison, Suite 1900, Chicago, IL 60602, USA

- All day-to-day operating responsibility was decentralized into three distinct service boards — The Chicago Transit Authority (CTA), Metra commuter rail and Pace suburban bus.

- A formula was established for allocating 85 percent of the RTA sales tax receipts directly to individual service boards and a new state funding source equal to 25 percent of the sales tax receipts was provided to the RTA for discretionary allocation to the service boards.

- As a condition for receipt of additional state funds, the regional transit system, when taken as a whole, must recover at least 50 percent of its total operating expenses from system-generated revenues.

The RTA System

The six-county area served by the RTA covers 3,700 square miles. Based on the 1990 census, the region has 7.3 million residents. The City of Chicago has a population of 2.8 million, or 38 percent of the regional total. Suburban Cook County accounts for 2.3 million residents. Of the collar counties, the largest is DuPage with 0.8 million residents. Lake, Kane, McHenry and Will counties have a combined population of 1.4 million. Between 1980 and 1990, suburban population grew by 9 percent while the City of Chicago experienced a net population decline of 7 percent. By 2020, the regional population is projected to be 9.0 million. While the city's declining population trend is forecast to reverse in the 1990s, suburban population is projected to grow but at a slower rate than historical trends.

Service Board Dimensions

CTA, Metra and Pace carried 529.0 million riders in 1997. CTA carried 418.8 million or 79 percent of the region's total riders. Metra carried 72.3 million passengers, representing 13.7 percent of RTA ridership. Pace carried 37.9 million passengers in 1997, or 7 percent of RTA system ridership

The CTA provides bus and rapid transit service to an area encompassing 220 square miles including the City of Chicago and 38 suburban municipalities. The CTA's operating budget in 1998 was $782.0 million with capital assets valued at nearly $15 billion. METRA provides commuter rail service connecting downtown Chicago with 68 other locations and PACE provides fixed-route bus, paratransit and vanpool services to 200 suburban communities and from suburban locations to the City of Chicago.

Evolution of the RTA's PRT Initiative

In response to its fiscal oversight, planning and funding mission as embodied in the Authority's enacting legislation, in 1989, the RTA sought and received authority from the state of Illinois, to issue $1 billion in capital improvement bonds. By 1990, the first $200 million of that investment towards rebuilding and improving the RTA system had been committed to the three service boards. In August of that same year, the RTA's PRT efforts began in earnest.

The Mandate

"The Authority and the Service Boards shall study public transportation problems and developments [and] encourage experimentation in developing new public transportation technology..."
Regional Transportation Authority Act, Chapter 70 ILCS
Section 2.09, Research and Development

In response to its statutory obligation, an experimental transit system-- a new and promising technology to improve suburban mobility in northeastern Illinois was unveiled.

The RTA Strategic Plan sets forth an experimental approach to serving the changing suburban markets in the region and encourages the adoption of appropriate new technologies. In 1990, there was increasing evidence where growing, yet dispersed, employment and population centers had been difficult to serve with traditional transit technologies and service designs. Several existing activity centers had densities and mobility needs high enough to produce severe highway traffic congestion, yet inappropriate for traditional rail transit service.

Throughout the 1980s, automobile usage had been increasing at an alarming rate. Between 1985 and 1993 Highway Lane Miles in the RTA region increased nearly 4.5 percent from 51,484 to 53,869 lane miles. For the same period, however, regional vehicle miles traveled increased by nearly 25 percent as the following figure illustrates.

Daily Vehicle Miles Traveled for Northeastern Illinois

Source: Transportation Facts, Chicago Area Transportation Study

The regional distribution of jobs was migrating outwards as well. In 1950, Chicago's suburbs accounted for 20 of every 100 jobs in the Northeastern Illinois region. By 1990 however, over 60 percent of the region's jobs were located in the suburbs. Work and travel patterns also changed dramatically. In 1990, more than half of all Chicago area workers commuted from suburb to suburb. At the same time, transit was weathering both a severe recession and a substantial loss of ridership as the following figure illustrates.

Public Transit Annual Ridership for Northeastern Illinois

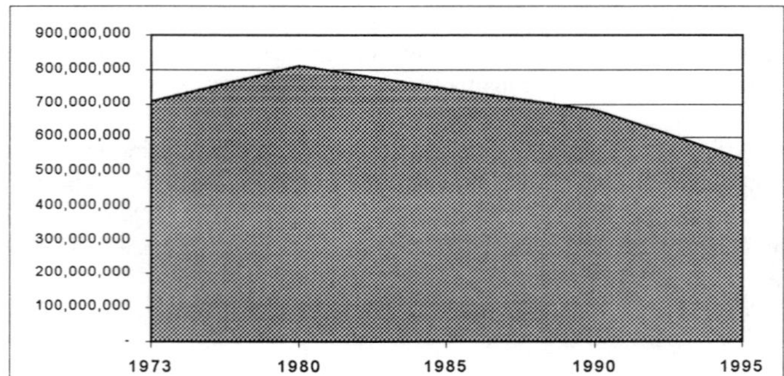

Source: RTA System Facts, Annual Report

The PRT Solution

The RTA's Personal Rapid Transit (PRT) development effort began as a vision. Guided by statutory mandate, staggering gridlock and the unique travel demands of emerging suburban activity centers--all against a backdrop of declining ridership on conventional transit modes, the RTA turned to technology for solutions. Known as Personal Rapid Transit, PRT attracted the attention of the RTA because it promised to be an innovative addition to the existing family of mass transportation services.

The preliminary research and discovery of the initial visioning process, had by 1990, evolved into a highly structured planning stage designed to evaluate both the feasibility of the PRT technology and the screening of suitable demonstration sites. With a new $1 billion capital program to help replace its aging infrastructure and the prospect of a new mode of transportation to augment traditional line-haul transit, the RTA initiated a conservative, three step PRT Program Development Plan. Phase I was the initial step and included System Design studies to evaluate the PRT concept. Phase II System Development and Testing, which was successfully completed in September of 1998, affirmed the technical and financial feasibility of the RTA's PRT system. Phase III or the final phase of the technology development plan consists of Site Demonstration and Deployment.

Goals of the RTA PRT Plan

- To conceptualize a new mode of public transportation--one that combines right-of way, technology and service in non-traditional, creative and unique ways;

- To evaluate this new mode and determine its applicability to suburban markets in northeastern Illinois and elsewhere;

- To determine if this new mode is safe, reliable, efficient, accessible to the disabled, and an effective complement to the existing family of transit modes; and

- To determine if this new mode can be deployed with a minimum of disruption to the community.

PRT Development Principles

The RTA's PRT initiative promised an innovative complement to the existing transportation network featuring:

Fully automated travel for 3 to 5 passengers
Direct origin-to-destination service
Mobility limited accessibility

Wait-times of three minutes or less
System availability > 99.7%

As a new technology, PRT offered a means of reversing the physical limitations of radial transit lines laid out nearly a century earlier. It promised to provide commuters with private, fully automated travel with wait times of less than three minutes. Shorter travel times would be possible because stations are off-line, and service is direct, Origin-to-Destination. PRT was considered a cost competitive technology because it could be built for a fraction of the cost of heavy rail. In addition, still reeling from the rising costs, fare increases and ridership loss spiral of the 1980s, as a fully automated system, PRT had residual benefits in that it was expected be relatively inexpensive to operate. Also, with its lightweight guideways and quiet cars, stated preference surveys forecasted almost universal acceptance in traditional suburban environments.

PRT Program Development

A PRT system design request for proposals was issued by the RTA in May 1990. Later that year, the RTA Board of Directors awarded two contracts for $1.5 million each to Intamin AG and Stone & Webster Engineering Corporation to conduct Phase I system engineering and design studies for two competing PRT concepts. Based on the results of Phase One, Phase (two) II, System Development and Testing and finally, Phase (three) III, Site Demonstration & Deployment completed the RTA's Program Development Plan.

With a development program underway, project planning turned to selection of potential sites. In March 1991, the RTA received expressions of interest from over 20 communities in the northeastern Illinois region. Of those, six communities submitted detailed written proposals identifying a 2 to 3 mile core alignment for the PRT demonstration, as well as potential expansion routes.

Site Selection

As a result of a comprehensive selection process, the list of potential sites was narrowed to four communities as shown below: Deerfield, Lisle, Rosemont and Schaumburg.

Four Suburban Sites for PRT Potential Deployment

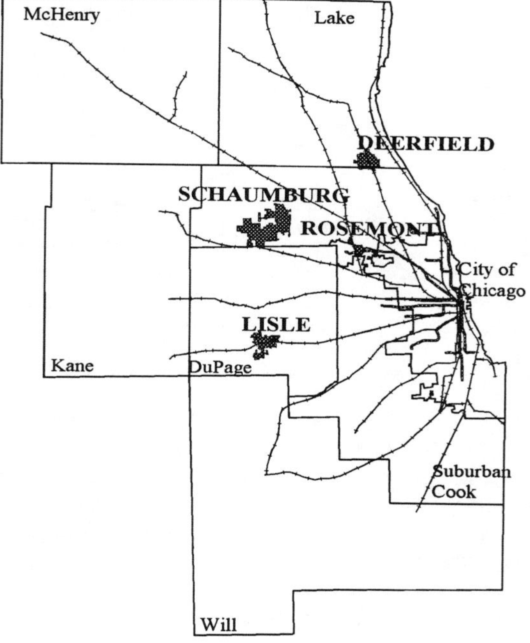

Ridership Potential

In March 1992, Wilbur Smith Associates completed a PRT ridership study for each of these four communities to determine potential ridership levels for their proposed PRT demonstration alignments. The main evaluation criteria included ridership, constructability and local commitment.

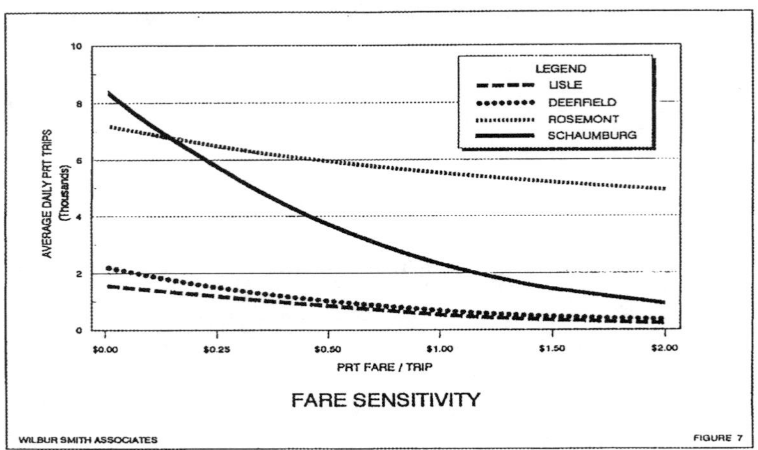

The graph above depicts the results of the PRT ridership study. It shows average daily ridership estimates for PRT as a function of fare for each of the four communities. A fare structure ranging from free to $2.00 per trip was tested for each community's proposed alignment. The communities of Rosemont and Schaumburg demonstrated the greatest PRT ridership potential--with Schaumburg showing the greatest sensitivity to fare (8,400 trips per day at zero fare and 800 trips per day at a $2 fare), and Rosemont being the least sensitive to fare (7,400 trips per day at zero fare and 4,800 trips per day at the $2 fare).

In April 1993, based on the results of this ridership study, an evaluation of the constructability of each communities' proposed demonstration alignment and local commitment, the RTA selected the Village of Rosemont.

Salient Characteristics of Rosemont are shown below:

- 3 Square Miles
- Population of 3,995 in 1990
- 1,726 Households in 1990
- 23,000+ Jobs
- Major Hotel and Convention Center
- Service Oriented Economy:
  - Business and Misc. Services 28%
  - Retail Trade 28%
  - Wholesale Trade 14%
  - Finance, Insurance & Real Estate 7%

Phase II PRT Partnership

In June of 1993, the RTA Board of Directors, approved initiation of Phase II, System Development and Testing, through formal partnership with Raytheon Corporation. This phase of the project, which was to be completed in November 1998, represents a joint venture, with the RTA providing up to $18 million in financing to be matched by a commitment of up to $20 million from Raytheon.

Phase II consists of detailed design, development, and testing of prototype hardware and software, integration of all elements into a functioning system, and the operation and testing of the prototype PRT system.

Both the RTA and Raytheon recognized the outstanding potential of this technology in the RTA region. A joint investment approach was adopted that allows each partner to balance the risk and share in the opportunities of this research and development project. Under the terms of the agreement, the RTA will share in 1.3 percent of all revenue generated from the future sales of the RTA-Raytheon PRT system for a 25 year period.

Phase II Program Status

The Phase II Program involved the design, development, and testing of a PRT system meeting performance requirements for deployed systems.

The Phase II Program was completed on September 30, 1998, on schedule and within the project budget of $18 million. It consisted of a two-step build process known as the engineering model (EM) and prototype model (PM) phases.

The EM Phase verified fundamental concepts for vehicle control on a single vehicle. Verification was done by a series of tests and demonstrations at the test track facility in Marlborough, Mass. Results of the EM demonstration provided a successful transition to the PM Phase. The PM Phase began with the integration and testing of multi-vehicle operations, followed by a demonstration. Raytheon's Independent Research & Development (IR&D) effort included system engineering and hardware and software development. In June of 1998, a successful PM demonstration was held at the test track facility in Marlborough, Mass.

Proposed Rosemont System

The proposed deployment of PRT 2000 in the Village of Rosemont is expected to provide an efficient and convenient transportation link connecting the offices and hotels in the heart of the Village of Rosemont, the Rosemont Convention Center, the Rosemont Theater and the CTA/PACE station. The

CTA/PACE station provides a direct connection to O'Hare International Airport, the City of Chicago, and neighboring communities served by the suburban Chicago bus system. By acting as a feeder and distribution system, the proposed Rosemont PRT system would compliment existing public transit services, while maximizing the accessibility of business destinations to hotels and other attractions. The system is designed for expansion, providing the transit infrastructure required to support additional development without incurring the traditional penalties of increased traffic congestion. The PRT system alignment, as currently perceived, is detailed in the following figure.

Proposed PRT System Alignment

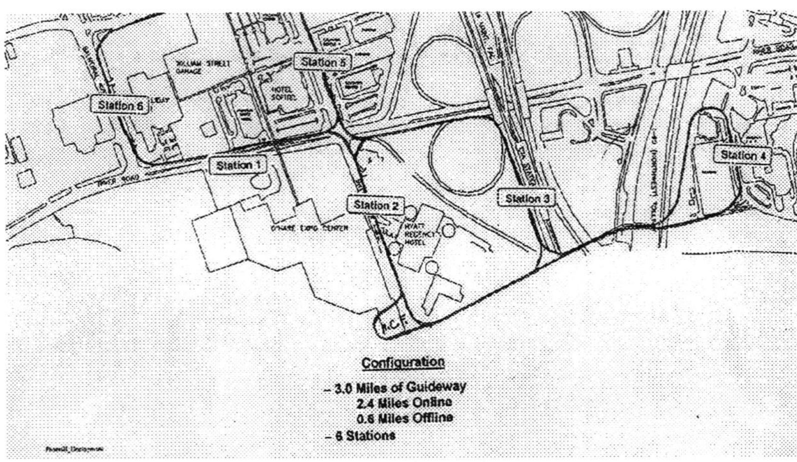

Rosemont Demonstration System Plan

An estimated 5800 people a day will travel among the six off-line, unattended stations, linked by 3.0 miles of one-way, elevated guideway. Seated passengers will travel directly to their destinations in comfortable, driverless vehicles, capable of seating up to four passengers. An automated fleet of 45 electrically powered vehicles will supply the capacity required to meet the projected ridership of the Rosemont system, and ensure that average passenger wait times remain below three minutes. The configuration allows non-stop travel between any two stations with an average one-mile travel time of under 3.75 minutes.

The six unattended PRT stations are standardized in design throughout the system to promote passenger efficiency through familiarity. Security is provided by a visually open design to permit high external visibility, by surveillance cameras located strategically within the station monitored by the System Control Center (SCC), and by two-way audio communications between the station and the SCC.

PRT system control, maintenance and system administration functions are performed in one integrated facility called the Maintenance and Control Facility (MCF). The MCF consists of the SCC, a maintenance area, vehicle test and storage track, and an administration area. The MCF is located at the rear of the Rosemont Convention Center.

The major hardware elements of the currently perceived Rosemont PRT system are listed in the following table:

Rosemont Personal Rapid Transit Major Hardware Elements

Item	Quantity
PASSENGER VEHICLES	45
Maintenance Vehicle	1
GUIDEWAY (MILES)	3.0
ON-LINE	2.4
Off-line	.6
Passenger Stations	6
Power Substations	3
Maintenance Control Facility	1

Current Status

The primary Public Transit concern in the northeastern Illinois RTA region continues to be focused on renewal and rehabilitation of the RTA's valuable infrastructure. With a capital improvement need approximating $4 billion, the RTA and its service boards have been concentrating all of their efforts at renewal of the state funded capital program initiative last authorized in 1989. The state's extensive network of roads and bridges requires extensive reinvestment as well. The Illinois legislature is expected to respond to these critical needs by June 30, of this year, when the state begins its next fiscal year.

In November of 1998, the RTA Board of Directors demonstrated its commitment to both the present and the future. Cognizant of its dual responsibility to secure capital funding for existing needs, and to continue technological research and development efforts, the RTA decided to defer progression to Phase III PRT Deployment and site Demonstration till spring of 1999.

SOCIO-ECONOMIC CHARACTERISTICS OF CURITIBA: Planning for Change through the Integration of Land Use and Transportation

LUCAS NIERI[1]

The city of Curitiba, called the "Environmental Capital of Brazil," has been awarded numerous environmental prizes including two awards from the United Nations. Like most cities in the Western Hemisphere, the story of Curitiba's development is a complex one, built on the confluence of natural conditions and human activity. This city's success and subsequent reputation is the result, however, of a number of specific interventions that have been carefully choreographed by the leaders of this city to the benefit of its citizens.

1. INTRODUCTION

Curitiba has distinguished itself for the last three decades because of its innovations in mass transit and urban planning. The city's transportation system is remarkable for its effectiveness and low cost. Its development is remarkable for its social and economic success and its environmental responsibility. In light of today's international economic instability, the Curitiba example is very important because it demonstrates that with determination and inventiveness a city can preserve existing qualities of value and create a positive future.

A further study of Curitiba's economic, social and political evolution demonstrates that people of every economic and social class can attain extraordinary results when they are united to a common purpose by a plan for urban development and public transportation.

Curitiba, the capital of the state of Paraná, is located in the south of Brazil near the Tropic of Capricorn at latitude 25°25', and has an area of 432 square kilometers (279 sq. mi.). The city's population in 1999 is approaching 1,800,000 inhabitants

---

[1]Lucas Nieri, Architect, Rua Angelo Sampaio, N:1037, Curitiba, PR, Brazil, CEP: 80250-120; Email: lanart@hotmail.com

within a metropolitan population of more than 2,400,000. Curitiba is strategically located in the most economically developed region of Latin America, being close to Brazil's Mercosul partner countries: Argentina, Uruguay and Paraguay. It is one of the principal international centers of business and investment in Brazil, and is the commercial and educational center of Paraná.

History

The city of Curitiba has always be a crossroads, growing in a series of waves with the changing tides of commerce. Founded in 1693, the settlement functioned as a center for the Portuguese extraction of gold from the rivers of the Planalto Curitibano. Curitiba's importance came from its location as a transportation crossroads. By the 17th century, the city was supplying convoys herding cattle from the pampas in the south to the booming gold and silver mining cities to the north.

Curitiba became the capital of the state of Paraná in 1842. Faster growth began in the middle of the 19th century with successive waves of agricultural exploitation of the vast high plain to the west. In the 1880's, first a granite cobbled roadway, and then a railroad, was cut from the plain, through rugged mountains of the Atlantic rainforest, to the Atlantic seaport of Paranaguá. Curitiba's location assured its success, being at the juncture between the agricultural land to the west and the road/railroad heading east. First matte (a tea herb), then timber, next coffee, and currently soybeans all made their way from source to market through Curitiba.

Immigrants came in large numbers in the second half of the 19th century and the beginning of the 20th to farm the fertile countryside of the state of Paraná. They came from many countries in western and central Europe—primarily Germany, Poland, Italy and the Ukraine, and also from the Middle East—primarily from Lebanon, and from Japan. Curitiba grew as well, reaching a population of over 130,000 in 1940. During the past 60 years, the population of the state has continued to grow; but as farm work becomes more mechanized there has been a large internal migration from farm to city. Much of this growth has concentrated in the state capital.

Curitiba functioned successfully as a commercial, educational and government center for the state of Paraná into the 1950's. The city realized, however, that these functions would not be sufficient to provide employment for the large number of people who were starting to arrive in the 1950's. A major expansion of the manufacturing sector, and further development of national and international commerce, would be necessary to meet the needs of the expanding city. Planning for this growth, so that the city would successfully meet the needs of all its citizens, began in earnest in the 1960's.

The continued growth of the city may be in part a result of the success of its plans for development; however, efforts are currently underway at the state level to improve the economic viability of remaining on the land, with the goal of slowing migration to the city.

Social and Economic Indicators

Curitiba's economic success is a function of its strategic location between producer and consumer, and the value added by its workers. Today, the city has one of the highest per capita incomes in Brazil. The median family income in 1990 was approximately US$4000. At that time, 5% of the city's families earned below the minimum monthly wage of US$75. This is similar to the statistics for the country's wealthiest state, Sao Paulo, whereas this figure is 18% for the country as a whole.

In 1994, there were 267 automobiles per 1000 residents. For the country, this was second only to Brasilia, the national capital, where most low-income residents live outside the statistical area. At the same time, 75% of all weekday commuters travel to work on public transportation. There were 350 unlinked (202 linked) transit trips per capita per year—these figures are nearly as high as New York City and Mexico City.

35% of the workers are employed in the retail/commercial/service sector and 20% work in manufacturing. Many international companies have established manufacturing plants in the Industrial City. This includes the vehicle manufacturers Volvo (trucks and buses), Renault (cars) and New Holland (harvest machinery and tractors), and vehicle equipment suppliers such as Bosch.

Curitiba is the site of the first Brazilian federal university, the University of Paraná. It has long been the regional educational center, and as such, a focus for intellectual debate and cultural activity. There are currently 35,000 university level students enrolled in the various institutions of higher education.

The statistics for 1990 show that 9% of the economically active population was unemployed, a figure that is low for Brazil. A number of social programs attempt to provide equitable access to education, jobs, housing, and health care. City services extend to nearly all households, often through self-help systems such as an innovative garbage collection/recycling program.

Curitiba has already interacted with and influenced its closest municipal neighbors to improve the quality of life for the metropolitan region, and this work is expanding. For example, the vision of an integrated mass transportation system for the metropolitan region, taking advantage of the evolution of the city's transit system itself, is resulting in improved conditions for the entire population. These efforts are a continuation of a long history of planning for urban development.

2. *PLANNING FOR URBAN DEVELOPMENT*

Since the last century, Curitiba has been privileged with plans of urban development which prescribe ordered growth. In 1855, the French urbanist Pierre Toulois made the first sketches.

Agache's Urban Plan

The first urban plan for the city was developed in 1943 by another French urban planner, Alfred Agache; at a time when the population of Curitiba had grown to 150,000 inhabitants. This plan assumed the dominance of the automobile. By means of circular boulevards and major radial arteries, created by the total reconstruction of existing roads, Agache's plan defined a modernist vision for the expansion of the city.

The Agache Plan proposed a strengthening of the center of the city and the development of four secondary centers: Portão, Bacacheri, Merçês and Cajurú, linked by the outer circumferential roads. The plan determined a hierarchy of streets for automobile circulation, and located open spaces and specialized centers.

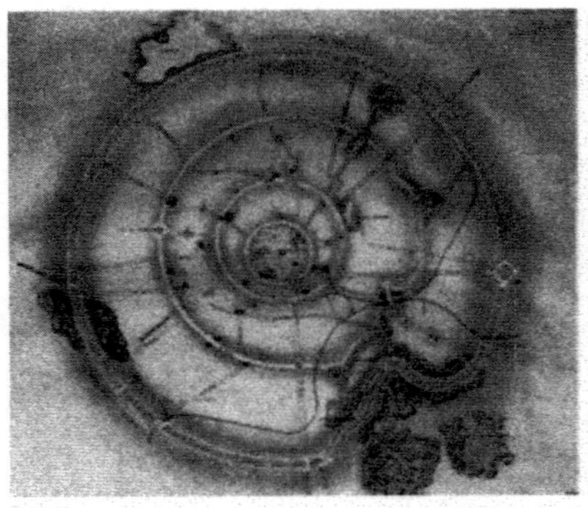

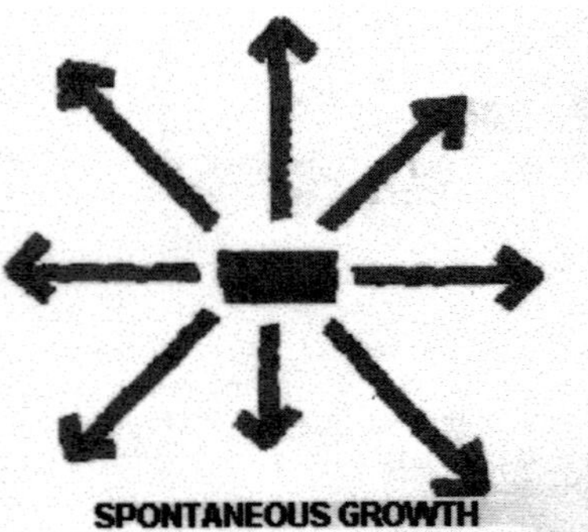

Agache's Plan

Drawing Of Spontaneous Development

Realization of this plan required a major restructuring of the city, to be achieved through massive demolition and public investment. Ultimately, little was carried out. Agache's plan did establish, and see realized, the Civic Center--an area of administrative buildings for the state and municipal governments. This area, representing mid-20th century planning and design of the modernist movement, stands out today as markedly different in character and scale from both older and newer parts of the city.

The Origin of the "Direction Plan" For Urban Development

After only two decades, the Agache Plan for the city's urban development was reconsidered. The Sarete Society of Studies and Projects, Ltd. created the "Preliminary Plan". This plan developed an analysis and socio-economic projections based on an urban proposal by the architectural office of Jorge Wilhem Associated Architects.

This work was financed by the Development Company of Paraná - CODEPAR (Companhia de Desenvolvimento do Paraná), formerly the Development Bank of Paraná - BADEP (Banco de Desenvolvimento do Estado do Paraná). In 1965, this plan won a competition sponsored by the Curitiba municipal government's Department of Urban Development.

Because the downtown center of Curitiba, at that time, was extremely dense, the plan proposed an expansion of linear growth, involving a system of peripheral avenues and a limited use of the center.

The principal directive of the Preliminary Plan of 1965 (subsequently known as the "Direction Plan") was to discipline the growth of the central area along one axis. The idea was to make the city linear, in order to increase the efficiency of circulation and the distribution of services. It proposed an organized infrastructure investment, and hoped to facilitate lagging urban development.

Old Photograph of Central Area

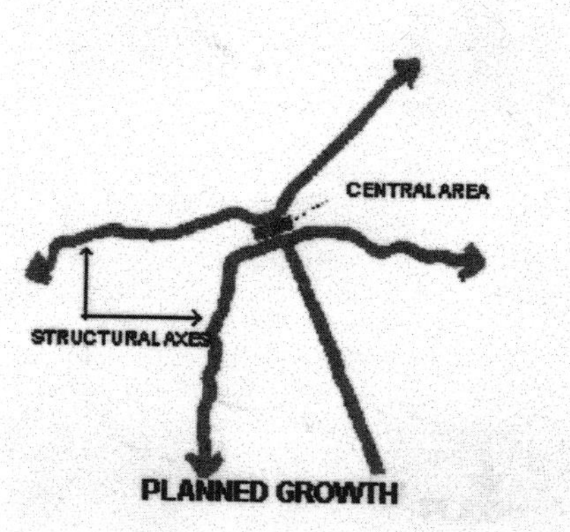

Drawing Of Planned Development

People believed that this plan would result in a continuous development without the difficulties associated with highly concentrated urban development. Secondary centers were considered unimportant, although some were predicted to grow in the future. This scheme, however, posed some technical problems: the inconsistency of the plan with reality, areas prone to flooding, low bearing capacity of the soils, and an awkward interlinking with the old interstate highway BR-2 (today BR-116).

The Three Basic Elements of the Direction Plan

One of the most important contributions of the Direction Plan was the definition of three elements that were the basis of the entire proposal. The three elements were: Land Use, the Road System, and Mass Transit. The evolution of these elements lead to the concept of the structural axes, the physical form around which the development of the city and its transportation network is now organized. This form, in turn, lead to legislation in the form of Land Use and Zoning that prescribed the desired growth of the city. The Direction Plan established the function of the structural axes as the focus for growth of the city. It also determined the character of the linking roads and of land use and conservation between the axes.

Thus, the linear center was realized. The functions usually found in the downtown were distributed along the structural axes (together with residences). This served to create an efficient transportation system, relieving the center of the city and allowing for its controlled expansion.

The Land Use element established a hierarchy of densities centered along each structural axis. The highest levels of residential and commercial development are concentrated in the two blocks at the center of the spine, with diminishing densities in the blocks to either side, thus preserving large areas for low-rise residential development in the sectors between axes.

The trinary Road System of each axis is made up of one central street with exclusive lanes for efficient public transportation and slow local access traffic lanes with parking. To either side of the central street are one-way arterial streets of traffic (fast lanes), headed into or away from the downtown.

The Road System was created through a re-definition of the existing streets. It was not necessary to resize the streets nor to construct overpasses. Expropriation, wholesale demolition and major construction was avoided, while major improvements to the efficiency of both transit and traffic were made. This avoided the destruction of the urban landscape that results when individual transport is intensified.

The Preliminary Plan had anticipated a transit system of street cars and subways. The Mass Transit system that was adopted in the Direction Plan, however, was the result of the search for a technology that was already in use and more accessible - the bus. This bus-transit system has a hierarchy of vehicle types to fulfill the different requirements of local feeder, line haul, and specialty routes.

In the Plan, one sector of the city was reserved for future growth. This area, with special land use regulations, is called New Curitiba. It is an area located between the Industrial City sector and one of the structural axes, and is crossed by three connecting arteries. This area has, in fact, grown with the increase in industrial jobs.

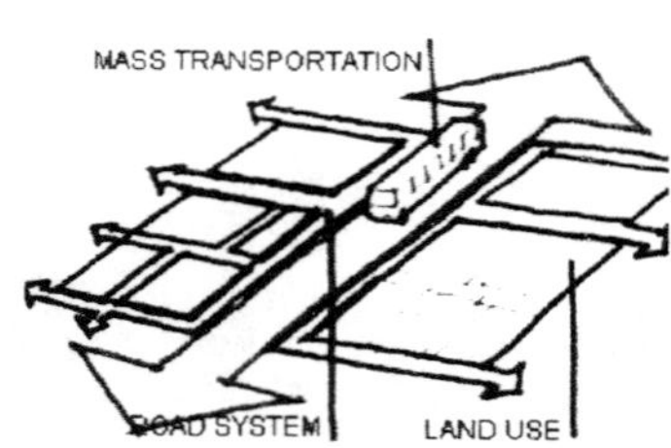

Diagram of the three elements

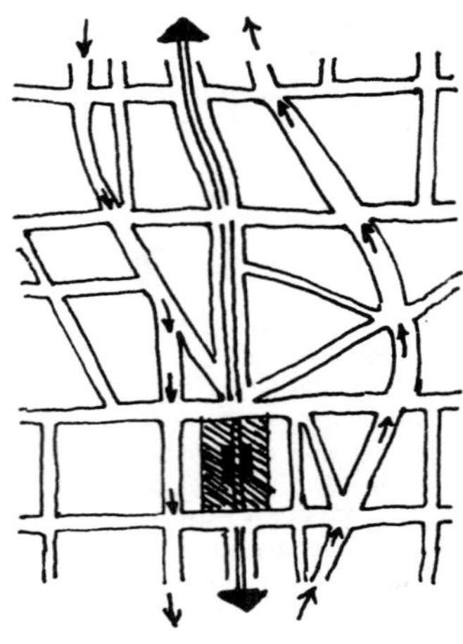

Illustration of the structural system

Implementation of the Direction Plan

An auxiliary to the city's executive branch of government was established on July 31, 1965 to implement the Direction Plan. First named the Auxiliary to Urban Planning and Research of Curitiba - APPUC (Assessoria de Pesquisa e Planejamento Urbano de Curitiba), this institution later became The Curitiba Institute of Urban Planning and

Research - IPPUC (Instituto de Pesquisa e Planejamento Urbano de Curitiba). This public agency was to be responsible for the development of a more complete plan and for carrying out the directives of the plan, while continually performing analysis and revisions.

Although the plan was approved in 1965, it was not immediately implemented. Its implementation began in 1971, when IPPUC's president at the time, architect Jaime Lerner, was appointed to be the Mayor of Curitiba. After three non-successive, but highly successful, terms as mayor, Mr. Lerner is now the two-term governor of the state of Paraná.

The basic responsibilities of IPPUC are not only to plan, but also to test solutions. This dual responsibility has been central to its success. As new plans are generated and accepted by the community, they are put into practice quickly.

The population began to trust the ideas of the Institute; and this trust has largely been responsible for the change of mentality of the city's inhabitants. The community began to believe in IPPUC and to support its ideas of urban transformation, to suggest improvements and to demand modifications.

It is important to highlight the evolution of urban planning in Curitiba concurrent with the growth of IPPUC. Thus the Direction Plan has become a "living guide" to the development of the city.

3. *INTEGRATION OF LAND USE AND TRANSPORTATION*

The plan for Curitiba has the objective of adapting the zoning and land use requirements to the socio-economic and territorial development of the city. Land Use is supported by the Road System and Mass Transit. It is a system that integrates the new structural axes of concentrated growth through a network of circular routes, connecting routes and penetrating routes. It generates jobs in regions close to residences and it develops new vocations.

Curitiba's Zoning and Land Use also implements new planning instruments that develop relations with the private sector, in joint action with the public administration.

A principal objective of Curitiba's Land Use plan is to coordinate the public mechanisms for the implementation of urban planning in a manner that guarantees ease of understanding by the community.

One very interesting example of this integration today in Curitiba--the fruit of a project that unites urban structure, zoning, land use, the road system, and private initiative--is the "Job Line", an avenue 34 km long, passing through 15 boroughs and taking ten job centers to the residents.

The places reserved for commerce and services are quite restricted, intending to reinforce the primary movement of people by public transportation, and to avoid reducing the efficiency of arterial streets in their role of serving the needs of vehicular circulation.

Locations for commerce and services are: 1) the Structural Sector (along the fast lanes), 2) the Central Zone, 3) "Connectors" in New Curitiba, and 4) the Service Zones.

The maximum area of commerce alternates from 100 to 400 m^2 per tract of land (except along connectors where up to 5,000 m^2 is allowed for commerce. In some cases more area can be acquired by approval of the Superior Council of Urbanism. In other places no commerce is allowed, such as along the penetration routes or in ZR1.

Heavy industries are concentrated in a large region in the west of the city called the CIC (Industrial City of Curitiba);

Zoning

The zoning of Curitiba is defined in conjunction with the Plan of Urban Development. It shows the following hierarchy today:

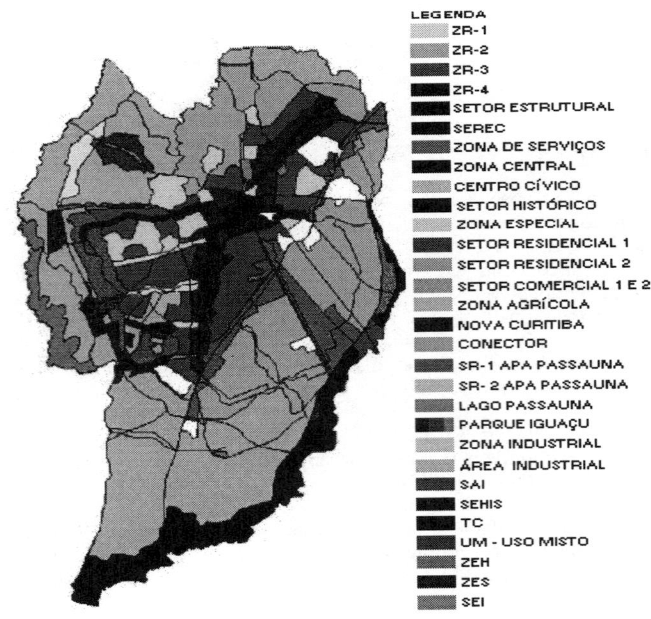

Plan and legend of Urban Zoning

HIGH DENSITY

Zone	Coefficient: floor area / site area	Height Limitation	Notes
Structural sector	4	unrestricted	
Central Zone	5	unrestricted	
New Curitiba	3	unrestricted	Coefficient 3 is only reached in conjunction with certain conditions

MEDIUM DENSITY

Zone	Coefficient: floor area / site area	Height Limitation	Notes
ZR4	2	6 floors	
ZR3	1.33	2 floors	
SEREC	2	6 floors	
Connectors	1.33	2 floors	
Transition-- New Curitiba	1.33	6 floors	

LOW DENSITY

Zone	Coefficient: floor area / site area	Height Limitation	Notes
ZR2	1	2 floors	
ZR1	1	2 floors	Only single-family housing

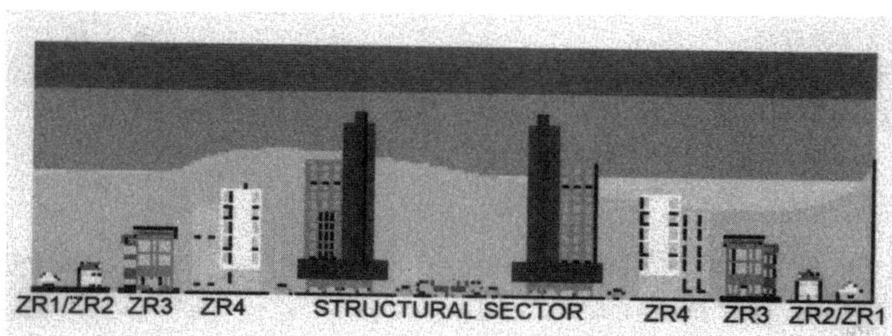

Schematic of the zoning vertical proportions

Several peculiarities of the land use legislation allow an increase in the constructive potential of each tract of land:

1- Incompatible Areas:
- Underground parking
- Fire escape stairs, elevator, balcony (6.00 m^2 / unit), mechanical areas
- Commercial foundation in the central routes of the structural sectors (can reach coefficient 1)

2 - Compatible Areas:
- First floors of the structural sectors outside the central routes (can reach coefficient 1)

Land Use and Benefits

Land Use and Zoning in Curitiba, through rigorous legislation, has brought important benefits to the residents of the city by preserving the undeveloped open spaces scattered throughout the city and transforming these spaces into public parks. This has enabled a considerable increase in the proportion of green area per inhabitant, giving preference to natural forests (some more than 100 years old) and giving the city a surprising index of 52 square meters of green area per capita (the UN recommendation is 16 square meters).

A critical feature of the Direction Plan has been the creation of conservation areas. Curitiba needed additional green areas to protect the city from floods, improve the water supply and promote basic sanitation. Instead of allowing the subdivision of critical large, empty parcels of land, the plan resolved to offer the owners of these properties either money, or more frequently trade of developable parcels, in order to conserve these critical pieces of open land.

Instead of permitting real estate speculators to subdivide large parcels into smaller lots indiscriminately, the real estate market wins in a different way. Consumers prefer to value the quality of life and live in a city in which the preservation of the environment is one of the primary administrative goals.

Of course, under close analysis of the road system and the circulation of mass transit, one can observe that much still must be done to improve and increase the system. It should be recognized that by defining vehicular circulation and the routes of the buses in such a way as to preserve green areas is an important element for the success of the city.

The Land Use and Zoning legislation also provides a clear policy of incentives which makes housing programs of social interest and preservation of historic buildings viable. With all these positive factors, the growing real estate market can profit with certainty because of the increase in demographic density and the quality of life in the state capital.

Constructive Incentives

In a city with few financial resources, many of the important benefits of the plan have been achieved by establishing incentives rather than by making cash payments. The primary incentive is to allow additional density in places where this is constructive to the planned organization of the city. Incentives are used to promote good things, and to prevent inappropriate things from happening.

Incentives allow a developer to increase the building coefficient (the ratio of floor area to site area) of a parcel of land, thus increasing the building height and thus the density of use. This provides a financial incentive at little direct public expense. Following are a few examples:

- Commercial use in the structural sector (can add to coefficient 1.5 in the commercial tower)
- Residential use in the tower linked with commerce on the ground floor in the central zone (can add up to two times the commercial area of the ground floor to the allowable area in the residential tower).

With these incentives some tracts of land can reach a constructive potential of 10 times the area of the tract of land (a floor area coefficient of 10).

Example:
Tract of land in the structural sector / central route / 1,000 m^2 / commercial / residential
Tract of land = zoning coefficient 4.0
02 underground parking provision = coefficient 2
Commercial incentive in the tower = coefficient 1.5
Commercial foundation = coefficient 2.0
Incompatible areas = coefficient 0.5
Total Allowable floor area ratio = coefficient 10

Created Land Project

In order to "create" land to be used for public purposes in particular locations, this project allows the height of potential construction in the ZR zones (medium and low density housing) to be increased by means of a payment, in cash or in tracts of land, to the city. The resources from this project go to humanitarian works.

Typical benefits to the developer are a 50% increase in density and an increase in the height limit by one or two stories.

Preservation of Historic Buildings

Owners of property which contains abandoned buildings can transfer the construction potential to other locations, through an analysis by a commission for heritage preservation called the Commission for Evaluation of Historical Patrimony (Commisão de Avaliação do Patrimônio Histórico). In this way, the construction potential of another tract of land can be increased, and the historic building can be preserved.

Green areas

The legislation concerning green areas in Curitiba is quite detailed, and has been changing in recent years. To preserve areas of vegetation on a particular parcel, the legislation, in some cases, allows a developer to increase the allowable building height. This promotes vertical growth so that green areas can be maintained without wasting construction potential.

Green areas have bee expanded by a hundred fold in thirty years. The purpose has been to enhance recreational opportunities concurrent with increased density of housing, and to protect areas from development that are prone to flooding. Of particular interest are the native pine trees of the city, which are all recorded by aerial photos. Cutting these trees is rarely permitted.

4. FUTURE PROPOSALS FOR THE ROAD SYSTEM AND MASS TRANSIT

At IPPUC and URBS (Urbanization of Curitiba, the bus authority), urban planners, architects and civil engineers are studying new ideas for improving the road system and mass transit. Included in current studies are electric vehicles, buses with diesel/alcohol fuel, a future underground rail transit line and underground parking.

Relation between population and circulation in the Curitiba region -- Predicting a gradual increase in the population of Curitiba, it is necessary to analyze the disastrous consequences that could occur if alternative solutions are not envisioned for the increase in automobile circulation in the city. The city is currently concerned about the air pollution index, the increase in volume of mass transit users, and the lack of space for parking in the downtown.

Electric Vehicles -- The city is developing, in collaboration with the new automotive factories in Curitiba, an alternative power system for automobiles: one which doesn't pollute (Most electricity in Brazil is hydro-electric). The current number of privately owned cars in the city of Curitiba is 580,000, which makes the development of electric transportation viable.

Electric Buses -- This power alternative can also be adapted for electric buses, serving the entire population, thus contributing to the preservation of the environment and the reduction in the number of cars in circulation.

Parking in Downtown -- The "Ecological Parking" project aims to create parking places surrounding the central area where the citizen can leave a gas-run vehicle and take an electric vehicle. In exchange, the electric vehicle user will the have easy access to all parking lots in the center of the city.

Underground Parking -- By means of a program which awards public services to private initiative, underground parking will be added to downtown. These parking structures will act as a support for the revitalization projects of the central region. The criteria for location will be a combination of technical feasibility, market, and the capacity of the environmental infrastructure.

5. *CONCLUSION*

Establishing the interest of the community, and creating the initial private investments necessary for the economic growth of a South American state capital is a difficult goal to attain for architects, urban planners and politicians.

The entire city must perceive the benefits of planned urban development that involves the most significant sectors of society: work, leisure, housing, transportation, health and education.

Without the help of Curitiba's inhabitants, it would be impossible to think of the development and of the response desired by the idealists of the Direction Plan, who searched for the identity of the community, its regional characteristics, and its economic potential.

With the success of the city's urban transportation system, it was possible for Curitiba to establish a relationship between land use in the city, its road system, and mass transit. Through these three elements, it has preserved the heritage of the city, enhanced its art and culture, restored the natural environment; and envisioned a future where an infinite number of ideas are yet to come.

OPERATION AND USE OF THE INTEGRATED PUBLIC TRANSPORTATION NETWORK OF CURITIBA, BRAZIL

CARLOS CENEVIVA[1]

In the worldwide search for intermediate transportation solutions between the bus and subway, the city of Curitiba, Brazil has achieved unusual levels of performance and quality. Curitiba has developed an integrated bus-transit network utilizing simple but efficient innovations: exclusive lanes, bi-articulated buses, direct lines, and tube stations. The efficacy of this system has been central to the city's ordered growth and high standard of living.

1. INTRODUCTION -- Urban growth and transportation

Curitiba hasn't escaped from accelerated urban growth and continued demographic concentration, a phenomenon that is characteristic of the medium and large sized South American cities.

In this region of the world, the twelve most populated countries (90% of the total of 300 million inhabitants) are growing at a rate of 1.5% per year (with the exception of Argentina at 1.2%). Nine of these countries, including Brazil with 250 million inhabitants, are growing at rates of between 2 and 3% per year.

Population growth is concentrated in the big cities. Today, with more than 50% of the population in the cities, the problem of mobility is aggravated, and has negative effects on industrial production and services. As a consequence, the GNP of these countries, generated for the most part in the cities, will demand proportionately more and more work hours, since productivity will tend to fall. No one produces well after wasting time and energy getting to work.

----

[1](formerly the Manager of Curitiba's bus-transit system) Architect, Ceneviva Planejamento Urbano, Rua Alberto Foloni 760, Ap.702, Curitiba, PR, Brazil, CEP: 80540-000; Email: ceneviva@bsi.com.br

The quality of life indexes, today unsatisfactory, can begin to change, beginning with the improvement of transportation. In order to reverse the current trends, it is perfectly viable (though not easily accomplished) to have a quick, transforming intervention by utilizing the public control over urban transportation. With the political will, and a small investment, it is possible to initiate a rapid process of reorganization in the transportation system, resulting in significant time savings and reduced operational costs.

It becomes clearer every day to all that, with some effort and organization, it is possible to eliminate waste and accommodate the growing demands with less work, lower cost and higher quality. The organs of finance and support for research in urban development play an important role in this process.

The search for solutions that are simpler, more efficient and more adequate to economic realities grows. The awareness of the need for a good urban transport system that promotes individual mobility takes form principally among the local governments. The concept of integration in the network, whatever the vehicle, takes priority over other technological innovations.

2. *THE CURITIBA EXPERIENCE*

In the city of Curitiba, the policy of transportation gives priority to collective over individual transport. The bus, the only economically viable means, incorporates simple, but very efficient innovations, and reaches unusual levels of performance and quality.

A municipal center of a metropolitan area with over 2.4 million inhabitants, Curitiba is the capital of the state of Paraná, in the south of Brazil. It is a center of commerce and services, and is becoming a major automobile, bus and truck manufacturing center of Brazil, second only to Sao Paulo.

2.1 Urban Planning as a Basis for Transportation

Curitiba inaugurated a new concept of urban planning and municipal public administration in the 1970's that has been essential to the city's transportation innovations.

Growth Structure -- Curitiba knows where to grow. The city's growth structure was defined and implanted at the beginning of the seventies. Since then it has experienced relative administrative continuity, in spite of the changes in priority and rhythm resulting in the alternation of public power.

Decentralization -- "Linear Centers" -- In the seventies, to the delight of the private interests, spontaneous growth gradually gave way to ordered growth, through an exercise of public power over three essential tools of urban planning: Land Use, the Road System, and Mass Transit. The growth of the downtown in a compact form was blocked and the "Structural Axes" were established as an alternative, using the concept of linear centers, extending in the directions of the outer boroughs.

Land Use, the Road System, and Mass Transit -- In these structural sectors, attractive parameters were designed for establishing commercial activities and residences of high density, thus circumventing the polarization of the downtown. The basic road system of each structural sector consists of three parallel streets. There is a central avenue, with 30 meters between the buildings, which before concentrated all the disordered traffic. Here, physical priority is given to mass transit by providing exclusive lanes for the Express Bus. The remaining space on this avenue is used for local traffic and parking for the commerce and services located along it. The voluminous through traffic was rerouted to two parallel streets, which were repaved and prepared to operate as efficient one-way arterials.

Without much investment or expropriation, the city now circulates better because of a new and efficient road system. Even today, this innovation meets the growing needs of collective transportation and the automobile.

2.2 The Basic Transportation Strategy

The local government has been promoting the expansion of an Integrated Transportation Network (ITN) that reinforces the global directions of the development of the city. The planners have been searching, with good sense and creativity, to find solutions most adequate to our reality. This has resulted in constant improvement in quality, without increasing costs and without the need to subsidize the operations.

I don't know another city, even among those that do subsidize their transit systems, that offers its inhabitants, for the same fare, the alternatives of transportation available in Curitiba.

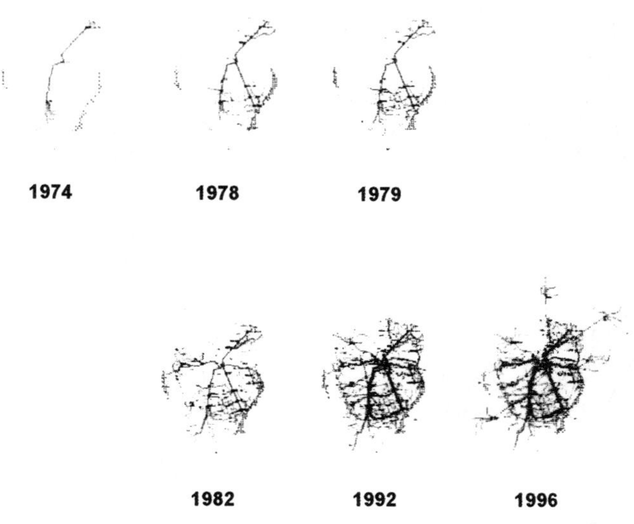

The Integrated Transit Network evolution from 1974 to the present.

The Integrated Transportation Network (ITN) -- Curitiba is in the final phase of a process of substitution of an old conventional system, a chaotic accumulation of superimposed lines with differing fares, with a complete network of interlinked lines, joined together by terminals of integration.

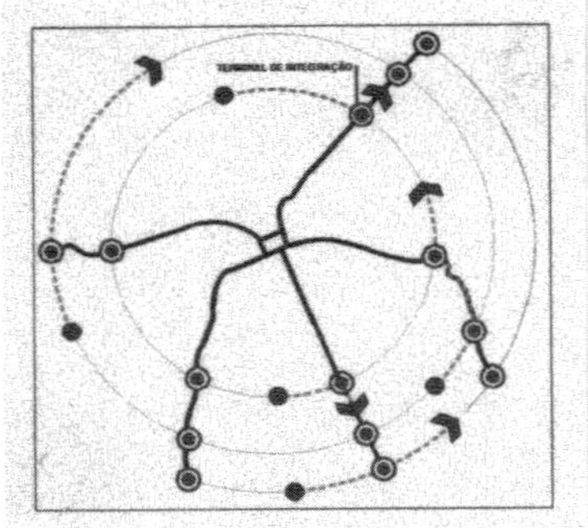

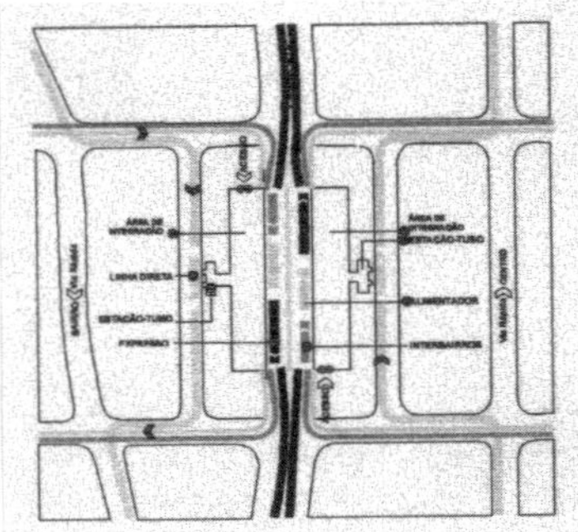

The Scheme for the Integrated Transit Network and the Terminals of Integration

Today the system operates under the concept of the Integrated Transit Network, permitting every passenger to compose his own path, utilizing various lines and paying only one fare. The result has been a better level of service and a more affordable fare.

Public Planning and Private Operation -- 193 interlinked lines form the ITN, while 86 more conventional lines are not yet integrated. 10 private companies (that have existed for many years) operate the lines under permits to provide services following rules pre-established by the public transit authority. These companies make all the investments in the fleet, and in facilities necessary for the operation and maintenance of the vehicles, according to measures determined by the authority. Drivers, fare-collectors, mechanics and administrative personnel are all employed by these private operating companies. The public pays for all investments in infrastructure: roads, stations, signage and the planning and control organization.

2.3 The Innovations

The transportation network is intimately interlinked with the general directives of urban planning, as defined in the city's "Direction Plan". This supports the city's growth within a pre-established urban structure, and optimizes public investment.

A repertory of innovations has been adding efficiency and increasing the capacity for transport as well as the life span of the system, in order to facilitate the use of the infrastructure for many years to come:

Exclusive lanes -- A strip seven meters wide is reserved exclusively for the operation of the Express Lines. The Express Lanes are located in the center of the structural axis. These lanes are delineated by concrete separators (raised medians) which prevent access by other vehicles. The first 20 km, the north-south structural axis, was planned in 1972, constructed in 1973, and began operations in 1974. Today there are five axes, with a total of 56 km of two-way exclusive bus lanes.

Express Lines -- (Red Bus) Established initially on the north-south axis, the Express Lines began operating with special buses for 110 passengers, with the majority of passengers standing. They run exclusively in the bus lanes, without conflicting with the rest of the traffic, and operate at a commercial velocity of 20 km/h. The Express Lines have normal stops every 500 meters, and Terminals of Integration every 4 km on average where they receive passengers from other types of lines.

Feeder Lines -- (Orange bus) These are short lines, operating with conventional buses. They interlink the neighboring boroughs with the Terminals of Integration.

Interdistrict Lines -- (Green bus) These are circular lines in two directions which interlink the boroughs without passing through the downtown area. They appeared in 1979 with the consolidation of the "Linear Centers." They diminished the polarization of the downtown area as the only destination. The growth of the structural axes generated tangential movements between the axes, creating a demand for the Interborough lines, and furthering even greater decentralization.

Integration and The Single Fare -- Until 1979, the fares reflected the operational cost of each line separately. Because they were less lucrative, the longer lines had a higher fare, and thus a high cost for the low-income population located at the periphery of the city. Since 1979, with the single fare reflecting the cost of the entire system, the short lines are able to subsidize the longer lines. Besides being socially just, the single fare facilitated the implementation of fare integration between different companies.

Concept of the Network / Terminals of Integration -- Following the implementation of the first Interborough line, the concept of network operation was imposed, interconnecting the lines from different companies. Between 1980 and 1982, 15 Terminals of Integration were constructed along the exclusive lanes, permitting the physical integration of fares to take place between the Express, Feeder and Interborough Lines, independent of the operating company. The rigidity of the system was broken with the countless possibilities of interchange between the lines. Demand is re-structured to new routes between the same points, providing operational benefits as well.

Remuneration per kilometer -- In 1986 the operating companies, which until then had received income directly from the transported passenger, changed to payment per kilometer. The municipal government takes a detailed measure (fleet, timetable, kilometers run, etc.), audits the execution, collects the daily receipts from the whole system, and pays the operators for services rendered in real costs. Detailed regulations establish the rights and obligations of the operating companies, define the faults and penalties, and seek to eliminate waste while constantly improving the quality of service.

The Tube Station and the Direct Lines (Speedy Bus) -- Created in Curitiba in 1984 and refused by another city, the project of the Tube Station had to wait for its opportunity until 1991. With the principal objective of facilitating embarking and disembarking, the innovation of the elevated, covered platform appeared; allowing for controlled access and safe and secure embarkation. The embarkation time fell to 1/8 of the previous time. People feel more respected. The innovation is applauded by the user. The design surprises and pleases everyone.

No matter which mode of transport is considered, approximately one third of all urban movement, at least in Brazil, can be classified as long trips in relation to the scale of the city. The automobile user generally chooses his daily route, opting for the quickest way, with more continuous movement and fewer stops, even if the path is longer. With the objective of saving time, this option can be offered to the bus user, through the adoption of direct lines between some of the more important points of the city where there is a concentration in demand. In Curitiba these points of concentrated demand are the Terminals of Integration.

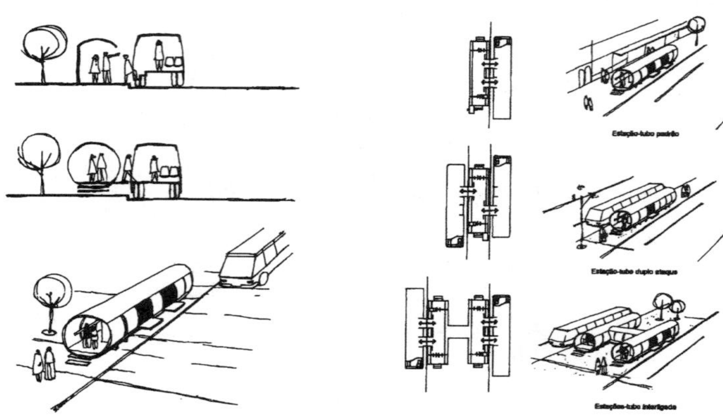

The Tube Station

Beginning in 1991, the Direct Lines (Silver bus), affectionately referred to as the "Speedy Bus" by the population, were connected to the Terminals of Integration. Direct line stops are separated by an average distance of 3.2 km and the lines run at a commercial speed of 32 km/h; favoring integration, and interlinking the city from point to point. They create new options, new routes, facilitate movement and, above all, save time for the passengers.

Table 1 -- ITN Curitiba

DIRECT LINE CAPITAL COSTS (US$)

		Quantity	Unit Cost	Total Cost	%
Public Investment -	Tube Stations	119	40,000	4,760,000	10%
Private Investment -	Vehicles	248	174,000	43,152,000	90%
Total Investment				**47,912,000**	**100%**

The biggest benefit offered by the Speedy Bus consists of the time savings for its passengers: one hour per day per person means more than 100,000 hours of time saved every day by citizens of the city.

An additional and important benefit of the tube station is the accessibility to disabled persons, including people confined to wheelchairs. They access the tube stations by means of a small lift. Embarkation on the bus is quick and safe, without delays or harm to the schedule, and without the need for special facilities that segregate disabled people.

Currently 13 Direct Lines are in operation, composing a network that is complementary to the previously existing one. They offer a new and attractive option for transportation that has already been adopted by a significant portion of the market. An analysis completed in 1992 showed that 28% of the Speedy Bus passengers were new users who left their cars at home. This means 30,000 fewer cars on the streets (around 7% of the fleet of 470,000 automobiles at that time).

Bi-articulated Bus ("Surface Metro") -- (Red bus) This is a bus in its most advanced level of development. It has a very large capacity (270 pass.), no steps or fare collection, travels in an exclusive lane, and incorporates the agility, comfort and security of level embarkation and disembarkation through the simple but effective technology of tube stations. This is a small "Surface Metro" with technology appropriate to the local necessities and possibilities.

The establishment of the elevated platforms at the terminals and the installation of tube stations at the bus stops guarantees that the bi-articulated bus has the necessary and sufficient operational conditions to support (with more vehicles) the growth in demand predicted for the next 12 years.

The first line was implemented in December of 1992, with 33 vehicles carrying 100,000 passengers per day on the south-west axis; with much greater comfort and a 6% lower operating cost than the previous fleet.

The second line began operations in August of 1995, on the north-south axis, with a fleet of 66 vehicles initially transporting 240,000 passengers per day. The daily demand has already reached 275,000, requiring an expansion in the fleet by 9 additional vehicles.

The Bi-articulated Bus concept

The adoption of the bi-articulated buses also has the advantage of unclogging the exclusive bus lanes. The wait time is now longer, as a logical consequence of the reduction to less than half in the number of vehicles. This allows for better regularity in the timetable and a better distribution of vehicles along the line. Scheduling is also improved by adjustment of the traffic signals, which in the near future will be triggered by the buses themselves.

Currently, the bi-articulated buses operate at an interval of two minutes. Tests in simulated operation proved it to be perfectly viable to operate at one minute headways, doubling the capacity of each line.

Table 2 -- ITN Curitiba

COSTS OF THE BI-ARTICULATED NORTH/SOUTH LINE (US$)

Public Investment			**7,160,000**	**17%**
Works	**quantity**	**unit cost**	**total cost**	
Adaptation of Terminals	6 terminals	200,000	1,200,000	
Repaving Exclusive lanes	20 km	160,000	3,200,000	
Stops (Tube Stations)	56 stops	35,000	1,960,000	
Central Stations	2 stations	400,000	800,000	
Private Investment			**34,275,000**	**83%**
Vehicles	**quantity**	**unit cost**	**total cost**	
Bi-articulated buses	75 buses	457,000	34,275,000	
Total Investment			**41,435,000**	**100%**

Financing -- The north-south Bi-articulated Bus Line constitutes the local contribution for the Curitiba Program of Urban Transportation, sponsored (financed) by the InterAmerican Development Bank. Together, this group of projects seeks to improve transportation and includes: paving the roads that constitute the bus routes, signage and road safety, computerized planning and control of public transportation, construction of Terminals of Integration, tube stations, bike paths, support centers for patrons, and finally communication with the user by means of a system of clear and current information. This program represents a total of investment of 200 million dollars, of which 120 million was financed by the InterAmerican Development Bank. Approved in October of '95, this program is in motion with 60% of the work already completed. It is entering now, with the new city administration, into its final phase of execution.

2.4 *The Benefits*

The analysis of the quantitative data in the previous and following tables, allows us to conclude that the adopted innovations have brought benefits and advantages to the city and to all the users of the Integrated Network.

Table 3 -- ITN Curitiba

SUMMARY OF OPERATIONAL DATA

Type of line	Color of vehicle	Pass per vehicle	Vehicle price x US$ thousands	No. of lines	Speed km/h	Fleet	km/day	Pass /km	Pass/day x thousands	Fares (revenue) /day x US$ thousands
Feeder	orange	90	91	163	16	621	150,348	4.5	681	366
Inter-district	green	110	170	8	16	159	34,829	5	171	88
Direct Line	silver	110	174	13	32	248	64,426	5.2	334	220
Articulated	red	170	252	6	20	73	17,041	5.8	155	94
Bi-articulated	red	270	457	3	20	95	17,772	16.3	342	207
Subtotal ITN				**193**		**1196**	**284,416**		**1,683**	**975**
Conventional	yellow	90	91	86	16	331	87,593	2.9	252	252
Totals				**279**		**1527**	**372,009**		**1,935**	**1,227**

The Direct Lines, for example, bring the following comparative advantages: pre-paid fare and level embarkation by means of the tube station diminish considerably the embarkation time per passenger, and consequently, the time that the vehicle must remain at the stop (dwell time).

The conventional lines have a commercial velocity of 16 km/h and a dwell time of 2 seconds/passenger. The Direct Lines have a commercial velocity of 32 km/h and a dwell time of 0.25 seconds/passenger. Comparatively, the speed is doubled and the embarkation time is 1/8. The shorter dwell times, achieved by pre-paid floor-level boarding at the tube stations, contribute very much to reducing the travel time for passengers. In addition, a smaller number of buses are required (with less private capital to remunerate) while the same number of people are transported.

On average, the Direct Lines save half the time previously spent on conventional lines. This represents one hour per day per person. As there are 334,000 passengers per day on the Direct Lines. Considering that each person takes an average of 3 trips per day, approximately 111,000 people per day are no longer wasting time (a savings of more than 100,000 hours) on the bus and can turn those hours to productive use.

Table 4 -- ITN Curitiba
TOTAL CAPITAL COSTS OF THE INTEGRATED TRANSIT NETWORK (US$)

Infrastructure (Public Investment)			**75,820,000**
Terminals	25 terminals	900,000 / terminal	22,500,000
Exclusive Lanes	56 km	800,000 / km	44,800,000
Tube Stations	213 stations	40,000 / station	8,520,000
Buses (Private Investment)			**188,504,000**
Bi-articulated	95 buses	457,000 / bus	43,415,000
Articulated	73 buses	252,000 / bus	18,396,000
Speedy Bus (Direct Lines)	248 buses	174,000 / bus	43,152,000
Inter-district	159 buses	170,000 / bus	27,030,000
Feeder Lines	621 buses	91,000 / bus	56,511,000
Total costs			**264,324,000**

Another example is the exclusive lanes on which the Express lines (bi-articulated buses) operate. The exclusive lanes total 56 km in length, serve the lines with highest demand and carry 497,000 passengers per day. These lines have a commercial speed of 20 km/h and save 25% of the time required by the old conventional lines. Considering an average of 3 trips per person per day, 166,000 people utilize the Express lines. Each person's daily travel averages 20 km, and thus one hour of travel time instead of one hour and 15 minutes per day. The time savings of 15 minutes per person per day for 166,000 people totals a savings of more than 40,000 hours per day.

Table 5 -- ITN Curitiba
COMPARATIVE SUMMARY OF OPERATIONAL COSTS (US$)

	Conventional line	Direct Line (Speedy Bus)	Articulated Express	Bi-articulated Express
1. Operational Cost (US$/km)	1.21	1.29	1.41	1.67
2. Administrative Cost (US$/km)	0.15	0.16	0.18	0.22
3. Cost of Capital (US$/km)	0.53	0.53	0.39	1.57
4. Taxes (US$/km)	0.06	0.06	0.66	0.04
Reimbursement (US$/km)	1.95	2.04	2.07	3.50
Passengers per km - IPK	2.9	5.2	5.8	16.3
Cost per Passenger (US$)	0.68	0.39	0.39	0.21
Vehicle Capacity - (pass.)	90	110	170	270
Vehicle Cost - (US$)	91,000	174,000	252,000	457,000

Still other benefits are the lower number of cars in daily use. 31,000 new users have been attracted to the Direct lines and 10,000 attracted to the Bi-articulated Express lines. This amounts to a diversion of 41,000 auto trips to transit. This represents almost 9%

of the automobiles in the city that are now staying in their garages; and proportionately less air pollution, less fuel consumption, reduced congestion, and a better quality environment for the entire population.

3. *THE NEXT TEN YEARS*

3.1 *Expansion of the ITN within the City*

New Direct Lines should be implemented, completing the full coverage of the network over the city. The "Speedy Buses" can serve 30 to 40 per cent of the demand of the collective transport, saving time and facilitating integration. This provides advantages for all passengers, including for those who don't use the Speedy Bus, by easing the congestion on the other lines, principally on the Express lines. In addition, they can attract still more new users.

Bi-articulated Buses will be implemented on the east-west axis, standardizing the operation in every direction on the exclusive lanes of the five structural axes; thus reinforcing the image of the system, and facilitating its understandability for the population.

3.2 *Metropolitan Expansion*

The municipalities surrounding Curitiba continue to grow at a rapid rate, and much of the travel demand is trips to economic generators in Curitiba. The construction of new automobile and truck assembly plants on the outskirts of Curitiba, with initial production anticipated within two years, demands some precautions in transportation as well.

The solution now being implemented consists of the expansion of the ITN to the greater metropolitan region. This extends the concept of network operations to all occupied areas whose demand for transportation has clear urban characteristics; in other words, daily round trip movement principally for the purpose of work.

Bi-articulated buses running in exclusive lanes will be extended, and Direct Lines will reach beyond the municipal limits to the neighboring municipalities, thus expanding and consolidating a Metropolitan Integrated Network.

Everything makes full use of the existing infrastructure (the road system, terminals and tube stations that are already installed). Large investments are avoided by increasing the use and efficiency of equipment already available; saving time and money, and promoting integration with a single fare--all characteristics peculiar to public transport in Curitiba.

4. *CONCLUSION*

4.1 *Time is life*

A quick view of urban transport in the developing world, in particular in the principal Latin-American cities, shows a desolate, almost chaotic, panorama. In every city, save a few exceptions, an enormous inefficiency, that mistreats the population, illustrates the difficulties that the public power has in fulfilling its mission to plan and control public transportation; a service so important that it effects the day to day life of nearly every person. Every day immense time and resources are wasted on a disjointed bundle of bus lines, precipitated by this lack of physical and operational organization.

Most people would agree that good cities require a good system of transportation. But to have good transportation these cities do not need to spend what they don't have--to acquire sophisticated and expensive technology (and which does not always resolve the problems as promised).

In spite of all the technological advances, the majority of people will only be able to count for their means of daily locomotion on a vehicle so common, of a technology so widespread and accessible, as the bus. It will continue to be the only means of transportation possible for the majority of our cities.

Improvements like those that Curitiba presents can raise the performance of the bus to the level of some of the most technologically sophisticated, hence much more expensive, systems of transportation.

This is what Curitiba has been doing for the last few years: searching for and finding these simple, low cost, but very efficient solutions with immediate benefits for the population.

The 270 passenger bi-articulated bus loads passengers from pre-paid high platforms

We have no doubt in affirming that this could be the solution for many other cities, even in more developed countries. The large number of visits and requests for technical collaboration that the city receives confirms that Curitiba is on the right track.

The support of the International Development Bank and BNDES (a Brazilian Development Bank), with important financing for a broad program of works for the whole city, aided by these low cost solutions, reinforces this conclusion.

More than 100,000 hours saved by the Speedy Bus, for example, constitutes an inestimable savings deposited day after day in the account of collective well-being and productivity of the working population of this city. Savings of time, which is much more valuable. Savings of time to be applied in rest, studies, and family life. Savings in time transforms into life.

4.2 A New Way

Curitiba has inscribed as much in its own urban landscape as in the international literature of cities: the current configuration of its Integrated Transportation Network, revitalized by the innovative solution of the Speedy Bus and the Bi-articulated bus. A world tendency has emerged in the search for intermediate solutions between the bus and the metro. Many cities in various countries are searching for new ways. Curitiba has already found, and continues to refine, hers.

CURITIBA: An International Perspective on the City's Bus-Transit Network

KENNETH E. KRUCKEMEYER M.ASCE[1]

Abstract

The bus-transit transportation network in Curitiba, Brazil, has received widespread attention for its innovations and effectiveness. This paper looks at those breakthroughs, and considers their applicability to transportation problems in other countries. The paper adds an international overview to the prior presentations by two professionals from Curitiba: Lucas Nieri's presentation on demographics and urban planning, and Carlos Ceneviva's description of the transportation system.

1. INTRODUCTION -- Experiences in Curitiba

The international visitor to Curitiba does not remember this city for extraordinary natural beauty nor for breathtaking architecture. The abiding impression of this place is one of friendly and energetic people, actively developing and caring for a city that works. The beauty of Curitiba is both on the surface, and in its soul.

One experiences Curitiba as a city that is easy to get around--whether by public transportation or by automobile. One finds great diversity--from the high level of activity at all hours in the downtown to the tranquillity of the metropolitan parks. And, one sees Curitiba as a great place to live—whether amidst the urbanity of high rise apartments or within the pleasantness of a single-family neighborhood. Things seem to work, during the day and at night, and for the rich and the poor. The organization and cleanliness of this city is certain to catch your eye; however, it is the high spirit of the populace that convinces the visitor that there is something very important to learn from Curitiba.

The statistics bear out the impression of a city that works: more than 70% of the residents get to work on public transportation. The transit system is chosen by people

----

[1]Research Associate, Center for Transportation Studies, Massachusetts Institute of Technology, 77 Massachusetts Ave., Cambridge MA 02139; (617) 267-2110, Email: kek@mit.edu

who have a car available (auto ownership is the second highest in Brazil) and it is actively used by those who have no car. The transit system operates frequently, safely and at a profit.

The stores and markets in downtown Curitiba are thriving. Because there is so much activity, one feels safe, day and night. Industry continues to move to Curitiba, and unemployment is as low as anywhere in Brazil. Over the past thirty years there has been a high level of public investment in visible and useful public facilities (parks, community centers and places for culture and celebration) and in less visible infrastructure investments (flood control, water resources and sewerage). Perhaps most importantly, there has been a great investment in the people of the city--in work programs and cottage industries, in schools and after-school programs, in adult education, health services and libraries.

A high-density structural axis and low density residential district seen from the Mercês tower

2. *OBSERVATIONS ON LAND USE AND TRANSPORTATION*

The prior presentations have described the transportation network and the land use patterns of Curitiba. From the Mercês tower, the structural axes of the city can be seen extending in each of five directions from the downtown to the edge of the city. (Structural axis is a term used to define the city's five linear spines of highest building density that center on the primary bus-transit trunk lines and parallel one-way arterial streets.) Between each of these five axes are large areas of low scale moderate to low density development. Four to six story medium density housing in a three to four block wide band adjacent to the spines is envisioned, but in most cases this conversion is yet unrealized.

The boldness of the plan is immediately visible. And the similarity between the plan and reality is striking. When one superimposes the transit and land use plans, one sees, in a cross section through the city, that the transportation and building densities coincide. Most remarkably, for a rapidly growing city, conformance with the plan by owners and developers is nearly universal.

Curitiba has developed and grown within its boundaries quite quickly according to this plan--the city's population has grown fourfold over the past thirty years.

Looking first at the structural axes, the evolution of individual parcels from low to high density means that some old single family houses remain adjacent to new high-rises. Over time, these homeowners choose to move (and can afford to choose because of increased property values) to single-family homes in other neighborhoods, or into units in the new towers. These new buildings, most often built as condominiums, are typically inhabited by the burgeoning middle class. This concentration of new residents along the spine is well situated to use the bus-transit for work and activities downtown. Most residents also have a car tucked in an underground garage for special occasions and use on weekends. It is part of the plan that these buildings are typically built by the private market, and without public subsidy.

At the center of the structural axis are the trunk bus line and the highest density development

The preserved and the newly developing areas of low density housing between the axes, while relatively homogenous in any one neighborhood, provide housing for people of all incomes. These areas are covered by the feeder bus network, which provides access to the spines and free transfer to the trunk lines. Direct bus lines connect concentrations of high transit demand within these areas, in order to reduce transfers as well as to reduce the volume on the trunks.

Curitiba has not recently tried to house its rapidly expanding low income population by building subsidized apartments. Not only has the money not been available to do so, but the city believes that low income people thrive best with their "feet on the ground," in the flexibility and adaptability provided by individual low-rise structures. The city provides "sites and services" to families who build their own housing, with an opportunity for gardens and home industry.

3. *UNDERSTANDING CURITIBA*

To learn from the Curitiba example, one needs to understand the realities of this Latin American city. Consider the following points that have a significant influence on land use and transportation.

Citizens -- Similar to the United States, southern Brazil is made up primarily of immigrants from European and Asian counties, many of whom arrived in the late 19th and early 20th centuries. The migration from farm to city has occurred more recently, and in greater numbers, in Brazil, than in the United States. Among Brazilians, the people of Curitiba have a reputation for being serious and hard working.

Economics -- People are generally poorer in Curitiba than in the United States or Europe, although they are better off than many of their Brazilian counterparts. The monetary situation in Brazil has, until recently, been extremely volatile as a result of frequent episodes of runaway inflation. One must, therefore, be very cautious about making financial comparisons with other cities and other countries. The lack of monetary stability has created a different relationship between capital expenditures financed by borrowing and operations financed by income and appropriations.

The Brazilian tax system provides little federal support of state and local programs, including infrastructure and transportation investments. With respect to public transportation, however, a federal law requires employers to subsidize an employee's cost of taking transit to work if it is greater than 6% of the worker's income. This provides an indirect subsidy to the transit system by private and public employers, although it does not help unemployed people afford transportation. It does allow a fare structure that is sufficiently high for operators to be able to make a profit.

Politics -- The volatility of Brazilian politics has created a greater awareness of the importance of rapid realization of public projects. In many parts of the globe, political change means that good ideas often die with a change in administration. In Curitiba, transportation strategies that are capable of speedy implementation have become a key to their successfulness. The political leadership of the city has changed frequently, in part because of a law that prevents a Mayor from holding consecutive terms, and in part because of the choice of the voters; but projects that can get completed in one term are up and running.

It is remarkable that urban planning and transportation has been a central focus of political debate in the city's political campaigns. The net result has been a successive improvement of the plans for the city with each administration.

Education -- Curitiba, the capital of the state of Paraná, is also the home of the University of Paraná, the first Federal University in Brazil. It has been an educational mecca for southern Brazil, and a center of intellectual leadership and ferment. The establishment of the Department of Architecture and Planning at the University around 1960 has created a group of leaders who have played key roles in making the city of Curitiba what it is today.

Planning -- The IPPUC (in English: the Curitiba Institute of Urban Planning and Research) was established in 1965, shortly after the graduation of the first class of Architects from the University of Paraná. This institution has not only been the center for effective planning in the city; it has also been the origin of many of the political leaders of Curitiba, and a foundation of their political power.

Growth and Change -- The growth of the City of Curitiba and its immediately surrounding municipalities has been exponential. The metropolitan population has increased from a half-million to over two million in the thirty years between 1965 and 1995. The success of the city in dealing with its challenges in a socially responsible way (good opportunities for housing, health care, education and employment) has further hastened this growth. In addition to growth there has also been change, with increased standards of living and greater expectations.

Is Curitiba's transportation system prepared for the future? Prosperity seems to inevitably mean more automobiles; and the growing number of automobiles will generate higher levels of traffic congestion. This can adversely effect the efficiency of street-running buses and of the trunk lines at intersections. It will likely result in increased pressure for greater accommodation of the automobile. The worldwide lack of courtesy of automobile drivers toward pedestrians, already at the extreme in Brazil, is likely to get even worse with greater congestion, whereas bus-transit riders need to feel comfortable when they are walking. Prosperity may pose a significant threat to the outstanding success that Curitiba has achieved.

4. STRATEGIES AND INNOVATIONS IN CONTEXT

The city of Curitiba has been exemplary in its achievements in many areas. This is especially true in light of the very real constraints of Brazilian political and economic volatility. Several achievements deserve further analysis in order to better understand their applicability to other cities and countries.

Land Use -- The land use strategy of Curitiba anticipates and forces extremes: very tall apartments with high densities, and small houses at low density, with moderate density walk-ups and apartments in between. The plan proscribes dramatic changes for formerly low density neighborhoods located along the structural axes that are planned for high-density development. The result, however, has been to assure the continued existence of other low density areas for rich and poor alike, and to further the preservation of large areas of open space. With the acceptance of aggressive land-use and density change comes a remarkable stability. Since the city comprises three-fourths of the metropolitan population, it has sufficient control to be able to determine both change and preservation.

Speed of Delivery of Public Infrastructure -- The leaders of Curitiba have made the conscious decision to choose strategies that can be implemented quickly. Some citizens complain that this means that things are not always built to last, and that there is insufficient attention to detail. This strategy is, however, a pragmatically appropriate one, given the rapid changes in Brazilian politics and economy; and incremental improvement is anticipated. For example, the early design for locally made bus stop shelters was quick and adequate, but not especially pleasant. Along the trunk lines, these shelters have now been replaced by the tube stations, also locally fabricated, which provide better service and are of the highest quality industrial design.

Financial Pragmatism -- Closely linked with the strategy of delivering projects quickly is that of dealing with financial realities. Building for the present, rather than building for forever, can reduce the immediate capital requirements of a project at the expense of a greater need for maintenance in the future. This may not be an undesirable condition, however, in light of a shortage of capital and a surplus of people needing jobs. In fact, many features of Curitiba's transit system emphasize the use of people instead of machines. Fares are collected by attendants who make change, thus improving the loading characteristics of local, express and trunk-line buses. Having a collector on local buses greatly reduces dwell times by speeding the process of entry. Fare collectors at the bus platforms for the express and trunk lines means that pre-paid passengers enter the bus in a manner similar to a rail-transit system. The collectors also

Collectors on the local buses (left) and at the tube stations (right) improve bus service

provide a continuous presence in the stations, improving security and reducing vandalism. The benefits to the system are greater efficiency and personal satisfaction, while at the same time benefiting the local economy by providing greater employment.

Incremental Implementation -- Speed of delivery and financial pragmatism are meaningless if the transit mode selected requires completion of the entire system before the first rider is carried. The Curitiba bus network has allowed individual parts of the system to be utilized as each portion is financed and constructed. It makes possible, as well, the upgrading of various components of the system as the needs require, as technology advances and as financing becomes available.

Integration of Public and Private Actions -- Many of Curitiba's public initiatives are intended to spur private action. This is true of housing: relying on private individuals and developers for upper income houses and buildings, and on self-help opportunities for low cost housing construction. In the transportation sector, the bus system is organized by URBS (Urbanization of Curitiba S/A), the public transportation corporation, which sets operating standards and owns the bus fleets. Private companies are contracted with to operate and maintain the buses.

Leadership -- A great deal Curitiba's success has to do with its remarkable leadership. A most unusual fact about this city is that many of its leaders have been Architects, Planners and Engineers, the best known being the former mayor Jaime Lerner, who is now Governor of Paraná. This fact has placed planning for the development of the city at the forefront of public debate. In the political ferment that has ensued, ideas about the direction of the city have benefited from the verbalization of different strategies. Even though with new administrations there has been some shifting of direction, this ferment seems to have resulted in a richer, more thoughtful plan at each step of the way.

Specific Innovations to Improve the Image and Effectiveness of Buses – Many people believe that only rail transit systems are sufficiently robust as to be able to generate high-quality dense urban development. The most notable feature of Curitiba's transportation network is that it is made up exclusively of buses; and buses have been central to the positive development of this delightful city. Curitiba's bus system has succeeded in providing excellent access and mobility, and thus is highly effective. The effectiveness of the system has been reinforced by a stylishness of design that has created a very positive image. And this positive image has engendered sufficient confidence in the system that it has defined the pattern of growth and development for the city, and engendered the high degree of compliance with the plan that is so evident.

The following two features of the bus transit system seem to have been essential for success in Curitiba. They may not, however, be sufficient to assure success in another city.

> 1) Reliable high-capacity buses running along trunk lines on the structural axes where the greatest population lives and works, featuring exclusive lanes with a limited number of cross streets and signal preemption, high platforms for entry and exit with pre-paid boarding; and
>
> 2) A complementary network of color-coded feeder and express buses, all with free transfer, providing dense coverage of the entire city.

5. *INTERNATIONAL ISSUES IN TRANSPORTATION*

Before the reader comes to any conclusions about the applicability of the Curitiba transit system, it is important to come to grips with several contemporary international issues in transportation.

The Dominance of Auto-centric Thinking -- Around the world, highest priority is normally given to automobile transportation; while at the same time, the lack of equity and the environmental liabilities of an automobile-based transportation system are frequently called into question. Ivan Illich wrote in 1973 that "high speed is the critical factor which makes transportation socially destructive... Participatory democracy demands low-energy technology, and free people must travel the road to productive social relations at the speed of a bicycle." Few have chosen to accept this advice.

The auto provides extraordinary utility and comfort (at considerable public and private expense). The challenges of providing equitable and affordable transportation in less developed countries are clearly documented in Charles Wright's book *Fast Wheels, Slow Traffic*. Public action is necessary to assure any equity in the transportation system. When there is a lack of decisions in the public sphere, power is put in the hands of private individuals of wealth, and thus the private automobile is allowed to dominate the roadways.

The Role of Government and Private Enterprise -- A number of more developed countries, particularly in Europe, have attempted to address the issues of equity and environmental impact. John Pucher's book, *The Transport Crisis in Europe and North America* documents the differences which government policy and aggressive action can make in affecting public action and personal choice in transportation. The issues of privatization (either ownership or operations) of the public transportation system is only a relevant debate once the public responsibility for the transportation network is established.

Profits and Subsidies -- Establishing a coherent transportation policy requires not only addressing the issues of service and equity, but also dealing with economics. It is difficult to have a productive debate on the financing of public and private transportation systems because we have developed separate languages for each. The Conservation Law Foundation, in its book, *Road Kill,* has attempted to clarify the terms of the debate.

It is standard parlance that the public expense needed to keep the highways functional for private vehicles (for which the cost of entry is owning a car) is called a "cost," while the public expense of operating a transit system (for which the cost of entry is paying a fare) is called a "subsidy."

The visible and invisible public and private expenses of operating private automobiles is difficult to compare with the expenses of organizing and operating public transit. Different people will use numbers to their own advantage to make their case for entirely different courses of action. For me, the bottom line is whether the ultimate benefits accrue to all members of society, or only to specific portions of society.

Timing with respect to Growth, Development and Affluence -- The ability to implement transportation policy in any society ultimately must have some public support. Even an aggressive shift in public policy must reflect contemporary realities in that society, since it is impossible to carry out a policy that is not accepted by the citizens.

Germany's high gasoline taxes were first instituted long ago when primarily the well to do owned autos, thus higher taxes were not seen as a burden by the majority. Curitiba's structural axes were a land-use and transportation strategy for a relatively low-density city that created high land value. The strategy was initiated when auto ownership was low, public transportation needs great, and prior to the traffic congestion that was anticipated with an exponential growth in population.

Change, Growth and Land Use Controls -- Every country differs in its legal and social aspects of land use controls. Private property rights are a complex interaction of laws, customs and mores. Japan builds airports on made land in the harbor and expressways over city streets to avoid taking private land. Germany has distinct boundaries for urbanized land within which all non-agricultural/forestry workers must live, thus protecting agricultural land, forests and open space. Typical United States zoning laws were developed in the 1930's without an understanding of the congestion produced by the private automobile. Every country, and every city, has its own reality that constrains potential innovation in transportation policy.

Public Policy, and the Effect of an Absence of Decisions -- An absence of public transportation policy results in favoring the automobile, and in putting all other modes at a disadvantage. By its size and speed, the private automobile seizes authority on the public right of way in all but the most rugged geography.

The proclivity of governments to pay attention to those with wealth, and the stylishness of following a western example, all mean that private transportation is more often satisfied than broader than public mobility needs. Examples can be found on every continent. It is more difficult to choreograph the public will and the financial resources to ensure an effective public transportation system than it is to let private actions take their course.

Frequent feeder buses with free transfers are integral to the success of the bus-transit network

6. LESSONS FROM CURITIBA

What happens when one takes the Curitiba transportation solution and attempts to apply it to the global transportation challenges described above?

The city of Curitiba has been able to establish a highly effective transportation network with economics that appear to be sustainable. This system appears to be more equitable, and to have fewer adverse environmental impacts than systems in other cities, especially when we look at the overall costs and benefits. This accomplishment seems to have occurred through the extraordinary leadership of enlightened, but pragmatic, politicians (who have been, and have employed, skilled and innovative planners and implementers).

The demographics of Curitiba continue to change. Development of the system continues as well. Will the transportation network be able to stay ahead of its success with continued high quality service as the population of the metropolitan area, especially that of the outlying cities, grows? Will the transportation network be able to accommodate and control the growing number of automobiles that greater affluence, and increased population, will inevitably produce? Is an increased level of safety for pedestrians and bicyclists possible? Curitiba is worth watching.

Is Curitiba worth emulating? Are the Curitiba lessons applicable?

This author believes that three aspects of the Curitiba system should receive greater consideration in more developed countries, particularly in Europe and in North America. These aspects are: a greater utilization of labor, acknowledging realistic limits on capital availability, and more effective public use of the existing public rights-of-way. These aspects may be important in developing parts of the world as well.

Employment of Labor -- The Curitiba system uses many people to do work that might be automated. This makes sense in Curitiba, but it also is important worldwide.

Unemployment remains in double digits in most European countries, and in minority and immigrant communities in North America as well. Putting more people to work can provide more efficient and more personal service, and can improve public security. These features are important to a good transit system, and they increase ridership.

Limits on Capital Availability -- Limited public capital resources, a fact of life in Curitiba, also severely constrain our ability to create comprehensive transit systems quickly in the United States. The process required in the United States to get federal support of complex and expensive new transit construction often adds many years to a project; and high-cost systems that are not extended to cover an entire metropolitan area leave many citizens underserved. Low-cost metropolitan systems, as realized in Curitiba, have the ability to more quickly and more effectively serve an entire metropolitan population--those without other means of transportation, as well as those who can choose.

Public Use of The Existing Public Right-of-Way -- Nearly all of Curitiba's transportation network, both the exclusive lanes of the bus-transit system and the parallel arterial streets, have been achieved by redefinition of the use of existing public streets. As automobile ownership grows, it has been a common occurrence throughout the world that every square meter of public space is taken over by the private car--parked on the sidewalks, and stalled in traffic on the streets.

In recent years, however, there are numerous examples around the world where the public right-of-way is being returned to public use. Italian piazzas that were filled with parked cars twenty years ago are now thriving pedestrian precincts, banned to autos. Curitiba has reclaimed a critical piece of its streets for its public transit system. This bold move can be accomplished in a matter of weeks rather than years, and can yield an extraordinary improvement in the speed and effectiveness of a surface public transportation network. Without such a move, surface systems cannot be effective.

Curitiba's effective public transportation system is the key to vibrant shopping districts

7. *CONCLUSION*

It is important for every person interested in transportation to know and understand the Curitiba example. Curitiba represents a highly successful response to a particular set of transportation problems in a specific context.

Curitiba has created its transportation system within the context of a rapidly growing metropolis with immediate needs and a low availability of capital. Its strategy has been to put an emphasis on labor rather than high technology, and to use a process of incremental improvements rather than attempt to create large new systems. This may seem merely pragmatic; however, it is a lesson that may be appropriate for many of us today.

The Curitiba bus-transit strategy might be used in many moderately sized but highly congested cities around the world. It might also be an integral part of a much more complex metropolitan network, providing efficient feeder service to, and extensions of, a rail transit system.

The "public will" to put public transport needs ahead of private demands (in this case, buses before cars) is not the normal or fallback condition. In Curitiba, a public will has flourished because of strong leadership; and it has been reinforced by the rapid realization of a low-cost, yet highly effective transportation network that serves a broad spectrum of the city's population.

Socioeconomic characteristics,
land use and travel patterns
of the Helsinki Metropolitan Area

Matti Pursula[1], Reijo Teerioja[2]

1. Socioeconomic characteristics

Population

The Helsinki Metropolitan Area consists of four cities: Helsinki, Espoo, Vantaa, and Kauniainen. Its area is 764 sq.km, the population is 920 000, and the population density is about 1 200 inh./sq.km. Table 1 indicates the distribution of the population by city at the beginning of 1998.

City	Population	Share (%)
Helsinki	539 419	58
Espoo	200 829	22
Vantaa	171 281	19
Kauniainen	8 507	1
Total	920 036	100

Table 1. Population distribution in the Helsinki Metropolitan Area by city at the beginning of 1998

----

[1]Professor, Transportation Engineering, Helsinki University of Technology, P.O.Box 2100, FIN-02015 HUT, Finland

[2]Head of Office, Transportation Planning and Research, YTV Helsinki Metropolitan Area Council, Opastinsilta 6A, FIN-00520 Helsinki, Finland

The population of the Helsinki Metropolitan Area has been growing constantly (Figure 1). There have been differences in the rates of growth. At the moment, we are living in a period of fast growth. By the year 2020 the population is estimated to reach 1,1 million.

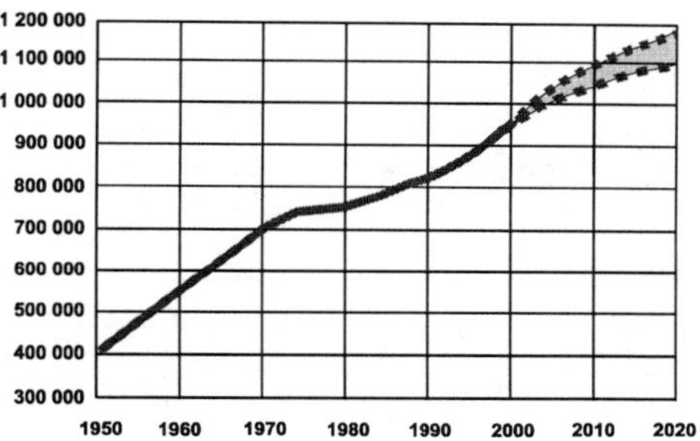

Figure 1. Population growth in the Helsinki Metropolitan Area 1950-1998, including predicted trends up to the year 2020

Helsinki is by far the largest city in terms of its population, while Espoo and Vantaa are in the same size class, and Kauniainen is the smallest.

Table 2 indicates the distribution of the population by age group as it stood at the beginning of 1996, and the table gives a prediction for the year 2020. The table shows a definite, large growth in the share of over 65-year-old population in 2020. The average household size was about 2,1 persons.

Age group	**Jan. 1, 1996**		**Jan. 1, 2020**	
	population	**%**	**population**	**%**
0-6	82 300	9.2	81 600	7.4
7-15	86 700	9.7	114 700	10.5
16-19	39 500	4.4	57 700	5.3
20-64	581 600	65.4	657 800	60.0
65+	101 000	11.3	184 400	16.8
Total	891 100	100	1 096 200	100

Table 2. Breakdown of the population of the Helsinki Metropolitan Area by age group at the beginning of 1996, and a prediction for the year 2020

The average taxable income of people living in the Helsinki Metropolitan Area was FIM 86 300 (USD 16 600) per person in 1994. The level of income per person was highest in Kauniainen, standing at FIM 124 500 (USD 23 900), and the lowest in Vantaa, where it was FIM 78 100 (USD 15 000). In Helsinki, income per person amounted to FIM 86 600 (USD 16 700), and in Espoo FIM 91 100 (USD 17 500).

Jobs

Figure 2 shows the trend in the overall number of jobs and a prediction for the year 2020.

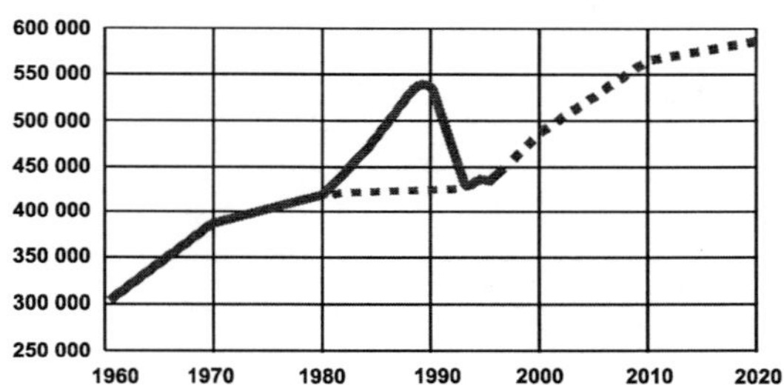

Figure 2. The trend in the overall number of jobs in the Helsinki Metropolitan Area and a prediction for the year 2020

The deep economic recession that beset Finland at the beginning of the 1990s shows up clearly in the diagram. The recession led to a considerable decrease in the number of jobs. The recession is still making itself felt, although production volumes are now reaching the level of the early 1990s. Table 3 shows the breakdown of jobs by city.

City	**Number of jobs**	**Share (%)**
Helsinki	292 800	65.9
Espoo	79 300	17.8
Vantaa	70 500	15.8
Kauniainen	2 300	0.5
Total	444 900	100

Table 3. The breakdown in the total number of jobs by city in the Helsinki Metropolitan Area in 1995

Workplaces were predominantly located in Helsinki city center. The number of jobs in Espoo and Vantaa has been rising dramatically. Table 4 shows jobs by field of industry in the entire area in accordance with international standards of classification.

City	Jobs				
	Primary production (A-B)	Processing (C-F)	Services (G-Q)	Unknown (X)	Total
Helsinki	446	42 940	249 797	4 749	297 932
Espoo	230	14 179	61 366	1 289	77 064
Vantaa	423	17 472	48 403	2 370	68 668
Kauniainen	11	191	1 895	76	2 173
Total	1 110	74 782	361 461	8 484	445 837

Table 4. Jobs by field of industry in the Helsinki Metropolitan Area at the beginning of 1996

The value added of fields of industry in the Helsinki Metropolitan Area in 1995 was totally 129 400 million FIM (25 400 million USD).

Figure 3 indicates the trend in car density in the Helsinki Metropolitan Area. Car density is expected to rise in the future as well and reach a level of 480 cars/1000 inh. in 2020.

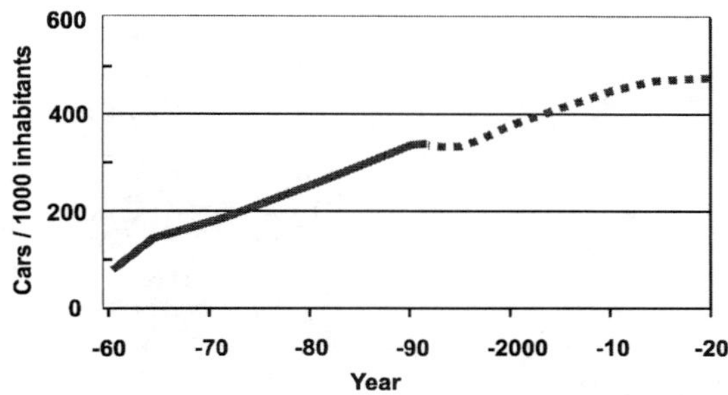

Figure 3. The trend in car density in the Helsinki Metropolitan Area and predicted trends up to the year 2020

About 60 % of households in the area has at least one car. The share of households without car is largest in the downtown of Helsinki. The number and density of cars by city at the beginning of 1997 is shown in Table 5. There are regional differences in car density. The density is smallest in Helsinki and largest in Kauniainen.

City	**Number of cars**	**Car density (cars/1000 inhabitants)**
Helsinki	165 579	307
Espoo	73 442	366
Vantaa	63 550	371
Kauniainen	3 198	376
Total	305 796	332

Table 5. Car density by city at the beginning of 1997

2. Land use

Residential structure

Figure 4 indicates the geographic residential structure of the area. The heavily populated areas are mainly located within the city limits of Helsinki: next to the Martinlaakso commuter train line, the metro (subway), and the main railway track, and in the environments of the western motorway.

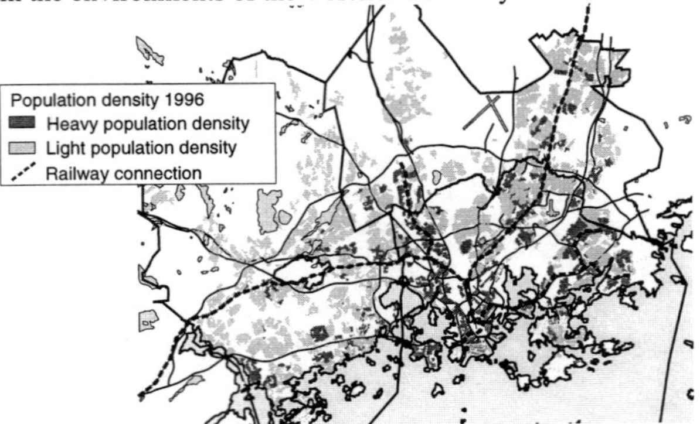

Figure 4. The geographic residential structure of the Helsinki Metropolitan Area in 1996

Jobs

The geographic distribution of jobs is presented in Figure 5. Most jobs are held within the Helsinki city limits. In addition, there are concentrations of jobs in the vicinity of western motorway, the railway routes, and the metro.

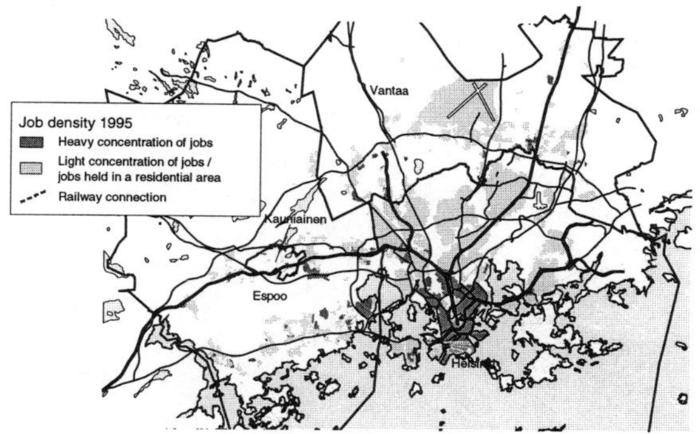

Figure 5. The geographic distribution of jobs in the Helsinki Metropolitan Area in 1995

Jobs and housing balance

Table 6 shows the theoretical jobs and housing balances for each city. The highest such ratio is in Helsinki, where it exceeds 100 % by a good margin. The theoretical jobs and housing balance is over 100 % for the entire region, which means that people are commuting to work from outside the area.

City	Resident workers	Jobs	Jobs and housing balance (100 x jobs / resident workers)
Helsinki	222 305	292 760	132
Espoo	88 500	79 304	90
Vantaa	77 558	70 479	91
Kauniainen	3 524	2 319	66
Total	391 887	444 862	114

Table 6. Job and housing balances for each city in the Helsinki Metropolitan Area at the beginning of 1997

Terminals

Figure 6 shows the locations of passenger traffic terminals of the area. The area has one airport, the Helsinki-Vantaa International Airport, which operates regular passenger flights. The two passenger harbors that are located in downtown Helsinki also handle international transportation. The central railway station is located in downtown Helsinki, as is the long-distance bus station. The figure shows the locations of the train and metro stations and bus terminals that serve the local bus routes.

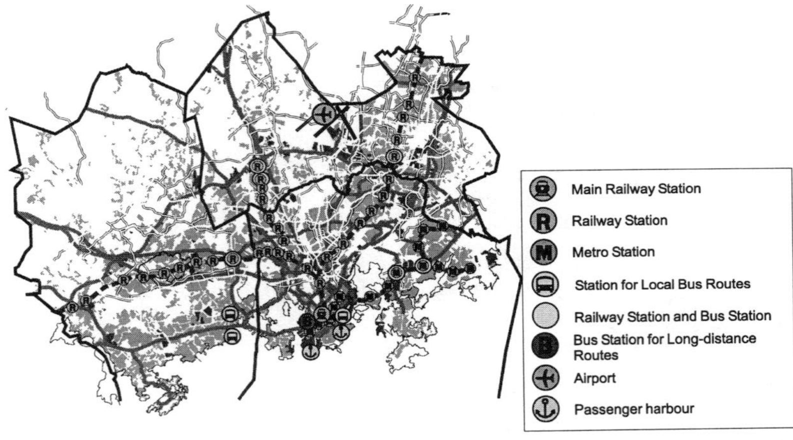

Figure 6. Major passenger traffic terminals in the Helsinki Metropolitan Area

Major transit destinations

The major passenger transit destinations are the main center of the area (downtown Helsinki) and the regional centers. The hospitals, university facilities, college-level institutions and vocational institutes that are located outside the centers also see large volumes of passenger transit. These destinations are laid out in Figure 7.

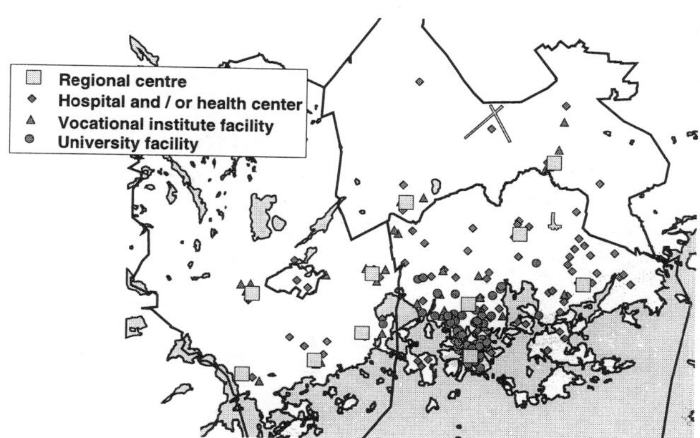

Figure 7. Major passenger transit destinations in the Helsinki Metropolitan Area

Industrial and warehouse areas

The largest industrial and warehouse areas that are located outside the centers are presented in Figure 8.

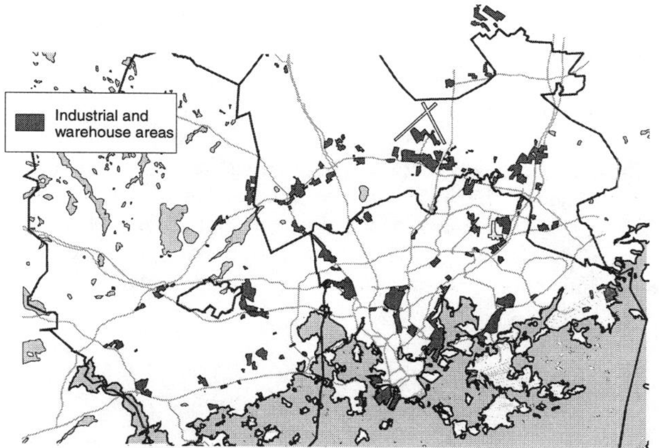

Figure 8. Industrial and warehouse areas in the Helsinki Metropolitan Area in 1997

Natural blockages occurring in the Helsinki Metropolitan Area are shown in Figure 9. The major blockages are the sea, numerous lakes, and the Vantaa River in particular.

Figure 9. Natural blockages in the Helsinki Metropolitan Area

3. Travel Patterns

Trip rates and modes of travel

The use of cars has increased by the increase of car density and population (Figure 10). The number of motorized trips in the area has more than doubled from 1966 to 1995. The number of public transportation trips has increased by 28 % and that of private car by 288 %. The correponding increase in population has been 38 %. Thus public transportation has lost market shares.

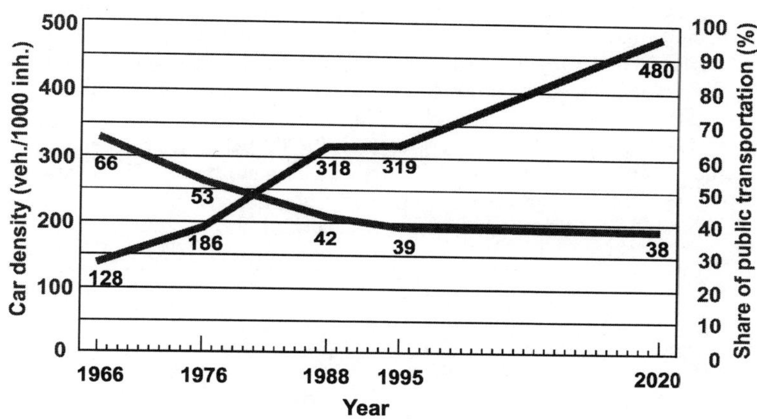

Figure 10. Trends in the share of public transportation and car ownership in the whole Helsinki Metropolitan Area

According to Figure 11 the use of public transport is highest among the inhabitants of the city center, and lowest in the cities of Espoo and Kauniainen. This is due to the socioeconomic characteristics of the people, and to the differences in land use patterns.

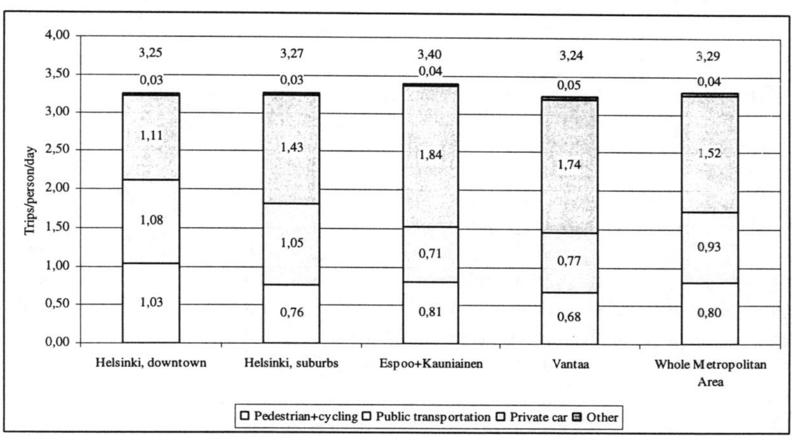

Figure 11. Daily trip rates in the Helsinki Metropolitan area in 1995

The public transportation modes available in Helsinki are bus, street car, metro, and commuter rail. The major travel corridors in the area (see Figure 6) are formed by six radial freeways and the adjacent arterials, four radial lines of heavy rail (three railway lines and one metro line) and two ring roads.

The highest traffic volumes, around 70 000 veh/day, are counted on the western motorway and on Ring I. Congestion is quite heavy also on the major arterials in the central areas of the city of Helsinki. Figure 12 shows the average travel speeds of private cars in the main road network.

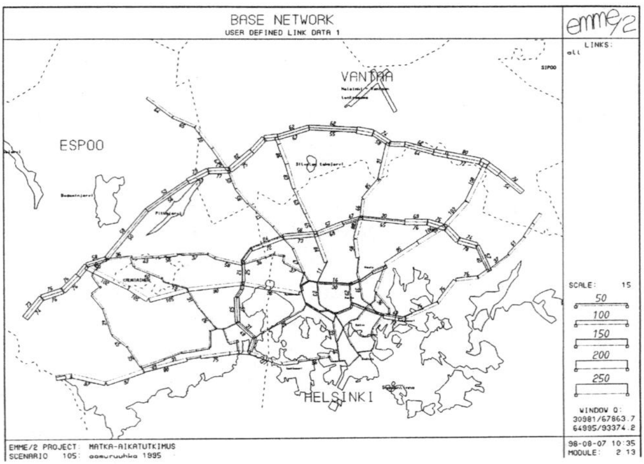

Figure 12. The travel speeds on main roads in the morning rush hour in 1995

Conclusions

The major travel problems in the Helsinki area are related to the increase in car traffic and population. Proper coordination of land use and transportation plans is essential to accommodate the great increase in population in the following 20 years. According to the Helsinki Metropolitan Area Transportation System Plan 2020 the increase in radial travel to the city center is 20-25 % but on ring road I 40 % and on ring road III 135 %. Correspondingly the share of public transportation in trips to the city center is 52 % (motorized trips only) and on transverse directions 33-36 % only.

Environmental problems caused by traffic are related to noise and emmissions. Noise barriers are under construction along the major roads, and many measures are taken to decrease the use of car and the level of emmissions. National air quality standards are usually exceeded only in winter time in certain weather conditions, and sometimes in the summer (ozone). Traffic safety in

Helsinki is at a good Nordic level, but much can be done to make the traffic safer for the pedestrians.

The promotion of public transportation has been a key element in the transportation policy in the Helsinki Metropolitan Area during the last 30 years. The high level of the usage of public transport has been achieved by strict parking policy in the city center, by prioritizing public transportation at traffic signals and by using bus and streetcar lanes and streets, and by investing in the development of rail traffic. According to the Helsinki Metropolitan Area Transportation System Plan 2020, about 92 million USD will be invested yearly in transportation system development in the area. These investments are divided about 50/50 between car traffic and public transportation. The cities also subsidize public transportation. In the city of Helsinki the subsidy is about 50 % of the operating costs including investments to rolling stock

An important part of the policy to promote public transport is also the common fare and ticket system in the area including free transfer between the vehicles of different operators in the area. The present law about public transportation gives the communities the possibility to take full responsibility about the planning and operations of the public transportation in their area. This possibility has been used by the cities and the Helsinki Metropolitan Area Council.

The lesson to learn from the experiences of Helsinki is that promoting public transportation is a never-ending task where the coordination of land use and transportation planning is of vital importance. Simultaneously, the promotion of public transportation must have high priority in the short range planning of traffic operations and control, too. In this work, the support of politicians and the general public are of utmost importance. In Helsinki, this is mainly the case.

References

Kaartokallio, M. (1997) Travel behaviour and its changes in the Helsinki Metropolitan Area (in Finnish). Helsinki Metropolitan Area Council Publication Series C 1997:7. YTV Helsinki Metropolitan Area Council. Helsinki.

YTV (1996). Travel time study 1995 (in Finnish). Helsinki Metropolitan Area Council Publication Series C 1996:5. Helsinki.

YTV (1998). Helsinki Metropolitan Area Transportation System Plan, Draft (in Finnish). Helsinki Metropolitan Area Council Publication Series C 1998:4. Helsinki.

YTV (1998). Transportation and land use database of Helsinki Metropolitan Area.

Public Transport in Helsinki

Seppo Vepsäläinen[1]
M. Sc. Engineering

Introduction

Traffic conditions are one of the factors influencing a city's appeal. In urban areas traffic is also considered a problem. Public discussion focuses on drawbacks related to the environment and safety.

An efficient transport system facilitates connections between different functions. However, urban sprawl can ensue if cohesion is lacking. The spread of cars can result in increased transport difficulties for many people.

Helsinki has chosen a transport policy which emphasizes public transport. Residents are also quite satisfied with the current level of public transport.

[1] Planning Director, Helsinki City Transport, Planning Unit, Toinen linja 7, FIN-00530 Helsinki, Finland

Customers

Public transport services are used by practically everyone who lives in or visits Helsinki. Only a few percent of residents do not use public transport, according to studies. Serving such a broad range of customers so that everyone is reasonably happy is quite a challenge.

The volume and direction of travel and which modes of transport are chosen depend largely on factors such as age, sex, social status and employment. Expectations regarding the quantity and quality of services also depend on these factors.

Young people, women and old people are the most faithful users of public transport. This is mainly due to a lack of alternatives. A small number of people have to use cars for one reason or another, at least for part of their journeys. Most people can choose between public transport and some other type of transport.

According to studies, satisfaction with public transport in Helsinki is generally quite extensive. There is a high demand for services and this demand is met in practice. Attitudes have become more and more favourable towards public transport. The increase in the number of cars is clearly more moderate in Helsinki than in other parts of the region and country. Car ownership peaked at the beginning of the 90s.

Different customer groups' expectations are systematically and regularly studied. The utilization of customer feedback has also been improved. Only by listening to customers can the use of services, which is already extensive, be further increased.

Women comprise 66% of users of public transport and men 34%.

There are also large differences in frequency of use according to age. 70% of people under the age of 20 use public transport daily. The figure falls with age, until only 30% of those over 60 are daily users.

Public transport's share of traffic

The Helsinki City Council decided in 1991 that the key goal of transport policy should be to increase public transport's share of passenger traffic. At the edge of the city centre during rush hours over 70% of traffic is handled by trains, trams or buses. This is quite high by international standards.

Public transport's share of traffic has not improved in the 90s, although passenger volumes have risen. There are two basic structural reasons for this.

Passenger traffic has continued growing throughout the 90s in spite of the economic recession. Growth has mainly involved traffic inside and between suburban areas. Traffic inside or to the inner city has not increased. Public transport's share of traffic inside and between suburban areas is clearly lower than its share of traffic inside and to the inner city.

The number of trips to work has fallen as a result of large-scale unemployment. Public transport is used much more frequently for this type of trip than for other travel. Growth has taken place in other categories, however.

Public transport's share of traffic to and from the inner city has remained at the same level throughout the 90s. Comparable annual data on traffic at other points besides the limits of the centre is not available.

Improving public transport's share of traffic is an important goal. There are considerable differences in the relative number of cars in different parts of the city. In the inner city about a third of households own a car. In some suburbs, on the other hand, over two-thirds of households are car owners. The level of public transport has clear significance in this regard. In the neighbouring cities of Espoo and Vantaa, car ownership is considerably higher than in Helsinki.

The latest studies show that improving the information available to passengers and residents leads to increased use of public transport services. There are many other means to increase public transport's share. The most important thing is to increase the share of traffic inside and between suburban areas.

Journeys

The people of Helsinki average about one journey a day and about 320 journeys a year per person using public transport. Around a third of these journeys include transfers.

Public transport volumes in Helsinki's internal traffic peaked in 1986 and then began to decline. This decline was due to intense growth in the number of private cars and car traffic in Helsinki and its neighbouring cities.

The volume of public transport again turned upwards in 1991 and has continued to grow ever since. This growth has been influenced by economic conditions and low ticket prices, the expansion of the Helsinki metro, improvements in quality and changes in parking policy.

Nowadays around 690,000 journeys are made on Helsinki's internal lines on the average weekday. (Here the term journey refers to each time a passenger boards a vehicle.) People who live in neighbouring cities account for nearly 7% of passengers.

Bus traffic has lost passengers in the past 10 years, while rail traffic has increased. The strongest growth has been in metro traffic. Rail traffic already accounts for 49% of the passenger volume in internal traffic. This includes commuter services provided by the railways.

Connections

Lines and the network form the service system for connections between areas. In Helsinki, as in most cities, lines mainly run from different areas to the city centre. In the inner city trams are the primary mode of trans-port. Cross-town lines supplement the network. The large capacity of the metro and commuter trains is put to good use with the help of feeder lines. The structure of the line network changes slowly.

In Helsinki the tram services are part of the traffic in the inner city. The service area and traffic volume have remained unchanged for a long time. The passenger numbers have increased during all the 90's amounting to some 55 million trips in 1997. The tram stock is being renewed. The trams, which will hold 121 passengers, were mainly bought in the 70's and 80's. The tramcars cannot be coupled together. For the time being, new low-floor trams are being purchased.

The metro services are mainly part of the suburban traffic. There are 6 stations in the inner city and 10 in the suburban area. The first metroline had six stations and was opened for traffic in 1982 whereupon the services have continued to expand slowly. In the 90's the use of metro has grown strongly: by 5-7%. In 1997 about 42 million trips were made by metro. The number of metrotrains will be increased. The present rolling stock bought in 1982 is functioning very well and running more than 120,000 kilometres a train annually.

Bus services' share of all public transport has decreased slowly while the rail network has been extended. So far every second of the trips within the city of Helsinki are made by bus and so are most of the trips in the whole Helsinki metropolitan area. The major part of bus traffic consists of transport between the city center and suburban areas. Also metro feeder and orbital routes are maintained by buses. The number of low-floor buses is growing rapidly. Some 15% of the services are run by articulated and bogie buses.

Local train services are maintained in three directions from the city center. The capacity of the network has increased strongly in the 90's. Annually about 25 million passangers go by train in the whole Helsinki metropolitan area. Low-floor trains are being purchased.

The development of supply focused on increasing metro traffic in the 80s and 90s. In 1990-95 bus traffic was reduced, but since then the supply of services has grown along with demand for every mode of transport. Bus traffic still accounts for around half (51%) of total seats, the metro 36% and trams 13%.

Service periods vary according to the line, but public transport generally operates from 5am to 2am on weekdays. There are also later night services during the weekend.

During rush hours the volume of traffic is almost twice as high as in the middle of the day. On weekdays the supply in the middle of the day is about the same as on Saturdays. On Sundays the supply is slightly less than half as large as during rush hours. The biggest fluctuations are in bus and metro traffic, while the supply in tram traffic is steadier.

Monthly fluctuations are visible in summer traffic, when some bus lines take a break and services are not as frequent during rush hours. During off hours the supply is nearly the same as in the winter. Metro traffic is just as frequent in the summer, but trains are shorter. In tram traffic reductions in summer traffic are smaller.

During rush hours bus traffic in July is about 80% of the figure in the winter. In total traffic all the reductions made during the summer correspond to about a week's traffic in the winter.

Purchases of trams, metro trains and commuter trains will allow the development of rail traffic during the next 10 years. The bus fleet has also developed at a rapid pace. Buses with low floors now account for 25% of services during rush hours and 50% of services at other times. This figure is rising rapidly and in a few years all buses will have low floors.

Stations, stops, terminals, parking at feeder stations

Helsinki has 13 metro stations and 15 railway stations. Metro stations include tunnel and surface stations. They have central platforms. There are two feeder terminals for metro traffic and two for other rail traffic. Large bus terminals are typical of the city. Most stops are equipped with shelters. The city has about 2,000 parking spaces at feeder stations.

Stations, stops and terminals form user interfaces for public transport services. It is very important for these interfaces to be not only efficient, but also userfriendly. In Helsinki the weakest length in the public transport chain is often stops.

Stations act as magnets and are landmarks in the city scene. Residents visualize the city with the help of such landmarks. For visitors in a strange urban environment they are even more important orientation points. Stations are just as important as squares, parks or markets.

Stations, stops and terminals are also major locations for travel information, where passengers can find out what they need to know during journeys. Adequate information is especially important at places where people make transfers.

Helsinki's largest bus terminals together with the number of departures (buses/day) and users (boarding passengers/day) in Helsinki's internal and regional traffic:

	departures	**users**
Kamppi	2600	62700
Itäkeskus	1100	41500
Railway Station Square	1550	39000

Helsinki has 240 tram stops and about 1700 bus stops. These stops are equipped with 1300 shelters, including 600 new glass shelters.

Service quality

Public transport's share of traffic can only be high if service is sufficiently appealing in the opinion of customers. Fast connections, short walking distances and intervals between services, easy transfers, adequate service periods and passenger comfort on vehicles and at stops are part of the equation. Friendly and skilled personnel are perhaps the most important corner- stone of appealing service.

The quality of public transport is being developed according to passengers' needs and expectations. The level of quality is evaluated with the help of customer surveys, in which passengers are asked to state their opinions regarding different service factors. Helsinki City Transport's quality programmes focus development measures so as to correct deficiencies.

Passenger feedback is collected in many ways. In addition to customer surveys, Helsinki City Transport has a feedback phone line and register. All feedback is taken into consideration. The quality of public transport can be divided into system quality and service quality.

The City Board has approved planning guidelines for system quality. These guidelines cover the following factors:
- the accessibility of stops and stations (walking distances)
- service periods

- line connectivity (transfers)
- passenger loads
- service intervals

Service quality includes the driver's driving and service skills, the condition of vehicles and passenger information.

Journey speed is also a key quality factor. In crowded traffic conditions the flexibility of public transport is ensured with the help of separate lanes and coordinated traffic lights. In this way public transport delays during rush hours have been kept reasonable.

Finance

Like many other cities, Helsinki has for decades supported the use of public transport with the help of economically priced tickets. About half of expenses are covered by ticket revenues, while the other half comes from taxes. The development of expenses has been favourable for some time. Thanks to efficiency meas-ures the level of costs has been reduced, avoiding the need for tariff increases in recent years.

Personnel costs account for about half of the expenses of public transport. These are covered by ticket revenues. The cost structure varies considerably according to the mode of transport. The metro is the most capital-intensive and buses are the most labour-intensive in this respect.

Expenses have developed favourably, thanks to cuts in traffic costs in bus traffic. Expenses apart from capital costs have also been reduced at a similar pace in rail traffic.

Ticket revenues have meanwhile risen as a result of increased passenger volumes. The relative prices of different types of tickets have remained roughly the same throughout the 90s. The price level has also remained stable.

Various ticket trials have been conducted. A reduced-rate tram ticket with no transfer has been a success. Cooperation is arising with the other cities in the Helsinki area, particularly along railway lines.

Infrastructure investments and maintenance (tracks, stations, stops and terminals together with other public transport costs) account for about a fourth of the public transport budget.

Slowness costs. If we compare unit costs for metro, bus and tram traffic, the differences are due mainly to differences in speed. Throughout the 90s investments have been made in speeding up traffic in order to improve cost development and service for public transport operating in the street network.

Organization

Flat fare is applied to public transport services in the four municipalities of the Helsinki metropolitan area. For the time being an enlargement of the ticket cooperation to a wider area is tested and planned. The major part of the regional services are maintained within the city of Helsinki and Helsinki City Transport is responsible for the planning, ordering and competition as well as mostly for the production of these services. Helsinki Metropolitan Area Council is in charge of the cooperation of the four municipalities and also of the planning, ordering and competition of regional services.

Metro and tram services are maintained by Helsinki City Transport, which is a business concern owned by the city of Helsinki. Today the Helsinki internal bus services are run by 8 companies, however, Helsinki City Transport is maintaining more than half of the production. The competition may change the division of labor in the next years. Two of the bus companies are international large-scale enterprises. The local railway services are maintained by the State Railways (VR) owned by the government. Helsinki Metropolitan Area Council makes the orders for train services.

Regional cooperation

Helsinki stretches beyond its own borders. Many of its suburbs are located in neighbouring cities. Traffic disregards borders. In order to maintain the attractiveness and economy of public transport, the four main cities in the Helsinki area have cooperated on tariffs and planning for the past ten years.

The people of Helsinki make part of their journeys on regional lines, using buses or commuter trains. The people of Espoo, Vantaa and Kauniainen likewise use Helsinki's internal lines on their journeys.

Helsinki residents' share of the use of regional services has grown throughout the ten years during which tariff cooperation has taken place. Most regional journeys made by the city's residents are on commuter trains, either inside the city or across its borders.

The annual costs of regional traffic are about FIM 448 million. Ticket revenues total about FIM 306 million. The difference is covered by subsidies from the municipalities. Helsinki's subsidy amounted to approximately FIM 28 million in 1996.

Rolling stock

In Helsinki rail traffic accounts for half of journeys and buses the other half. The metro went into operation in 1982, using new Finnish trains. Tram traffic is mainly handled by articulated trams purchased in the 70s and 80s. These were also built in Finland. Buses were purchased in the late 80s or are low-floor models purchased in the 90s.

Buses

Helsinki presently has 552 buses in internal traffic. This includes 102 articulated buses. Buses with low floors total 140 or 25% of the fleet. Two of these are articulated and four are bogie types. The fleet includes three gas buses.

The measurements of a typical low-floor bus are:

	2-axle	**bogie**	**articulated**
length	12m	14,5m	18m
width	2,5/2,6m	2,5/2,6m	2,5/2,6m
seats	36	55	49
standing room	34	39	61

The percentage of buses with low floors is rising rapidly. All new buses are of this type.

Rail vehicles

Helsinki has 105 trams, including 82 articulated trams and 23 old 4-axle trams. The metro fleet consists of 42 two-car trains. Measurements are as follows:

	4-axle tram	**articulated tram**	**metro train**
length	13,3m	20m	44,2m
width	2,3m	2,3m	3,2m
seats	29	40	130
standing room	59	81	270

Rolling stock is being renewed.

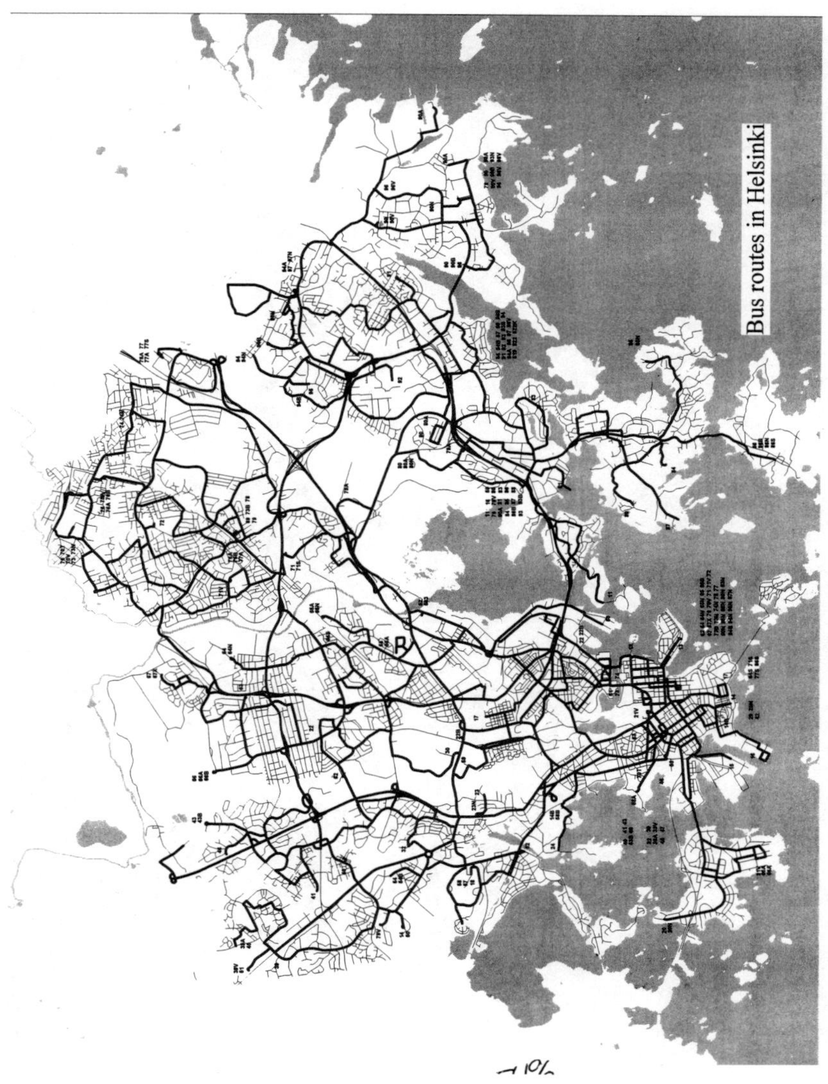
Bus routes in Helsinki

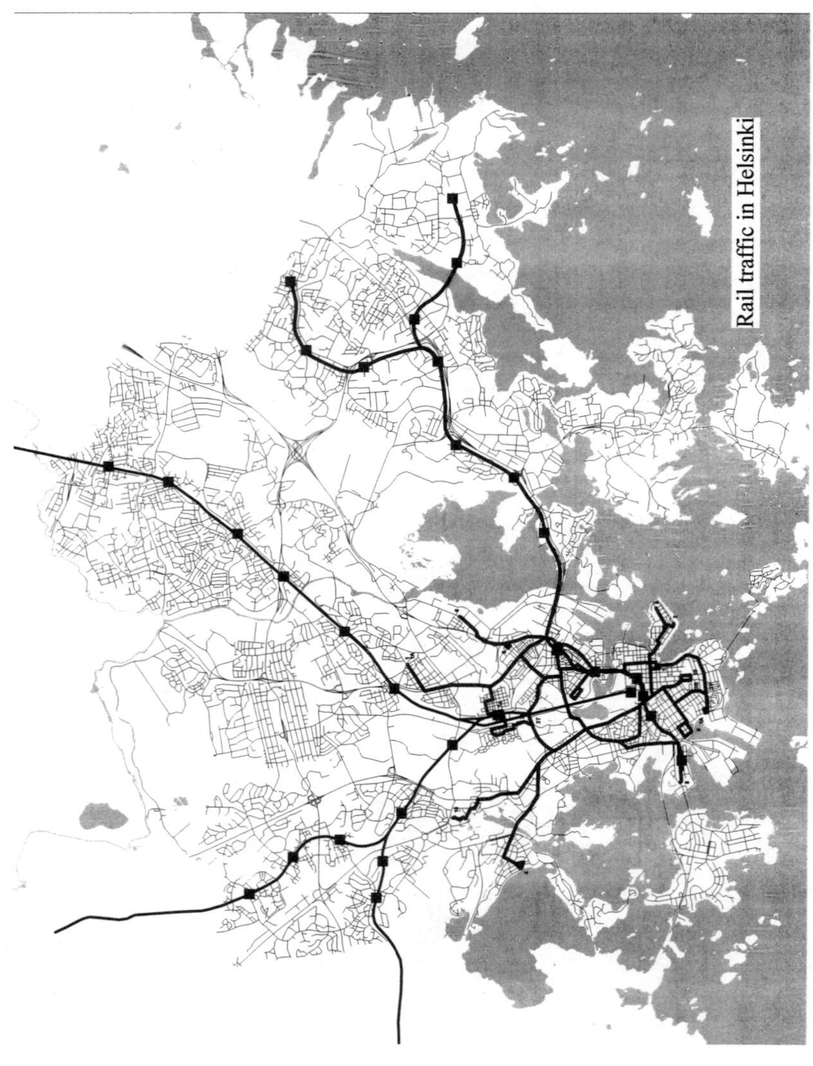
Rail traffic in Helsinki

TRAVEL CARDS IN THE YEAR 2000

The Helsinki region ushers in the smart card era

Frank Burmansson[1]

Abstract

Once the travel card system for the Helsinki region has been fully implemented, it will be one of the most advanced ticket sales and fare collection systems around. Passengers will carry remote contactless travel cards. They will be able to load the cards with travel rights for a certain period (a season ticket), or with stored cash value that can be used to pay for trips (a stored value card). The card may also combine both attributes.

When passengers use these cards as season tickets, they will flash their cards at a remote reader when boarding a vehicle. If a passenger's card is valid, the reader beeps and a green light comes on to show that his card has been approved. Take another case: here a passenger wants to use a stored value card to pay for her fare. She presents the card to the remote reader and presses a button to indicate which kind of ticket she wants. The card then registers the purchase.

The data transmission, back-end and front-end systems will be faced with a formidable challenge, as the system as a whole will process some 2 million transactions each day.

The system that will be introduced is only a basic version. It can still be enhanced in a variety of ways to produce results and to match up to the expectations people have of it.

---

[1] Head of Travel Card Project, YTV Helsinki Metropolitan Area Council, Travel Card Project, P.O. Box 521, FIN - 00521 Helsinki Finland

Travel cards in the year 2000

The Helsinki region ushers in the age of smart cards

1. The system's clients

The new travel card system was commissioned by the Helsinki Metropolitan Area Council (YTV), Helsinki City Transport (HKL), and VR Ltd. (Finnish Railways).

YTV purchases bus transportation services linking the different cities within the Helsinki Metropolitan Area. HKL is the primary provider of internal bus, tram, and subway transit in Helsinki, the capital of Finland, while VR Ltd. supplies commuter train services in the Helsinki Metropolitan Area and its environs.

The region in which the new system will be launched contains 16 independent cities or municipalities populated by total of 1,300,000 people. Some 700,000 trips to work are made each day.

2. The objectives of the travel card system, considered from the points of view of different user groups

2.1 The passenger's perspective

The new system enables a passenger to use one card to travel on any vehicle, and to do so in the entire region in which the system has been implemented.

He or she may freely select the days on which the card will be valid, and purchase passage for these days in advance.

2.2 The driver's perspective

The card makes drivers' lives easier by offering a simple and easy way to sell single tickets. As drivers will no longer need to carry cardboard tickets, their fare pouches will become lighter. It will be simpler and faster to process fare payments when passengers do not need to buy tickets from drivers.

When the travel card and its stored cash value comes into public use, a part of the fare collection work that drivers now have to take care of will be handled by remote readers. The pricing of single tickets will be used to support this development: it will be cheaper to purchase a single ticket from a remote reader using the stored cash value of a travel card than to buy the same ticket from a driver.

Driver security will be improved, as they will not have to carry as much money as before. When the new system is implemented, passengers will most likely purchase less tickets from drivers than they do today. At the end of the day, the driver will also be able to deposit all the money he has collected during the day, as fare deposit machines will be installed in the depots.

Drivers need not keep an eye on the fare collection units' sale times and pricing scales: the system takes care of this automatically.

2.3 From the clients' and society's perspective

For the clients who commissioned the system, it offers a fast and virtually flawless method of collecting transit information and income. At the same time, the system makes it possible to direct ticket revenues to the right parties, and separate surveys on ticket use will not be required anymore.

The image of public transit will also be improved when it deploys a flexible and modern fare collection system allowing it to offer new and cheaper products to passengers. It will be possible to have fare types that are valid only on the times of day or weekdays on which the fleet has capacity, and new passengers will not cause additional costs.

Demand during rush hours can be controlled by means of the pricing of single tickets. The new equipment will enable public transit to exercise an influence over which modes of transportation passengers select, as different types of transit can be priced in different ways.

The system will also produce comprehensive data on passenger volumes and statistics on transit times. Use of this data will make planning more efficient, and public transit will become cheaper and more reliable.

At some later date, the collected data can be employed to set up quality benchmarks for public transit.

3. System tests

During 1992 - 1992, YTV tested different types of travel cards. The results were positive. The design of the system got underway in 1995.

The first Finnish travel card based on smart card technology was implemented in the city of Oulu, which has somewhat under 100,000 inhabitants. After this, travel cards have been introduced all over Finland. Once the present project is seen to completion, the Helsinki Metropolitan Area will have such cards as well.

4 The travel card system

4.1 The suppliers of the system, the schedule, and the procurement cost

The procurement agreements were signed at the end of 1996. The procurement itself consists of two parts: the back-end and the front-end. The front-end system will be delivered by Wang Global Oy. Buscom Oy and Western Systems Oy are acting as subcontractors. The front-end system comprises the sale and loading of single tickets and remote cards at service outlets, newspaper stands, and vehicles, as well as their use on buses, trams, the subway, and trains. The procurement included all the equipment needed on board vehicles and at points of sale as well as the required computers, programs, and cards.

All installation and commissioning work falls outside the scope of the agreement.

The back-end system will ensure hitch-free use of remote cards for passengers. It transfers data on fare types and prices to the card readers, points of sale, and automatic vending machines. It handles closed card lists and other card information. The back-end system will also process the data collected by the front-end system. The back-end system will be delivered by Andersen Consulting Oy.

The automatic vending machines and their software will be supplied by Kauppatalo Hansel Oy, with Lancro Oy and Dildata Oy as its subcontractors. A datacommunications system will act as the link between the front- and back-end systems. The required user services will be produced by Tieto Corporation.

The system will be set up in two stages. During the first stage, single ticket sales systems will be installed on trams and buses, and these will be equipped with the data processing software required. The first stage can be fully launched on March 31, 1998.

The second stage comprises the sales systems, complete with sales and loading points for the remote card, remote readers for the vehicles and stations, and the distribution of the remote cards themselves. The second stage could be launched on October 31, 1998. According to the agreement, the distribution of travel cards to public transit passengers should begin on that date.

The procurement price, as specified in the agreement, should be $25 million.

Travel card system

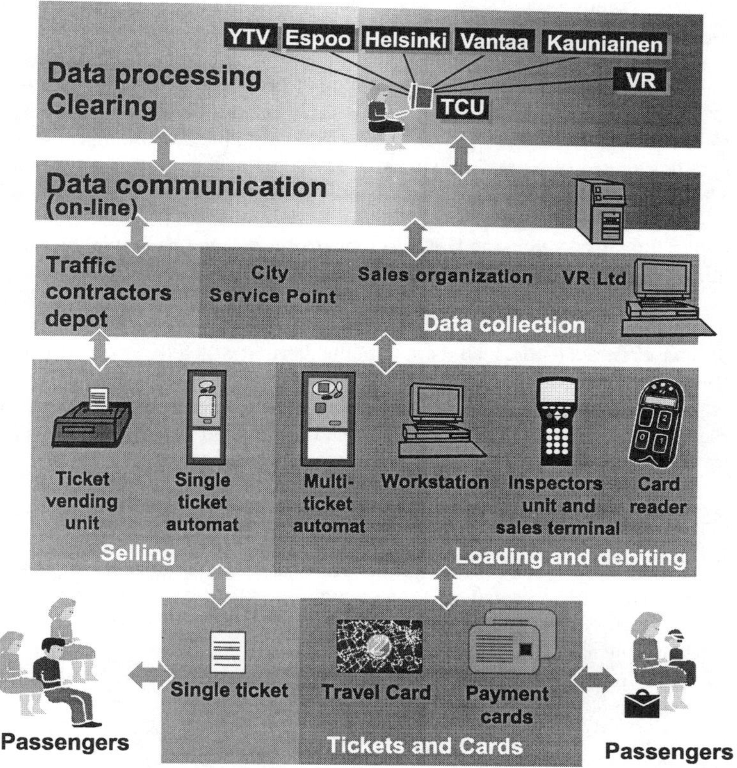

Figure 1. Travel card system

4.2 *The scope of the system*

The travel card system will include:

- over 300,000 reloadable remote cards for passengers
- 2500 remote card readers for vehicles and stations
- 1500 single ticket sale devices for drivers
- 100 inspection devices
- 30 depot machines for bus and tram depots
- over 100 sales machines for loading tickets and purchasing single tickets
- over 200 point of sale workstations for loading and selling remote cards.

The travel card system will cover some 1350 buses, 25 depots, and about 2500 bus drivers. There are about 100 trams at the three depots, and they are manned by some 300 drivers. The system encompasses 150 local train units, 56 railway stations, and some 300 conductors. There are 19 subway stations. Equipment is required at the three ferry terminals.

All in all, over 4000 people will be responsible for the operation of the system. They must be trained to serve in different capacities under the new system, which will be introduced gradually. During the transition period, which is rather long at two years, they must use both the new and existing systems. The deliveries as ordered include training packages. The installation of the system and the training of personnel will be carried out at the same time as the public transit system in the Helsinki Metropolitan Area serves over half a million passengers a day.

Passengers will also have to be taught how to use the new system, purchase remote cards, load them with the days on which they will be valid (their periods of validity), and to show them in vehicles and to the remote readers at stations. They will be able to purchase their trips with the cash value loaded into their cards. It will take from six to nine months to distribute the travel cards.

4.3 The system's travel cards and tickets

Passengers can purchase either personal travel cards or cardholder tickets. The same personal card can include a season ticket valid for a certain period (14 to 365 days) and/or loaded cash value. A person traveling with a loaded value card can purchase a single ticket, pay the fare of someone traveling with him, or pay the additional fare that in certain situations is required of people who have season tickets.

Two periods of validity will fit on a single card: the period that is currently valid and a new period. Passengers may purchase the new period when it suits them best.

A family can purchase user-specific cards (cardholder tickets) for the adults in the family and for the kids alike. All employees of a company can use the company's own cardholder tickets to travel.

A travel card does not expire. The only restriction is the card's duty life, which is estimated to be from 3 to 4 years.

In addition to the travel cards, the system includes the single tickets which will be printed out on paper by the drivers' sales devices and the automatic vending machines.

The system controllers will hold control cards. These include "main user cards", which can be used to make driver, repairman, management, and salesperson cards. Drivers will use control cards on board their vehicles to open and close the driver's sales device. The card will be used to transfer fare sales data: this data is read from the card at the fare deposit point, and the driver hands over the corresponding amount of Finnish marks to the recipient of the deposit.

The salesperson cards will work in the same way as the driver cards.

Repairmen will need these cards when replacing equipment and ensuring that they function.

4.4 The vehicle equipment and its functions

The remote readers on board vehicles will have four buttons numbered 0, 1, and 2, with one button reserved for future needs.

If a passenger wants to buy a single ticket with a loaded value card, and will not transfer to another vehicle, he places the travel card on the remote reader and presses "0".

If a passenger wishes to transfer to another vehicle within a single municipal zone, she places the travel card next to the remote reader and presses "1".

A passenger traveling in numerous municipal zones presses "2".

The system positions itself automatically on the basis of the mileage indicator reading and the data on distances between stops which is contained in the vehicle's CPU. This system automatically takes care of zone crossings and fare changes. The driver will correct any possible errors in positioning.

Drivers will sell single tickets. The vehicle system starts up when the driver inserts her control card in the sales device. When she is done with her shift, she once again inserts the card into the sales device. Data on the fares sold during the shift is then transferred onto the card. The card has enough capacity for 50 periods.

Drivers, depending on the deposit instructions given to them, will deposit all the fares they have sold once a week, for example, and the money is transferred from their accounts to the transit operator's account.

A single driver's sales information is transferred to the depot system via the deposit point, and from there to the back-end system.

When a vehicle arrives at a depot, its CPU contains sales information which has been transferred from the fare collection equipment and remote readers. The depot system contacts the vehicle and transfers the data from the vehicle CPU to the depot system. Correspondingly, the depot system updates the vehicle's system, such as by uploading current closed card lists.

The back-end system compares the sales information from the vehicle system and the deposit point to ensure that there are no errors in the sales information.

The back-end system collects data from the depot systems once a day and then begins to process them, transferring the processed and updated data to the depot systems by the following morning.

Money handling, the transfer of sales data, and the management of card stocks between different points of sale and the back-end system will work in a similar way and at the same pace.

4.5 Distributing money between the parties involved in the system

The money is collected and then distributed to each client party with the aid of the information collected from the system. YTV's Travel Card Unit will perform summary batch runs of the daily sales information twice each month, and will send sales and money distribution reports to the municipalities, ticket vendors belonging to the system, and to the transit operators.

Vendors will deposit money into the account of one of the client parties on the basis of the sales report. This client will keep the share indicated in the money distribution report and deposit the moneys owed to other parties into YTV's account. YTV in turn will retain its share and distribute the rest of the moneys to the other cities. The distribution of money must have been completed by the tenth of each month, when the cities shall pay transit operators for the transit services ordered.

Somewhat under $200 million will be processed by the system during one year.

Financial statements will be drawn up once a year, complete with review and summary batch runs.

5 A situation report on the project

Setting up a travel card system has proved to be considerably more difficult than was initially thought. The estimate of the amount of actual work required both on the clients' and suppliers' side has been too small. It has not been possible to find additional, professional personnel to work on the project quickly enough, and as a result the project has fallen behind its original schedule.

The three client parties' differences of opinion during the system's specification phase have also significantly slowed down progress.

The fact that no single supplier can take total responsibility should have led YTV, as the party ordering the system, to make major outlays on coordination. However, YTV had not been prepared for this, and the personnel resources required were not available.

In the summer of 1997, an external company audited the project and proposed changes and revisions to the control, administration, and organization of the project. Changes were made, but personnel resources, cooperation between the ordering parties, and the coordination of the suppliers' functions remained wanting.

The schedule of the project was not revised after the audit.

New and considerable problems in data communications cohesion surfaced in the spring of 1998. Another external audit was commissioned, and the auditor's report stated that the project would run into major problems if this were allowed to continue.

Following this report, the project was reorganized and its management group and project manager were changed, while the new management group was given a greater power of decision and influence. In accordance with the new draft schedule, the first stage of the implementation will be get the go-ahead in the spring of 1999. The distribution of customer cards will most likely begin at the start of the year 2000, and the travel cards will have been distributed and put into operation before the end of the year.

6. *What should have been done differently*

The project is about a year behind schedule at the time of writing. The question is, what should have been done differently so that this situation would not have arisen? The main factors to be considered are the following:

* The owners of the coming system should have specified their objectives, and formulated a schedule of objectives, and stage breakdown in greater detail.

* A single representative should have been selected from the clients' ranks to act as the system's owner and to represent the clients as a whole, and this person should have been given sufficient discretionary powers.

* The clients should have had a sufficiently large project organization as early as during the specification phase. Its members should have included people who will use the system and were at that time performing tasks similar to those which the system will take over. This organization should have come up with a functional description of the system and, insofar as possible, carry out the specification work in detail, complete with screen displays. This work should have been completed for the most part before the invitations to bid were submitted. At the same time, the organization should have gained a better understanding of the amount of work that setting up the system calls for.

* As the implementation phase got under way, the clients should have used a project organization that had enough information on the new system as well as the required personnel resources. They should have had an adequate knowledge of information technology, transit operators, financial administration, and customer service, along with knowledge and practical experience of sales systems. In this way, the specification review phase following the submission of the order would have been completed in the agreed time.

* One supplier should have set up the card system, taking complete responibility for this work. The other suppliers should have cted as subcontractors. A responsible supplier would have set a schedule for the system set-up process as well as coordinated and tested it. The original order should have included all those components - such as systems management, databases containing detailed information, and database searches when looking into error situations - and total packages which should have been ready when the pilot work on the first stage got underway.

* The implementation of a system as large as this one always requires a pilot phase, in which all the components of the system are used in field conditions. Pilot phases have been added to implementation stages 1 and 2. They also stretch out the new schedule.

Socio-Economic Characteristics, Land Use and Travel Patterns

Development Coordination of New Towns in Hong Kong

S C Lo[1]

Introduction

The total land area under the Hong Kong Special Administrative Region (HKSAR) is small, comprising only 115,000 hectares. The surrounding territorial waters cover 180,000 hectares. Hong Kong's mainland portion consists of the urban area of Kowloon and a portion of the New Territories. Hong Kong Island, across Victoria Harbour from Kowloon and about 19 km east of Lantau Island, houses the seat of government and the chief business district, known as Central.

Despite Hong Kong's small size, the topography is varied and rugged because it is largely comprised of folded mountains. Less than 15 percent of the land is developed because of the rugged terrain. Land reclamation schemes began in the mid-19th century and they continue to be important means of acquiring new land for urban development.

Hong Kong has an extensive network of roads in the New Territories, in Kowloon, and on Hong Kong Island. Road-base public transport includes franchised buses, 16-seat mini-buses and taxis. The road network is supplemented by the Kowloon-Canton Railway (KCR) and the Mass Transit Railway (MTR).

The KCR serves the central Kowloon and new towns in the eastern New Territories. It also provides an important facility for people travelling between Hong Kong and the Mainland.

---

[1] Deputy Project Manager, Hong Kong Island and Islands Officer, Territory Development Department, Hong Kong SAR Government, Hong Kong

There are three MTR lines operating between Hong Kong Island, Kowloon and Tsuen Wan. Two latest lines serve respectively the Tung Chung New Town and the New Chek Lap Kok International Airport. The MTR serves more than 2.5 million passengers daily.

A 13-km long electric tram line operates on Hong Kong Island. A modern Light Rail Transit system is implemented in Tuen Mun New Town connecting it to the Yuen Long New Town.

Ferries shuttle between Kowloon, Hong Kong Island, and all other major outlying islands. Ferry services now form an important transportation mode to link Hong Kong with the estuary and coastal areas in Mainland.

Hong Kong's most ambitious transportation project, Chek Lap Kok International Airport to replace the previous Kai Tak Airport at Kowloon City, was constructed on an inlet off Lantau Island. The entire project costs more than $US20 billion. Highway and rail projects form the core transport projects of the new airport. The new airport is expected to accommodate as many as 87 million passengers annually by the year 2040.

The General Historical Background of New Town Development

The Hong Kong New Towns Programme commenced in 1973 with the objective of providing housing for 1.8 million people with proper community facilities. The sites of the early new towns were in the rural hinterland. Unlike the urban areas, there was greater flexibility in implementing massive development to reasonable planning standards.

The continuing economic growth in the past decades enabled better living environmental standards and the Programme was expanded to benefit more people. The number of new towns has increased from an initial count of three to nine in 1999, and the design population has also doubled. To date, nearly 2.8 million people are living in the new towns representing some 42% of the total population. It is anticipated that by the turn of the Century, more than half of the Hong Kong population will be accommodated in new towns.

The development of a new town is not only concerned with the provision of living quarters for the people, but also providing them with adequate supporting facilities such as roads and drains, public transport facilities, shops and markets, schools, clinics and hospitals, police and fire stations, etc. The Territory Development Department, being responsible for new town development, believes that a balanced development of population and facilities within the town at any time does not only provide comfortable living places for the residents but would also minimize social costs.

This paper attempts to describe the various aspects and changes in the emphasis in the coordination role for New Town Development in Hong Kong. Some important and interesting features will be highlighted.

The Territory Development Department

To achieve the target set by the New Towns Programme, the New Territories Development Department was set up in 1973 and charged with the responsibility of building the new towns. It was re-titled as the Territory Development Department (TDD) in 1986, when its responsibility embraced a number of regional Development Offices, each comprising multi-disciplinary teams of engineers, town planners, architects, and landscape architects under the overall management of the Project Manager.

In around 1990, several new towns had been substantially completed and the remaining new towns were well advanced in the construction stage. With the creation of the Planning Department, the development team was re-organized with mainly engineers remaining in the Department. They are responsible for the development coordination and consultants management. Development emphasis tended to shift towards reclamation, Tung Chung and Tai Ho New Town development in Lantau, and urban renewal.

The Development Process

The process of development commences with planning and engineering studies. They enable the drawing up of development plans and programmes. The implementation of the plans in accordance with the programme starts with the acquisition and clearance of land. Site formation works then follow with land becoming available for the building of public and private housing, factories, commercial undertakings and community facilities. Provision of roads and drains as well as other infrastructure proceed in parallel. The aim is to provide, as far as is practicable, a suitable mix of public and private housing, job opportunities and community facilities both at the end of each stage of development and on completion of the total programme.

Planning and engineering studies are preceded by a Preliminary Project Feasibility Study report that ensures that the development proposal is well defined and viable. In the study report, the Environmental Protection Department has to confirm that the project is environmentally acceptable with or without mitigation measures. The Lands Department has to confirm availability of staff to meet the land acquisition and clearance requirements. Other supporting departments have also to indicate any additional resources requirements for the project. The report also contains the estimated recurrent costs. The Secretary for Planning, Environment and Lands, as head of the policy bureau for land formation, will approve the report whereby the project will be in Category C of the Public Works Programme (PWP). The project can then compete for funds in the Resources Allocation System. Successful candidates are upgraded to Category B of the PWP. Funds will only be available for commencement of works after the project has been upgraded to Category A. In the year 1998/99 the Government of the HKSAR estimates spending $US350 million on new town development.

The "Package" Approach

In order to achieve a balanced development at each and every stage of a new town development, a "package" approach is adopted. With this approach, the development programme for a new town is derived by grouping works into a series of "packages". These packages, as far as is practicable, include all those works required to produce a balanced development of housing, industry, and community facilities. In certain instances, it has not been possible to achieve this concept in a particular package but it remains a primary objective to provide a balance, in relation to population build-up, on an overall town basis by ensuring that the various types of land use areas are developed concurrently.

Resources for works in packages are contributed from different policy areas. For example, funds for the site formation works come from the Planning, Lands and Environment policy area; those for the construction of a school come from the Education policy area, while the Secretary for Security has to provide funds for the construction of the police and the fire stations. It is vital to have the consensus of the heads of these policy areas on the development proposal and their commitment to providing resources for the packages.

Departmental Coordination

The overall coordination among government departments, utilities companies, and public agencies on the development progress is overseen by the TDD. The Department maintains a Development Programme for each new town to monitor the whole process from planning to completion. Before the Programme is publicly announced, internal consultation is carried out to solicit support and confirm the availability of resources for the Programmes. This will ensure that all the essential infrastructure and facilities are in place at the intake of population for each stage of the development of a new town.

Team Coordination

The setting up of a department for the new town development was intended to have closely working multi-disciplinary teams for the development of each new town. This is important to ensure the timely completion of all the necessary facilities before population intake. The team consists of Engineers, Town Planners, Architects, Landscape Architects, and their technical supporting staff.

Although TDD employs consultants to carry out feasibility studies, design and construction supervision, the team must handle administrative procedures, departmental liaison and consultation as well as consultant management. Engineers, forming the hub of the system, draft consultancy briefs and agreements, proposed contractual arrangements and documents, process necessary statutory procedures as well as coordinate planning, environmental, electrical, and mechanical engineering aspects.

Coordination Emphasis

Despite the fact that our general principle in new town development is to have a balanced development in a new town in each and every stage of its development, the need for coordination differs from town to town and from time to time. The paper will not go into details of coordination work. It will highlight some important and interesting aspects requiring coordination in the new town development in Hong Kong.

All New Towns are Different

The development strategy for each new town is different, mainly pre-determined by their geographical conditions and constraints. Others may involve major site formation work over rural land or by reclamation from the sea. Each of these new towns also has its own characteristics. Size and population are the basic differences. For example, the smaller new town, Tin Shui Wai, has a development area of about 460 hectares for an initial design population of about 150,000; and Tsuen Wan, one of the bigger new towns, is 2,500 hectares in size with a population already stabilized at about 700,000. Difference in proportion in the various types of land uses in the new towns is also a common feature. Some new towns have incorporated areas for industrial estates or special industries, while others have provisions only for light industries and offices. The following two new towns are typical examples of special purpose new towns.

The "Container Port" New Town

Container Terminals (CT) Number 1 to 9 centre around the Kwai Chung and Tsing Yi areas, which are within the Tsuen Wan New Town. All CTs except for CT9 have been completed and are in operation making Hong Kong one of the world's busiest container ports. A lot of the construction works and infrastructure provisions in this new town were for the port development that created many job opportunities.

The "Airport" New Town

Tung Chung and Tai Ho areas on Lantau Island are being developed into a new town as a supporting community for the new Chek Lap Kok Airport. The first phase of the new town was completed in mid 1997 to accommodate 20 000 residents. The new airport was brought into operation in July 1998. The Phase 2 development is in full swing and scheduled for completion in 2002. The remaining development is intended to meet the medium and long term housing demand in 2006 and 2011. With the upward adjustment of population prediction, the required population level of this strategic growth area is about 50% more than that previously anticipated.

In the Early Days

When new town development first took place, the most important mission was to create new population centres in order to thin out the population density in the existing urban areas. This first batch of new towns was developed by converting, upgrading and expanding the original traditional rural market towns. They already have road networks connecting them with the urban area.

People in those days had lower expectations. What is considered as a small and substandard flat today was the dream of most families. The bread earners of the family could easily find a job near their home. The new towns were very attractive and successful.

TDD took care to conserve and enhance important heritage features. During the planning and design stage, measures were taken to retain as much as possible the original local features, such as the local indigenous villages. In general, higher planning standards as compared to the old urban area have been adopted in order to increase the attractiveness of a new town to people looking for a better quality of life in respect of living environment. In order to encourage people to move into these new centres, public housing estates were first developed with the necessary supporting facilities. This is because these estates provided better homes for the lower income group. After some public housing tenants moved in and settled to form a community, land sale for private development then proceeded. This brought into the new town a new batch of people while further facilities were being provided and more commercial activities attracted into the new town.

Development for Mobility

A good transportation system is one of the important factors attracting people into a new town. This factor is taken into consideration in the choice of development sites. Areas with existing roads or a railway linked to the urban areas have been developed first. In order to cope with the increased demand as a result of the development, these existing roads were widened and new roads were built. The

Kowloon-Canton Railway was double tracked and electrified in early 1980s to permit development of new towns along the railway. Construction of the West Kowloon Expressway, the North Lantau Expressway, the Lantau Link, the Airport Express Line and the Lantau Line form the core transport projects for the New Chek Lap Kok Airport and the supporting Tung Chung New Town.

Local roads are built in parallel with or immediately after the formation of sites. They are required to allow access as well as to accommodate utilities. Cycle tracks are now normally provided along major local roads on top of footpaths. Trees are grown along an amenity strip. When these footpaths/cycle tracks cross major distributors, they are grade-separated for convenience and safety.

Public transport termini are provided for buses, mini-buses as well as coaches, if required. Bus companies are required to maintain five-year development programmes to cope with the future development. Taxi stands are also required. Most of them are established on the roadside. However, to cater for locations with anticipated high demand for taxis, specific taxi stands are provided. They are planned at strategic locations in a new town with the "Route Development Plan" for buses formulated and agreed with the Transport Department and the franchised bus companies.

The railway systems are operated by public funded corporations. Close liaison between the government and these companies are maintained to ensure that the public is well served by them and that their development has acceptable patronage.

The need for ferry services is also considered at the early feasibility stage. Their popularity cross the harbor has been observed to diminish. However, they are still the major public transportation mode for residents on the islands such as Cheung Chau, Ping Chau and Lamma Island.

Efforts are made to integrate all modes of transports by providing convenient interchanging points. Public transport interchanges are normally developed at the early phase of a new town.

What About the Environment?

After the housing demand had been met to a certain extent and the environmental awareness of the public increased, people inquired about the environmental aspects of new town development. The Environmental Protection Department monitors and assesses the environmental aspects. The Department takes part in new town development. In the feasibility study of new towns, an environmental impact assessment study is carried out as part of that study. This is to ensure that the environmental standards regarding air quality, noise, and liquid and solid waste disposal are met.

For instance, industrial areas are segregated from residential development areas. Air polluting industries are not allowed in new towns situated within airsheds, and breeze paths are planned to facilitate air movements. Particular attention is paid to the siting of sewage treatment plant and the selection of treatment process to safeguard the quality of the receiving waters. Landfill sites are only allowed on the fringe of some new towns to accommodate solid waste to a certain level, and after reaching the designed capacity, the sites are rehabilitated and landscaped.

Green New Towns with Landscaping

Better living environment of a new town is made possible through the inclusion of landscape design as a planning element. Landscape architects in TDD design and construct parks, recreation areas, open spaces and sitting out areas of various sizes.

Public Participation

In the Eighties, District Boards were formed as local consultative bodies. Their function is to act as a bridge between Government and the public. In order to encourage public participation in the new town development process, the Department presents and explains development projects to the District Boards. Comments and proposals from them are taken on board for improvements and thus minimizing any adverse impacts they conceive. Some may look upon the Department as taking a public relations role. However, this forum allows explanation of proposals and considerations in order to allay fears and to understand the feelings and aspirations of the local people through feedback.

Filling Material and Dumping Area

New town development also faces engineering problems and constraints. Land acquisition and resumption have proved to be a very time consuming exercise requiring close liaison with the Lands and the Housing Departments.

In order to achieve rapid completion of reclamation and early settlement, sand is the best filling material. At one stage, dredging of sand from the seabed within the Territory was carried out. Special dredging areas were designated in order to minimize any adverse environmental impacts. It was soon found that dredging had a negative effect on the environment. Alternative sources had to be identified. Currently most sand is imported from the Pearl River Delta.

Vertical seawall is the most popular type of structure used along the seafront. It requires a very substantial foundation. Dredging of the seabed was required for the construction of the foundation. Mud so removed needs to be disposed off properly in order to maintain the environmental standards. This is especially required for mud having excessive heavy metal content. Such contaminated mud is required to be dumped at special pits in the seabed that had been dredged. The pits with contaminated mud are covered with suitable material.

In order to reduce the requirement for mud disposal, mud within a reclamation area is to be retained as far as possible. Early consolidation of the mud layer below is achieved using wick drains and other methods such as surcharge or preloading.

Problem with Construction and Demolition Waste

The continued development in Hong Kong produces more and more building debris. Some had to be disposed of in landfills. This consumed the capacity of the landfills for solid waste. The constructive utilization of construction debris is to make use of it for land production – reclamation. It is not a good reclamation material, as its size and composition varies and land formed with it is susceptible for large settlement. On the other hand, it is the cheapest material, available and would otherwise have to be disposed of at a cost.

The use of public dumping material for reclamation is in conflict with the intention of speedy land formation. The rapid growth in Hong Kong requires reclamation sites be available as early as possible. As it is difficult to predict or even control the rate of industrial waste, as a combination of both dumping materials as well as normal filling material is sometimes used. It would be even better if they could identify areas for reclamation which are only required in the long term. These reclamation areas could then be used as a long-term public dumping ground. It is now a policy to plan reclamation projects maximizing the use of public dumping materials.

People Coming and Going

Although there has been concern about "brain drain", the influx of residents from the Mainland has always presented a problem to Hong Kong. Most of the new arrivals need public housing. As a result, they are putting pressure on the public housing development programme. The statistics also show that immigrants are relatively young and require education provisions for their children. The number of primary and secondary schools in new towns will have to be reviewed frequently.

At present, the arrival rate is being controlled to 150 per day. The current population estimation by Year 2010 is around 8.1 million and may go up to above 10 million. This may put great pressure on the housing supply and thus land production in Hong Kong in the medium to long term.

Where to Develop further?

To enhance the role of Hong Kong as an International City and a Regional Centre of business, financial and information, tourism, entrepot activities and manufacturing, controlled reclamation in the Victoria Harbour has to be carried out to provide more land for further development. However, the public is concerned about the narrowing down of the harbor resulting in higher flow velocity to affect navigation safety but lower volume of flow to reduce the harbor's self-cleansing ability. The alternative of further development in the New Territories was raised by a group of people. This group of people considers the Harbour as natural heritage and should no longer be used for land reclamation.

Others take a more balanced view and do not object to reclamation works in the Harbour as long as they can maintain the existing environmental standards. They are aware of the fact that development in the New Territories will pose other environmental impacts. Development constraints include international convention/bilateral agreements, statutory control, administrative control, physical and infrastructures constraints. These development constraints can be resolved by concerned departments working together.

The Government of the HKSAR, in its Territorial Development Strategy Review (TDSR) recommends a mixed development proposal in the urban area as well as in the New Territories. Strategic growth areas have been identified and recommended. Subsequent feasibility studies for these development proposals will have to address and confirm the preliminary findings in the TDSR. In order to win the support of the public on the proposal, presentation packaging for public consultation requires special attention.

Redevelopment of Housing Estates and Urban Renewal

On top of creating new land for development, increasing the development intensity and re-development of some existing areas are also being considered.

Aging urban areas with obsolete buildings where people live in crowded conditions have been a continuing cause of concern and embarrassment to the Territory, which is ranked among the most developed economies in the world. The need to upgrade living standards and replace slums with new and modern buildings is increasing. This triggers the possibility and need of rejuvenating dilapidated areas through a widespread urban renewal programme.

Such redevelopment proposals will demand more infrastructure, utilities, and community provisions. The Department has to take a new role in meeting with this challenge.

The Challenge and the Way Ahead

After 25 years of new town development, the Territory Development Department continues to contribute towards the vitality of the Territory. Development of new towns to meet the housing need as well as improving the general living environment has proved to be a success. As Hong Kong is a small city with high population density that is still increasing at a significant speed, decentralization to the New Territories will continue to occur while the importance of the Metro area as a place in which to live and work will also continue to increase. The adequate provision of transport facilities, in particular public transportation, has proved to be vital.

Through the above brief description of the various coordination aspects for development we have seen that the Department:

- emphasizes on its coordination roles during the planning, design, and implementation stages,
- takes note of the differences among new towns,
- takes note of the need and desires of the people through public consultation,
- reviews its approach to development taking into consideration factors such as the special characteristics, variations in development pattern and maturity of the town,
- reviews its development proposals, at each stage of a new town's development to cope with the changes in the Territory as well as those in the Regional context,
- makes use of the latest technological advancements to meet the increasingly stringent demand

The adaptability of the Department to changes and challenges is perhaps one of the most important factors in its contributing towards the ever continued development in the Territory.

Hong Kong's Public Transport System

Harmonising Choice, Diversity and Urban Integration

Christopher J. McCarthy[1]

Introduction

Hong Kong's rapidly growing population, currently around 6.3 million, but with a trend over 30 years of a seemingly inexorable increase of 1 million each decade (with growth expected to continue) benefits from a wide range of public transport opportunities. These include heavy and light rail; city bus routes; mini-buses; trams (streetcars); a large regulated taxi fleet and a network of ferry services.

Car ownership has remained at less than 15% of the population, in part because of fiscal and other financial influences on the cost of car ownership but arguably also because of the good and affordable alternatives readily available to most people in this geographically small (about 1,000 sq. km) place. Hong Kong is fortunate that these conditions support and push growth in its exceptionally diverse and dense network of passenger transport modes under different operators. Many modes compete, either as a result of Government licensing policy or (as with taxis and some types of mini-buses) by their very nature. The resultant choice available to the travelling public in terms of convenience, comfort, cost and other factors is evident and helps keep the better operators on their toes with service improvements, while poorer operators fall (or are pushed!) away from time to time.

The heavy rail network currently has over 112 route kilometers, with 61 stations, and is operated by two companies each with reasonably defined spheres of influence. Several new heavy rail lines have recently been authorised by the Government and both rail operators have recently let initial contracts for construction works. These new lines

[1] Town Planning Adviser, Mass Transit Railway Corporation, MTR Tower, Telford Plaza, Kowloon Bay, Hong Kong

will add some 56-route kilometers to the total heavy rail network. Both the rail companies are wholly owned by the Government as sole shareholder but are legally tasked to, and do, operate according to prudent commercial principles in the same way as any private sector business.

There is one major light rail network (operated as a division of one of the heavy rail operators) of around 32 kilometers located as the primary transport mode in one of the new towns built in the last 15 or so years. This network is also currently being extended as the town grows.

In the downtown area of Hong Kong, a local network of double deck street running trams (streetcars) provides a light rail type local service for short distance trips in the crowded urban area. This network is more than 70 years old but still provides a useful service.

Bus transport, with routes franchised by Government, predominates in terms of passenger boarding and use. Three main private operators compete in many of the crowded downtown areas, but have more discrete areas of influence moving away from downtown. Services are very extensive and complex in routing with short headways and commensurate wait times of, at worst, only a few minutes for most passengers. The dense network ensures that almost everyone in Hong Kong has at least one bus route service very close by. Some all-night services and special, additional, routes are operated as required. The operators have recently been reequipping rapidly with enhanced new fleets of super-large air-conditioned double decker buses that have brought new standards of comfort and service to users. These are providing strong competition to rail and other modes. Less well-patronised bus routes try to compete by changing to air conditioned single deck fleets.

Two forms of mini-buses provide a public transport function. Each holds 16 people and are flagged down at the roadside as required. One type (called Maxicabs) operates on fixed routes at regulated fares. These provide services to low demand areas or corridors or to residential zones where full-size bus service is not viable or would be infrequent. They also provide short distance convenience travel as an alternative to regular buses and railways. The Maxicab network is growing with urban expansion.

The other mini-bus (called Public Light Bus) is an unregulated, “entrepreneurial” provision which runs on such routes and charges what the operator thinks the market will bear at any particular time. Fares change by time of day and day of the week. This service tends to be more expensive than other road forms but is often convenient and fast (often recklessly so!). It sits between bus and taxi in terms of its market position, but is gradually being superseded by Maxicab and other transport services.

Licensed taxis are widely available throughout Hong Kong and there are three categories, generally distinguished by geographical license zone, although there is some overlap. Catching a cab in Hong Kong is very common across society and is, by most world comparisons, quite cheap and reasonably competitive with other modes.

Hong Kong also has extensive franchised ferry services due to the many intervening water bodies and its more than 200 islands. These services tend to be less than one-hour duration and are largely distributed from the downtown area. One main operator predominates, although other operators have recently captured the franchises for some routes. Many types of craft are used, ranging from conventional ferryboats, through hovercraft to jet catamarans. There is also a more informal network of local boat links in the outlying areas where franchised services are not justified.

As evident from this description, Hong Kong is well served by public transport. Increasingly, customer service improvements, such as access for the disabled, are being provided on all modes, especially rail, bus and ferry. To date, however, ticketing has been largely independent to each operator, although some common tickets have been available, especially between the two heavy rail operators. In the last two years, however, a pre-paid "smartcard" system has been introduced by the two rail operators and increasingly now fitted to buses and ferries. This is expected to have almost universal transport application within a few years and already overwhelmingly dominates ticketing in terms of passenger usage because of its convenience and ease of use. It is understood that it is currently the most widely used system in the world.

Integration of public transport with land use and development

As Hong Kong's urban development has expanded, so too has the public transport network. Government planning policy has in recent years placed much emphasis on the benefits of using fixed track systems, primarily heavy rail, to form the backbone for the new growth areas and their links back to the downtown and other parts of the existing city. It has also been recognized that improved transport has a role to play in urban renewal for the older parts of the city.

A good example of this is the way the Mass Transit Railway has been used as a key tool for urban expansion and induced restructuring. The role of this transport operator, how it relates to integrated land use/transportation policy agendas set by Government planning, and how it has risen to this challenge whilst still operating on mandated "prudent commercial principles" is discussed in the remainder of the paper.

Background to MTRC Involvement

The Mass Transit Railway Corporation (MTRC) is the main passenger heavy rail operator in the urban area of Hong Kong and has over 20 years of experience in planning and implementing the integration of railway facilities with associated land uses and very large scale related property developments.

The Corporation was established in 1975 as a statutory body. The Hong Kong SAR Government holds all the equity but despite Government ownership, the Corporation was conceived, and is run, as a commercial enterprise with the philosophy that it would not become a subsidised public transport system. It was seen to also provide a flexible organisational structure, capable of responding to changes in market demand expected of a modern railway and operating with minimum Government involvement.

Within its important statutory obligation to operate on prudent commercial principles the Corporation, unlike many other railways elsewhere, determines its own fares, negotiates its own borrowings and debt, and does not ask the Government for subsidies to operate the railway. Indeed, it has been making an operating profit such that in 1997 it was able to pay its shareholder a dividend of US$160 million. In conjunction with railway construction, the Corporation has undertaken many joint venture developments of key residential and commercial properties above stations and depots; manages completed estates, and retains commercial property for investment purpose.

Since opening its first line in late 1979, with gradual extensions until 1986, the Corporation has completed and operated a 3-line railway system of 43.2 kilometres route length in the main urban areas of Hong Kong. Its electrified system (1500v dc overhead line) has comprised a total of 38 stations (28 underground and 10 above ground) and 3 maintenance depots. From Hong Kong's around 6.3 million population, the Mass Transit Railway (MTR) carries about 2.2 million passengers each weekday and has a market share of about 25% of the daily total of around 8.6 million fixed route public transport trips made in the territory. Its urban lines each have a design capacity of around 85,000 persons per hour one way flow.

A new 34 kilometre Airport Railway route was completed in 1998 and opened with two types of rail service, partly sharing tracks and infrastructure. The first, the Airport Express business class dedicated feeder train service into Hong Kong's new airport from the city's CBD and key downtown areas, is purpose designed and one of the world's first to be fully integrated into the airport function, with in-town-check-in for baggage and issue of airline boarding passes and other features, in effect bringing the airport downtown.

The other key function of this route corridor is to provide a high-capacity rail link to enable new strategic development areas to be opened up as part of the Hong Kong Special Administrative Region Government's plans to ease the overcrowding and pressure on housing in the city's existing urban areas, where anticipated population growth up to a total of around 8 million people in the next decade (a more than 20% population increase on the present) is likely to be unsustainable. The line will also ease current peak hour train overcrowding on some sections of the existing railway. The 5 initial railway stations on the new route have been targeted as the focus for major, intensive, property developments. Other stations may be built later as strategic development plans evolve.

Characteristics of the public transport system

The following table summarizes the main characteristics of Hong Kong's public transport:

Table: PUBLIC TRANSPORT MODES

TOTAL PUBLIC TRANSPORT		
Fixed Route Public Transport Operators		
Passenger Journeys (pax)	258 mill/month	8.6 mill/day
All Public Transport Operations	321 mill/month	10.7 mill/day
BY MODE		
Franchised Bus		
Passenger Journeys	116 mill/month	3.9 mill/day
% share of total trips	36%	
No. of vehicles	5840	
Heavy Rail		
Passenger Journeys	91 mill/month	3.0 mill/day
% share of total trips	28%	
No. of cars	1270	
Light Rail		
Passenger Journeys	17 mill/month	0.6 mill/day
% share of total trips	5%	
No. of cars	280	
Maxicabs		
Passenger Journeys	30 mill/month	1 mill/day
% share of total trips	9%	
No. of vehicles	2180	
Private Light Buses		
Passenger Journeys	24 mill/month	0.8 mill/day
% share of total trips	7%	
No. of vehicles	2150	
Taxis		
Passenger Journeys	41 mill/month	1.4 mill/day
% share of total trips	13%	
No. of vehicles	18000	
Ferries		
Passenger Journeys	5 mill/month	0.2 mill/day
% share of total trips	0.2%	
No. of vessels	110	

Notes: 1) Source: Hong Kong Government Transport Department August 1998
2) Numbers rounded by author

Relating To Urban Planning Strategies

The Government's urban planning strategies have as one of their aims to capitalise on rail transport assets, in part to reduce pressure on road infrastructure but also to benefit from the convenience and reliability of rail services. In those strategies (the latest being titled "Territory Development Strategy Review") it has taken the bold and strategic planning decision to focus much of the predicted future development of the new urban expansion areas around nodes based on rail station "walk-in" catchment areas. This concentration conversely provides for lower density, more space extensive, land uses to "fill in" between the station nodes. This essentially simple but radical strategy fundamentally changes the pattern of land use and activity where it can be applied.

From the railway perspective, the concept is especially attractive. It locates high-density zones at railheads and ensures that rail captures a high percentage of trips made from those areas. This gives long term assurance of ridership and greatly helps forecasts for operational development and revenue/financing strategies for the rail operator. Other, road based, public transport operators can also benefit with the certainty that the station focal points give both convenient interchange opportunities and also a widely recognised "feeder" requirement that simplifies route planning and gives greater robustness to their revenue forecasts. This feeder role does not, however, mean that the bus companies do not also compete with the railway for line haul passenger traffic! Government licensing has ensured that rail and bus are in keen competition on all main routes.

Of course, it is only a minority of the urban area that is, or will be, capable of being developed on new land where such concepts can be directly applied. In the dense existing urban area, current land use patterns and diverse interests preclude this elegant planning framework from being imposed, although over time the existing city has tended to adapt to the railway through, mainly private sector, urban renewal projects induced by the location advantages given by a nearby station.

A resulting effect is to reinforce the role of public transport as the main focus for travel in Hong Kong, notwithstanding that GDP growth trends have made car ownership more affordable across society.

This approach is having a fundamental influence on the land use and transportation patterns of Hong Kong and is a significant contributor supporting the society's transition into a service economy. It is helping to redraw the map of urbanisation and with it the patterns and distribution of socio-economic activities in the territory.

As densely populated cities (perhaps most particularly in Asia) mature, mass transit hub locations will increasingly become key factors for patterns of urban development and the distribution of key economic functions. This is being recognised in several major cities and it seems clear that the Hong Kong experience in taking forward railway and urbanisation in this well planned and integrated manner has resonance elsewhere.

While the principle of railways leveraging urban development and capturing resultant enhanced value is as old as the railway industry itself, the focus in most countries of the world over more than the last half-century has clearly been for the growth of cities to rely on road-based transportation solutions. This has led to the loss of a wider vision, leading to under-investment, an inexorable spiral of decline in public acceptability, and consequential lack of political support for urban rail. The result has been a long, sterile and over-simplistic, "rail versus road" contest with rail demonstrably the loser. Until fairly recently, rail was way off the public agenda in most places!

A combination of explosive world growth in cities and conurbations, leading to public reaction in many places against perceived increasingly unacceptable congestion and environmental impacts, is attracting renewed interest in using transport as an active urban management tool. The reintroduction of urban rail, both light and heavy rail, within the portfolio of techniques has gained wide attention, and evidenced in part by this Conference. There are now very few crowded cities where single-mode transport provision is advocated, even by a mode's own lobby groups, or where land use and transport planning are not converging. Greater acceptance of a multi-modal approach dovetailed with city planning policies seems evident, albeit institutional restraints and the "baggage of history" means that much is still in the talking and paper stage but with less under way in practice! For various reasons touched on in this paper, Hong Kong has been able to progress a more integrated multi-modal urban development framework where rail is acknowledged as perhaps the driving influence.

How Has MTRC's Experience Been Gained?

The MTR network of rail lines is seen by most Hong Kong people as the essential "backbone" to the city and it has followed that station related development sites are highly sought after by developers and end user occupiers. This has led to significant "value capture" for the Corporation from both development receipts and enhanced railway patronage revenues. Together, these have made a major contribution to its financing strategies and are enabling it to maintain and enhance the quality of its public transport product as well as supporting active plans for new rail network expansion.

On the earlier MTR lines, a total of 18 developments (comprising in round figures 31,500 domestic apartments, 200,000 square metres of office space and 250,000 sq. m. retail/commercial space) have been built using joint venture arrangements with private developers. MTRC continues to manage many of the domestic estates and has retained for its own investment three prime shopping centres built over the railway depot sites in order to generate recurrent rental income. In 1997 these shopping centres' rental income and the residential estate management charges produced a useful 10% (approximately) of MTRC's total revenue. This in fact represented 21% of the Corporation's net profit for the year, reflecting the higher potential profit margin of those businesses. Of course, year on year, the contribution of the retail centres is rather more affected by the vagaries of achievable rents depending upon overall economic conditions than the residential estate management sector, which is somewhat less susceptible to this factor.

It may be interesting to note that a further 12% of Corporation revenue is gained through other non-farebox sources such as poster advertising sites; use of tunnels as easements for commercial cabling; providing facilities for pager and mobile telecom services in stations and trains (while travelling in the underground sections); retail concession kiosk shops and mini-banks in station concourses, etc.

Taking both this and the property income stream into account, farebox revenue from rail tickets in 1997 represented only about 78% of total recurrent income and less than 58% of the net profit. In addition to recurrent income, capital inflows from property development profits also contribute significantly to the viability and cash flow of the Corporation, with valuations of properties retained for investment contributing to a strengthening of its asset base.

The Attraction Of Railway Station Locations

Experience in Hong Kong for over 20 years has clearly demonstrated the power of the MTR stations as major determinants of attraction, both for users and in terms of value generation to adjacent development. Location close to an MTR Station has become a significant influence in the development strategies of many companies and individuals because in Hong Kong's crowded metropolis, no transport system offers the travel reliability and certainty that the urban rail services can give. From Government's viewpoint, this locational interest has been welcomed because it has induced a high level of urban renewal around stations.

From the beginning, the route and construction of the railway lines themselves created more opportunities by way of new land formation which could be used for development. Additionally, it was recognised that an obvious way of generating more rail ridership was to make it especially convenient for people to get to the stations - what more convenient way than to build large scale high density residential estates directly above the stations themselves? This focus widened the vision of many parties to the opportunities created by the new lines and rapidly identified a range of other development prospects, large and small, associated more or less directly with stations. There began to be appreciated, arguably for the first time in Hong Kong, the direct causal relationship between convenient, reliable, public transport and the willingness of the general population to consider locations at some distance from traditional core areas of the city.

The understanding accordingly led entrepreneurs to identify related development opportunities and led to a rapid growth in what are now acknowledged as Hong Kong's leading property companies. Their focus on developments close to MTR stations, and their heavy promotion to the general public of the locational advantages of this, gave great success to those companies. Of wider planning importance, however, it established a climate of opinion in both the public and in the investment community that accessibility and, very important, reliability of journey time is of vital significance in this congested territory. Led by the example of the residential schemes, over time decentralised office functions have gradually crept ever outwards along the MTR towards non-traditional locations.

For MTRC, these tendencies have been helpful because better distributed locations for high density workforces, such as are typically found in these new office areas, can spread ridership more widely across the MTR network and take advantage of spare capacity in counter-flow situations during peak demand periods. Clearly this is in everyone's interest, and enables both remarkably high utilisation of the MTR asset and extends the time until key links will finally reach their absolute maximum peak hour capacity.

The consequences of induced change to land values and locational attractiveness which are attributable to railway stations has been a lesson which did take some years to be fully appreciated as a potential tool for public policy urban management at the wider scale of planning. It follows that the emphasis given to this effect in plans has evolved over time as confidence in the theory has grown and been shown to be warranted. For MTRC, too, there has been an evolution in the scale and nature of the railway related development schemes as it has learned to balance development (that is to say market) opportunities with risk avoidance/minimisation techniques as behoves a public corporation.

Leveraging Benefits - and Risk Management

Hong Kong has thrived, and its success greatly based, on its mobility both at the business and personal level and notwithstanding one of the highest ratios of electronic communications aids in the world, personal contact remains important. Unlike most metropolitan areas, the peak hours for travel in any day are not so distinctive from the off peak and this will be quickly apparent from a journey on the MTR at almost any time of day, when busy trains are the norm. There is a great deal of movement around town for business purposes during the day and for leisure and family reasons in the evenings and at weekends.

This has been, and largely remains, Hong Kong's traditional pattern of activity for the city where the main urban centres always seem busy throughout both day and night, with living, working and leisure functions existing all together.

This point is made because the strategy of focusing large scale, mixed use, urban development projects on MTR station locations is perfectly attuned to this now accepted pattern of economic and social activity, whose vitality is arguably one of Hong Kong's greatest assets. The new sites are planned so as to continue this and not to become sterile, single use, zone. Mixed use within large scale schemes does also provide some hedge against cyclical movements in market conditions since it is common for, say, residential, office and retail demand cycles to be different. It can thus also be a prudent strategy from a property development viewpoint.

The comparatively higher level of confidence that a development located close to an MTR station would be a commercial success, whether it is residential, office or other has certainly made these prime locations. In many cases, the Corporation itself has been able to directly benefit from the enhanced property values that its own presence has brought about. The resulting capital receipts obtained from property joint ventures have made a very significant contribution to the Corporation's funding ability for continuance of high standards of system maintenance, customer service enhancements and ability to consider new railway extensions. Retaining a limited number of selected developments as a longer-term investment so as to give revenue diversification has also shown demonstrable benefits to the balance sheet.

A careful (and continuously reviewed) balance between reserves, borrowings, shareholder's funds and property contributions to finance the Corporation's activities does, however, needs to be struck. Over-reliance on property contributions must to be avoided as it could expose the Corporation to unacceptable levels of risk and weaken its financial status, albeit that its joint venture arrangements are structured to lay-off all direct market risks onto the JV party, with the Corporation recovering its direct enabling cost outlays for the scheme, the land costs and other exposures "up front". All development costs, including financing and construction costs, are carried by the other party.

In this way it has operated in a risk averse manner but with the capability of sharing surpluses (net of such costs) in proportions, and ways, contracted with JV partners who bid for the development rights through open tender. The financial status of tendering JV parties is carefully reviewed during pre-qualification. In their tender, they specify the share and nature of the surpluses that they are willing to return to the Corporation as its profit from the venture. These may include, for illustration, cash payments up front, percentages of sales proceeds upon staged completion of the development, or sliding scales of income distribution from property rental assets retained by the developer for investment. There are several possible formulae and combinations.

Other than in exceptional circumstances the MTRC will, of course, time its tender announcements to maximise its likely return from the venture and it follows that the details of its scheme designs are always kept under continuous review, with adjustments to ensure they remain attractive and affordable to the development industry.

Recent Developments

As Hong Kong's society has moved ahead into the post-industrial era, the MTRC's "mixed use" development concept and design emphasis of the new Airport

Railway station developments are focusing even more on creating fully integrated communities, with appropriate land use and a wide range of supporting facilities. The concepts have been fully backed-up by suitable Government urban planning strategies and area-wide public infrastructure programmes. This latest railway extension has provided for about 3.3 million square metres of development which includes some 24,000 apartments; office, hotel and other commercial towers (together around 990,000 sq. m.) and 5 major shopping centres (about 350,000 sq. m.) together with open space and other amenities. All are planned to the highest standards and most are now well into the construction phase in association with the well-tried joint venture arrangements. The development value of these schemes has been estimated at more than US$20 billion.

The latest railway extension is to the new satellite town of Tseung Kwan O, where work on the railway project is commencing around now, and this has 4 development sites integrated with stations and maintenance depot. These will give rise over a period of some 10 years to about 29,000 apartments housing around 80,000 people, 3 shopping centres and a local office development.

Together these many developments will contribute significantly towards improving the standards of living and working for hundreds of thousands of people as well as giving them greater convenience and choice. Further new railways with integrated urban development plans are also in the pipeline for the longer term.

The scale and nature of the new developments offers a huge challenge to the planning and management skills of the Corporation and for integrating these with Government's own plans and works programmes not least because, in terms of the planning of Hong Kong, the MTRC's sites are a crucial component in the urban structure and therefore have a wider significance.

Conclusion

This paper has just touched very briefly on some of the benefits of integrating railways with land use planning and active urban development and why this has been rather successfully achieved in Hong Kong. Increasingly, as cities become more congested and the service sector comes to dominate their economic activity, places such as Hong Kong will only prosper if given suitable infrastructure tools within a closely related urban land use planning policy. The two are indivisible.

In Hong Kong's experience, the urban railway when dovetailed with land use strategies and visionary development concepts (which must be appropriate to the community's needs and aspirations) has a key role to play as a particularly significant influence in supporting growth and structural economic changes in society and at the same time helping meet people's aspirations for improvements in the quality of their lives.

Not the least, the value leveraged from the concept can, if circumstances allow, make a useful contribution to the balance sheet of the transport undertaking in terms of both higher and more certain ridership generation (or market share) and farebox income, capital injections from development surpluses and support from property management and other ancillary revenue streams. Together these further support its ability to maintain a quality transport product and a competitive position in the market place.

The Mass Transit Railway Corporation has certainly benefited in this way, with the result that it has been able both to sustain its "benchmarked" position as one of the leading urban passenger railway systems in the world but also support and actively participate in Hong Kong's growth and development as a post-industrial society.

The people of Hong Kong derive great practical benefits from the diversity of public transport modes available to them, whether it is the railway, buses, or the various other types of transport. The combination of Governmental controls through licences and route franchise arrangements, taken together with the less controlled secondary modes, has worked well for many years. As Hong Kong moves forward with a growing population it is important that the validity of the current arrangements are kept under continuous review to ensure they remain relevant, affordable to society and viable - and continue to be accessible in both convenience and cost to the individual users. This is the challenge for Hong Kong's public transport as it continues its harmonisation with the changing city.

The Transportation System for the New Hong Kong International Airport

Stephen Wa-Kwai Lam[1]

Abstract

The decision by the Hong Kong Government in 1989 to relocate the city airport at Kai Tak had resulted in the largest infrastructure project ever undertaken in Hong Kong in recent times. The Airport Core Programme consisted of the construction of a new replacement airport and the associated road and rail link, reclamation and a new town, an investment totaling to some US$ 20 billion. The programme was completed in eight years and the new airport opened on July 1998 as scheduled. With the tight time frame, the size and complexity of the project, the ACP was one of the most challenging engineering projects in the world.

Introduction

Hong Kong is situated at the Pearl River Delta in the southern part of the Mainland China. Hong Kong is a major financial centre and is the world's seventh largest trading economy. Its per capita gross domestic product (GPD) reached US$24,500 per year.

Hong Kong has a land area of 1,094 km^2 with a population of 6.6 million. With the current rate of net migration from the mainland China, it is anticipated that the population will grow to over 8 million in year 2011. At present, about 65% of the population are accommodated in the urban area and the remaining 35% in the new towns and the New Territories. The terrain in Hong Kong is hilly and this is especially so in the urban area where land for development is scarce. Much of the built-up area in the urban part of Hong Kong has been reclaimed and the closest point between Hong Kong Island and Kowloon peninsula separated by the Victoria Harbour is now some 800 meters apart (Figure 1).

[1] Airport Authority Hong Kong, 8 Chun Yue Road, Hong Kong International Airport, Lantau, Hong Kong

Background

The old airport at Kai Tak was a city airport, situated next to the populated residential area at the eastern part of Kowloon (Figure 2). The Kai Tak airport was the third busiest airport in the world in terms of air passengers and the busiest in term of international cargo handling. Kai Tak had a single runway and there was an overnight curfew on flights between the hours of 00:00 - 06:30. It handled more than 30.2 million passengers annually, and over 60 airlines operated scheduled services there, with over 400 aircraft arrivals and departures, one aircraft movement in every two minutes in peak hour.

Kai Tak had a finite capacity because of the physical and operational constraints at the site. With the increasing demand on air traffic, there was a need to build a replacement airport that would provide sufficient capacity to support the anticipated growth in air traffic in the future years. The new airport would play a vital role in supporting the economic growth of Hong Kong and in maintaining its leading position in aviation and tourist industries.

While the study of a replacement airport commenced as early as in the 70s, it was in October 1989 when the Government formally announced the plan to construct a new airport at Chek Lap Kok, a small island off North Lantau, as part of the airport and port development strategy. Together with the network of associated road and rail infrastructure, the overall airport relocation exercise was referred as the Airport Core Programme (ACP) projects, a construction investment totaling to some HK$155.3 billion (US$20 billion).

The ACP, consisted of 10 interrelated projects which constituted some 200 works contracts, was amongst the largest infrastructure development in the world (see Figure 3). The magnitude of the programme has created opportunities for the private sector participation. The ACP projects were implemented by the Hong Kong Government, the Airport Authority (AA) and Mass Transit Railway Corporation (MTRC) - two statutory bodies wholly owned by the Government, and a franchised company Western Harbour Tunnel Co. Ltd. The Government contributed about HK$109.9 billion (US$14.5 billion) by direct funding and implementation or in the form of equity injunction into the AA and MTRC for the airport and railway works.

Airport Core Program

Chek Lap Kok is approximately 28 km (17 miles) from the main city centre. An efficient transport system is required to serve the air passengers and other users requiring access to and from the new airport. The development of this transport link formed the core of the ACP works. The cost of ACP is summarised in Table 1 below.

Table 1: Airport Core Programme Cost

ACP Works	HK$ (billion)	US$ (billion)
Airport	70.7	9.1
(Airport Authority)	(49.8)	(6.4)
(Government facilities)	(5.5)	(0.7)
(Franchisees)	(15.4)	(2)
Government works	47	6
Airport Railway	34	4.4
Western Harbour Crossing	6.5	0.8

Source: NAPCO

The Airport Transport Link

To support the operation of the new airport at Chek Lap Kok, an efficient transport network is necessary to ensure that the air passengers and other airport users can easily access to the airport. To achieve this aim, a dedicated highway and railway system of some 34 km (21 miles) was therefore built to provide a high-speed link to the airport. This transport link not only meets the transport demand of the airport users, but also creates opportunity for future development along the transport corridor. The airport transport link includes the Airport Railway and five major highway projects - North Lantau Highway, Lantau Link, Route 3 (part), West Kowloon Expressway, and the Western Harbour Crossing (Figure 3). These are described below.

The Airport Railway

The Airport Railway, costing HK$ 34 billion (US$ 4.4 billion), is the first purpose-built railway to serve an international airport. It was built and is now being operated by MTRC, the body operating the existing urban mass transit system in Hong Kong. The Airport Railway provides two services. The Airport Express Line (AEL) is a fast-dedicated line linking the airport with Hong Kong Central, with two intermediate stops at Kowloon and Tsing Yi. The Tung Chung Line (TCL) is a domestic service, designed to provide mass transit community service between Tung Chung New Town and the urban area.

The AEL Airport Express trains offer "business-class" rail services targeted at air passengers. The stations have also been designed to match airport standards. To enhance the service to air passengers, in-town check-in facilities for baggage and seat allocation are provided at both Hong Kong and Kowloon Stations. The express train can travel at speeds of up to 135 km/h (84 mph) and a journey from Hong Kong to the new airport would only take 23 minutes. Construction of the Airport Railway commenced in November 1994 and works were substantially complete in June 1998. The Tung Chung Line in fact commenced services on 22 June 1998.

The Airport Express operates 19 hours daily from 06:00 - 00:45 hours, and it was expected that more than 36,000 passengers each day would make use of this high quality rail service in the first year of airport opening.

To increase the competitiveness of the airport express, the fare has been reduced by 50% from the originally planned fare to HK$100 (US$12.8). (The fare for an airbus is around HK$50 (US$6.4), and taxi fare approximately HK$350-400 (US$44.9-51.3) from Hong Kong.)

North Lantau Highway

The 12.5 km (7.8 miles) dual 3-lane North Lantau Highway (NLH) connects the new airport to the Lantau Link along the northern coast of the Lantau Island. It was built by the Government, costing HK$ 5.8 billion (US$ 0.74 billion). The construction began in 1992 and the highway was opened to traffic in May 1997. In addition to serving the new airport and Tung Chung, it also provides the road link connection to the future port facilities on South Lantau and other property development along the highway corridor in future.

A 2-lane utility road is also provided along the NLH and this can be used for traffic diversion in case of traffic blockage on the highway.

The Lantau Link

The 3.5 km (2.2 miles) Lantau Link, the largest of the transport infrastructure projects in the ACP, provides the railway and road access to the Lantau Island and the new airport from the urban Kowloon and Hong Kong (Figure 4). The Link consists of the Tsing Ma Bridge, the Kap Shui Mun Bridge and the Ma Wan Viaduct. The US$1 billion Tsing Ma Bridge is a steel suspension structure of 2160m with a main span of 1377m. The Kap Shui Mun Bridge, costing US$300 million, is a cable-stayed bridge with a main span of 430m long. The Tsing Ma Bridge is the longest suspension bridge in the world that carries both road and rail traffic.

The Crossing carries 6 traffic lanes (3 in each direction) on the upper deck and two railway tracks on the enclosed lower deck. In addition, two traffic lanes are also provided on the lower deck for emergency use so that traffic can be diverted in severe weather conditions.

As Hong Kong is subject to typhoons, and Lantau Link is the only land link to the new airport, it is important for the bridge to remain open as long as safety permits so as to minimise the impact on the traffic to and from the airport. To assist in effectively managing the operations of the Link, a sophisticated Traffic Control and Surveillance System (TCSS) has been designed to facilitate the implementation of various traffic management measures safely and expeditiously under different operating conditions and especially during severe weather conditions.

The construction of the Lantau Link started in May 1992 and opened to traffic on 21 May 1997.

Route 3

Route 3 is a strategic road which provides a direct road access from the North West New Territories to Kowloon. The two sections of Route 3, Tsing Yi and Kwai Chung Sections, form part of the ACP as they carry the Lantau Link traffic to the distributor road network in Kowloon.

These two sections of Route 3 are about 7 km (4.0 miles) in total, and were constructed by the Government. They run through a twin-bore dual 3-lane tunnel and an elevated dual 4-lane structure. Together with the West Kowloon Expressway and the Western Harbour Crossing, they were opened to traffic on 30 April 1997.

West Kowloon Expressway

The HK $2,266 million (US $290 million) West Kowloon Expressway (WKE), built by the Government, connects to the Western Harbour Crossing and is also a strategic link to the road network serving Kowloon. The WKE is a dual 3-lane expressway of 4 km (2.5 miles) long. The WKE is built on reclaimed land, the northern section of which is elevated with the airport railway running underneath it.

Western Harbour Crossing

The Western Harbour Crossing (WHC) has a dual 3-lane twin-bore tunnel under the Victoria Harbour, providing a linkage between the western part of Kowloon and the Western District of the Hong Kong Island. This is the only one of the 10 ACP to be privately funded as BOT project by Western Harbour Tunnel Co. Ltd. under a 30-year franchisee agreement with the Government.

The crossing is 2 km (1.2 miles) in length, 1.36 km of which is of immersed tube tunnel and the remaining 0.64 km cut-and-cover approach tunnels. The WHC is the third tunnel under the Victoria Harbour and has a design capacity of 180,000 vehicles/day.

Reclamation

New land is required to facilitate the provision of airport transport infrastructure. The HK $1.47 billion (US $0.188 billion) West Kowloon Reclamation provides an area of 334 hectares (825 acres) of land which increases the size of Kowloon Peninsula by one-third. This is the largest single reclamation project ever undertaken in urban area.

This reclaimed land becomes the site of the West Harbour Crossing, the West Kowloon Expressway, and the Airport Railway Stations. Moreover, the reclaimed land is also used for other much needed development such as housing, recreational and other infrastructure development.

The Central Reclamation provides 20 hectares (49 acres) of new land for the Airport Railway's Hong Kong Station and commercial development. Both reclamation works were carried out by the Government.

Tung Chung New Town Phase 1

Tung Chung was a small fishing village at north Lantau adjacent to the new airport. In order to support the airport operation, a new town in Tung Chung has been created to provide supporting facility such as housing, shopping centre, and office development. This initial phase of new town development is aimed to provide airport employees with an opportunity to live nearby their workplace.

Phase 1 development is designed to cater for a population of 20,000 people initially. Subsequently, in the future, the new town will be extended progressively along the coastal strip of reclaimed land alongside the North Lantau Highway.

The New Airport At Chek Lap Kok

The Airport Authority

In April 1990, the Hong Kong Government established the Provisional Airport Authority (PAA) as a statutory organisation to plan, design and construct the new replacement airport. The New Airport Master Plan was commissioned by PAA and it basically set out the development scheme for the implementation of the new airport with one runway initially, and subsequent expansion into a two-runway airport operating 24 hours a day.

The Airport Authority was formally established on 1 December 1995, tasked with the obligation to develop, operate and maintain the new airport in a prudent commercial manner.

The Airport Plan

The master plan was published in March 1992. It provided a programme for the phased development of the airport to meet the traffic forecasts and demands until the year 2040. Important assumptions which influence the planning and design of facilities were made in the formulation of the airport development plan:

- The former airport at Kai Tak would cease to operate when the new airport opens.

- The provision of a multi-modal surface transport system with a high priority given to rail; with full rail, road and ferry accesses be made available on airport opening.

- All passengers are considered international, thus defining the security, immigration and customs requirements of the airport facilities.

The airport is a phased development and the Airport Site Plan is shown in Figure 5. The airport was designed with an initial operating capacity of 35 million passenger and 3 M tonnes cargo annually in phase 1, and this can ultimately be expanded to design capacity of 87 million passengers and 9 M tonnes of cargo annually. The main development features of the 1st phase of airport plan can be summarised below:

- 2 parallel runways separated by 1525m, each of 3800m long and 60m wide.
- Passenger terminal building (PTB) of 1.3km long with a GFA of 515,000 m^2, served by an underground Automated People Mover, and a baggage handling system capable of processing 1920 bags per hour.
- Ground Transportation Centre (GTC) to house the Airport Railway station and all the ground transport modes.
- A high speed rail and road network to facilitate an efficient access to the PTB through the GTC and other airport supporting facilities at CLK.
- Airport support facilities such as air cargo, aircraft maintenance, aircraft catering, aviation fuel supply, ground support equipment maintenance, air traffic control, emergency and rescue facilities etc.
- Commercial and non-commercial facilities such as the 1,100-room airport hotel, a freight forwarding centre, general business aviation, heliport facilities and Government Flying Services Department's facilities.
- Other ancillary facilities that will enhance the commercial viability of the new airport including cargo village, industrial park or a business park.

The airport was opened with a single runway, a passenger terminal building with 38 gates and 27 remote aircraft stands. The second runway will be brought into operation in mid 1999, bringing additional 10 frontal gates provided by the extension of passenger terminal building concourse. Table 2 provides a comparison overview of the facilities at the old Kai Tak airport and those at the new airport.

Table 2: Airport Facilities at Kai Tak and Chek Lap Kok

Airport Facilities	Kai Tak Airport	New Airport on CLK Phase 1(a)
Site Area	333.8 ha	1248 ha
Passenger handling annually		
actual passengers in 1996 (excl. transit)	29.5 million	
design capacity (phase 1)		35 million
eventual design capacity		87 million
Air cargo handling annually		
actual cargo in1996	1.56 million	
design capacity (phase 1)		3 million
eventual design capacity		9 million
Runway	1	1(2)*
Runway length	3393 m	3800 m
Passenger Terminal Building (PTB) area	66,000 m^2	515,000 m^2 (550,000 m^2)*
Aircraft gates:		
frontal	8	38 (48)*
remote	56	27
air cargo	5	13
Baggage reclaim units	6	12
Check-in counters	210	288
Immigration control counters	170	224
Automated People Mover	0	1.
Retail outlets	40	120
Airport Railway	0	1

* Provision with the completion of the second runway and extension of the terminal concourse.

The approximate cost of the development is given in Table 3 below:

Table 3: Airport Development - Cost Summary

Airport Facilities	HK$	US$
Reclamation/site formation (1248 ha)	9 billion	1.2 billion
Passenger terminal building	13.9 billion	1.78 billion
Airfield	3.8 billion	0.49 billion
Transport facilities and roads	3.5 billion	0.45 billion

The Airport Platform

The replacement airport at Chek Lap Kok was built on a platform of 1,248 hectares (3,083 acres) of land formed by reclamation of 938 hectares (2,317 acres) from sea and the leveling of the original Chek Lap Kok Island and Lam Chau Island (Figure 6). The reclamation work began in Dec 1992 and was complete in Jan 1995, costing HK$ 9 billion (US$ 1.2 billion). The Airport Authority was responsible for the main operating facilities and all the infrastructure provision on the platform, with the exception of certain facilities such as the Government facilities including air traffic control tower, and the franchised development such as air cargo, catering and aviation fuel supply.

The Surface Access System at the New Airport

The new airport demands the provision of very highly efficient transport system to serve the large number of air passengers and to efficiently support the core operation of the airport and other businesses at CLK. The transport system at the new airport comprises a road and rail network, a Ground Transportation Centre and a ferry terminal.

Roads

It was forecast that some 70% of airport users (including air passengers) would access the airport by road transport modes. The 34 km of road network at the airport basically consists of a primary access road designed to expressway standards and a network of distributor roads serving other development on the airport platform. The dual 3-lane primary access road and the Airport Railway form the backbone of surface transport system for passenger access.

In the planning and design of the road network, particular emphasis has been given to traffic management and operation requirements such as the provision of alternative routes and priority to high occupancy vehicles.

The primary access road becomes the main access route to the passenger terminal building and other main operation facilities at CLK. To maintain a high quality of service to the airport users, it is extremely important that the road network is managed efficiently and effectively such that delay or disruptions to the users are minimised. To assist the management of the road network at CLK, an airport traffic control and surveillance system (ATCSS) has been installed with the following main features: CCTVs, Automatic Incident Detection System, Traffic Diversion System, Roadside Emergency Telephones, Dynamic Car Parking Guidance System, Dynamic Departures Guidance Signage System to provide direction to the motorists in advance the correct drop-off zone at the departures kerb, Taxi Staging Control System, and the associated control centre installation.

Ground Transportation Centre (GTC)

The GTC is the focal point of all surface transport modes at the passenger terminal building (Figure 6). The GTC contains the station for the Airport Express Line and other public transport facilities such as taxi rank and stations for franchised bus, tour coach and travel industry vehicles. Table 4 compares the transportation facilities of the former airport at Kai Tak and the new airport at CLK. At Kai Tak, the taxi was the major provider of transport access because of its proximity to the urban area. At the new the airport, the demand for taxi service has been substantially reduced as expected.

Table 4: Airport Transport Facilities and Modal Split

Modes	Size of Provision		Modal Split	
	Kai Tak	New Airport	Kai Tak*	New Airport**
Rail	n/a	1	n/a	40%
Bus	6 bays	17 bays	7%	17%
Tour Coach	12 bays	18 bays	8%	7%
Hotel Vehicles	60 m	110 m	7%	3%
Taxi	14 bays	24 bays	58%	19%
Private cars	1732 spaces	3100 spaces	20%	14%

* Air passengers only ** air passengers, and greeters and well wishers

The GTC building and the associated transport facilities have been designed to facilitate movements of passengers with emphasis on efficiency, speed, comfort and safety. The centre is essentially comprised of 4 levels of passenger transfer/interchange activities. The departure curb at the top level is a vehicular drop-off area, which connects to the passenger check-in concourse with 4 gentle down ramps. The departure kerb has double-kerb arrangement, with high occupancy vehicles such as buses and coaches on the inner kerb, and private cars and taxis on the outer kerb.

The second level is the departure platform of the Airport Railway. Passengers arriving by rail can travel along the connecting ramps and moving travellators to the check-in concourse in the terminal building.

The next level of the GTC is the Airport Railway's arrivals platform that is immediately adjacent to the passenger terminal building's meeters and greeters hall. Again, arrivals passengers can access the rail platform without the need of level change.

The ground level of the GTC is occupied by the pick-up facilities of other road transport modes, and can be easily accessed from the PTB by gentle down ramps and escalators. The ground transport facilities are segregated and the design of which is to ensure that the passenger can access their choice of transport with ease. The pick-up facilities include bus terminus, tour coach station, limousine pick-up area, taxi station and car parks, all of which are within easy reach of the PTB.

Ferry

The ferry terminal is provided at the east of airport island, approximately 1km distance from the GTC. A ferry service is provided to serve Tuen Mun. A public transport interchange is also provided at the ferry pier.

Airport Relocation

With the completion of the new airport, the Kai Tak airport ceased to operate. The relocation of the operating airport at Kai Tak to the new Hong Kong International Airport was a major and complex task. At 00:00 hours on 6 July 1998, the Kai Tak airport closed down, and at 04:00 am on the same day, the new airport at CLK was put into full operation. The overall airport relocation required extensive planning, organisation, and coordination of many bodies and organizations. Through careful planning and by concerted efforts of all involved, the relocation exercise was conducted and concluded smoothly.

The Airport Operation

The new Hong Kong International Airport opened on 6 July 1998 as scheduled, and all facilities are now fully operational. Like all other airports or public transport infrastructure in the world, the operation of the new airport shall be continuously refined to improve its services to the air passengers and to cater for the ever changing circumstances. Airport planning is a continuous process. Expansion programme and improvement works are under constant review, ensuring that the airport facilities shall meet the increase in passenger demand in the future.

Conclusion

The Airport Core Programme is the largest ever infrastructure project undertaken in Hong Kong. The completion of such important infrastructure shall enhance the position of Hong Kong as a leading international financial, trading and business centre in the region.

Hong Kong's economy depends greatly on the provision of an efficient transport system. The new airport, together with the rail and road transport facilities, will support Hong Kong's continued economic growth into the next century and beyond.

The airport and associated works were amongst the largest and most challenging engineering projects in the world in recent times. It is the devotion of the individuals, who have been involved in the whole programme from planning to construction and operation, can make it possible to achieve the completion of the ACP works on time and within budget.

Acknowledgments

The author would like to thank the Airport Authority for permission to publish this paper.

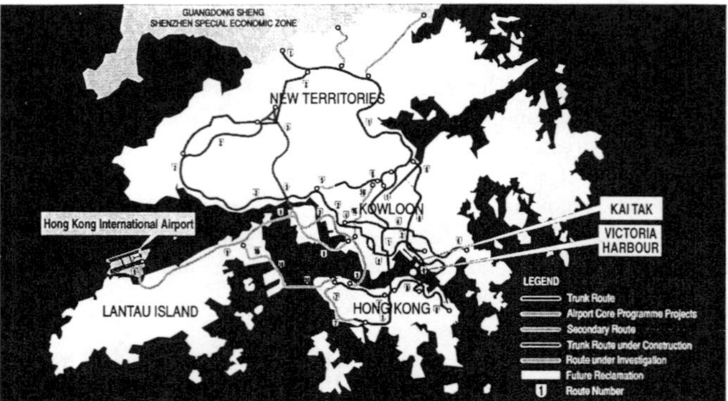

Figure 1: Hong Kong

Figure 2: The Former Airport at Kai Tak

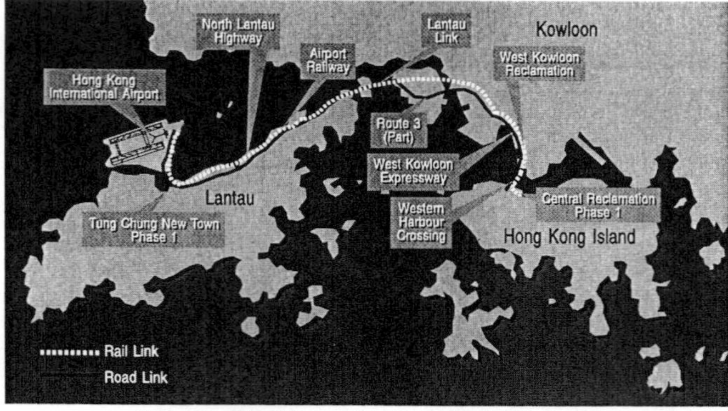

Figure 3: The Airport Core Programme Projects

Figure 4: The Lantau Link

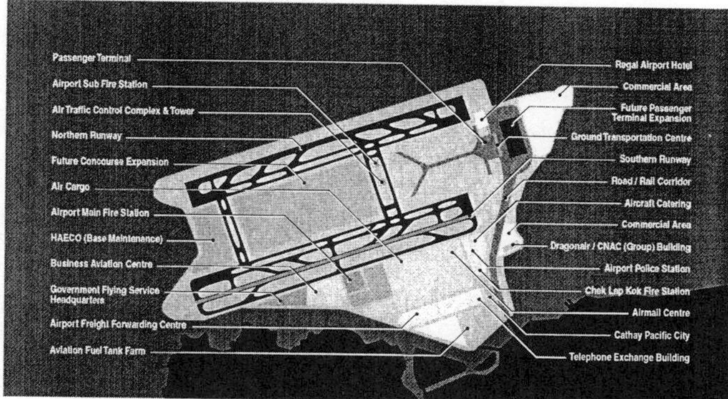

Figure 5: The Airport Site Plan

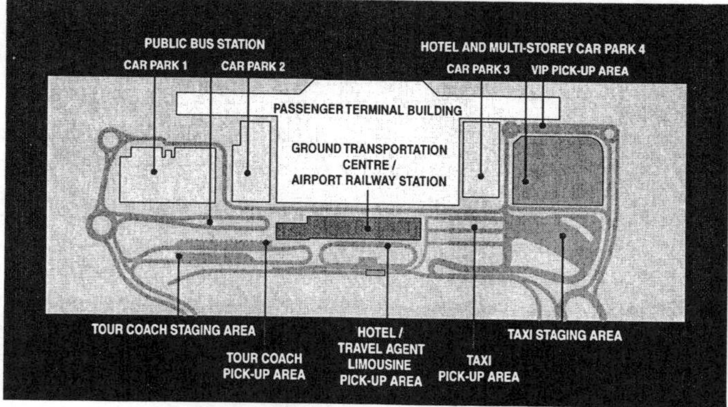

Figure 6: The Ground Transportation Centre

LONDON - SOCIO-ECONOMIC, LAND USE AND TRAVEL PATTERNS

POSITION STATEMENT, ISSUES AND SOLUTIONS

Professor Stuart Cole[1]

THE LONDON CONTEXT

The major cities of the European Union are facing a congestion crisis as a result of inadequate investment in both road and mass transit infrastructure over the last forty years.

London in particular has over the last ten years faced a crisis of such proportions that the Government commissioned a series of studies (DOT, 1989). However it is clearly looking for agreements on rail funding where most of the cost is being met by passengers through fares increases and market growth; by OPRAF franchised train operating companies; by property owners who would benefit from development gain; through joint funding using the Private Finance Initiative (HMT, 1993, HOC, 1993) (already used in the Central Line, Metrolink, Northern Line and Jubilee Line Extension) or in the case of Railtrack through private investment by a private company. Some schemes will also require public sector support (from local authorities, OPRAF or DETR) to ensure viability through financial payments, land allocations or capital equipment included in funding packages (Railtrack, 1997) to for example Railtrack or London and Continental Railways (the Consortium set up to build the CTRL).

The development of the road network is also under consideration. The ringways planned in the 1960s are evidenced so far in the M25 and parts of the inner ringway (R1) at Shepherds Bush and Stratford. Their construction would involve large scale demolition and controversy of the type which followed the publication of the London Assessment Studies Options (DOT, 1989) which contained both rail and road proposals with the line of route shown in general terms, with homes blighted as a result. There have also been demonstrations against housing demolition (on the M11 extension in Walthamstow) and about environmental issues (the A40 improvements in west London, where homes were bought and demolished, and the scheme was cancelled on grounds which appeared to be environmental but which are more likely to have been Treasury-led.

[1] Professor and Director TRaC University of North London, 277 Holloway Road, London N7 8DB, England, Great Britain; Professor of Transport, University of Glamorgan, Business School, Pontypridd, Wales, Great Britain.

THE CONGESTION PROBLEM

It has long been suggested that London should have a strategic transport authority - in 1905, in 1925, in 1943, in 1980, in 1989, in 1996 and in 1998 and in most years in-between.

The Confederation of British Industry (CBI, 1989) produced what they saw as solutions for London's transport congestion. It was almost the twenty-first anniversary of the `Transport in London' White Paper (MOT, 1968) which proposed an overall transport planning authority for London with wide powers of control over London Transport, British Rail and highways. These powers would be held by the Greater London Council and the Minister of Transport.

Consider three separate statements:

First: `London should be a transport priority zone under the authority of a government minister ..... who would have sufficient powers to provide real and obvious cohesion between future rail, road, Underground and land developments to ensure that all modes of transport complement each other.'

Second: `The cause of the London traffic problem, that is to say the want of proper and adequate streets, is not primarily finance, nor the growth of London; it is the want through the centuries and at the present time of some controlling authority with comprehensive power such as has existed in Paris, Berlin and Vienna.'

Third: "Transport in London has a proud tradition of achievement and innovation - it had the first underground railway and its red buses and black cabs are recognised around the globe. But at present the capital's public transport is under heavy strain. Services are poorly co-ordinated and the system suffers from years of under-investment in some areas London needs a clear transport agenda."

It might at first be thought that all three statements were made in the last few months. The second however was a statement of the underlying cause of congestion in London both on the roads and public transport identified by Sir Lyndon Macassey as President of the Institute of Transport in 1927. There have been similar calls by the GLC in 1980, the CBI (1989), (the source of the first statement), the last Government's policy document Transport - The Way Forward (DOT, 1996), the new Government (DETR, 1998) (the source of the third statement), and by planners and academics before and since.

The new Greater London Authority will deliver an integrated and sustainable transport strategy in London; unify the presently fragmented responsibility for transport in London by creating a body which can tackle issues at a London-wide level; and define clear boundary lines between the responsibilities of government, the GLA and boroughs. This is further emphasised in recent reports from London Transport (LT, 1995, 1996) and from London Pride Partnership and from the Government Office for London (LPP 1996, GOL 1996) which further indicate concerns about London's position as a world city and its role as a major financial centre from which many jobs and billions of dollars (US$) of foreign exchange earnings are derived.

The fact is that no past provision has been made for traffic growth (Fig 1) and no consideration of the best solution for congestion (Fig 2). The major highway and railway authorities have always had separate objectives although they were controlled by the Department of Transport or by metropolitan councils. Before the present Government's Integrated Transport Policy White Paper, the only indication that a more cohesive decision-making process might be on the way is in the Central London Rail Study (CLRS, 1989) the recommendation of the London Assessment Studies which contain road and rail solutions, and the A470 Study in the Cardiff area (WO, 1993). The CLRS took a refreshing approach and estimated the economic benefits of each rail scheme to public transport users, road users and railway operators, which is favoured in the last Government's "A Transport Strategy for London", and is at least a starting point from which an Integrated Transport Policy can be developed.

Fig 1 London household car ownership

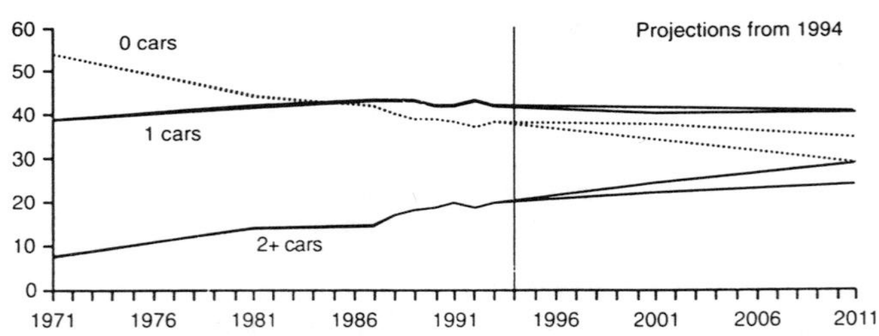

(Source: Department of Transport GOL 1996)

Fig 2 Trip Density (am peak)

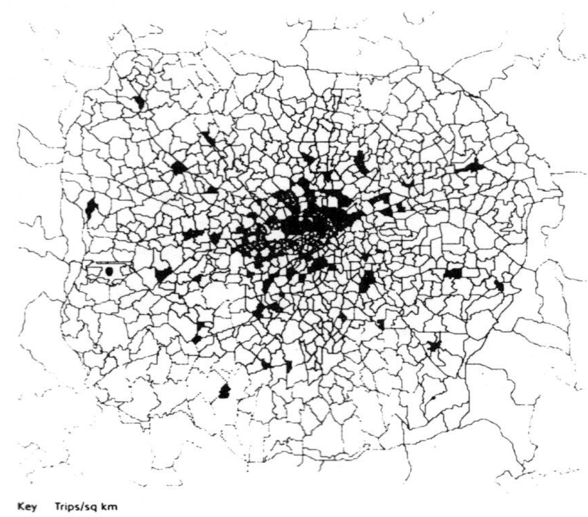

Source: Department of Transport, London Transport (LT 1995)

A finer grouping of policy objectives one would be hard pressed to find. However, few decisions have been made on major funding into rail, metro, light rail or bus infrastructure, particularly since most of those operations are in the private sector and would require extensive negotiations prior to commencement it would have been expected that discussions on proposals would by now have been well advanced.

Causes of Congestion

The causes of road congestion are insufficient capacity, greater demand and a low level of investment in new roads or road improvements. Congestion in mass transport is caused by a significant increase in demand since 1981 on a system which has suffered from decades of low investment.

There are two main reasons for this increased demand. The upsurge in economic activity in London since 1982 (Fig 3) and the cheap fare policies of the Greater London Council between 1980 and 1983 and the advent of the Travelcard which led to a change in public awareness of mass transit, and brought a major shift by commuters from car to rail (LT, 1994).

Fig 3 Population in London (and south east England) 1981-1994

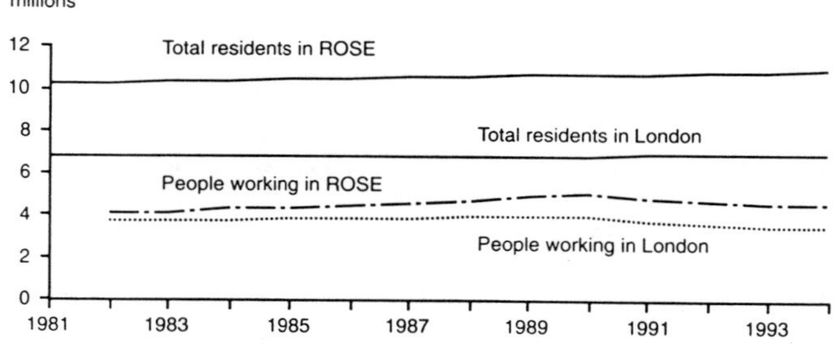

Source: Office of National Statistics (GOL 1996)

Road Congestion

The London Assessment Studies commissioned by the Department of Transport over ten years ago in 1986 examined traffic problems in four main London corridors. The South Circular Road study (DOT, 1989) showed that private cars constituted on average over 80% of total vehicles on that road, light vans used by delivery firms and repairers represent 12% while heavy goods vehicles constituted only 2% of the total.

To achieve any major impact on road congestion therefore it is clear that reductions in car flows will play a key role. By coincidence passengers are also the customers for whom other alternative modes are available. Traffic speeds also need to be considered in parallel to modal split; car ownership has risen and those speeds have fallen to an average of 18.4 kph (11.5 mph) in the peak (CIT, 1992) and speeds on some main roads are little different to what they were in Sir Lyndon Macassey's day.

London's peak traffic problems may be divided into three types - commuter movements to/from the central business district (the City and the West End), orbital movements, and local work or school traffic. Added to this are the `off peak' movements in the central area when even then, a minor incident can cause serious traffic jams. (Fig 4).

Fig 4 Daily journeys by each mode in London

	Millions	%
British Rail	1.4	5
Underground/DLR	2.6	10
Bus	3.0	12
Taxi	0.2	1
Car and motorcycle	10.8	42
Walk and pedal cycle	8.0	31
Total	26.0	

Note: Journeys involving a change to the same mode are counted as one journey

Rail Congestion

Congestion on the London rail mass transit system has followed increased demand since 1981, with a low level of renewal investment and no major increase in capacity other than the Docklands Light Railway.

Demand has increased on the mass transit systems operated by the London area train operating companies (TOC's) and the London Underground as a result of the upsurge in economic activity in London since 1980. In the sixties and seventies the transfer of jobs out of London as a result of Government policy led to reduced usage on Underground and heavy rail lines. This reduction in usage was also the result of increased car ownership and the availability of free on-street parking in the early 1970s as near to the Central Business District as South Kensington. Even after parking restrictions were imposed between 1970 and 1980 car usage remained at its existing level. This decline in demand coincided with the construction and opening of the Victoria Line in the 1960s; as a result its passenger figures fell short of expectations and the Treasury became sceptical about further Tube line proposals. By 1987 this overcrowding was particularly bad in certain sections of the network and this scepticism was only finally removed in London following the 1989 Central London Rail Study and subsequent evaluation studies. In consequence two out of four major schemes are now in progress or approved.

1. The Jubilee Line Extension from Green Park to Stratford via London Bridge and Waterloo has all tunnelling completed (October 1996) but it will be mid-1999 before trains begin operating. Current cost estimate is $....

2. Thameslink 2000 (Metro) is to be part-funded through a financial arrangement between the Treasury/Department of Transport (HMT 1993) and the privatised

Railtrack plc with supplemental track access charges or OPRAF agreements. Construction costs are currently (October 1998) estimated at $1200m (£700m) (excluding new trains) (Railtrack, 1998) to provide a metro-type frequency of up to twenty-four trains per hour in each direction. But this will not open to traffic until 2005.

THE QUESTIONS FOR LONDON

1. London's Transport Needs

The basic requirement for the successful operation of any major urban area is an efficient transport network for people and goods. There is deep concern that London no longer has such a system and that in consequence as a financial and commercial centre London may be the loser.

These relate to the role of London as a major employment centre (where 84% of the economy is service-related), as the centre of government for Great Britain and as a world city for finance and commerce competing with Paris, New York, Tokyo and Frankfurt (LPAC, 1991).

The transport needs (Fig 5) may be examined in relation to three modes:-

a) Rail: Radial routes in peak periods are overcrowded in the main direction of travel (and whose CBD users contribute to an estimated 20% of foreign earnings) but have relatively low load factors at other times and are relatively unimportant for overall travel. The issues of an outer orbital line to provide links between radial routes; the construction of new lines (eg Hackney-Chelsea; Jubilee Line and Crossrail) and the construction of the Channel Tunnel rail link completed in France but not started for the London end.

b) Road: Major links, both radial and orbital, have heavy traffic flows and consequent low speeds for much of the day (0700-2000). More local routes have peak congestion and the issue of preferential movements for buses and trams through traffic management has to consider its benefits (to public transport) and disbenefits (to other road users). The question of traffic management and urban road pricing has also to be seen within the context of an Integrated Transport Policy.

c) Airport: The options are to construct Terminal 5 at Heathrow Airport, or expand the airports at Gatwick/Stanstead. Any further development at Heathrow Airport will increase the present traffic flows on M25/M4 and require the Heathrow Express. Can the system cope with landside traffic or is there a need to further increase service/frequency or the public transport network (rail/buses) into the south east region from all three airports?

Fig 5 Travel patterns

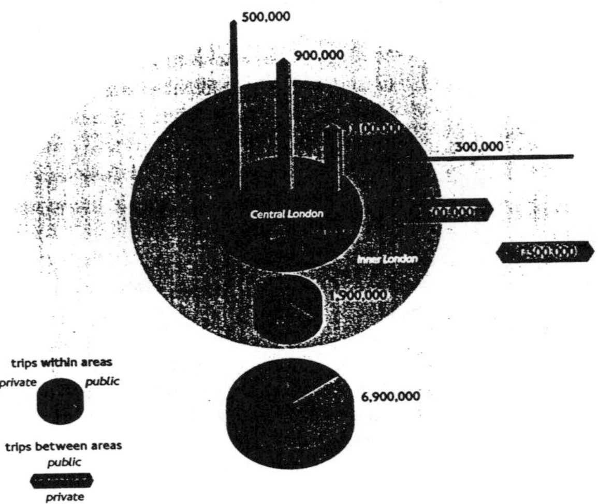

Source: London Transport (LT 1995)

The functions of a financial and tourism international destination have to be part of an integrated transport policy. Landside links from London Heathrow Airport to the central business district are currently available by metro as a heavily used commuter route (Piccadilly Line) or by bus, car and taxi via the M4 motorway while routes to all other areas are solely available by road. The development of the Heathrow Express brings additional capacity and a high speed route to the central area - or at least to the edge of it at Paddington. This development cuts the journey time to Paddington to fifteen minutes; had the East West Crossrail Scheme been built at the same time to connect Central London with Heathrow Airport and provide relief for overcrowded radial routes and both schemes been seen as an integrated network then significant reductions in journey time from airport to CBD would have been achieved (50 minutes to 20 minutes).

In considering the issue of integrated transport policy in the capital, there is a prerequisite: to establish the investment position currently and that investment level (of $ 25 bn / £15bn) needed to achieve what Sir Wilfred Newton, past Chairman of London Transport, described as a "decently modern metro". The London Underground metro system and the OPRAF/Railtrack main line railway network form the only possible way of moving London's millions of radial route travellers efficiently (in terms of speed and cost) and with minimum environmental impact.

London's rail investment requirements may be set out as follows:

Major Rail Network Extension Projects	$ million	£ million
Jubilee Line Extension	3200	1900
Channel Tunnel Rail Link	4600	2700
CrossRail	3400	2000
Chelsea-Hackney Line	2600	1500
Thameslink 2000	1100	650
Piccadilly Tube Extension to T5	120	70
East London Line Extensions	190	110
Croydon Tramlink	260	150
DLR Lewisham Extension	190	110
Woolwich Rail Crossing	260	150
Croxley Link (Watford end of Metropolitan Line)	45	25
Intermediate mode projects	500	300
Estimated costs of the major rail schemes:	9,655	16,445

Source: London Transport, Planning London's Transport (LT, 1995)

2. Predominant issues

a) Environment

How are pollutants to be reduced?
Government policy set out in Sustainable development - the UK Strategy (1995) states:

> *"A transport policy that is compatible with sustainable development objectives is one which strikes the right balance between serving economic development and protecting the environment and future ability to sustain quality of life."*

Is this balance achievable? How is it to be done? Are the environmentalists right in saying "the balance is not a part of a sustainable policy only environmental improvement is" (FOE, 1996). Cars are currently 75% of London's vehicles on major routes - are these to be the first target for removal? What role will public transport play; what will be the effect on pollutants?

b) Congestion, efficient movement and journey time

The public transport and road networks are overcrowded on certain sectors at particular times of day. This causes delays to people and to freight.

What type of investment is to be undertaken to reduce congestion?

What are the most effective ways of moving people?

If the car is inefficient and polluting how are people to be persuaded to use them less? If the public transport network is also seen as overcrowded, old, dirty, slow and infrequent (the most quoted quality factors for not using metro/buses/trains) what has to be done to that network? Why has public transport (and bicycles) become acceptable in Amsterdam for all socio-economic groups? What does the Dutch experience tell us about integrating road, public transport and traffic management policies?

3. Policy Scenario 1: Cheaper, short-term solutions

Are there cheaper solutions since no solution in the London context would be cheap? Perhaps in relative terms to the $ 40 bn (£23 bn) envisaged by London First, a series of park and ride car parks on major radial routes, would with link roads to motorways and main roads and depending on whether at grade or multi-storey construction was intended, cost between $340 m and $850 m (£200m/£500m) - modest indeed!

At a series of locations based on major radial corridors beginning where land was in public or Railtrack ownership with sufficient space for 5000 cars, construction of such interfaces could reduce traffic flows into London in peak periods by 50,000 cars - about 30% of current flows. The example is found in Oxford where free secure well-lit car parks are serviced by low fare, frequent public transport (buses - provided by private companies) but where parking in Central Oxford is expensive even by London standards and where few car users are subsidised for car parking charges as occurs on a large scale in London. Consequently similar prerequisites making park and ride fiscally advantageous would encourage usage in London.

But these provide only short-term solutions and might cause additional traffic flow problems near such parks. However the four locations suggested by London Transport using its own land and from which it hopes to generate additional parking and train revenue, are a start. But a particularly valuable opportunity may have been missed at Hillingdon in West London where a new station and track bridge were built to accommodate the new M40 motorway section to bypass Hillingdon. Direct access from road to car park provide a perfect example of such a park and ride link adjacent to a major radial route and brings the discussion back to the principles of an integrated transport policy where the car parks would be free, security controlled, safe and well lit; the fares policy for all operations would be at considerably lower levels than at present and train frequency would rise. Some but not all of the increased capital and revenue support costs would be retrieved from higher load factors and higher revenue, but providing directly for the car user to travel into London would incur prohibitive costs, worsen the congestion in the central area and create an increase in the level of pollutants. Potential savings are therefore considerable.

Policy scenario 2: - integrated transport policy

In its fullest, and, some would say, only form there is an overall public body which determines expenditure priorities, trade offs between options and sets the framework for the integration of:-

(a) public and private transport (walking; park and ride; cycle and ride)

(b) public passenger transport (seamless interchange and common ticketing for train/metro/bus/tram/boat (ferry)/air (where ticketing is excluded)

(c) electronic information systems to give comprehensive data leading to greater choice and optimum journey time and convenience.

The trade offs in expenditure would be between expenditure on:-

- road investment
- road maintenance
- traffic management
- rail investment
- bus investment
- cheap fares

This was suggested as far back as 1980 by the House of Commons Select Committee on Transport (HOC, 1980) which pointed to a Metropolitan Transport Authority responsible for roads and public transport (through a partnership) with financial and strategic planning powers.

A transport authority of this form would provide for a trade-off in transport expenditure between capital expenditure on new roads (eg urban motorways), traffic management schemes (for example through fewer parking spaces, greater resources spent on preventing illegal parking, road pricing - as a means of parking restraint, and area control) and the use of public transport (through subsidies, investment, improved service quality and integrated ticketing).

Fig 6 Average gross weekly earnings 1995 male full-time (£)

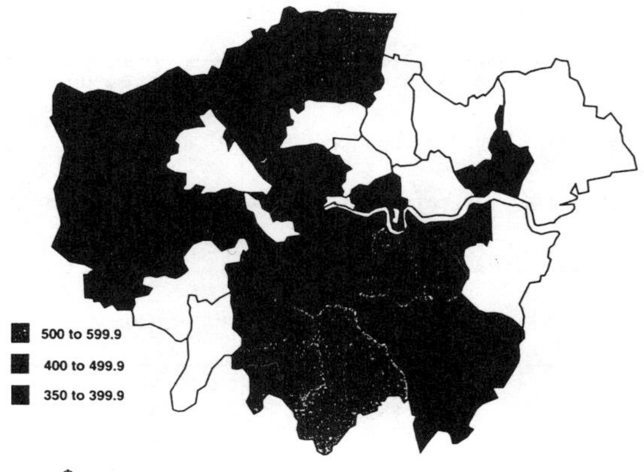

£	$
500-600	850-1020
400-500	680-850
350-400	600-680

Source: GOL 1996

Policy Scenario 3: Market forces and an integrated transport policy (The Single Market)

The cheap fares policies of the GLC between 1980 and 1983 led to a change in public awareness of rail mass transit. There was a major shift by commuters from car to rail as a result of fares reductions. This was paralleled by the introduction and development of the London Transport/BR Travelcard which made regular travel more convenient and the card's multi-ride facility led to additional trips during the day and for evening or weekend leisure travel which the cardholder perceived as `free'. The growth in popularity of London as a tourist destination has also led to increased usage of the central sections of the Underground.

In the 1980-1985 period the lowering and raising of fares (Fig 7) showed a cross price elasticity of demand over the total commuter market which indicated that the market was not separate segments - car, bus, tube, rail, but a single integrated market within which travellers would move from one mode to another as they responded to changing relative charges of each mode of transport.

Fig 7 Index of London fares and earnings

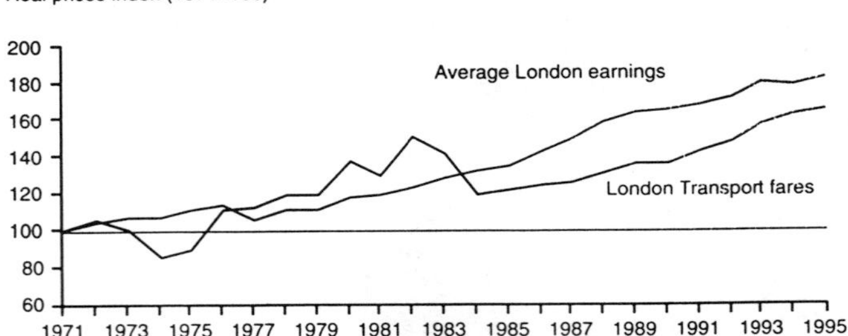

Source: Office of National Statistics; London Transport

Table 1 - Effects of LT fares charges on peak central area commuter traffic (%)

	Fare	Underground	Year on Year change Car
	%	%	%
1980-81	-38	+ 6	- 6
1981-82	+82	-13	+14
1982-83	-27	+11	- 9
1980 base figure (000's)	-	435	184

Source: London Regional Transport: The LT Fares Experience, 1984 (Economic Research Report 259 (LT, 1984)(LT 1997)

The existence of cross-price elasticity indicates a need to consider all the available transport modes in determining the most efficient solution. This applies to movements into Central London, orbital journeys, local journeys in the peaks and often throughout the day. Such a consideration can only be achieved if there is a single authority responsible for all the strategic transport decisions for the whole of London and with control of all aspects of the strategic transport budget.

Policy Scenario 4: Evaluation and financing of railway and road investment within an integrated transport policy

This brings the argument back to its most difficult point. The Central London Rail Study contained a development strategy for improving services to rail passengers, forecasts of demand; suggestions on improving existing resource utilisation, a list of strategic choices and packages of measures whose costs are justified in terms of revenue and external (to the railway) benefits. It is this last factor which gives some hope for those wish to see an Integrated Transport Policy where the proposals consider congestion on London's railways together with the options for attracting car users onto the network. But the study will also need to be put in a context of the total funding to be provided and the priorities for expenditure.

The evaluation of new roads is at present carried out on a cost benefit basis where the benefits (or returns) are relief of congestion, reduced journey time, reduced vehicle operating costs and reduced accident costs. Railway investments on the other hand are evaluated on a cost revenue basis. Inter-City rail services until 1996 received no government grant and it might be argued that currently these payments are to sustain the profitability of the privatised railway (OPRAF 1996). Previously it had an objective of earning a 5% financial (cost revenue) rate of return on assets it uses, while long distance radial (or, indeed, orbital) motorways are evaluated on a cost benefit basis.

In investment terms the primary financial criteria for the old British Rail as a whole was that investment should yield a test discount rate of return of at least 7% compared with the minimum cost alternative. In the subsidised sectors (eg the London Connection rail network), the only difference is that the appraisal is conducted in relation to the most cost efficient way of keeping the existing service running rather than developing it. Other grants (eg Section 56) can be set against the costs.

With one financing authority, the use of the same investment criteria would be possible. These would be applied to fares (and this achieve cross-price elasticity effects), improved service quality (with for example payments for maintenance of stations and train interiors), wage rates (to assist in achieving required staff levels) and capital investment in new trains, signalling, station refurbishment, and new track. These could be compared with the use of Government funds for new roads and road maintenance. The basis for the comparison should be cost benefit analysis with the inclusion of externalities such as environmental factors, journey times, congestion and comfort of passengers. It should not involve any form of cost revenue or profitability analysis since this would make such a direct comparison impossible.

An integrated system would provide better value for money since the most appropriate solution to any particular problem could be selected. At present the various highway and public transport authorities in London can only consider solutions within their own statutory limits. There is no overall authority to decide whether a road or railway would be the best of resources in that particular case.

Conclusion

This paper sets out the current socio economic, land use and travel patterns in London. On this basis it discusses the issues which arise from these patterns and poses questions and suggests solutions.

Two essential policy requirements emerge:

1. The need for adequate funding in advance of demand for both capital investment and revenue support as the services achieve higher ridership.

2. The introduction of an integrated transport policy which links land use planning to transport infrastructure (existing and planned), and which creates seamless interchanges between modes with easily available, comprehensive information systems for public transport services to enable users to achieve the optimum rout in terms of generalised cost (time and cash).
Note: Exchange rate = £1 = $ 1.7 US (May 1998)

REFERENCES

CBI (1989) Traffic Congestion in London; Confederation of British Industry, London.

CIT (1992) Chartered Institute of Transport, London's Transport - The Way Ahead (Bames R, Cole S, Finney N, Niblett R, Smith R), CIT, London.

CLRS (1989) Central London Rail Study Department of Transport, British Rail, London Regional Transport, London.

Cole, S (1998) Applied Transport Economics, Kogan page, London

DETR (1998) Mayor and Assembly for London, The Government's proposals, Department of Environment, Transport and the Regions, London.

DOT (1989), Four Assessment Studies were commissioned - South London, East London, West London and South Circular Road. The Stage 1 reports were produced by Consultants in December 1986 identifying the problems and issues. In January 1989 Stage 2 reports included recommendations. The studies covered reducing congestion, lorry movements, road safety, the environment, and access. Department of Transport, London.

DOT (1996) Department of Transport, Transport - The Way Forward (CM 3234), HMSO, London.

FOE (1996) Friends of the Earth Less traffic, better towns, FOE, London.
Fournier, Jaques (1991) The Railways in Europe SNCF today and tomorrow, Franco-British Chamber of Commerce (Additional data from J-P Loubinoux, Direction de L'Economie et finances, SNCF).

GOL (1996) Government Office for London, A Transport Strategy for London, HMSO, London.

HMT (1993) Private Finance Joint Ventures, British Government Treasury, London.

HOC (1980) House of Commons Select Committee on Transport, Transport in London (HC 127 Session 1980-81).

HOC (1993) Joint public-private sector financing of infrastructure problems (93/11/201/E/as; House of Commons Library, London.

LPAC (1991) London World City, London Planning Advisory Committee, London.

LPP (1996), London Pride Partnership, London's Action Programme for Transport: 1996-2010.

LT (1984) London Transport, The LT Fares Experience Economic Research Report R259 (M. Fairhurst), London.

LT (1995) London Transport, Planning London's Transport, London.

LT (1996), London Transport, To win as a world city, London.

LT (1997), London Transport Traffic Trends 1970 - 1995, Research Note m(97)71, LT Marketing, London

Macassey, Sir L (1927) The Problem of London Traffic, Institute of Transport Journal (November pp 14-21), Inst. of T., London.

MOT (1968) Transport in London White Paper (Ministry of Transport), HMSO, London.

OPRAF (1996) Press release on the Great Western trains franchise, The Office of Passenger Rail Franchising, London

Raltrack plc (1998), Network Management Statement, London

SNCF (1990), Rapport D'Activite 1989, Paris.

UKG (1990) UK Government, This Common Inheritance, Britain's Environmental Strategy (CM 1200), HMSO, London.

WO (1993) A470 Corridor: approaches to Cardiff (Ove Arup) Welsh Office, Cardiff.

Roadspace re-allocation increasing bus use

Kevin Gardner, BSc, CEng, MICE, DMS[1]

Abstract

This paper briefly outlines the history and current development of the role of the bus and how roadspace re-allocation particularly through bus priority is used to increase public transport demand in London. Recently changed transport policy is examined particularly encouraging increased use of public transport plus bus and pedestrian priority. Implementation of measures on the London Bus Priority Network (LBPN) and Priority (Red) Route Network (PRN) using previous policies and strategies is outlined along with their potential development using the new policies. The paper concludes with a discussion on the part Intensified Bus Priorities and advanced technology applications could play in the future within a package of complementary traffic and parking restraint measures. The bus requires far higher levels of priority than is presently the case to meet the changed policy objectives and to produce the mode shift from private to public transport and traffic reduction.

Introduction

The increase in urban traffic in the UK has given rise to increasing problems for bus operators, particularly in terms of delays and the ability to provide a reliable service. In London and in a number of UK urban areas since the 1970's major road schemes have been abandoned and balanced transport policies introduced targeting more efficient modes of transport. The introduction of traffic management measures including bus priority are key elements of these balanced policies, which has seen car commuting drop and bus passenger journeys increase over a period of 15 years in London. Other public transport provision including the benefits from the multi-modal Travelcard ticketing system and road pricing are not discussed in detail in this paper, which develops and updates a recent paper describing the history and current development of bus priority (Gardner K and Cobain P;1997).

[1] Head of Bus Priority and Traffic, LT Buses, Bus Priority and Traffic Unit, 172 Buckingham Palace Road, London, SW1W 9TN, UK.

Transport policy

In 1960 the Government set up the strategic London Traffic Management Unit to develop "total" traffic management with large one-way systems and controlled parking zones (King G;1986). These were intended to be short-term measures, increasing road capacity in advance of the construction of urban motorways and extensive off-street car parks. The schemes achieved the capacity objectives but at the considerable expense of the environment where communities were severed by heavy volumes of traffic created by the one-way systems and banned turns which required traffic to divert onto previously quiet residential streets. Buses were forced to follow the one-way systems removing buses in one direction from main passenger objectives, thus making them difficult to find and use. The urban motorway proposals, generated enormous opposition and were abandoned by the strategic traffic authority the Greater London Council (GLC) in 1973. By the early 1980's a road network had developed primarily to meet the needs of cars. To rectify this, GLC policy in the 5 years preceding abolition in 1986 was directed to "Changing the Balance". The key element of this policy was the provision of efficient and relatively cheap public transport. This improved mobility for all those who did not have access to private transport and also enabled a start to be made towards the objective of reducing the increases in non-essential traffic by persuading some drivers to change to public transport where convenient services were available. A balanced transport policy, updated by the 1989 "Strategic Guidance for London", continued until recently.

Nationally, the UK's integrated transport White Paper "A New Deal for Transport: Better for Everyone" (DETR;1998a) and within London "Traffic Management and Parking Guidance" (GoL;1998) provide new policy guidance addressing increasing concerns about impacts of transport on the environment and how far forecast traffic growth is sustainable. The integrated transport White Paper includes policies proposing:

- more and better buses;
- better interchange and better connections;
- enhanced public transport networks with simplified fares;
- more reliable buses through priority measures and reduced congestion;
- improved personal security when travelling; and
- easy access public transport.

Revised London "Traffic Management and Parking Guidance" was issued by the Government Office for London in February 1998. The guidance provides a new framework for traffic management and parking implementation and is directed towards London's various highway authorities. It also gives a very positive change of emphasis with regard to the provision for buses, with the opportunity to obtain further bus priority and traffic management measures assisting buses. Local highway authorities are urged that bus priority measures should be concentrated along complete bus route corridors to assist bus operations and passengers. The importance of effective enforcement is recognised as essential for the bus priority to fully succeed and local authorities should give priority to removing vehicles hampering buses. It is also acknowledged that the community benefit resulting from bus priority means some disbenefits to frontager businesses. Finally the application of Intensified Bus Priorities, which would result in delays to other traffic, has been highlighted as an area where innovation and experiment would be appropriate subject to resources. One of the core principals of the revised guidance indicates that the management of traffic and roadspace should be based on the movement of people and goods (rather than vehicles). Examples of the revised aims/guidelines include:

- to facilitate the movement of people and goods.................;
- to support reduced car commuting, especially into/across inner and central London;
- to assist measure to reduce local traffic; and
- to provide priority for buses so as to achieve their efficient movement.

Implementation and benefits of bus priority

Bus lanes, protecting buses on the approaches to congested junctions, and other traffic measures to assist buses first appeared on London's roads in 1968; other schemes followed primarily between 1972-79 and 1982-86. Additionally during the latter period, to support the recently agreed balanced transport policy, the introduction of active bus priority at traffic signals began, thus complementing bus lanes by giving the bus priority through, as well as on the approach to, junctions. Between 1968 and 1990, some 240 bus lanes were introduced and some 56 signal junctions provided with Selective Vehicle Detection (SVD) to give buses priority.

Despite a decade of balanced transport policies, until 1990 bus priority measures were spread thinly across the bus network and, as a result, most of the key bottlenecks remained. In August 1991 the "Green Routes" report (MTRU for LT and LPAC;1991) on bus priority recommended that the vast majority of bus routes in London would benefit from end to end bus priority. In addition, at that time, a lot of London's rail network was running at full capacity hence a modal shift to bus offered a low cost solution and one which could be achieved in a fairly short timescale. The current bus priority initiative began in 1992 with three bus route length demonstration projects in South and West London (Blitz R and Yates M;1997), the Uxbridge Road (Gardner K and Metzger D;1997) and North Docklands, a total length of approximately 50 miles. The development of measures involved a close partnership between the 11 boroughs involved, London Transport (LT) and the bus operators and the physical bus priority measures were introduced between 1993 and 1998. Bus route 220 (South and West London) and routes 207/607 (Uxbridge Road) were selected as the benchmark routes because they have the most extensive set of physical bus priority schemes following consideration of measures on an end-to-end basis. On Route 220 the average bus journey time savings over all peak periods showed an improvement of over 8 minutes, a journey time reduction of 14.5%. Cash paying patronage has showed an increase trend of around 10% per annum since 1993. This is much higher than the recorded 'control' routes where there has been no bus priority. Similar improvements in bus speeds have been seen on the routes 207/607. Patronage data shows a general trend of increased usage of about 15% over and above any changes in the control routes over a two year period. Good economic rates of return have been achieved on all the demonstration projects.

Following the success of the demonstration projects, in 1994 approval was given to develop a London Bus Priority Network (LBPN). This 540 mile network, covers around 65% of the worst traffic trouble spots for buses. 30% of the remaining problems are being dealt with in the 314 mile Priority (Red) Route Network (PRN) programme. Once the networks are in place it is anticipated that there will be a measurable improvement in bus service reliability and a significant reduction in bus journey times commensurate with the results from the demonstration projects. The time savings achieved from the two networks, when bus route length implementation has been completed, will be ploughed back into improving bus services further. The LBPN incorporates nearly all bus routes within the Inner Ring Road and most others with 8 buses per hour (one way peak) or more. Implementation of LBPN bus priority measures started in 1996 and will progressively increase over the next few years with completion in 2003, at a total cost of £80 million. The Red Route Network, covering London's strategic roads, aims to improve road conditions for all users not just buses. There is a no-stopping regime, with exemptions for buses at stops and designated parking or loading bays, identified by special signs and red lines which replace the waiting and loading signs and yellow lines. Stricter enforcement by police and their traffic wardens on these roads, higher parking fines and additional bus priorities are expected to assist buses along Red Routes, with a target 10% bus journey time reduction. Bus routes operate on about 250 miles of the Red Route network and it is presently proposed that the network will be finished by the year 2000.

"Active" traffic signal bus priority utilises Selective Vehicle Detection (SVD) within various traffic control strategies such as vehicle actuation, fixed time Urban Traffic Control (UTC) and traffic responsive SCOOT. At vehicle actuated sites, the system pays for itself in about 15 months, with bus delays at the junctions reduced by 32% (9 seconds per bus per junction) and a reduction in variability of 22%. Active bus priority within UTC systems and has been developed as part of the EC DRIVE II project PROMPT (Bowen G, *et al*;1994). Results from the PROMPT trial in Camden Town and Edgware Road (Hounsell N and McLeod F;1995) plus further implementation at a further 50 installations indicate that BUS SCOOT bus delay savings average between 20%-30% (4 seconds to 8 seconds per bus per junction) and that overall benefits to all traffic produced benefits to buses repaying over 70% of system costs within the first year.

Until now the introduction of the bus priorities, have at best protected buses and their passengers from the increases in traffic congestion and slower traffic speeds. They are however, a key component in the implementation of balanced transport policies which has achieved a reversal in the spiral of decline of bus travel in London. Since 1982 there has been an increase in local bus passenger journeys, compared with a fall in the deregulated metropolitan areas of the UK (see Figure 1), additionally priorities have played their part in reducing road based commuting into Central London by private car. Today's balanced transport and bus priority initiatives, introduced from 1990, are the first to really tackle the challenge of protecting buses from traffic congestion on a comprehensive basis and are:

- supported by all the main authorities;
- starting to treat complete bus route corridors comprehensively; and
- starting to assess the people-moving capacity of roads, rather than vehicle capacity.

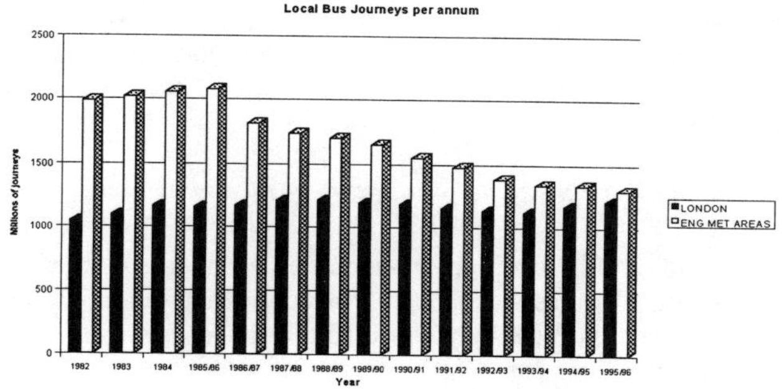

Figure 1

As of March 1998 there are 525 bus lanes in London, 126km in length (see Figure 2); 24 bus advance areas and 349 traffic signal bus priority junctions operating. London-wide implementation of traffic signal priority is proceeding at some 1000 signal junctions and at some pelican crossings.

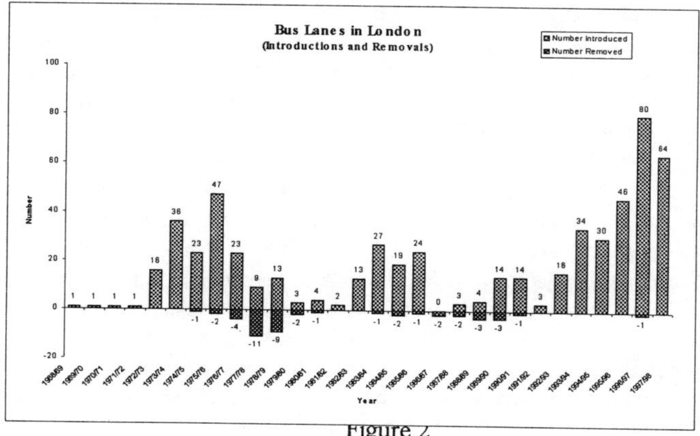

Figure 2

The revised objectives set by the integrated transport White Paper and Traffic Management and Parking Guidance emphasise the need for intensification of the LBPN and PRN in relation to roadspace reallocation. Results from both initiatives, as judged by the cost/benefit evaluation of the demonstration and pilot schemes, indicate the measures have been extremely successful with impressive first year rates of return, often in excess of 100%. They have also had the important additional benefit of attracting higher bus patronage. There is little evidence however, from either study that significant mode shift has occurred from the private car. There is limited evidence from a small scale attitudinal survey undertaken for the South and West London Bus Priority Demonstration project, which indicated that the increased bus patronage from bus priorities resulted in a mode shift of 3.7% from car to bus. However for the PRN, with the aim of improving the flow of traffic, including achieving as far as practicable a 10% reduction of general traffic journey times, there is little encouragement for the private motorist to use public transport, walk or cycle.

Shifting the balance through greater roadspace re-allocation

Looking to future roadspace re-allocation using urban traffic management systems and parking measures, it is firstly worth considering the "do-nothing" scenario. Potentially the increase in urban traffic will continue and give rise to ever increasing problems for bus operations particularly in terms of delays and the ability to provide a reliable service. The spiral of decline of buses will re-commence, and it is against this scenario that all potential actions should be judged.

Better understanding of the relationships affecting travel behaviour is opening the way to policies giving greater priority to public transport on London's streets. It is now recognised that there is a relationship between the level of congestion within a large city such as London and the speed and quality of public transport. People with a car and parking space available will tend to use it unless congestion is such that they can reach their destination more readily by the next best alternative. Until recently, it has been assumed that transferring present roadspace to free buses from congestion would be impractical; other traffic would either divert and intrude into unsuitable roads or the whole network would become gridlocked. There is now growing recognition that such fears have been exaggerated. Analysis of road users' reactions to changes in road capacity (MVA *et al*;1998) suggests that, far from causing excessive congestion, when reducing capacity for cars (for example through pedestrianisation schemes, traffic calming, bus priorities, roadworks, bridge closures, trip end restraint and disaster effects) people readjust their travel patterns more frequently and more easily than previously thought. While there are some disbenefits for those making precisely the same journeys as before, changes in behaviour result in traffic congestion remaining broadly constant. The preliminary view is that some 25% of traffic "disappears" from the area network in response to highway capacity reduction and the predicted traffic chaos does not materialise.

If the bus is to make any significant contribution to reducing private vehicle use it must become an attractive mode, particularly to car users. This has given rise to studies being undertaken which aim to create more sophisticated bus priority measures than those previously implemented. These may be thought of as "Intensified Bus Priority" which requires the segregation of buses from general traffic along parts of the bus route and maximum junction priority. Along the sections not physically segregated, fully enforced measures and metering techniques are required to ensure that the bus is not affected by congestion. Central to many of these schemes is a willingness to shift the balance of roadspace in favour of sustainable transport strategies. Typical Intensified Bus Priority measures are described below:

Busways: Segregation can be achieved by introducing busways which allow buses exclusive rights of way and are physically separated from general traffic. The complete separation from other traffic gives the same freedom from congestion enjoyed by rail systems. This increases speed of operation and reliability of services by freeing buses from the variability of traffic delays.

> Signal system and timing strategies; traffic metering plus system integration: Active traffic signal bus priority via SVD is now possible within most signalling systems. Presently this provides benefits to buses without significantly disbenefitting other vehicles, however, once systems are fully implemented the signal timings and priority parameters can be used as the tool to provide the levels of roadspace re-allocation to public transport, pedestrians and cyclists as determined by individual transport policies. Traffic metering signalling techniques enable general traffic delays to be transferred from narrow sections of road to wider sections. Bus lanes provided on the wider sections allow buses to bypass the relocated queues. The metering reduces congestion on the narrower sections of road allowing traffic to flow more freely. The bus gains from the bus lane on the wide section and the freer flowing traffic on the narrow section. Developments are in hand to integrate priority traffic control systems, and bus Automatic Vehicle Location (AVL) management systems. This approach will be part of a system providing bus operating data on vehicle headway, timetable and bus route to the signal system to target improved bus reliability, by giving greater signal priority for late running buses.

Enforcement: The importance of enforcement cannot be overstated. If all violations of bus lane facilities were eliminated, the efficiency of bus lanes could improve by up to 50%. The Traffic Director for London, is implementing an area scheme which places enforcement cameras at roadside locations and on buses, with the aim of providing evidence suitable for prosecutions. The London Traffic Control Systems Unit (TCSU) has developed complementary roadside CCTV camera systems which will be used for enforcement early in 1999.

Bus advance areas: A bus advance area is a scheme which permits buses to advance into an area of road, cleared of traffic, before the main signalled junction. Traffic signals ("pre-signals") in advance of the major junction always control traffic entry to the advanced area; a bus lane is provided up to the pre-signals. The objective of the pre-signal and advance area is to re-order vehicles such that buses may be given priority to reach the main junction and is particularly useful for right turning buses. The traffic metering technique can be enhanced by using a bus advance areas and is particularly applicable on Town Centre approaches where because of constraints such as roadwidths and loading requirements, bus lanes or busways cannot be introduced. The technique has the advantage that bus journey time benefits can be obtained throughout all periods of the day, 7 days a week and may be operated only when congestion problems occur.

Bus lanes: Though widespread, potential to improve the traditional design exists. For example route length bus lanes would give high levels of priority. In the UK most bus lanes end prior to signalled junctions at what is known as the setback area. The purpose of the setback is to maintain junction capacity, usually the main constraint in maintaining link capacity. However, were roadspace re-allocation to be promoted the bus lane setback could be shortened or removed and the hours of operation extended.

Bus gates and pedestrian/bus only roads: Bus gates are short sections of bus only, or bus and other permitted vehicle only sections of road. Bus access to pedestrianised streets gives buses considerable advantages over the car by being able to deliver passengers directly into shopping areas.

Bus stop boarders and clearways: New low-floor accessible single, and double-deck buses are being introduced throughout London. To achieve the maximum access on and off the bus, stops need to be free from parked vehicles and buses need to be able to stop directly at the stop. This can be achieved by provision of bus boarders, kerbing and bus clearways.

Traffic management and restraint measures: Restraint by parking, congestion charging and physical restraint, the latter to ensure induced through traffic does not result, would re-allocate roadspace to buses. This would make driving into town centres less attractive and have a positive effect on bus operations. Speeds and hence patronage would increase. A modal shift towards public transport from private cars would be encouraged as public transport becomes relatively more attractive. The extra passengers will produce additional revenue which would allow more frequent services to operate, providing a virtuous circle of improvement.

Demonstrations and modelling of Intensified Bus Priority

A European Community project, INCOME (INtegration of traffic Control with Other MEasures) is presently being undertaken within London and other European cities to study some of the essential elements of Intensified Bus Priority (Hounsell N. *et al*;1997). The work centres on use of UTC systems to provide "active" bus priority and traffic restraint/metering plus the potential of providing greater selective bus priority to individual buses targeting reliability. The trials have taken advantage of the implemented physical bus priority measures along the Uxbridge Road bus priority demonstration project corridor, while network traffic metering has been undertaken in, and on the approaches to, Twickenham Town Centre, which has a number of bus lanes introduced as part of the LBPN. Preliminary results from the project (Southampton University *et al*;1998) indicates that the traffic control bus priority provides a further 4 seconds per bus per junction benefit along the Uxbridge Road UTC network surveyed. With over 50 signal junctions along the corridor this value-added benefit will be in addition to the 4 minutes bus benefit resulting from the physical bus priorities. The synergy of traffic metering in Twickenham has enabled a further 2.2 seconds per bus per junction benefit in the morning peak period when the gating strategy operates. Simulation of integrating UTC with the bus Automatic Vehicle Location management system has indicated that benefits to bus journey time and reliability, by providing enhanced priority for late buses (based on headway), can potentially be doubled.

It should be noted that the INCOME trials study only certain aspects of potential Intensified Bus Priority measures. In particular they do not provide the level of road based public transport segregation essential to provide a high quality public transport system. INCOME has fully demonstrated the benefits to be obtained by introducing active bus priority and traffic metering on town centre approaches where roadwidth and other constraints preclude complete segregation of public and private vehicles. Recent studies including those by the Cross River Partnership and the LT study of Intermediate Modes in London provide pre-feasibility indications of potential benefits of the segregation of buses as part of an Intensified Bus Priority scheme and an integrated transport initiative. The Green Areas study (MTRU for LT Buses *et al*;1996) involved examining an area of West London with a view to progressing the idea of providing an efficient, sustainable, multi-modal transport system, as a key contribution to improving the economic, social and environmental fabric within a designated area. The report builds on the new generation of bus priority measures. The key concepts of the report explain the need for an objectives led approach focussing on: total journey quality; transport as part of the nature of place; a mutually supportive, multi-modal approach; sustainability and local transport; reducing the need to travel; rebalancing the network; and partnership, participation and leadership.

The London Transportation Survey (LTS) model was used to provide advice contained within the Traffic Management and Parking Guidance. This work has attempted to investigate different London-wide effects of such individual strategies as parking restraint, LBPN/PRN implementation and Intensified Bus Priority, together with a combination of individual strategies. The findings from the strategic appraisal modelling of LBPN and Intensified Bus Priority implementation are very encouraging, particularly when combined with parking restraint. These results indicate a reduction in some 700 million highway vehicle kilometres per annum (3%) and bus passenger journey kilometres rising to over 5,500 million passenger kilometres per annum (a rise of over 30% from the base 2001 LTS estimate). The Intensified Bus Priority plus parking restraint tests also indicate:

- 1.8% PT mode share increase (including a significant shift from rail to bus)
- 3% more PT passenger kms including 32% more bus passenger kms;
- 3% less highway vehicle kms;
- 0.6% reduced fuel consumption;
- highway vehicle speeds unchanged; and
- 3.6% fewer annual personal injury (PI) accidents.

High quality public transport

As well as improving buses' performance through roadspace re-allocation and Intensified Bus Priority, a high quality bus service is also required. LT Buses, who procure bus services within London through a tendering regime, are addressing this by the introduction of *Bus 2000.* Initiatives comprise:

Improvements to the bus network: Including new inner London orbital routes; new outer London routes providing better links to town centres and rail stations from residential areas; higher levels of off-peak service provided on key routes; and routes operating 24-hour services increased from roundly 50 to 70.

Vehicle and service quality improvements: Bringing forward delivery of already contracted new vehicles, including low-floor double decker buses; negotiating new vehicles on to contracts not yet re-tendered; retrofitting wheelchair ramps to non-fitted low-floor buses; refurbishing older vehicles; customer care courses for bus operating staff; more new shelters plus lighting fitted to all shelters; and increasing staffing coverage at bus stations.

Fares and ticketing changes: Including changing from existing bus fare structures, which are partly zonal and partly distance based, to a system of flat fares; plus a trial of two-stream boarding. All initiatives are aimed at speeding boarding of passengers on to and off buses to reduce delays at stops.

Information and marketing upgrades: Including new local maps and diagrams at key locations; new style departure lists at each stop replacing timetables; more Countdown "real time" passenger information at bus stops; "next stop" signs on buses; extensive marketing of this improvements package; and wider based Travel Awareness and Green Commuting initiatives.

Major projects combining aspects of all of these: Bus demonstration projects are currently planned to include 5 key bus routes. These routes will receive new low-floor fully accessible double-deck buses, as part of new contract awards, over the next two years; while the traffic authorities have also agreed to incorporate these services into "whole route bus priority" schemes over the same timescale, incorporating either second generation, enhanced priority based on new guidelines or intensified priority.

Organisation

Largescale implementation of bus priority schemes and complementary traffic and parking measures to assist the bus have been made both by the strategic authority, the GLC, and more recently by individual authorities co-ordinating package LBPN implementation. The proposed arrangements, as contained within the strategic Greater London Authority (GLA) White Paper (DETR;1998b), whereby the GLA will be the highway and traffic authority for the strategic London road network and will have both positive and negative reserve powers for the LBPN implementation plans, which the individual authorities will implement. The proposals provide the ideal organisational opportunity to maximise bus priority implementation and achieve roadspace re-allocation. The importance of setting up a specialist bus priority team co-ordinating such an initiative is emphasised in the conclusions to a "Bus Priority Review" paper (Dean J;1983): "for others contemplating the implementation of a bus priority policy, an approach whereby staff are specially recruited and trained for the work is strongly recommended".

Conclusion

The spiral of decline of bus travel in London was halted during the early 1980's. The implementation of bus priority measures combined with balanced transport policies has been an essential element in the reversal. However, if the increase in urban traffic continues, this could be halted or reversed.

There is now considerable evidence that bus priority measures provide substantial benefits to bus passengers and bus operators with no significant disbenefits to other vehicles. When implemented over a complete bus route length, network efficiency benefits result. Cost benefit analysis indicates that these measures represent excellent value for money with typical first year rates of return often in excess of 100%. The introduction of bus priority measures involve elements of roadspace re-allocation, and the important question to be resolved is the amount of roadspace re-allocation required to achieve sustainable transport policies and objectives. The implementation of the 540 mile LBPN and 315 mile PRN, introduced up to the revised policies and objectives in early 1998, will lead to a relatively small level of roadspace re-allocation. The networks should improve bus journey times by up to 15%, which in turn should lead to patronage increases of around 10%. However, the amount of mode shift from car to bus is expected to be limited, without higher levels of roadspace re-allocation.

A greater, medium, level of roadspace re-allocation should result from the new policies in the integrated transport White Paper and Traffic Management and Parking Guidance. The revised objectives, particularly for the LBPN and PRN, could be enhanced further in the future by targets as well as decreasing bus journey time (and reliability) to also include an improvement in the bus journey time relative to other vehicle speeds. The additional benefits derived from such measures would be quantified by the mode shift from car to public transport, along with the social, environmental and accessibility benefits, as measured by an objectives led approach.

In order to attract substantial numbers of car users to buses the total experience of bus travel will have to improve, involving a close partnership between highway and public transport authorities, their agents and operators. The "Green Areas" report provides a clear lead to the way forward and a framework to assess the introduction of Intensified Bus Priority in conjunction with bus initiatives such as *Bus 2000* plus other traffic management and parking roadspace re-allocation measures and strategies. Intensified Bus Priority will require extensive separation of the bus from other traffic giving rise to the sort of journey times and reliability obtainable with light rail systems. Where this cannot be achieved because of roadwidth or other constraints such as essential loading and access requirements, particularly through town centres, then traffic metering protecting buses on the approaches to town centres can be successfully used. The INCOME EC project has successfully proved the value added benefit of combining traffic control bus priority and physical priority plus the synergy of traffic metering, potentially doubling the bus benefits of the Bus Priority Demonstration project along the Uxbridge Road and targeting improved bus reliability.

If Intensified Bus Priority is implemented and the private car is less attractive, due to congestion or specific restraint policies, the highest levels of modal shift to the bus are likely to be achieved. Such a package approach provides the most appropriate strategy to achieve an efficient, sustainable, multi-modal transport system, a key contributor to improving the economic, social and environmental fabric within a designated area. Such high levels of roadspace re-allocation have not been attempted previously in the UK and their effect would depend on a number of factors such as public acceptance and traveller willingness to change their travel patterns, which may take a number of years to achieve. However the bus improvements will improve the image and will attract more passengers making the roadspace re-allocation more acceptable to the public. To support the further development of roadspace re-allocation and bus priority it is important that a specialist, dedicated team is provided within the proposed strategic GLA and that they develop a series of area wide trials, evaluated over a period of time to test empirically a high level roadspace re-allocation strategy.

REFERENCES

Bowen G, Bretherton R, Landles J, Cook D-Active Bus Priority in SCOOT, IEE, 1994

Blitz R and Yates M – Bus priority:the South & West London experience; PTRC 1997

Dean J - Bus Priority Review; PTRC April 1983

Department of the Environment, Transport and the Regions (DETR): A new deal for transport, better for everyone – integrated transport; July 1998a

DETR: A Mayor and Assembly for London; March 1998b

Government Office for London – Traffic Management & Parking Guidance; Feb 1998

Gardner K, Metzger D – Uxbridge Rd bus priority demonstration project; PTRC 1997

Gardner K, Cobain P - Bus priorities: A solution to urban congestion; Proceedings of the Institution of Civil Engineers, Transport; November 1997

Hounsell N, and McLeod F - Field Trial Implementation and Evaluation, Deliverable No 19, University of Southampton for DRIVE II Project PROMPT; 1995

Hounsell N, Landles J, Bretherton R, Gardner K – Intelligent systems for priority at traffic signals: the INCOME project; 4th World Congress on ITS, Berlin, October 1997

King G: Traffic Management in London–state of the art at GLC abolition; PTRC 1986

Metropolitan Transport Research Unit (MTRU) for LT, LPAC and 6 London Boroughs - Green Routes: Report on Bus Priority; August 1991 (Unpublished)

MTRU for 3 Borough and LT Buses - Green Areas; May 1996. (Unpublished)

MVA and ESRC Transport Studies Unit (University College London) for LT and DETR - Traffic Impact of Highway Capacity Reductions; February 1998

Southampton University, LT, TCSU, TRL – Final report on results of UTC/PTS strategy implementation & evaluation for London, INCOME Deliverable No 18; 1998

SOCIOECONOMIC CHARACTERISTICS, LAND USE AND TRAVEL PATTERNS IN LOS ANGELES COUNTY

James de la Loza* and A.S. Narasimha Murthy**

Introduction

The ability to move people and goods, particularly during peak periods, is a challenge to both administrators and traffic engineers in Los Angeles County. There have been several major capital investments in transportation infrastructure such as the subway, light rail lines and freeway high occupancy vehicle (HOV) lanes. The Los Angeles County Metropolitan Transportation Authority (MTA) has been designated as the congestion management agency for the County and commissioned to plan and program all transportation improvements into the 21st Century.

The MTA was established in 1993. It is the nation's second largest provider of public transportation (New York City is number one). The MTA is responsible for developing an integrated Metro system for the County, which includes improvements to light and heavy rail transit, commuter rail, bus and paratransit services, carpool and bus lanes. The MTA has made considerable progress in creating a unified public transportation agency. It continues to construct and operate a safe, cost-efficient and reliable public transportation system for Los Angeles County.

Socioeconomic Elements

Los Angeles County, established in 1850, is located in Southern California (State of California) on the West Coast of United States of America. Los Angeles County is the State's most populous county. The county covers over 10,400 km^2 (4,000 square miles of which 7,426 km^2 (2,856 square miles) is unincorporated. There are 88 incorporated cities within the County with ten cities over 50,000 population. The average population density is 912 persons per km^2 (2,372 persons per square mile). In 1995, the County population was 9.4 million people. By 2020,

*Executive Officer, Regional Transportation Planning and Development, Metropolitan Transportation Authority, Los Angeles, CA 90012

**Project Manager, Regional Transportation Planning and Development, Metropolitan Transportation Authority, Los Angeles, CA 90012

the population is projected to increase by nearly 2.9 million people, an increase of 31%. Table 1 shows the historical population growth in Los Angeles County, Los Angeles City, and State of California.

The City of Los Angeles is the largest city in the County and covers 1,209 km^2 (465 square miles) with a population of 3.7 million. The City has more kilometers (miles) of street channelization than any other jurisdiction in the world with 10,240 street km and 256 freeway km (6,400 street miles and 160 freeway miles). Also, it has the second highest number of city traffic signals in the United States. The City is surrounded by several other cities and over 100 neighborhoods. Los Angeles City Department of Transportation (LADOT) has an annual budget of $77 million with 1,450 employees. The City also administers a transit program.

The county is one of the most ethnically diverse metropolitan areas in the world. In the 1980's, the Latino population of the County increased by 1.5 million and the population is spreading eastward in the Los Angeles basin. People from over 140 countries live in the County and traffic reports in seventeen foreign languages can be heard in Los Angeles. Table 2 shows the ethnic groups equal to or over 100,000 population in the County.

Los Angeles area per capita income exceeds that of the rest of the nation, but it is offset by the high cost of living. During 1980's and 1990's the median household income in the county rose by 18.8% to $34,965. The household expenses, estimated by U.S. Department of Labor for the county, are $33,480. It is also estimated that on an average the transportation cost per annum is $6,000 for those owning a car. A 1989 study by the University of Los Angeles (UCLA) indicated that 15.6% of the local population live in poverty.

TABLE 1: POPULATION TREND FROM 1781 TO 1990

Year	City of LA	County of LA	California
1781	44	-	-
1830	770	-	-
1845	1,250	-	-
1850	1,610	3530	92,597
1860	4,385	11,333	379,994
1870	5,728	15,309	560,247
1880	11,183	33,381	864,694
1890	50,395	101,454	1,213,398
1900	102,479	170,298	14,850,53
1940	1,504,277	2,785,643	69,500,00
1960	2,481,595	6,038,771	15,863,000
1970	2,811,801	7,055,800	20,03,9000
1980	2,967,000	7,477,503	23,78,0100
1990	3,485,390	8,769,944	29,976,000

TABLE 2: ETHINIC GROUPS

Mexican	2,519,514
African-American	950,000
White	350,000
Salvadoran	350,000
Filipino	350,000
Armenian	250,000
Japanese	200,000
Chinese	170,000
Guatemalan	100,000
Greek	100,000
Thai	100,000

Industry took root in the Los Angeles basin in the 1890's and peaked during and after the Second World War. The majority of manufacturing originated at the rail yards in downtown Los Angeles and spread towards the harbor (south). During the 1920's and 1930's, industries such as movie making, oil refining and aircraft construction were prominent. According to records during 1940's and 1950's, many light industries such as pottery, flour mills, meatpacking, soap manufacturing and milk processing took root in the basin. In the 1960's to present day, the largest employers have been the film industry, aerospace, ship building, apparel and garment textiles, tools, scientific instruments, industrial machinery, banking, painting and publishing trades. In 1995, the County employment base was estimated at 4.2 million jobs and is expected to increase to 5.8 million by 2020.

Land Use Pattern

Single family detached housing is an indelible part of Southern California. The City of Los Angeles in 1980's was said to contain more private houses per capita than any other city in the world. The Los Angeles City Planning Department estimates that out of 1.4 million housing units possible, nearly 1.3 million have already been built. One hundred forty-four developers spent a total of $5.4 billion in 1985 to erect new homes in the County. Since 1995, local jurisdictions began to report building permit activity (construction and demolition) as part of the Countywide Deficiency Plan process or MTA Congestion Management Program (CMP). From June 1994 through May 1997, permits for 26,563 dwelling units, and 3.4 million square meters (37.9 million square feet) of non-residential permits were issued. The non-residential category included commercial, industrial and office space.

Travel Patterns

The transit modal split in the county is 7.7% for home to work trips and 2% to 3% for all other trips. There are over 10,000 signalized intersections, 10,400 freeway km (6,500 freeway miles) and 8,000 major arterial km (5,000 major arterial miles). In 1996, the top six most congested urban areas in the United States included: Los Angeles, Washington D.C., San Francisco, Oakland, Miami and Chicago. Each day, over 280-million vehicle km (175 million vehicle miles) are logged or the average vehicle km (miles) per vehicle is equal to 45 km (28 miles). The annual average delay is estimated at 63 person-hours per eligible driver. The cost of congestion is estimated at $8,620, per year, per eligible driver. Also, Los Angeles has the highest Roadway Congestion Index (RCI) of 1.52 among fifty urban areas in the Country. RCI is a ratio of existing daily vehicle-kilometers of travel per lane (DVKT) to calculated DVKT values identified with congested conditions. An RCI value of 1.0 or greater indicates that congested conditions exist areawide.

The number of registered vehicles in the County was over 6.2 million (1996). The number of vehicles per household in the County is 1.68 (1996). The percentage of households in the County without autos was 9.2% (1996). The average a.m. peak period speed, by 2010, is expected to be 24 km per hour (15 miles per hour) without improvements made to the present transportation system. In some rapidly growing areas, speeds could drop to as low as 16 km per hour (10 miles per hour) during a.m. peak hour per MTA transportation model.

The percentage of Single Occupancy Vehicle (SOV) trips has reduced from 80% in 1994 to 78% in 1996. The average commute distance has reduced from 24 km (15.3 miles) in 1994 to 23 km (14.6 miles) in 1996 (MTA transportation model). The average commute time remains the same at 30 minutes for both years. Also, the commute satisfaction rate remains the same at 6.6 (on a scale of 1 to 9, with "1" the lowest and "9" the highest). The work trip transit and the carpool & vanpool mode share was estimated at 7.4% and 19.3% respectively in 1996 per the MTA model. The percentage of employment that could be reached within 60 minutes by transit was only 16%. The estimated transit person speed and bus speed were 19 km per hour (12 miles per hour).

Freeway and Highway System

Freeways and highways have unified and defined the physical structure of Los Angeles since 1960's. Streets and freeways are the backbone of the County transportation system. There are over five freeway interchanges that carry over 500,000 vehicles per day. Table 3 shows the average daily traffic on selected freeway interchanges. The Los Angeles County freeway system is shown in Figure 1.

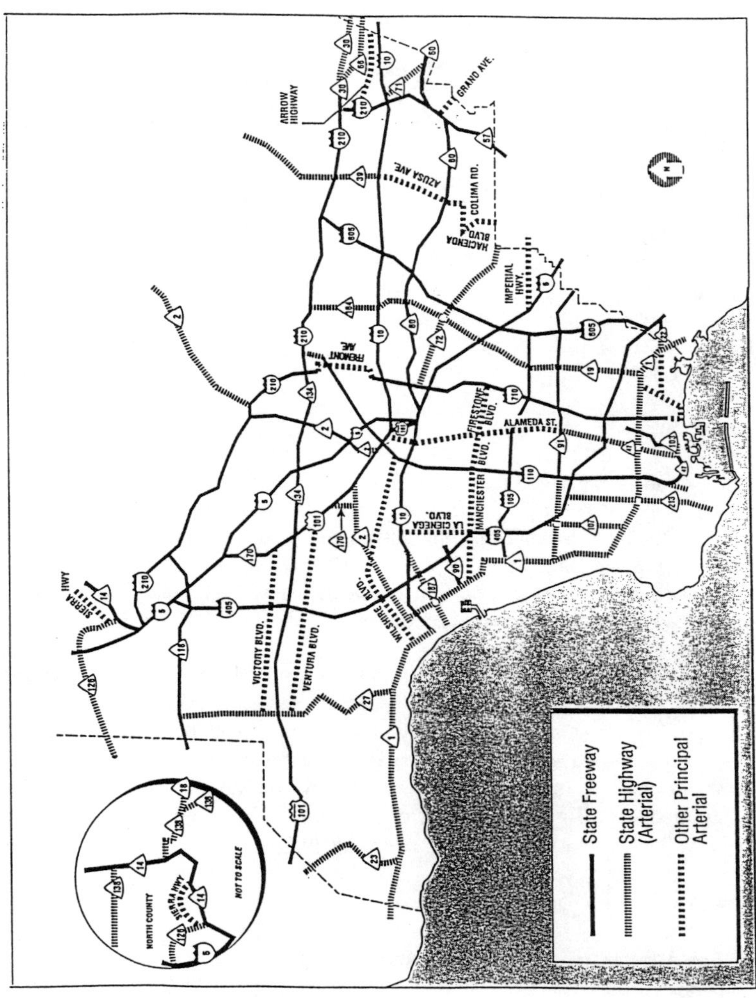

Figure 1: Los Angeles County Highway and Roadway System

TABLE 3: AVERAGE DAILY TRAFFIC AT SELECTED FREEWAY INTERCHANGES

(in thousands)

Interchange	1972	1982	1992	1995
US 101 at I-110	165	192	232	233
US 101 at I-405	187	238	273	300
I-405 at I-10	208	201	313	313
I-10 at I-210	75	116	181	188
I-5 at I-10	127	164	228	285
I-5 at I-605	122	139	192	238
I-605 at I-210	58	77	95	105
I-605 at I-5	133	165	222	238
I-605 at I-405	12	28	35	34

Source: Caltrans

Los Angeles has added the Glenn Anderson (I-105) Freeway in 1994 leading to the Los Angeles Airport. The next major freeway segment, Route 30, is anticipated in 2002. Following this, the completion of the remaining I-710 Freeway segment is not expected until after 2015. Due to right-of-way and construction costs, land constraints, and concerns about environmental impacts, no additional freeways are programmed for construction in the County.

There is an extensive program for adding carpool lanes or High Occupancy Vehicle (HOV) lanes. Over 205 km (130 miles) of freeway carpool lanes have been added in the County since 1992 with funding programmed by MTA. With the capacity to move three times as many people as a regular lane, carpool lanes make more efficient use of our already over-crowded freeways, and are critical to maintaining mobility. The total number of miles of HOV is 264 km (165 miles) in 1998 and it will increase to 480 km (300 miles) by 2010.

Earthquakes

The Los Angeles transportation system must cope with natural phenomenon such as earthquakes, mudslides, wind storms, sand storms and wild fires. Earthquakes occur as frequently as 30 times a day in Southern California, ranging from 0.5 to 3.5 on the Richter Scale. Three major faults run through the Los Angeles area (the San Andreas, the San Gabriel and the Newport- Inglewood faults). Table 4 shows major earthquakes in the Los Angeles area. Due to the frequency of major earthquakes in the area, it is necessary to retrofit old bridges and columns to new seismic design requirements.

TABLE 4: MAJOR EARTHQUAKES IN LOS ANGELES COUNTY

Epicenter	Year	Richter	Time
Tejon	1857	7.7+	8:00 a.m.
N.W. LA	1893	6.0+	11:40 a.m.
Long Beach	1933	6.3	5:54 p.m.
Brawley	1940	7.1	8:37 p.m.
Techchapi	1952	7.7	4:52 a.m.
Sylmar	1971	6.6	6:00 a.m.
Imperial Hwy.	1979	6.4	4:16 a.m.
Whittier Narrows	1987	5.9	7:42 a.m.
Landers	1992	7.4	-
Northridge	1994	6.7	4.31 a.m.

Source: LA Times

Since 1990, the local jurisdictions, on their own or in partnership with MTA, have been responsible for adding 1,400 lane km (875 lane miles) of major roads within the County. This addition of new roads is responsible for accommodating 1.6 million vehicle km (a million vehicle miles) of travel daily.

Local agencies have improved traffic flow, since 1990, by participating in projects to synchronize traffic signals along 2,560 km (1,600 miles) of roads. This effort has tremendous benefits in terms of the travel time saved for motorists and bus riders, as well as reducing air pollution. MTA has funded between 1991 and 2001 over $300 million towards Signal Synchronization and Bus Speed Improvements in the County through Call for Projects (CFP).

Climate and Air Pollution

Southern California has two seasons. One is a long, dry spell, moderately warm from May to November and the second is a short, wet spell, moderately cool, but hardly bitter cold from November to May. Rainfall is normally scarce on a yearly basis, especially between October and April. Due to the warm weather throughout the year, the traffic load on freeway and highway network is constant.

The topography and climate of Los Angeles County combine to make the Los Angeles Basin an area of high air pollution. It is the only city in the county designated as "extreme" by the EPA for air pollution. The area is bounded by the Pacific Ocean to the west and the San Gabriel, San Bernardino and San Jacinto Mountains to the north and east, creating a pocket for pollutants to remain in the area. In summer months, the warm upper layer (inversion layer) does not allow the cool marine air (mixed with pollutants) to disperse and it is therefore trapped by the mountains. The abundant sunlight throughout the year triggers a photochemical reaction, creating smog. Table 5 below shows the cancer risk in the Los Angeles area.

TABLE 5: CANCER RISK IN THE COUNTY

City	Cancer risk per million people
Los Angeles	470
Burbank	483
Long Beach	323

Note: Goal of Clean Air Act is to reduce the cancer risk to 1 per million
Source: Los Angeles Times, March 9, 1999.

Transit System

While Los Angeles is known worldwide for its extensive freeway and roadway systems, there is also a comprehensive public transportation system provided by MTA and several local transit operators. This section describes the existing transit system in the County. The MTA operates one of the largest bus systems in the United States, with a service area covering over 3,640 km^2 (1,400 square miles) and providing on an average 1.3 million passenger trips per day. MTA's transportation partnership includes twelve fixed-route operators who receive regional formula-based funding. These operators are Antelope Valley Transit, Commerce, Culver City, Foothill Transit, Gardena, Long Beach, Los Angeles, Montebello, Norwalk, Santa Monica, Santa Clarita and Torrance. In addition forty-two cities provide their own community shuttle service.

An extensive rail system, in the County, has been under construction since 1990. The current system includes 384 km (240 miles) of light rail, subway and commuter rail services within the County.

Metro Rail lines span a total of 78 km (49 miles) to date and serve over 100,000 boarding passengers each weekday. The Metro Blue, Red and Green lines are branches of the Metro Rail system being developed by the MTA for the County to provide an alternative mode of travel.

The Metro Blue Line was the first operational segment of the Metro Rail system and provides 35 km (22 miles) of light rail service between Downtown Los Angeles and Long Beach. The average ridership on this line has increased from 30,000 in 1990 to approximately 47,000 in 1997. In July 1998, the average boarding was over 50,000.

The Metro Red Line is designed to be the backbone of the Metro Rail System. It began operation of its first segment in early 1993, providing the city with modern, heavy rail subway service from Gateway Plaza/Union Station, a multi-modal transit center to the Westlake/MacArthur Park Station. With the recent addition of the extension to Wilshire/Western Station, the Red Line is a 12-km (7.7-mile) line encompassing seven stations. More than 37,000 passengers ride the Red Line daily, a 69% increase over the previous year. Also, the 7th Street/Figueroa Street station provides the required connection to the Metro Blue Line. The Red Line is being currently expanded to include an additional four stations by summer 1999.

The Metro Green Line was added in 1995 and operates between Norwalk in Eastern Los Angeles County and terminates in Redondo Beach. Constructed mostly along the median of the I-105 (Glen Anderson Freeway) Freeway, this 32-km (20-mile) east-west light rail system serves over 10,000 passengers. At Redondo Beach terminus passengers can take local shuttle buses that connect to Los Angeles International Airport (LAX). These connections provide the necessary foundation for a multi-modal system in the County.

Metrolink is the Southern California Regional Rail Authority's (SCRRA) commuter rail system and connects commuters living and working in six counties: Los Angeles, Orange, Riverside, San Bernardino, San Diego, and Ventura. Metrolink commuter rail service began in late 1992 to Downtown Los Angeles from Moorpark, Santa Clarita, and Pomona. This original commuter rail service covered 179-route km (112 miles) and carried approximately 8,000 passengers daily. Today the Metrolink regional rail system covers 666 route km (416 miles), with an increase in ridership to over 25,000 passengers daily. SCRRA operates as a joint power's authority with funding provided by the above listed counties. Together, on an average weekday, the above mixtures of rail transit systems provide countywide service to over 124,000 passengers.

Air Transport

The County has eleven airports, including its major airport, the City of Los Angeles Airport (LAX). Table 6 shows expected growth in the County between the years 1995 and 2020.

TABLE 6: ESTIMATED PASSENGERS AT AIRPORTS
(Million Annual Passengers)

Airport	1995 Actual	2020 Projections		
		Low	Medium	High
Los Angeles	53.9	90.7	94.2	101
Burbank	4.9	8.4	9.2	9.7
Long Beach	0.4	2.6	2.8	7.3
Palm Dale	0	0.1	1.7	1.8

Source: Draft 98 RTP. Community Link 21. November 1997. SCAG.

The air cargo is expected to grow from 2.7 million tonnes (3.0 million tons) in 1995 to 8 million tonnes (8.9 million tons) in 2020. This additional air cargo will put more truck traffic on the transportation network. It is suggested that new airports be added along with improvements to the transportation network to serve the projected air cargo.

Maritime Transport

In Los Angeles County there are two major seaports, the Los Angeles Port and Long Beach Port. In 1542, Julian Cabrillo identified the LA Harbor as "an excellent harbor" located about 25 miles south of downtown Los Angeles. The Los Angeles Harbor was officially created in 1899. In 1914, the Long Beach Harbor was commissioned. Both harbors create over 500,000 local jobs. The Alameda Corridor project is a 32-km (20-mile) train corridor from the port of Long Beach to the port of Los Angeles through Downtown Los Angeles. This project is expected to facilitate goods movement via rail between the ports and alleviate highway traffic to some extent by reducing truck traffic. This project will consolidate railway lines between the ports.

Funding Sources

A common thread that connects all transportation activities is funding. The MTA receives funding for itself and other transportation improvements within the County. The MTA fund includes funds from Federal, State and Local taxes. The Transportation Development Act (TDA) was a one-quarter of a cent sales tax that provided $216 million in 1996-1997 for transit use. State Transit Assistance (STA) is based on population share and base revenue. Los Angeles County received approximately $25 million from STA in 1996-1997. Federal Section 9 subsidies were $21 million, 1996-1997 and Congestion Management Air Quality funds were $50 million per annum. The enactment of the Transportation Equity Act for the 21st Century (TEA-21) on June 9, 1998 was hailed as the largest public works bill ever in the history of U.S. It is expected that TEA-21 will provide about $329 million in 1999 for transportation projects.

Conclusion

1. Expected population and employment growth in Los Angeles County and the City of Los Angeles by 2020 will exacerbate existing transportation and transit system. Since freeways are built out in Los Angeles, new methods to provide mobility and accesses are to be developed, including methods to change travel behavior.
2. One solution would be to restructure existing transit services to support multi-center hubs and find ways to increase ridership. Also, to effectively utilize the existing transportation network of major arterials and highways throughout the County to improve transit mobility. Any solution to the Los Angeles transportation system should include a more effective use of this transportation investment.
3. Flexible transportation funds are required to provide required infrastructure and services and also to maintain the existing infrastructure. Also, Cooperation

among various agencies is required to share transportation funds and resources to build an efficient transportation and transit system based on long-term needs.

4. There is a need to set aside funds for education, training and operation, and maintenance of investments. Sharing of information among agencies is more vital than ever to provide feasible solutions and to be cost-effective.
5. Advanced technology should be used to provide information, mobility and accessibility in many areas of transportation. All transportation projects also should address the issue of clean environment for its citizens.

Acknowledgements

The authors wish to express thanks to MTA for the study.

References

1. 1997 Congestion Management Program for Los Angeles County, Metropolitan Transportation Authority, 1997, Los Angeles, California.
2. Community Link 21, Draft 98 Regional Transportation Plan, Southern California Association of Governments, 1997, Los Angeles, California.
3. State of the Commute, Southern California Association of Governments, 1996, Los Angeles, California.
4. "Ongoing Study Reveals Most Cities Losing the Congestion Battle," Urban Transportation Monitor, Volume 123, Page 1-5, October 1997.
5. "Appendix B- 1990 Census Population," A to Z an Encyclopedia of the City and County of Los Angeles, 1996, University of California Press.
6. Taylor, Wendy. "The Gateway to L. A's Future," Black's Guide. Summer/Fall 1995, Los Angeles, California.
7. "Urban Roadway Congestion—1982-1994, Volume I: Annual Report." Texas Transportation Institute, Austin, Texas.

LOS ANGELES COUNTY PUBLIC TRANSPORTATION SYSTEM ITS OPERATION AND USE BY CUSTOMERS

Gary Spivack[*]

Transit Operations

There are nearly 1.2 million daily boardings on nearly 200 bus routes and rail services in Los Angeles County. Bus services provided by the Los Angeles County Metropolitan Transportation Authority (MTA) are provided directly and through contracted service providers hired by the MTA. Nearly 1,400 square miles (3,640 square km) are covered by transit in Los Angeles County. Table 1 presents the projected rail and bus service levels for the 1998-1999 fiscal year. It also shows the amount of service provided by direct contract to the MTA.

The MTA fleet consists of 2,115 active standard transit busses and 104 active railcars (30 heavy rail and 74 light rail). Additional cars are being delivered and are in making ready status at this point. An additional 74 heavy rail cars are on order, although some of these might be leased to other properties, given the slowdown in the heavy rail construction schedule. An additional 52 light rail cars are on order for the Metro Blue and Green lines.

The MTA Operations Division is organized into four basic units – Metro Bus Operations, Metro Bus Maintenance, Metro Operations Support, and Metro Rail Operations. Metro Operations provides telecommunications, facility maintenance, and other agency wide services. Metro Rail Operations provides operations and maintenance services for each of the three rail lines. Figure 1 shows annual bus boardings.

[*] Director, Transit Operations Support & Services, MTA, Los Angeles, CA 90053-2952

TABLE 1: MTA BUS AND RAIL OPERATING CHARACTERISTICS PROJECTED FOR FY 1998-99 (IN THOUSANDS)

	REVENUE SERVICE HOURS (000)	REVENUE MILES (000)	PROJECTED BOARDINGSS (000)
BUS			
OPERATED DIRECTLY	6,099	73,120	327,540
PURCHASED	579	8,260	15,350
SUB-TOTAL	*6,678*	*81,380*	*342,890*
RAIL			
BLUE LINE	75	1,605	16,120
RED LINE	27	541	11,450
GREEN LINE	48	1,443	6,560
SUB-TOTAL	*150*	*3,589*	*34,130*
TOTALS	**6,828**	**84,969**	**377,020**

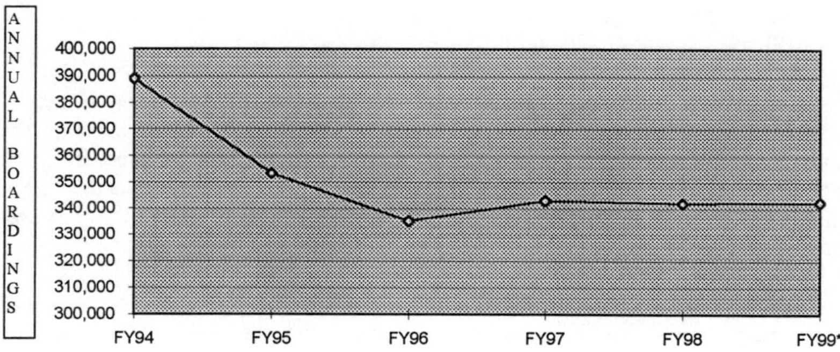

FIGURE 1. BUS ANNUAL BOARDINGSS

All of the MTA bus fleet is wheelchair accessible. The age of the fleet, however, has added significant costs to the system. The average age of the bus fleet is approximately 9 years. Many of the older buses scheduled for replacement are 18 or 19 years old. Twenty, thirty-five foot coaches (10.5-m) are among the oldest in the fleet. The average system speed is approximately 23 km/h (14.3 miles per hour). During peak periods and depending on the location, surface bus speeds can range from a low of 12 km/h to 48 km/h (8 miles per hour to 30 miles per hour).

Rail Services

The first local rail line from Los Angeles to San Pedro (harbor) was implemented in 1869. In 1901, the Pacific Electric Railway Company developed an Interurban rail network of 1000 miles (1,600 km), which served much of Southern California. The rail cars were 50-foot (15 m) Red Cars made of wood and steel. Some cars, called Yellow cars, were used for short routes within downtown area. Recorded travel speeds were 40 to 50 miles per hour (64-80 km/h). This extensive rail network was used by commuters for both work and recreation.

During the 1930's, automobile traffic increased, resulting in a decline in rail commuting. One major cause for the decline in commuters by rail was travel time due to reduced speed. Another major cause was the buying of transit lines by giant automobile corporations and abounding them later. In 1961, the last Red Car rumbled through downtown LA.

Today, the MTA operates three rail systems in Los Angeles County. Opened in 1990, the Metro Blue Line travels some 22 miles (35 km) from downtown Long Beach to downtown Los Angeles. The Red Line (Heavy Rail) has opened in segments beginning in 1993 and runs from the heart of downtown at Union Station to Wilshire Boulevard and Western Avenues. In June 1999, it is expected that the system will continue into Hollywood.

By the year 2000, the Red Line will open to North Hollywood, a system of approximately 27 km (17 miles). The newest of the rail systems operates in the median of the Glenn M. Anderson Freeway (I-105), running east to west from the City of Norwalk to Los Angeles International Airport. The line intersects the north-south Blue Line at Imperial Station that also serves as the Rail Operations Control Center. The Green Line opened in 1996 and runs 27km (17 miles).

Rail system speeds are governed by the distance between stations and interference by grade crossings or street running segments. Figure 2 shows the annual rail boardings.

Other Transit System in the County

Sixteen Municipal operators and two local cities are funded by the MTA as part of the regional funding program. The Municipal systems are all bus systems, operated by Foothill Transit, Long Beach Transit, Santa Monica Municipal Bus Lines, Culver City, City of Los Angeles (LADOT), Torrance Transit, City of Gardena, Culver City and the City of Montebello. Santa Clarita and the Antelope Valley (north Los Angeles County), operate systems within their local communities as well as over the road services into the Los Angeles basin. In addition to these operators, there are 22 additional fixed route service providers funded through local sales taxes. In total, there are 38 fixed route service providers.

FIGURE 2: RAIL ANNUAL BOARDINGS

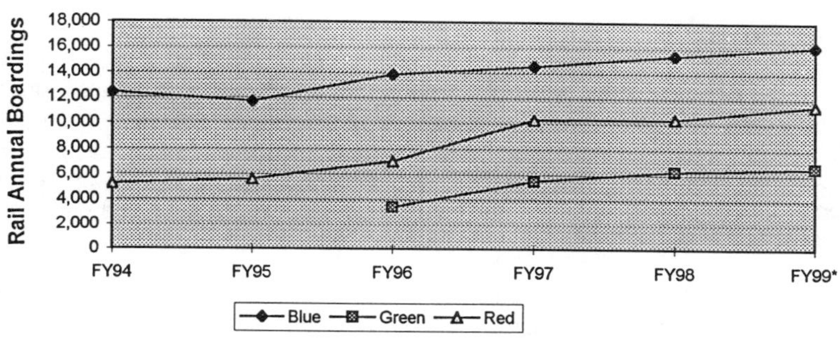

Access Services Incorporated is a newly formed operator for those individuals that are unable to use the regular bus and rail system options provided in Los Angeles County. Beginning as early as 1988, the MTA and its predecessor systems made a commitment to ensure that the basic transit system was fully accessible. Under the Americans with Disabilities Act passed in 1990 by the United States Congress, disabled or wheelchair customers were to be given access to public transportation.

Fares in the County

The following Table 2 presents the adult cash boarding fares for operators in the County:

TABLE 2: CURRENT FARES IN LOS ANGELES COUNTY BY OPERATOR

Access Services Inc.	$1.50	Antelope Valley	$0.80
Arcadia	$0.75	Claremont	$1.25
Commerce	$0.00	Culver City	$0.75
Foothill	$0.85	Gardena	$0.50
LADOT	$1.10	La Mirada	$1.00
Long Beach	$0.90	Montebello	$0.75
LACMTA	$1.35	Norwalk	$0.60
Redondo Beach	$1.00	Santa Clarita	$0.75
Santa Monica	$0.50	Torrance	$0.75

The MTA is working to develop an integrated "universal" fare collection system that would make transferring between operators "seamless" from the perspective of the transit user. Currently, customers can transfer between operators for a $0.25 fee. Local operators do not accept passes unless a special fare agreement is reached between them. For example, LACMTA has fare agreements between Foothill, LADOT, and Long Beach. Pass holders on the MTA system must pay a full fare or a transfer fee to board a foreign operator's service.

Customer Base and Usage Trends

MTA bus and rail boardings have followed separate trends since 1994. On the bus side, boardings decreased because fares rose and the economic recovery from the last recession was extremely slow. The local aerospace industry was hit hard by layoffs resulting in out-migration of major population segments. Jobs in the service sector that depended on these highly skilled people also dried up and only recently has the Southern California economy began to come back.

Privatization of services began in the late 1980's with the formation of the Foothill Transit Zone. This spun off a large section of the agency's service area, taking with it a large segment of the MTA's market. Since that initial experiment, cities funded by local tax initiatives and by grants provided by the MTA began to develop their own systems for intra-community travel, but in addition, absorbed some of the higher cost MTA express services into these programs. Today, there are 16 Municipal Operators in Los Angeles County. Two of the newest operators, Foothill Transit and the City of Los Angeles have developed large bus operating systems.

City of Los Angeles buses are operated by the Los Angeles City Department of Transportation Department (LADOT). MTA provided $2 million to conduct transit-restructuring studies of the entire City. The City at the behest of the MTA and the County has completed the studies in cooperation with other transit operators in the area and has assessed how to provide better bus service to the public.

In addition to the introduction of new bus systems that absorbed portions of MTA services, the Southern California Regional Rail Authority began operations in 1992. The MTA provided commuter rail services at first to only Los Angeles and Ventura Counties, but today serves Orange, Riverside and San Bernardino Counties as well. Their service network carries approximately 25,000 people daily. In total, the MTA bus and rail system makes up nearly 80% of the total customer base for the Los Angeles County region.

Bus boardings reduced more drastically, clearly revealing the impacts of the changing service picture. Use of the MTA system has declined with the formation of new operations and local transit initiatives, which allow cities to provide services that overlay MTA routes.

Regionally, people perceive transit travel as slow. The travel time perception held by transit users is shown in Table 3. Both transit users *and* non-transit users estimated that using public transit to commute to work takes more than twice as long as driving alone.

TABLE 3: TRAVEL TIME PERCEPTION OF COMMUTERS

Respondents' Mode to Work	*Respondents Estimated Time "Drive Alone"*	*Respondents Estimated Time via "Public Transit"*
Drive Alone	25 minutes	**66 minutes**
Public Transit	25 minutes	**56 minutes**

The bus average trip length of 3.7 miles (5.95 km) for the MTA has not changed substantially for many years. The Blue Line from Long Beach to Los Angeles and the Green Line traversing the County from Norwalk to Los Angeles International Airport have average trip lengths of 8.16 miles (13.12 km) and 7.99 miles (12.86 km) respectively. The Red Line has the shortest trip length at 1.4 miles (2.25 km). This is due to its current 7-mile (11-km) length. Extension to Hollywood and North Hollywood will result in added ridership and longer trips. A part of the new extension to Hollywood will be opened in June 1999. Table 4 shows the MTA system ridership by time period:

TABLE 4: DISTRIBUTION OF SYSTEM BOARDINGS BY TIME PERIOD

TIME PERIOD	% (of daily riders)
A.M. (4 HOURS)	27
Midday (6 HOURS)	25
P.M. (4 HOURS)	37
OTHER (10 HOURS)	11

Who Rides Transit?

The majority of our bus customers are female (55%), while on the rail system, males make up 52% of the customers (Table 5). The Hispanic market dominates the customer base with 51% of the bus rides and 35% of the rail riders (Table 6). The African American population makes up the next largest segment on the system (21% bus, 29% rail). This population is followed by those classified as "white" (12% bus, 24% rail), Asian (8% bus and 8% rail), and American Indian (1% bus and 1% rail). Tables 5, 6 and 7 show the distribution of ridership by gender, age, ethnicity, vehicle availability and income for bus and rail users.

Fifty-one percent of the rail riders indicated that they had a vehicle available, but chose to use transit (Table 6). Bus riders, alternatively, indicated that only 20% had access to another mode of transportation

TABLE 5: GENDER AND AGE DISTRIBUTION OF RIDERS

Gender

	Bus Users	Rail Users
Male	43%	52%
Female	55%	48%

Age

	Bus	Rail
Less than 18	8%	5%
18-34	39%	41%
35-54	33%	45%
More than 55	15%	9%

TABLE 6: ETHNIC DISTRIBUTION AND AUTO AVAILABILITY

Ethnicity

	Bus Users	Rail Users
Am. Indian	1%	1%
Asian	8%	8%
Black	21%	29%
Hispanic	51%	35%
White	12%	24%
Other	3%	3%

Vehicle Availability

	Bus	Rail
Yes	20%	51%
No	80%	49%

TABLE 7: 1995 HOUSEHOLD INCOME

Income (BEFORE TAXES)	Bus Users	Rail Users
Less than $14,999	48%	35%
$15,000 to $34,999	14%	16%
$35,000 to $49,999	4%	10%
More than $50,000	3%	21 %
No Response /Not Sure	30%	18%

In the Household Telephone Survey of Residents of Los Angeles County, respondents were asked their usual mode of travel to work, to shop, to school, to visit, and to medical appointments. Of the 2,440 respondents 18 years of age or older who are employed full or part time, 68% drove alone to work and 14% used transit. The Household Telephone Survey respondents who used bus or train to travel to work have the following characteristics:

- 55% are foreign born
- 52% have limited access to a vehicle
- 50% have no driver license
- 31% own no car or other vehicle

Most of the reasons given for not using transit (Table 8) in the past 30 days include many factors that are outside the MTA's control. Only about one-third of the factors that the LACMTA may be able to *partially* influence.

TABLE 8: REASON FOR NOT USING TRANSIT

Rank	*Reasons*	*Frequency of Response*	*LACMTA Influence Possible*
1	Own/bought car	42%	
2	No need to ride transit	14%	
3	Transit is too slow	9%	•
4	Not convenient	9%	•
5	No access	8%	•
6	Need car	4%	
7	Too expensive	2%	•
8	Carpool	1%	
9	Too dangerous	2%	•
10	Disability	1%	•
11	Walk/Bikes	1%	
12	Uncomfortable/Conditions unacceptable	1%	•
	Other responses	6%	
	Total	100%	

Note: Data are from a multiple-response item and are percentages of responses (rounded).

The perceptions of Los Angeles County residents towards transit riders is given in Table 9. The residents perceive that bus riders predominantly belong to *low socioeconomic status*. However, the residents have a somewhat better perception towards train riders than bus riders.

The household survey concluded that the limitation in service coverage was not a substantial barrier to getting people to their ultimate destinations. However, only 19% indicated that they were solely dependent on transit for their travel needs. The rest of the population surveyed (81%) indicated that that had other options for travel.

TABLE 9: PERCEPTIONS OF RESIDENTS TOWARDS TRANSIT USERS

Bus Riders Are (public's perception)	*Rank*	*Rail Riders Are (public's perception)*
Low income	*1*	Working Class
Working class	*2*	Business/Professional
Average people	*3*	Average people
All kinds of people	*4*	All kinds of people
Students	*5*	Middle class
"Vehicle-less" household	*6*	Low income
Senior citizens	*7*	Friendly / considerate
Crazy/strange	*8*	People who dislike driving
Middle class	*9*	Students
Friendly / considerate	*10*	Upper class
Commuters	*11*	Safe to be around
Gang members/criminals	*12*	"Vehicle-less" household

Table 10 shows the residents' weighted importance of service attributes and their weighted perceptions towards the service attributes of MTA. The largest discrepancy between expressed importance and perception was for travel time. The second largest discrepancy was for "schedule information."

Challenges and Opportunities

The MTA is one of the leading and innovative transit providers in the nation. That being said, the debt service requirements are restricting the agency's ability to provide new capital facilities. A great number of communities are disappointed given that some areas are not forecasted, any longer, to receive fixed guideway services in the future. This has prompted calls for studies to break up the regional system into smaller parts, increase the rate of privatization, and even divest the MTA of its operating arm.

TABLE 10: WEIGHTED IMPORTANCE AND PERCEPTIONS OF RESIDENTS TOWARDS SERVICE ATTRIBUTES

Service Attribute	*Weighted Importance*	*Weighted Perception*	*Difference*
Schedule Information	***4.6***	3.5	**-1.1**
Cleanliness of Bus	***4.5***	3.6	-0.9
Cleanliness of Train	***4.5***	4.2	-0.3
Travel time	***4.4***	3.0	**-1.4**
Cost	***4.2***	3.5	-0.7
Vehicle Exterior	***3.6***	3.9	*+0.3*

Importance Scale: 5 is Most Important, 1 is Least Important
Perception Scale: 1 = very bad, 2 = bad, 3 = neither good nor bad, 4 = good
5 = very good

Table 11 shows the survey respondents' answers regarding safety. Bus and rail transit during daylight hours is perceived to be equally safe. However, the rail is perceived as safer than bus after dark.

TABLE 11: SENSE OF PERSONAL SECURITY ON BUSES AND TRAINS

Time/Location	Very Unsafe	Unsafe	Neither	Safe	Very Safe
Bus during daylight hours	5%	16%	4%	55%	17%
Train during daylight hours	2%	6%	3%	49%	24%
Bus after dark	27%	33%	4%	22%	5%
Train after dark	11%	21%	4%	33%	10%
At bus stop	24%	33%	9%	27%	4%
At train station	11%	20%	6%	39%	9%

Source: FY96-97 Survey of Los Angeles County Residents (Weighted for Non-Telephone Households). Note: Percentages do not total 100 percent due to rounding and omitting "does not know/refused."

MTA Operating Budget

The fiscal year operating and capital budget for the MTA are over $2.5 billion. The MTA is more than just an operating entity. It provides substantial other services to the Los Angeles Metropolitan Region. As shown in Table 12, the MTA budget is divided into three programs: 1) MTA Operations, 2) Bus and Rail Capital and 3) Regional Programs.

TABLE 12: FY 1998-99 MTA BUDGET

ELEMENT	$ millions	$ millions
1. OPERATING FUND SUB TOTAL		$1,183
BUS AND RAIL OPERATIONS, PLANNING, PROPERTY MANGEMENT	$717	
ADMINISTRATION	$42	
SPECIAL REVENUE FUND	$75	
DEBT SERVICE	$338	
BENEFIT ASSESSMENT	$11	
2. BUS AND RAIL CAPITAL		$701
3. REGIONAL PROGRAMS		$642
MUNICIPAL OPERATORS	$119	
ADA/ACCESSIBILITY	$39	
LOCAL PROJECTS	$451	
COMMUTER AND INTERCITY RAIL	$33	
TOTAL 1998-99 MTA BUDGET		$2,526

Conclusion

The following are the main conclusions of this study:

1. Transit program should focus on using urban street system more systematically and provide for "people movement" over vehicular movement.
2. Transit terminals should be designed to be secure and areas of activity.
3. It is important that all equipment's used should be well researched and tested under extreme conditions to reduce breakdowns.
4. A balance needs to be found between the needs of transit customer, the merchant and non-transit commuter

Acknowledgements

The author wishes to express thanks to LAC MTA for the study.

Disclaimer

The opinions expressed in this paper solely reflect the views of the author.

References

1. Los Angeles County Metropolitan Transportation Authority, Fiscal 1998-1999 Adopted Budget, June 1998.

2. Los Angeles County Metropolitan Transportation Authority, Short Range Transit Plan, Fiscal Years 1997-2000, June 3, 1996.

3. Los Angeles County Metropolitan Transportation Authority, Service Planning Market Research Project, Phase I Summary Report, April 1988.

4. Los Angeles Department of Transportation, City of Los Angeles, 1997.

5. Meet the MTA, Los Angeles County MTA, 1997.

INNOVATIVE SOLUTIONS TO IMPROVE THE TRANSPORTATION SYSTEM IN LOS ANGELES COUNTY

A.S. Narasimha Murthy* and James de la Loza**

Abstract

Providing bus and rail service to over one million boardings each day is not an easy task, yet it is accomplished each day in Los Angeles County (County). In a County, where over 175 million vehicle miles are logged each day, smog from vehicle emissions is a great concern. The Los Angeles basin is the only area to be designated as an "extreme" non-attainment area for air pollution by the Federal Government. Finding solutions to provide an efficient transportation system is a challenge under present conditions in the County. Los Angeles is one of the premier centers for innovative solutions developed and implemented in all sectors of transportation.

The Los Angeles County Metropolitan Transportation Agency (MTA) has made considerable progress in developing solutions with the help of several agencies. This paper discusses various techniques and programs developed by MTA and its partners in improving the transportation system. Some of the solutions include: Gateway Transit Center, Advanced Technology Transit Bus (ATTB) as shown in Figure 2, Fuel Cell Buyers Consortium, Automated Traffic Surveillance and Control (ATSAC), Fuel Cell Consortium, Smart Corridor, Alameda Corridor and High Occupancy Vehicle (HOV) Lanes System. Also discussed are the Transitway Project, Rideshare 2000, Bike Stations and Transit childcare Center, Metro Art, Televillage and Project IMAJINE under Intelligent Transportation System (ITS).

Los Angeles County has been a national leader in providing accessible service to all individuals. The following sections briefly describe various, adaptations,

* Project Manager, Systems Analysis, Regional Transportation Planning and Development, Los Angeles County Metropolitan Transportation Authority, Los Angeles, CA 90012.

** Executive Officer, Regional Transportation Planning and Development, Los Angeles County Metropolitan Transportation Authority, Los Angeles, CA 90012.

innovations and programs developed by the MTA in partnership with other jurisdictions to improve transit system in the County.

Gateway Transit Center

The Gateway Transit Center (GTC) continues the transformation of downtown Los Angeles into Southern California's commercial and transportation hub. The center at historic Union Station connects the County through regional and local bus service. It also, links Los Angeles to neighboring cities and counties by providing easy connections for commuters using Amtrak, Metrolink, Subway and the El Monte Busway. The GTC acts as a real multimodal center by integrating bus and rail services while offering park-n-ride options for auto users. The MTA, along with several municipal bus operators and the Orange County Transportation Authority (OCTA) serves the center's Patsaouras Transit Plaza. The plaza can accommodate 100 buses every hour and more than 114,000 transit passengers daily. In addition, the center includes a 25-story building that serves as the MTA's headquarters and houses 1,700 employees. Figure 1 shows the layout of the Gateway Transit Center with the transit plaza.

Advanced Technology Transit Bus (ATTB)

ATTB is a national program initiated by the MTA in 1992, with FTA funding. The ATTB has been selected as the best point of entry for bringing fuel cells into the transportation market. The objective of the ATTB project is to develop a lightweight, low floor, low-emission transit bus using proven and advanced technologies of the aerospace industry. This vehicle is being designed to meet Federal, State and Local axle weight and clean air requirements. The bus will use a low, flat floor and a simple ramp system that is more reliable than the current wheelchair lift technology. Prime contractor Northrop Grumman is developing the ATTB. Some of the ATTB specifications are:

- Maximum unit price of $300,000 ($ 1992)
- Meet California Air Resource Board's Low Emission Vehicle requirements for urban buses with 2.5 gm / bhp-hr Nitrates, 0.05 gm/bhp-hr of particulate matter
- Curb weight 4,500 kg (10,000 lbs.) less than the equivalently configured current buses
- Meet American Disabilities Act (ADA) requirements for disabled boarding plus 2 wheel chair positions
- 12-meter (40-foot) bus that accommodates 43 seated and 29 standing passengers
- Use vehicle management system to achieve driver-commanded high quality performance
- Ergonomically designed operator's station and maintainer-friendly modular design
- Low maintenance and operating cost

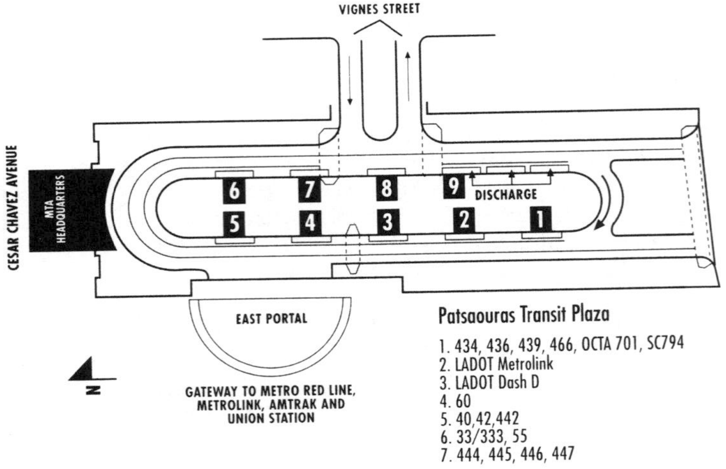

Figure 1: Gateway Transit Center

Figure 2: Advanced Technology Transit Bus (ATTB)

The ATTB program has finished the design and fabrication of the first prototype. Currently, field-testing is being conducted by the MTA in the City of Los Angeles. ATTB is the nation's newest advancement in heavy-duty public transportation.

Fuel Cell Buyers Consortium

MTA is a member of the Fuel Cell Buyers Consortium (FCBC) which is a coalition of thirty local and state government entities, public utilities and transit agencies. These agencies are working together to create efficient transportation and clean energy markets through commercialization of fuel cells. A fuel cell is an electrochemical engine. It converts the chemical energy of fuels directly into electrical energy that can power an electric vehicle. The consortium has agreed to develop a single product to a single specification, fuel cell bus such as ATTB. Another important objective of the consortium is to link environmental issues with the development of buses to create jobs and open markets for fuel cells.

Automated Traffic Surveillance and Control (ATSAC)

The City of Los Angeles is the second largest city in the United States. With a population of over 3 million people, the City holds about two-fifths of the population of the County. Transportation is an essential element contributing to the quality of life in Los Angeles. In the city, there are 256 km (160 miles) of freeway and 10,240 km (6,400 miles) of surface streets.

The ATSAC system was developed and managed by the City of Los Angeles Department of Transportation (LADOT). ATSAC is a national award winning interconnected and coordinated traffic signal system that automatically monitors and manages surface street traffic. With this sophisticated system, detectors monitor traffic throughout the network, traffic surveillance cameras and various timing programs that are automatically or manually implemented in response to fluctuating traffic demands.

The ATSAC system was first put into operation during the summer of 1984, during the Summer Olympics in Los Angeles. The system currently handles over 2,500 intersections. The system is a powerful tool for reducing delay and minimizing traffic congestion. The system is being currently adopted to include multijurisdictional operation of major transportation corridors such as Santa Monica Freeway Smart Corridor demonstration project. ATSAC has also installed a fiber optic communications network throughout the City.

ATSAC is indispensable for responding to nonrecurrent events such as parades, freeway closures, special award ceremonies, street sports events and major emergencies such as earthquakes. The ATSAC system can accommodate up to 64 different timing plans. The system allows to restoration of mobility, which is essential for transit to provide better service in the County. ATSAC was completed with funding from Federal Highway Administrations, MTA and other agency's assistance.

Santa Monica Freeway Smart Corridor

The Santa Monica Freeway Smart Corridor is one of the most innovative and visionary Intelligent Transportation System (ITS) projects in the nation. Utilizing a coordinated series of measures that include advanced vehicle detection technologies, sophisticated computer programming, and proven traffic management strategies, the project significantly improved traffic flow in the corridor. By maximizing the ability of the freeway and surface street network to carry traffic, congestion and delay to motorists and transit were reduced. The revolutionary aspect of this project is that it introduces an unprecedented level of collaborative decision making and operation of participating systems.

The Smart Corridor extends from Downtown Los Angeles to Santa Monica along the Santa Monica Freeway. It includes the five parallel arterials of Washington, Venice, Pico, Olympic and Adams Boulevards and fifteen major cross streets. The Smart Corridor uses a three-pronged approach to improve traffic flow:

- coordinating the existing traffic control and monitoring systems of the participating agencies to maximize the ability of the freeway and surface street network to carry traffic;
- providing dynamic, real-time traffic information to motorists traveling the corridor, and;
- providing timely and effective accident or incident management within the corridor

The project is expected to improve mobility, air quality and increased safety. The project cost was estimated at $40 million. Figure 3 shows the Santa Monica Smart Corridor.

Alameda Corridor Project

There are two major seaports in the County, the Los Angeles Port and Long Beach Port. These seaports have national and international links and thus are a primary load factor for sea cargo in the Western Hemisphere. The proposed Alameda Corridor project is a 20-mile train corridor from the ports of Long Beach and Los Angeles to Downtown Los Angeles. This project is expected to facilitate goods movement through rail between the ports and alleviate highway traffic to some extent by reducing truck traffic. Also, it is expected to consolidate existing rail lines in the corridor. Figure 4 shows the project location. MTA has provided funds for this project and is a member of the Alameda Corridor Task Force. The completed rail corridor would contain two main line railroad tracks and a maintenance access road.

The project would improve traffic flow along Alameda Street and selected locations. The project is estimated to cost $2 billion and be completed before the end of 2000. It will also improve traffic congestion, air quality and reduce vehicle delays and noise along the corridor.

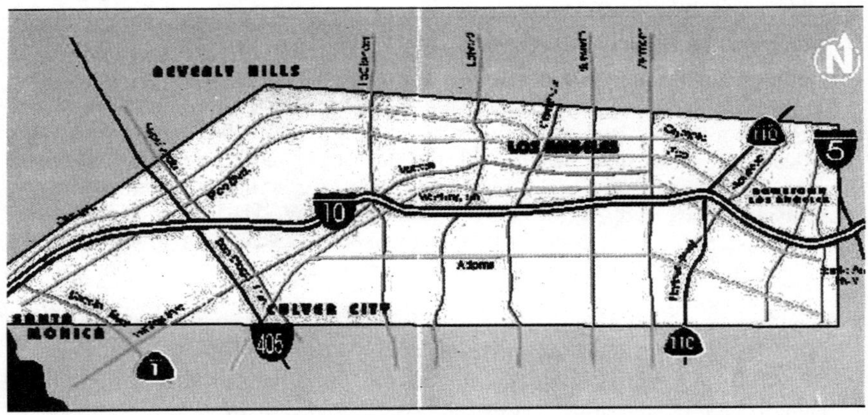

Figure 3: Santa Monica Smart Corridor

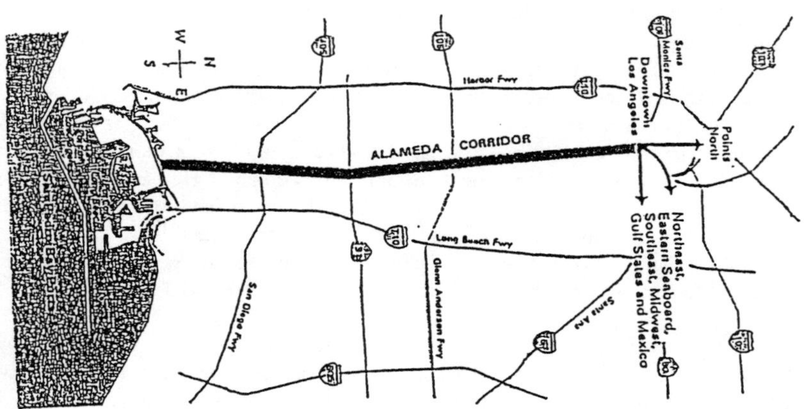

Figure 4: Alameda Corridor Location Map

High Occupancy Vehicle (HOV) System

The purpose of the HOV system in the County is to enhance mobility for all County residents. By providing dedicated HOV lanes, the County encourages the use of transit and carpooling. The HOV will also support other countywide objectives of improving air quality, trip-reduction and efficient movement of persons and goods.

A HOV System Integration Plan was developed to provide MTA with the technical input necessary to evaluate the cost and sequencing of HOV projects in the County. The MTA has provided funds to add over 208 km (130 miles) of HOV to the highway system since 1992. With the capacity to move three times as many people as a regular lane, carpool lanes make more efficient use of our already over-crowded freeways, and are critical to maintaining mobility. The total number of miles of HOV is 264 km (165 miles) in 1998 and it should increase to 300 miles in 2010. The majority of HOV lanes will have direct freeway to freeway connectors without needing to exit.

The County's HOV lane system is already one of the largest in the country, and will become the largest and most heavily used by the end of this decade. A million person trips each day is expected to be served by the HOV system in the County. The HOV system is assisted by the construction of Park-and-Ride and Transit Stations within the HOV corridor at convenient locations.

Transitway Study and Rapid Bus System

The transitway feasibility study was conducted by the MTA to analyze the feasibility of constructing and operating an exclusive bus transitway on three or four major corridors across the County. These corridors would provide access for welfare recipients to find employment within a reasonable time frame within the County. The Welfare to Work Program by Federal Transit Administration encourages construction of such corridors. Currently, MTA has chosen four major corridors: Los Angeles Downtown Connector, Exposition Corridor, San Fernando Burbank-Chandler and Crenshaw-Los Angeles Airport Corridor. The idea is to have exclusive bus lanes constructed along railway right-of-ways (ROW) that are owned by MTA and along mixed flow lanes. This was to develop transitway corridors similar to Brazil's Curitiba Model. Figure 5 shows the preliminary corridors selected for study.

The transitway project is expected to relieve congestion on parallel freeways and primary arterials. It also provides a viable transit usage of rail right-of-way (ROW) purchased by MTA. The existing MTA facilities can house and service vehicles required for the Trasnitway. The project provides the opportunity to reallocate duplicative services to areas with unmet transit needs. The estimated cost of Transitway for all four corridors selected is approximately $700 million.

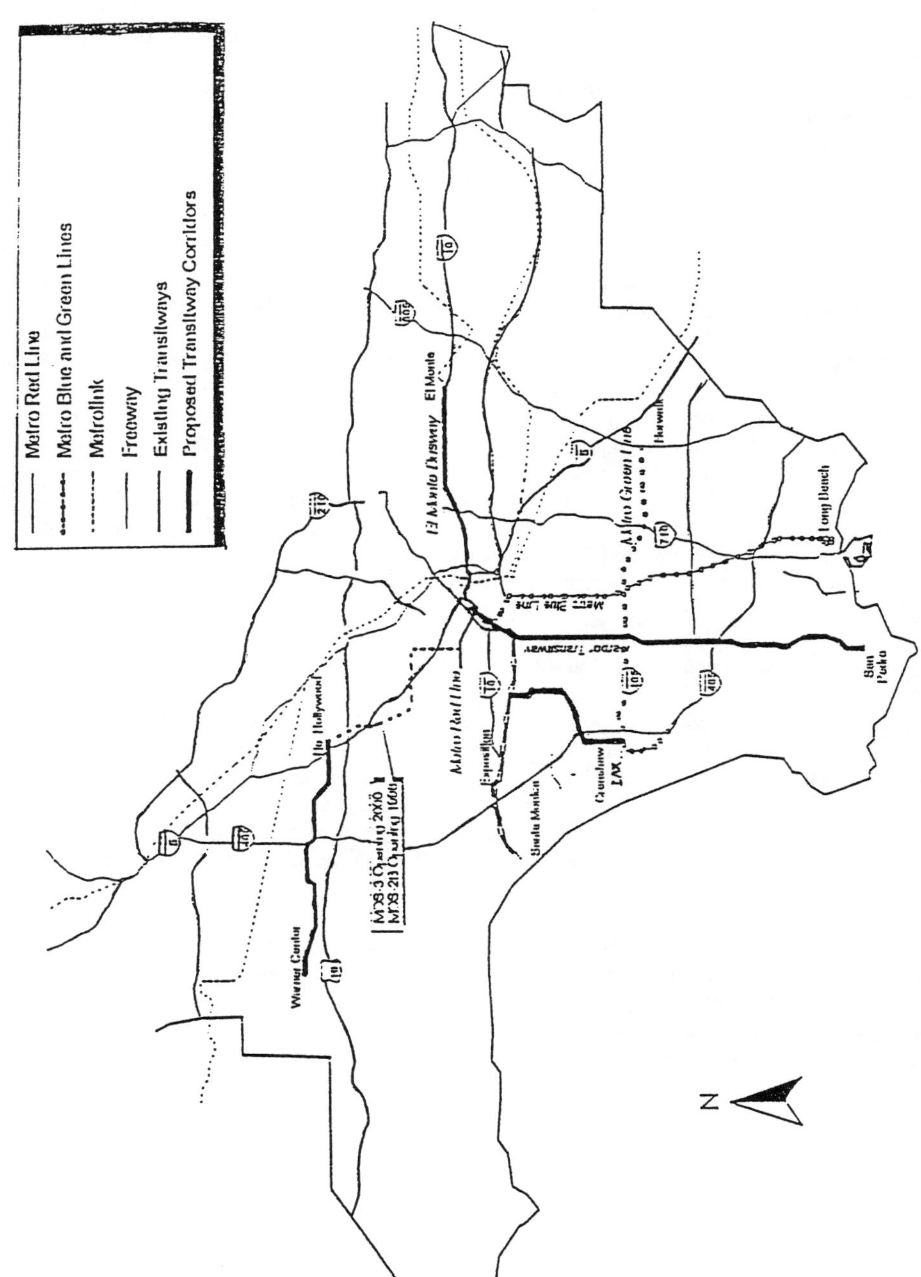

Figure 5: Proposed Transitway Corridors

A "Rapid Bus" concept that is currently operating in Vancouver, British Columbia, Canada was identified as a feasible solution for Eastside Los Angeles County. Under the Rapid Bus concept, it is possible to maximize people's movement through mixed traffic flow in busy corridors. Travel speeds are increased and dwell time is reduced by limiting stops to major intersections and implementing signal priority, low floor buses with multiple doors and a pre-paid fare system. Four east-west routes have been identified as candidates for Metro Rapid Bus operations.

Rideshare 2000

Rideshare 2000 is a Transportation Demand Management program developed by MTA in cooperation with Regional Transportation Agencies. The objective is to get more people to use carpool, vanpool, buses, metro, bike, walking and telecommuting to work. The program is designed to benefit employers and employees while it also benefits by relieving congestion and improving air quality. To qualify for the program, a workplace must have fewer than 250 employees and be located within the Los Angeles County boundary. The Rideshare 2000 encourages employees to earn $120 in financial incentives. It's easy to join and you can earn $2 a day during the three-month trial period. If the participants rideshare five days a week for three months they will be rewarded with $120 by the program. The incentives are paid in the form of gift certificates to be used in local groceries, restaurants and super stores. Also, merchant discounts up to 10% to 20% are available through the "Club Metro" card.

Transit Bike Stations and Child Care Centers

The advertisement for bike stations says "Quality Daycare for Bikes." The first Bikestation was opened in the First Street Transit Mall in the City of Long Beach. It is a full service bike transit facility opened in 1996 as a demonstration project to complement the downtown corridor's bus, light-rail and pedestrian modes. The purpose of the bike station is to encourage trips by bicycle and on public transportation in an effort to reduce vehicle emissions and congestion and foster more livable community.

To further promote the use of public transit and ridesharing in the County, childcare services are available to parents using bus and rail transit centers. The Transit Tots Center is operated by Children's Discovery Centers, a state-licensed and nationally accredited provider. There are two centers now offering this service (Chatsworth and Sylmar/San Fernando hubs). The objective is to provide a program for commuting parents who vanpool, carpool or use public transit.

Blue-Line Metro Televillage

One of the primary purposes of the Televillage is to provide access to a cross section of the population using transit. The program will provide computer and telecommunications technology that is rapidly changing the way people learn, communicate and conduct business.

The City of Compton is the first community to have a facility of this kind. The facility is located inside the Martin Luther King Transit Center at the Compton Station on the Metro Blue Line. The Televillage uses state-of-the art computer and telecommunications technology to be used by public. It is a center for people to work, take classes, visit after school, obtain information and hold meetings. The on-line kiosks include an on-line Housing Authority's database, job data base and consultant contracts.

Metro Art

To make the trip to a transit center or metro station more pleasant and interesting, MTA has always included artwork that reflects local history. For example, "The California Dreams" is the theme of the North Hollywood Station, which will honor generations of immigrants who have come to the San Fernando Valley to fulfill their aspirations. At the Universal City Station, the focus of the artwork will be on the historic significance of the adjacent Campo de Cahuenga. At this place the Articles of Capitulation were signed and Mexico relinquished control of California to the United States. The Hollywood/Highland Station will show dramatic images projected on surrounding walls by an overhead theatrical lighting element. This artwork and others are expected to depict the local history to residents as well as tourists.

Priority Corridor and Project Imajine

The federal government along with state and local governments has recognized Southern California as a Priority Corridor. This corridor travels through the urban areas of Ventura, Los Angeles, Orange, San Bernardino and San Diego counties and only four national ITS priority corridors. The theme of priority corridor is to "design once and deploy often."

In Los Angeles and Ventura Counties the priority corridor projects are: Rural Surveillance, which deploys "smart" call boxes to provide rural traffic surveillance capability; Rail Crossing, which integrates traffic management centers with rail operations to improve safety at highway rail grade crossings; Transit Radios, Assess viability of converting current MTA's radios to Global Positioning System (GPS) technology and IMAJINE, The Intermodal and Jurisdictional Integrated Network

Environment project will integrate freeway and arterial street operations in South East Los Angeles County.

Project IMAJINE

This Intermodal and Jurisdictional Integrated Network Environment project will include the synchronization of local and state signals and the adjustment of signal coordination to allow transit vehicles through with minimal delay.

Bus Priority Pilot Project

This is a MTA funded project with a budget of $4.8 million to design, develop, implement and evaluate a Bus Signal Priority Pilot project. Also, the project will develop a countywide guideline for Bus Priority. The project plans to conduct studies along three or more high volume bus corridors, identified using criteria such as headway, traffic volume, congestion, length and accessibility. The project is now underway and expected to be completed by the end of year 2000.

Other projects not discussed above are: Santa Monica Transitway, Smart Card, Signal System Technician Training Program and Career Development and Training Center.

Conclusion

With more than 18,000 bus and rail stops serving over one million boardings a day, the Metro System (buses and rail) operated by MTA is an extensive network. MTA is making strides in improving the service reliability and quality by taking major steps to improve transportation and transit in the County through innovative solutions with limited transportation funds. MTA is looking forward to being a leader in the 21st Century by operating a safe and efficient transportation system in the nation.

Acknowledgements

The authors wish to express thanks to LACMTA for the study.

References

1. Access to the Future: Televillage. Los Angeles County MTA. 1994
2. ATSAC Evaluation Study Report. City of LA DOT. June 1994.
3. Chain Letter: Bikestation. Los Angeles County MTA. Spring 1998
4. Community Link 21. Draft 98 Regional Transportation Plan. Southern California Association of Governments. 1997. Los Angeles, California.
5. Fuel Cell Buyers Consortium: Briefing Report. Fuel Cell Consortium. 1997.
6. Los Angeles Department of Transportation. City of Los Angeles. 1997.
7. Meet the MTA. Los Angeles County MTA. 1997.
8. Los Angeles County Metropolitan Transportation Authority. A Recommended HOV System for Los Angeles County. October 1996.
9. Nuttall, Ian. "Smart Neighbors." Traffic Technology International. Pp.37-40. 1997.
10. Rideshare 2000. Los Angeles County MTA. 1998.
11. Show Case. Southern California Priority Corridor Steering Committee.1995.
12. The Community Connection. Los Angeles County MTA. Vol3, No.1, August 1998.
13. Transit Tots Childcare Center. Los Angeles County MTA. 1997.

SOCIOECONOMIC CHARACTERISTICS, LAND USE AND TRAVEL PATTERNS

A PROFILE of MIAMI-DADE COUNTY

Jose L. Mesa[1] and Frank F. Baron[2]

History

Miami-Dade County is a large metropolitan area located near the tip of the Florida peninsula along its Southeast coast. While Miami and the Beaches are a world-class conurbation in the midst of the sprawling, modern, urban area that is Miami-Dade County, the County's beginnings were relatively recent and rather modest. But the metropolitan area has grown explosively over only the past century to become a significant metropolis on the American landscape.

With roots in Paleo-Indian prehistory of some 12,000 years ago, the locale was more recently a small Tequesta Indian habitation in the 1400s. Ponce de Leon sailed Biscayne Bay in 1513. Spain established a mission settlement on the Miami River in 1567; though abandoned a few years later, it was the first non-native establishment in what is now Miami-Dade County. Creek Indians fleeing from settlement pressures in the Southeast pressed into Florida in the 1700s, and became known as Seminoles. Also in the late 16th and early 17th centuries, Bahamian squatters were granted land by Spain and became the first permanent non-native residents if the area.

The U.S. acquired Florida from Spain in 1821. In 1836, Richard Fitzpatrick, head of the territorial legislature, established Dade County. Hostilities with the Seminoles began that year; the U.S. responded by founding Fort Dallas at the mouth of the Miami River. In one of the quiescent periods of the ensuing two Seminole Wars, Fitzpatrick sold his sizable holdings to William English. English laid out a village on the south bank of the river in 1843. Thus was Miami born.

The Dade County seat was moved to Miami in 1844, Florida attained statehood in 1845, and joined the Confederacy in 1861. No substantive growth occurred in the area until the Brickell and Sturtevant families came to Miami c. 1870. Cocoanut Grove, a small village 8 km (~5 mi.) southeast of Miami, attracted the area's first tourists c. 1875-80. Other small settlements were

[1] Staff Director,
[2] Transportation Systems Specialist, Miami-Dade County MPO, 111 N.W. 1st Street, Suite 910, Miami, FL 33128, USA

established in the surrounding woods of the demanding wilderness in the ensuing years, and many lent their names to present day neighborhoods or grew into present day municipalities in their own right.

Miami and surrounding communities of Dade County remained isolated and relatively unknown at a time when other Florida cities such as Key West, Tampa, and Jacksonville grew and prospered. In this pre-auto era, the railroad was the vector for development, but none had made it to Miami. Julia Tuttle, local civic presence, successful realtor, and visionary, persuaded Henry Flagler to extend his railroad to Miami from Palm Beach. Miami handily survived the disastrous freezes of 1894-95, which disrupted agriculture farther north. With completion of the railroad, development accelerated and the area grew.

Miami incorporated in 1896. Since then, sunshine, beaches and palm trees have been constant features, and the constant promotion, resulting bountiful tourism, and ensuing growth have been synonymous with Miami-Dade. Started by hardy pioneers, distinguished by mild winters, opened by the railroad, fueled by rampant land speculation, and shaped by the automobile, the County and its communities developed through the first half if the 20th century. Interrupted only temporarily by events such as great hurricanes, the land bust of the late 20s, and the Depression, Miami and Dade County grew steadily through World War II.

Following WW II, tourists began to truly flock to Miami Beach; many returned to live here, either as working adults or retirees. Regular exposure on national '50s and '60s media, and the advent of widespread household air conditioning, only served to further encourage greater domestic migration. The County then expanded explosively, and became one of the dominant urban areas of Florida by the early 1960s.

In 1959, what had been a modest trickle of emigrants, from South and Central America and the Caribbean Basin, became a flood with the overthrow of the Cuban government by Fidel Castro. It was a development that would permanently and profoundly transform the complexion and character of the metropolis. The influx of Cubans first changed the face of Miami, and then the County; it paved the way for other immigrants to come to the County; and Hispanics eventually became the dominant cultural influence and ethnic majority of Miami-Dade.

Florida's rapid growth has propelled it from a modestly populated state with a largely agricultural and tourism-based economy to the fourth largest state in the Union. State growth was led in large part by growth in South Florida, and particularly the growth of Miami-Dade. That growth has continued through the present, when its immediate impacts and long-term effects are locally being seriously questioned, debated and challenged for the first time.

Land Use and Urban Form

Miami-Dade is flat, with only a few small (<~30 ft./10 m.) oolitic limestone ridges offering minimal elevation change across its expanse. Elevation topography has offered no resistance to Miami-Dade development. More importantly, the county is divided by Biscayne Bay, the Miami River, and numerous canals. The Bay separates the offshore barrier islands of the Beach communities from mainland Miami (the city) and Miami-Dade County. Biscayne Bay covers approximately 611 km^2 (~236 mi^2), widening from its narrower confines in the north where it separates the mainland from the Beaches, to a wide, substantive shallow water aquatic nursery and fishery in the south. Commercial shrimping, and intense recreational boating and fishing occurs throughout the bay.

The Miami River drains southeasterly from central northwest Miami-Dade diagonally to the shoreline of the Miami CBD. A number of canals cut through the county, significant remains of efforts to drain the swamps, create dry land, and promote development and growth. Except where bridged they pose barriers to local development, local streets, and local mobility and accessibility.

Miami-Dade's general urbanized area extends south some 56 km (~35 mi.) from Broward County on the north to urban fringe scrub and everglades. From the offshore string of barrier islands of the beach communities, separated from the mainland by Biscayne Bay, the urban area extends about 40 km (~25 mi.) west, to agricultural and horticultural farmland and then the everglades.

Of the ~6400 total km^2 (~2460 mi^2) within its borders, land covers some 5035 km^2 (~1945 mi^2). However, only 1190 km^2 (~460 mi^2) –about 19% -lies within the Urban Development Boundary (UBD), which prescribes the (remaining) urban function-developable land. The western 60% of Miami-Dade is basically everglades savanna, currently prohibited from development. About 311 km^2 (~120 mi^2) additional land external to the UDB are designated and developed for agricultural uses, also constrained, at least presently, from further urbanization.

Because of its location and topography, and that the area grew in the early and mid-20th century, and is now beginning to reach an intermediate maturity in the late 20th century, Miami-Dade's patterns of growth and development are excellent examples of those of the prototypical contemporary 'sunbelt city'. It is generally low density, dispersed and increasingly polycentric, and development patterns and resultant travel are most assuredly auto-dominated Miami sprawls.

Transportation Networks

On the mainland, Miami-Dade streets are generally laid out in a rectilinear grid. Avenues run north and south, spaced 10 per mile (1.6 km), while streets run east and west, and are spaced 16 per mile (1.61 km); roadways occurring at the 1-mile intervals are usually major arterials. Mainland Miami-Dade is divided in four unequally sized quadrants, the SE, NE, NW, and SW sections, which intersect near the coast in mid-Miami CBD.

The general countywide grid is interrupted by the Miami River and the canals, and by manmade (or man-designated) obstacles. The gridded street system is also suspended by a number of municipalities that have instituted unique street designations, and non-gridded street configurations, though most cities retain county street conventions. Miami Beach and the beach communities are numbered separately without congruency with the mainland numbering system.

Miami International Airport (MIA), located in central Miami-Dade, presents a sizable obstacle interrupting the grid. A number of larger parks and Metrozoo, two other smaller feeder airports, and a number of institutions occupying significantly sized sites, such as Florida International University (FIU) in central West Dade, the University of Miami (UM) in Coral Gables, and Miami Dade Community College's various campuses also interrupt the grid.

The roadway network includes a limited expressway/freeway system, the southernmost extension of the Florida Turnpike, and numerous primary major arterials of the grid system.

Miami is connected to the Beaches by a series of seven causeways that provide these densely developed barrier islands' only transportation links to the County mainland; two other more southerly islands, Key Biscayne and Virginia Key, are separately connected south of Downtown Miami via causeway as well. The beaches are connected with Broward County to the north by State Road (SR) A1A. Also known as Collins Avenue, this is the only north-south thoroughfare connecting the barrier island communities along their entire length.

A number of major non-expressway facilities serve the metropolitan area. These acted as the original conduits for early development in the pre-Interstate era. Foremost among these is US 1. Roughly paralleling the coastline, it spans the length of the county from north to the Monroe line in the south; variously known as Biscayne Boulevard and South Dixie, US 1 was the area's 'main street' in the period preceding the county's Interstate era of rapid suburbanization. Other significant arterials include the Tamiami Trail, SW 8th Street, which traverses

Miami-Dade and travels west out of the county. Before the interstates, this was the route to Florida's west coast; it is the contraction of TAMpa and MIAMI.

A variety of other major north-south avenues, east-west streets, and other irregular roadways provide a hierarchy of roads for intraurban travel. These facilities were the roads that provided transportation, and perhaps just as importantly, directed development in Miami-Dade prior to 1961-62, when the area's first expressway, SR 826, and its first Interstate, 1-95, were reaching completion.

I-95 run north from Miami's CBD towards Broward, through Florida, and thence to Maine. SR 826, the Palmetto Expressway, is a circumferential highway belt that encircles the central core of the developed County on the north and west sides; the Bay and the Ocean form the east side of the County core, and Kendall Drive, about 5-1/2 miles south of the CBD, forms its south side. SR 836, the Dolphin Expressway, is the only limited access roadway bridging the County from east to west; it connects Miami Beach in the east with the Turnpike in the west. It passes through the northern third of Downtown Miami, skirts the southern edge of Miami International Airport, and travels about a mile north of FIU before linking to the Turnpike about 14 miles east of the Atlantic Ocean.

Florida's Turnpike forms a second beltway, ringing the older, earlier waves of suburban development; new suburban growth now radiates even farther out from the Turnpike. Four other, shorter expressways interconnect the four fundamental limited access facilities described above.

Of course, the development and completion of even the current limited expressway system and the Florida Turnpike in Miami-Dade increased accessibility and shaped the form of Miami-Dade during its emergent period of development. Unsurprisingly, development both anticipated and followed major expressway openings in the late 1960s through the 1980s. The expressway network synergistically augmented and amplified the rapid suburbanizing urges first realized decades before their completion, originally shaped by the major arterials.

Transit Network

The Miami-Dade Transit Agency (MDTA) operates the vast majority of mass transit systems and services in Miami-Dade. MDTA's Bus Division, Metrobus, provides the main public mass transit service to County residents; routes are aligned with most major arterials, and more than 70% of Miami-Dade Countians are within walking distance of a Metrobus route. Metrorail is the central backbone of the system, and connects the near SW with the CBD via an alignment which follows US1 north from (now) close-in Dadeland suburbs to Miami. A 'stair step'-type of alignment is then followed away from the CBD to the NW section, first to the Civic Center, and then through southern Liberty City neighborhoods to Hialeah. From the main CBD station, an automated people mover, Metromover, provides an elevated circulator accessing all three parts of Downtown Miami: Brickell in the south, the central CBD, and the Omni section in the north.

Augmenting conventional publicly-provided transit services are jitney vans and minibuses. These privately operated services are restricted from a general run of the streets to specified routes; operations must meet a number of threshold criteria.

Miami International Airport

The area's major air facility, Miami International Airport (MIA), occupies a landlocked, and development- and access-constrained rectangle of approximately 30 km^2 (~10 mi^2) in central Miami-Dade. MIA is one of the world's most active airports, ninth busiest nationally, and second for international passengers. All enplanements continue to increase, but international passenger transfers between arriving and departing flights is increasing beyond travel originating in or destined for Miami-Dade. The east side passenger terminal is directly accessed by from LeJeune Road/NW 42 Avenue; and access to LeJeune is made via SR 112 or NW 36th/41st Street on the north, or SR 836 on the south.

The burgeoning west side cargo area is accessed via NW 12th, 25th, and 36th/4lst Streets. MIA is the county's second leading cargo airport; it is the U.S. and North American air gateway to Latin America and the Caribbean, and an important connecting node to Europe, and somewhat surprisingly, to the far east. MIA cargo carried is increasing even faster than passenger traffic is increasing. For both passenger and cargo activities, airside operations are crowded onto three runways; this is shortly anticipated to be insufficient to handle increasing flight loads. Both cargo and passenger surface accessibility is likewise becoming exceedingly constrained; multimodal projects are being planned to relieve traveler access problems, and improved and possibly truck-only grade-separated highway access solutions are planned to relieve freight transport congestion.

The Port of Miami

The area's other major transportation facility is the Port of Miami. The Port is located on an island immediately east of the northern midsection of the Miami Downtown. Expansion is limited, and access is highly constrained, the latter limited to a single six-lane bridge connecting the Port with the CBD, through which access is forced because there is no direct expressway or ramp connection.

Heavy trucks, large buses and motor coaches, and private vehicles must all access the Port across the bridge, either via the northern or central portions of the CBD. Like MIA, the Port is a primal economic engine driving the Miami-Dade metropolitan area economy, and like MIA, congestion threatens its, and the County's, economic vitality.

Population

Miami-Dade County has a population of about 2.14 million living in 30 municipalities and in unincorporated Miami-Dade. The City of Miami is the County's largest municipality, with a population of about 370,000, followed by Hialeah (210,000), Miami Beach (93,000), North Miami (50,000), Coral Gables (40,000), North Miami Beach (35,000), and Homestead (27,000). None of the other 23 municipalities are larger than 20,000. The unincorporated area is home to about 55% of all county residents: as befits a sprawling sunbelt metropolis, more people now live in the suburbs and exurbs than in cities and smaller municipalities. While a number of traditional city downtowns remain indeed major nodes of activities, led by the City of Miami, non-municipal urban activity center aggregations popularized as 'edge cities' may be found dispersed through the metropolitan area, growing in size, economic importance, and social significance. In combination with grassroots political dissatisfaction of many suburban subareas and non-municipal communities in unincorporated portions of Miami-Dade, the development of such edge cities has contributed to the incorporation of a number of these areas. A former edge city now in fact becoming the 'downtown' of a bona fide municipality with the opening of city hall offices and functions within -or in proximity to -the environs of such suburban conurbations.

Of the County's population registered in the 1990 Census, about 49% was Hispanic or of Hispanic descent, around 21% was African-American, and the remaining 30% Non-Hispanic White. The percentages have changed in the near decade since the census; now there are approximately 55% Hispanics, with corresponding decreases in percentages, though not absolute numbers, of Blacks and Non-Hispanic Whites.

County population age profiles developed from census data since 1970 indicate that the median age of county citizens has remained relatively stable. However, the distribution of ages has changed, as more residents are found in the age 20-44 cohorts in 1990 than previously. While percentages of those 65 and over slightly increased, those aged 65-74 have decreased. The increases in school-age children, and the immense pressures on the county school system to reduce crowding, construct more facilities, and build them farther out in suburban areas indeed indicate a somewhat younger and dispersing population.

The growing 'Hispanisization' of the county has introduced a certain dissatisfaction between the other two major population segments, and has led to at least two phenomena in response. Members of the African-American community have lobbied extensively for increased attention, much along lines of support for increased political representation and economic development programs for the Black community. A number of Non-Hispanic Whites, on the other hand, appear to have 'emigrated' to neighboring Broward County, bordering Miami-Dade on the north, in search of a more prototypically classic suburban lifestyle, without the more pervasive Hispanic-influenced or -laden, culture. In large part, the former has led to political redistricting at the county level, with a one County Commissioner one County Commission District representative form of county government, increasing the size of the County's governing body and replacing the at-large election of all commissioners with district election of specific commissioners, and increased attention to programs aimed at economic development for poorer areas of the Black community. The latter has led to changes in transportation and travel patterns, as many if not most former residents who move to Broward retain employment in Miami-Dade, thus increasing demands on the regional transportation system, increasing travel times and distances, and increasingly highlighting the need for the three Southeast Florida counties of Miami-Dade, Broward and Palm Beach to recognize and act on regional transportation issues and needs as well as those more parochially oriented to each county's internal travel and transportation concerns. Employment in the county occurs mainly along the central east-west axis defined by the Port of Miami, Downtown Miami, the Civic Center, MIA, and Airport West. There are many other important employment nodes scattered throughout the county. These have occurred with greater frequency as the county, like its counterparts nationally, has experienced the second great wave of suburban dispersal, the movement of employers and jobs to the suburbs.

Employment

West of MIA is a burgeoning area of warehouses, import-export and international trading companies, and light industrial firms, appropriately known as Airport West, where great economic activity is agglomerating. In Hialeah, northwest of the Miami CBD and north of MIA, is located a considerable portion of the county's light industry. Southwest of the Miami CBD lies Coral Gables, a city marked by its above average income status, beautiful Mediterranean style architecture, and a concentration of offices which house many internationals and headquarters for the Latin American divisions of large domestic firms. Miami Beach is the focus of the local tourism industry, and during the past decade or so, the home of numerous modeling talent agencies, and sites of movie, catalog, magazine, and commercial shoots. Just northwest of the Downtown lies the legal and medical centers collectively known as the Civic Center, home to a large agglomeration of courts and legal firms. The Civic Center includes the largest medical center in the Southeast, home to a major medical school, several large hospitals, a large research center, and numerous medical offices and clinics.

The Miami-Dade school system is fourth largest in the U.S., with approximately 350,000 students in public schools; 31 high schools are located in Miami-Dade alone. The school system was the largest public employer in 1997, with more than 33,600 workers on its payroll. The county itself employs more than 28,000 and is the second-leading public sector employer in Miami-Dade. State and federal employment in Miami-Dade surpass 17,500 each. The largest private sector employer, in contrast, was American Airlines, who reported 9,000 workers in 1997; that is expected to increase as AA creates an international hub at MIA. The three largest health care facilities (one private and two public institutions) reported a combined employment of about 13,300 in 1997, while the three largest secondary educational institutions (a community college, a public university, and a private university) combined to employ about 13,000. In 1996, unemployment was about 75,000 out of an employment base of some 1,030,000 workers, exhibiting an unemployment rate of about 7.3 percent in 1996; probably it is currently higher.

Growth and The Geography of the Land

Miami-Dade County encompasses 6382 km^2 (~2,464 mi^2) in total area, 5,038 km^2 (~1,945 mi^2) on land, but only about 1,502 km^2 (~580 mi^2) are developable, including agricultural acreage presently excluded from any more dense development. The Urban Development Boundary (UBD) currently encompasses 1189 km^2 (~459 mi2), and will be expanded by approximately 34 km^2 (~13 mi^2) by the year 2020. Agricultural uses account for about 310 km^2 (~120 mi^2) external to the UBD. The remaining 3324 km^2 (~1,245 mi^2) are located in environmentally sensitive areas to the east and south of the urbanized area, mostly southern coastal wetlands and the Everglades to the west. The mainland portion of the county borders Biscayne Bay on the east, a sizable area of 611 km^2 (236 mi^2), which splits the barrier islands comprising the Beach communities from the mainland north of the Miami CBD; Miami Beach and its neighboring municipalities to the north front the Atlantic Ocean. The county is largely a low-lying land mass; elevations rarely exceed 20 feet above mean sea level.

Unsurprisingly, the county is becoming increasingly urbanized along the suburban axis, expanding west from the coast, as has been the trend for the past half century. However, because of a number of factors, this tendency is expected to slow over the coming 20 years.

The first consideration curtailing indefinite suburban expansion is the limit of developable land. Bordered on the east by water and on the north by Broward County, any potential expansion is limited to southerly and westerly directions. However, Everglades National Park occupies a significant southwestern section of Miami-Dade, and remaining land to the south and west has been designated so environmentally sensitive as to preclude development. Furthermore, development is proscribed beyond the UDB unless approved by amendment to the Miami-Dade County Comprehensive Development Master Plan (CDMP), and that such an amendment receives approval from the South Florida Regional Planning Council and State Department of Community Affairs. Development in most agricultural land within the UDB is also CDMP constrained, but growth pressures sometimes preclude the preservation of agricultural interests and zoning variances and changes may allow suburban expansion.

There is considerable public sector urban planning interest in redirecting development towards the core of the county at increased densities by promoting infill development. This planning not only removes some pressures from continuance of suburban sprawl at or near the inland UDB periphery, but also away from the coastal high hazard areas near the Bay and along the ocean, areas subject to hurricane surge inundation and associated tropical storm flooding.

Finally, aiding in controlling and redirecting growth are Florida's concurrency laws. Essentially, the concept of concurrency stipulates that infrastructure for proposed development must be in place when the development occurs. Concurrency encompasses virtually all public sector areas of urban infrastructure, most notably water and sewer lines, storm sewer drainage, educational facilities, and transportation facilities. Capacities are determined, and the proposed development's size and estimated impacts are assessed. If there is not sufficient capacity to serve the proposed development, the size of the development must be reduced, the capacity of the various deficient infrastructure elements increased, the project postponed or abandoned, or an exception granted.

Despite these constraints, Miami-Dade appears to be an area which will continue to suburbanize in the near future, characterized by continuing dispersion of population, employment, and activity centers.

The Transportation System

The Texas Transportation Institute (TTI), under contract to the US DOT, has measured traffic and congestion in the nation's largest cities for the past decade. Miami now ranks as the fourth most congested metropolitan area nationally. The area's overall mode split is about 4%, with the CBD-oriented work trip mode split roughly around 15%. Roadways are increasingly congested throughout the day, and like many larger U.S. metropolitan areas, Miami-Dade is experiencing peak period spread, Traffic in many suburban areas rivals or sometimes surpasses that of more developed areas such as the downtowns of Miami and other larger satellite city CBDs in the county. On weekends, traffic in the suburbs can reach central county weekday peak period proportions.

While trip lengths have increased some over the past few years, average trip times have increased more. More people are driving more widely available personal vehicles more often for more purposes to greater numbers of destinations. Major travel corridors along 1-95, and all the expressways, are clogged with peak period traffic, frequently resulting in stop-and-go movements on 6- and 8-lane freeways. Incidents stop traffic. Daily travel volumes approach 200,000 on many of these facilities; arterials are becoming more stop-and-go congested for longer periods as well. With the advent of planned unit developments (PUDs), the countywide grid has been interrupted even more, as large-scale residential areas are developed with few points of access, a doubly SOV travel alternative compromising situation. PUDs force more traffic on major arterials because there are few or no other through streets available, and constrain bus routes because of the pattern of winding streets and cul-de-sacs designed into the developments. Transit use is further hampered because walk access is also highly constrained by the PUD street patterns; homes which may be located a few hundred feet from a bus route may actually require a trip of a half-mile or more to access the nearest stop, and this is an effective deterrent to using transit. Suburban commercial development also militates against transit use when campuses of office buildings are created. Surrounded by areas of parking, without numerous land uses within such campuses, it becomes a necessity to drive to go to lunch or access other services. The contemporary design paradigm for both commercial and residential development encourages or demands increased private vehicle use, and because of the combination of segregated land use and the dispersal of people and jobs in this suburban landscape, SOVs become the overriding choice of travel mode. Of course, social inclinations to own single family detached housing, cheaper in the farther suburbs, supported by or necessitated by greater numbers of private vehicles that may be bigger and more prestigious models, are pervasive as well.

More people live in the county, and auto ownership has increased faster than population growth. VMT and VHT both increase. But roadways fill to capacity, and there is little room for expansion of physical capacity improvements in many areas. In the suburbs, the main roadways are now routinely built as wide 6-lane divided arterials, often with dual left turn lanes and right turn lanes at the intersection throats, making even more difficult for pedestrian movement.

County transit service is hard-pressed to keep up with increased travel demand, much less making any inroads into a diminishing mode split, despite having the country's first truly fully multimodal transit system. A bus fleet of over 500 buses (peak service), serving routes covering virtually the entire county with a general 20-hour service span, offers traditional big city bus transit, supplied by the Miami-Dade Transit Agency (MDTA), a county department. The bus system is augmented by Metrorail, a heavy rail system which follows the US 1 corridor from the Kendall edge city of Dadeland in the south to the CBD, travels northwest to the Civic Center, and continues in a stepwise fashion farther northwest through southern and central Liberty City and Hialeah to its present termination in central west Hialeah. An extension will shortly begin to extend the rail to the Palmetto expressway at NW 74th Street, the first such extension since the system was inaugurated in 1984-85. Metrorail connects to the central Downtown district, the Miami CBD, the southern downtown district, the Brickell financial center, and the northern Downtown district, the Omni area, via Metromover. Metromover is an automated guideway system (AGT), which loops through the CBD, with relatively recently completed spurs south to Brickell and north to the Omni. Metromover was developed to better serve the Downtown. Because of difficulties in positioning an elevated heavy rail line through the already built-up CBD, the AGT was selected to act as a collector-distributor for this largest Miami-Dade activity center. In South Miami-Dade, an exclusive bus-only facility, the South Dade Busway, runs between Dadeland at the southern terminus of Metrorail south to Cutler Ridge, a South Miami-Dade suburb. In addition to large buses, the busway utilizes small buses to transport patrons along the US 1 corridor it closely parallels. A variety of Special Transportation Services (STS) are offered for specialized E&H transportation needs, also operated by MDTA.

Additionally, there are a variety of jitneys offering private sector small bus and van transit in select corridors. Now more closely regulated than in the recent past, they provide relatively more flexible services than does the more rigidly scheduled MDTA, and offer a non-SOV travel alternative to travelers. Jitneys have been particularly successful in catering to residents of immigrant communities. Such travel was a common travel mode in the counties of origin for residents of the Hispanic and Haitian communities which most often utilize their services. In the early 90s, basically unregulated jitneys proliferated in such numbers, encroaching upon established high-patronage bus routes, that estimates of up to 25,000 daily riders were being diverted from Metrobus. However, since regulations have been enforced, this siphoning of MDTA ridership has considerably lessened.

Concluding Observations

Transportation is becoming an extremely costly undertaking, with large new highways and non-bus transit alternatives costing in the billions of dollars. However, it appears that to most of the public, transportation is akin to the weather -much to be said, little to be done. Unless and until a specific project interferes with a particular individual's or community's normal existence, the public seems to merely accept increasing congestion and proceed with their lives, grumbling, but taking no proactive stance. Transportation issues locally take a back seat to issues of crime and public safety, education, and other civic concerns. The mindset of suburbia seems to be one that excludes non-auto travel choices and opportunities; most residents seem to prefer the suburban lifestyle, and given the opportunity, most gladly move to the suburbs. And above all, despite their professed support of transit, it seems that the public supports transit for others, not themselves, and indeed do not vote with their wallets, as county referenda to create a dedicated local source of tax support for public transit have been thrice defeated over the past decade by increasing margins. Most believe that highways are completely paid for via gas and other car taxes and road tolls, and that transit should be self-sufficient. Efforts to educate the public on more fully allocated costs of all modes of transportation, especially private cars, appear to fall on deaf ears, and there seems to be no meaningful cognition of either energy use or air quality impacts of continued and increasing reliance on private vehicles.

Unless and until congestion worsens to unacceptable levels, transit and other personal vehicle alternative modes are seen as socially and culturally acceptable means of travel, acceptance of increased residential densities occurs, more alternative mode- accommodating urban design becomes the norm, and the public is willing to pay for multimodal transportation improvements, there is little expectation of substantive change. ITS, novel transit approaches, travel pooling, revisited urban design paradigms, increased residential and commercial densities, and other approaches will succeed only marginally until transportation and its impacts are taken more seriously more often by a significant plurality of Miami-Dade Countians.

Urban Public Transportation System in Miami-Dade County and Its Operations and Use by People

Carlos F. Bonzon[1]

Introduction

Miami-Dade County is an urbanized area comprising over two million residents. It is the fourth most congested metropolitan area in the United States. Thus, there is no doubt that an efficient transit system is critical to the area's economic prosperity. Nevertheless, inadequate funding for transit and the lack of a dedicated source of funding at the local level are beginning to take a toll. This paper briefly describes the existing transit system and planned expansions.

The Miami-Dade Transit Agency (MDTA) is one of the largest departments within Miami-Dade County government. It is responsible for the planning and provision of all public transit services in the Greater Miami metropolitan area.

MDTA's integrated transportation system carries over 50% of the transit passengers in the entire state of Florida and consists of four major components: the Metrobus fleet, with some routes running almost 22 hours per day, connecting most areas of Dade County; Metrorail, an electrically powered, elevated rapid transit system stretching 33.6 km (21 mi.) from Dadeland to Hialeah; Metromover, a 7.04 km (4.4 mi.) elevated and automatic people-mover system that serves the downtown central business district of Miami, including the Omni and Brickell areas; and Paratransit Services, designed to meet the needs of the disabled rider who cannot use regular transit services. Annually, MDTA carries over 80,000,000 passenger boardings on the three conventional transit services.

---

[1] Miami-Dade Transit Agency, 111 NW First Street, Miami, FL 33128

General Outline of Bus and Guideway Systems

Metrobus

The Miami-Dade Transit Agency provides bus service throughout Miami-Dade County 365 days a year. The Metrobus system intersects with Metrorail and Metromover, and serves all major business, shopping, entertainment, and cultural centers, as well as major hospitals and schools. Metrobuses travel over 41.6 million kilometers (26 million miles) throughout Miami-Dade County each year. The fleet consists of 590 full-size buses, including 66 articulated, 46 mini-buses and 15 vans. Service throughout Miami-Dade County is provided along 72 routes, up to 22 hours each day plus Park-and-Ride service to special events. Metrobus averages over 202,000 weekday boardings. The total annual ridership for FY 1997 was 61,925,029.

Busway

This 13.1 km (8.2 mi.) roadway was built just for Metrobuses. Buses travel on a two-lane, at grade roadway to and from Metrorail. A one-way trip between Cutler Ridge and Dadeland South Metrorail station takes only 25 minutes.

Both full-size and minibus routes operate on the busway and in adjacent neighborhoods, entering the exclusive lanes at major intersections. Six different bus routes use the busway. They offer local and limited-stop service between Florida City, in far south Miami-Dade County, and Dadeland South Metrorail station. Park-and-Ride facilities are provided at two strategic locations. Figure 1 depicts the South Miami-Dade Busway with Metrobus feeder routes.

Metrorail

As shown in Figure 2, Metrorail, the elevated, electrically powered, heavy rail portion of Miami-Dade County's transit system, provides service to 21 stations on a double track, 33.76 km (21.1 mi.) line. The last major addition to the system was the construction of a station which provides integrated transit service with Tri-Rail, the commuter rail line operating in Palm Beach, Broward and Miami-Dade Counties.

The system was opened in 1984 at a cost of $1.03 billion. The fleet consists of 136 cars with a normal capacity of 164 passengers per car. Metrorail operates at a top speed of 93 kph (58 mph) with an average speed of 50 kph (31 mph).

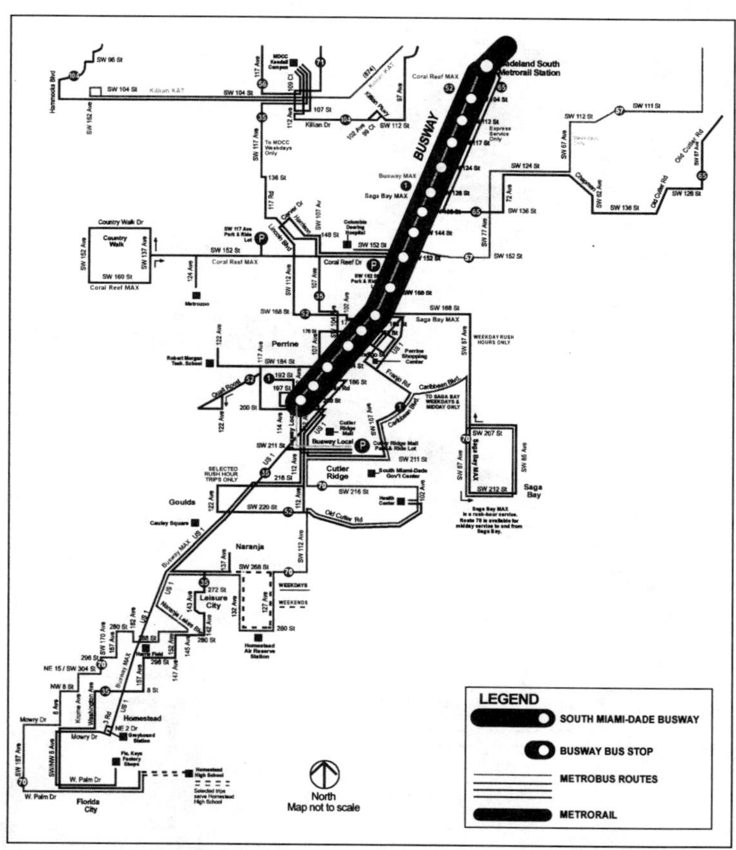

SOUTH MIAMI-DADE BUSWAY WITH METROBUS FEEDER ROUTES

Figure 1

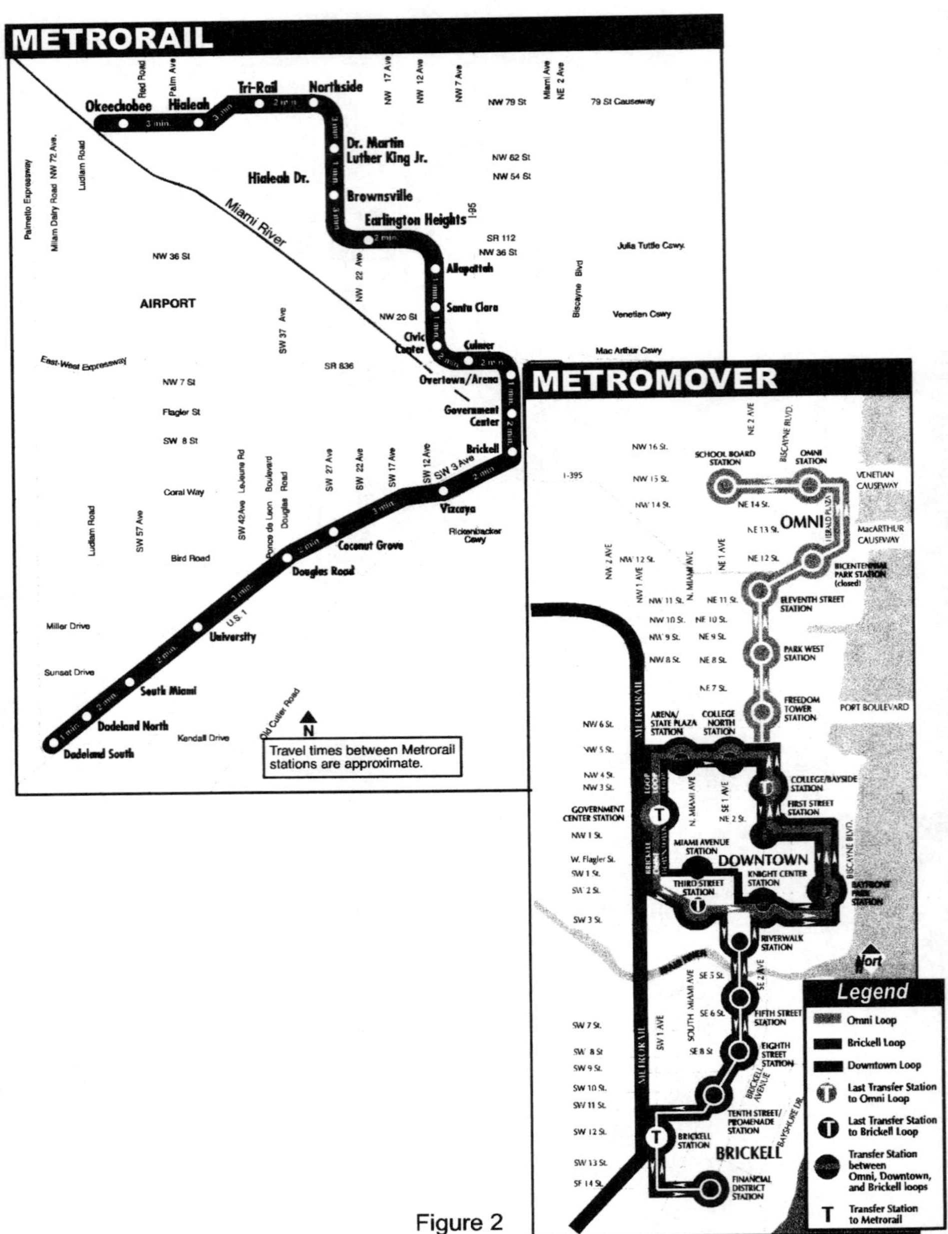
METRORAIL
Okeechobee
Hialeah
Tri-Rail
Northside
Dr. Martin Luther King Jr.
Hialeah Dr.
Brownsville
Earlington Heights
Allapattah
Santa Clara
Civic Center
Culmer
Overtown/Arena
Government Center
Brickell
Vizcaya
Coconut Grove
Douglas Road
University
South Miami
Dadeland North
Dadeland South
Red Road
Palm Ave
NW 17 Ave
NW 12 Ave
NW 7 Ave
Miami Ave
NE 2 Ave
NW 79 St
79 St Causeway
NW 62 St
NW 54 St
Palmetto Expressway
Miliam Dairy Road NW 72 Ave.
Ludlam Road
Miami River
I-95
SR 112
NW 36 St
Julia Tuttle Cswy.
AIRPORT
NW 22 Ave
Biscayne Blvd
NW 20 St
Venetian Cswy
SW 37 Ave
Mac Arthur Cswy
East-West Expressway
SR 836
NW 7 St
Flagler St
SW 8 St
SW 42 Ave
LeJeune Rd
Ponce de Leon Boulevard
Douglas Road
SW 27 Ave
SW 22 Ave
SW 17 Ave
SW 12 Ave
SW 3 Ave
Coral Way
Ludlam Road
SW 57 Ave
Bird Road
Rickenbacker Cswy
U.S. 1
Miller Drive
Sunset Drive
Kendall Drive
Old Cutler Road
N
Travel times between Metrorail stations are approximate.
METROMOVER
SCHOOL BOARD STATION
OMNI STATION
VENETIAN CAUSEWAY
MacARTHUR CAUSEWAY
OMNI
BICENTENNIAL PARK STATION (closed)
ELEVENTH STREET STATION
PARK WEST STATION
FREEDOM TOWER STATION
PORT BOULEVARD
ARENA/ STATE PLAZA STATION
COLLEGE NORTH STATION
COLLEGE/BAYSIDE STATION
FIRST STREET STATION
GOVERNMENT CENTER STATION
MIAMI AVENUE STATION
DOWNTOWN
THIRD STREET STATION
KNIGHT CENTER STATION
BAYFRONT PARK STATION
RIVERWALK STATION
FIFTH STREET STATION
EIGHTH STREET STATION
TENTH STREET/ PROMENADE STATION
BRICKELL STATION
BRICKELL
FINANCIAL DISTRICT STATION
METRORAIL
BRICKELL AVENUE
BAYSHORE DR.
BISCAYNE BLVD.
Legend
Omni Loop
Brickell Loop
Downtown Loop
Last Transfer Station to Omni Loop
Last Transfer Station to Brickell Loop
Transfer Station between Omni, Downtown, and Brickell loops
Transfer Station to Metrorail

Figure 2

Metrorail runs from 5:08 a.m. to midnight, with extended hours provided for special events when warranted. Trains arrive at a station every 5 minutes during maximum peak hours, every 10-15 minutes midday and Saturday, and every 20 minutes nights and Sunday. Metrorail averages 47,000 daily boardings.

Metromover

The Metromover system is a fully automated guideway people-mover. It includes a $153.3 million, 3 km (1.9 mi.) loop which opened in April 1986 and serves the core of the downtown area, and two extensions: one north to the Omni area; the other south, serving the Brickell area. The extensions, which opened in May 1994, added twelve stations to the original nine, 4 km (2.5 mi.) of guideway and seventeen additional vehicles at a cost of $228 million.

The service span is from 5:30 a.m. to 10:30 p.m. everyday, with extended hours provided for special events when warranted. The frequency of service during the peak hours is 3 minutes. During the off-peak hours, as well as weekends and holidays, service is 5 minutes.

The system connects with Metrorail at Government Center and Brickell stations and with Metrobus at various ocations throughout downtown Miami with a major bus terminal at the Omni station. Figure 2 also depicts the Metromover system.

The fleet consists of 29 vehicles. Each car is designed to carry 88 standing and 8 seated passengers. Metromover averages 11,400 daily boardings.

Paratransit Special Transportation Services

In addition to the three fixed-route modes, MDTA operates a demand-responsive service, called Special Transportation Service (STS). STS is a shared-ride, curb-to-curb transportation service for the disabled and mobility impaired riders who are unable to utilize the fixed-route modes. STS service is available for anyone who is certified by a physician as being unable to use conventional public transit in accordance with the American with Disabilities Act. The service area includes all of urbanized Miami-Dade County. Service is provided by sedans and lift-equipped vans, seven days a week from 4:30 a.m. to 2:30 a.m. the following day to match the service hours of Metrobus.

Presently, there are 196 sedans and 85 lift-equipped vans available for STS service. They are provided through a contract with a broker. There are in excess of 18,000 clients enrolled in the STS program; of which 15,000 are ambulatory and 3,000 are non-ambulatory. STS averages over 3,000 weekday boardings.

In addition, MDTA administers the state-subsidize Medicaid program for transporting low-income persons to and from medical facilities. The service is performed by a private contractor.
Under this program, over 1,700 weekday medical trips are provided free of costs to eligible persons.

Eligible riders in both the STS and Medicaid programs must make trip reservations no later than 24 hours prior to traveling.

Interfaces with other Transit Providers

Broward County Transit (BCT)

Broward County is the county to the north of Miami-Dade. Broward County Transit (BCT) operates four local routes serving northern Miami-Dade County. MDTA in turn operates three local routes serving southern Broward County. Transfers between the two systems are accepted at designated points where routes meet.

Tri-Rail

Tri-Rail is the commuter rail line that runs 107 km (67 mi.) within Palm Beach, Broward and Miami-Dade Counties. Tri-Rail is operated under the direction of the Tri-County Commuter Rail Authority (TCRA) with operations being managed by a private company.

There is a single track, with passing sidings, shared with Amtrak long distance passenger trains and CSX freight trains. A second track is being installed to permit higher frequency commuter service. Trains stop at eighteen stations, five of which are in Miami-Dade County. One of the Miami-Dade County stations provides a direct connection to Metrorail.

Weekday service starts at approximately 4:30 a.m. and operates until midnight. Trains run with headways of approximately 60 minutes during service hours. Tri-Rail also provides service on Saturdays, Sundays and for special events, when warranted.

Metrobus provides two feeder routes for Tri-Rail: one to serve Miami International Airport to and from Miami Airport Station; and one to serve the Doral/Koger Center area to and from Hialeah Market Station. Two local bus routes serve the Tri-Rail Metrorail Station, three local bus routes serve the Opa-locka Station and five local routes serve the Golden Glades Tri-Rail Station. The Tri-Rail fare system is comprised of six zones, and ticket prices are determined by the number of zones through which a passenger travels.

Jitneys

Private jitney operators provide for-profit service on numerous roads that supplement, or partially compete with, Metrobus services.

Presently, eleven approved jitney operators provide jitney services along 13 routes.

Any problems with unlicensed and unregulated jitney services operating along MDTA bus routes are addressed through enforcement of the Miami-Dade County Ordinance regulating the jitney industry.

<u>Fares, Revenues and Operating Costs</u>

Approximately 33% of the MDTA operating revenues come from fares. Transit fares are paid with exact change coins, dollar bills (buses only), tokens or a Metropass. Many qualified elderly, persons with disabilities, and youths (with permit and ID) are eligible for reduced fares on conventional transit as indicated below:

	Full Fare	Reduced Fare
Bus/Rail	$1.25	$0.60
Express Bus	1.50	0.75
Metromover/Shuttle Bus	0.25	0.10

Certain qualified persons are eligible to ride for free on conventional transit.

Transfers cost $0.25 ($0.10 for reduced fare passengers) and one can ride Metrobus, Metrorail and Metromover any day, anytime, by purchasing a $60.00 monthly Metropass ($30 for reduced passengers).

STS fares range between $2.50 to $4.00 per trip based on the fare charged for a similar trip made on conventional transit.

MDTA also receives revenues from advertising and joint development as well as significant subsidies from County (49.9%), State (7.5%) and the Federal government (1.9%). Operating revenues from the local gas tax amounts to (2.8%).

Transit Support Facilities

Bus Garages

The Miami-Dade Transit Agency operates three maintenance bus garages to serve a fleet of 590 full-size buses, 46 mini-buses and 15 vans. These garages, located throughout the County, provide services such as inspections, paintings, body work, major overhauls, preventive maintenance, transmission and air conditioning repairs.

Rail Maintenance

The Metrorail fleet of 136 rail cars is stored and maintained at the William L. Lehman Center facility located at the north end of the line in northwest Miami-Dade County while the Metromover fleet of 29 cars is supported by the maintenance facility located in downtown Miami.

Park and Ride Facilities

The Miami-Dade Transit Agency has close to 10,000 park-ride spaces available. An average of 55% of these spaces are utilized on weekdays. However, actual parking usage is highest on the southern portion of the Metrorail line, and at the Metrobus Golden Glades parking lot.

Bus Fleet Replacement

One half of the Metrobus vehicles have run in excess of 800,000 km (500,000 mi.) and/or are over 12 years old and are due for replacement. To achieve a regular replacement cycle, fifty replacement buses are programmed to be acquired annually through FY 2005.

High Capacity Improvements

To provide for the anticipated mobility needs of Miami-Dade County residents and visitors in the years to come, high capacity transit system improvements have been identified in the Long-Range Transportation Plan and are in various stages of planning and implementations. They are briefly described below.

Palmetto Metrorail Extension

A 2.2 km (1.4 mi.) extension of the Metrorail system will be built westward from the current north terminal at Okeechobee Station and a new station will be built just west of the Palmetto Expressway (S.R. 826).

The station will have approximately 800 regular, 20-handicapped, and 9 pick-up and drop-off parking spaces. Parking facilities will be at-grade and provide improved auto access to Metrorail for commuters using S.R. 826 to and from northwest Miami-Dade and Southwest Broward. Construction of the extension has started and the scheduled completion date is in the year 2002. In addition, bus feeder services to the station have been programmed.

East-West Corridor

Multimodal solutions for severe traffic congestion along S.R. 836, the principal east-west expressway in central Miami-Dade County, were examined. As a result, an integrated system of highway improvements, high occupancy vehicle (HOV) lanes, a new rail transit line, bus system enhancements, an intermodal transfer point at Miami International Airport, and pedestrian and bicycle facilities have been proposed. The study considered a 38 km (24-mi.) rail line (heavy and light rail) connecting some of the region's most important economic generators, with direct transfers to the Metrorail, Metromover, Tri-Rail, Amtrak and eventually high speed rail. As proposed, the project links the suburban areas west of Miami, Florida International University, the Miami International Airport, downtown Miami and the Port of Miami, with a separate light rail line from downtown Miami to Miami Beach Convention Center. Continued work on this project is contingent upon a dedicated funding source.

North Corridor

This corridor extends northward, approximately 5.2 km (9.5 mi.) from the existing Metrorail line, along N.W. 27 Avenue to the vicinity of the Pro-Player Stadium and the Broward County line. The possibility of a future extension into Broward County was also studied. The North Corridor project would also serve the North Campus of Miami-Dade Community College, the City of Opa-Locka and would interface with highway facilities at Gratigny Parkway, Palmetto Expressway and the Homestead Extension to Florida Turnpike. Continued work on this project is contingent on a dedicated funding source.

Kendall Corridor

A Major Investment Study (MIS) for this corridor is currently being conducted. The MIS is examining transportation solutions in terms of appropriate modes (e.g., heavy rail, light rail, busway), alignment and options. The corridor consists of two segments. The first segment extends west to east from S.W. 147 Avenue to the Dadeland area via Kendall Drive. The second segment of this corridor would provide a northerly extension from approximately the Dadeland area to the Miami International Airport, connecting into the proposed Intermodal Center. Such a connection would reduce traffic along the Palmetto Expressway, one of the most congested expressways of South Florida, and provide additional service to and from major employment centers in the vicinity of Miami international Airport. Continued work on this project is contingent on a dedicated funding source.

Northeast Corridor

This project is a proposed 21 km (12 mi.) transit way facility connecting northeast Miami-Dade and southeast Broward areas with downtown Miami generally along the existing Florida East Coast (FEC) railroad right-of-way corridor. Funding is currently being sought to conduct a Major Investment Study for this corridor.

Future Outlook

Miami-Dade County has sought and intends to continue seeking funding to enable the implementation of critical projects and expansion of the existing bus service that are necessary to serve the mobility needs of citizens and visitors and is trying to enhance their quality of life. The County is trying to provide for the continued development of Miami-Dade County as an international center of tourism and commerce.

Development of Innovative Transit Services Using Market Research Techniques

Danny Alvarez[1]

Abstract

The private sector has used market research techniques to determine the characteristics of their markets and marketing to develop and encourage the use of their products and/or services. The public sector transit industry is starting to use these techniques to draw and retain passengers. The Miami-Dade Transit Agency (MDTA) has been applying such conventional business market research and marketing techniques to design services, facilities, and a marketing program for those services and facilities.

This paper discusses two cases where those functions played a vital role in the design and operation of new and successful suburban transit service in Miami-Dade County, Florida.

Introduction

The private sector has used market research techniques to determine the characteristics of their markets and marketing to develop and encourage the use of their products and/or services. The transit industry in the United States was primarily a private sector industry until the middle of this century. As the result of the proliferation of the automobile and the facilities to support the automobile, the elements of the transit industry that needed to exist for the public good became the responsibility of government.

1. Director, Miami-Dade Transit Agency, 111 NW First Street, Miami, FL 33128

It has been especially difficult to provide transit service to the suburbs, areas where the automobile dominates and where land uses are designed to serve the automobile. As in other metropolitan areas, MDTA was looking for ways to effectively provide transit service particularly in the growing southwestern suburbs in Miami-Dade County.

The Agency started to use market research and marketing techniques that have been used successfully by the private sector. Such applications had only recently come into vogue in the development of transit services and in promoting those services. This paper discusses two cases where these functions played a vital role in the design and operation of transit in Miami-Dade County, Florida.

Background

The MDTA provides multi-modal transit services primarily in the urbanized portion of Miami-Dade County. However, the growth of the suburbs has created traffic that, by public policy, can only be relieved by improving transit service. The State of Florida has made available several sources of funding to the urban areas of the State. Two of those funding sources, that are meant for urban corridors and service development, have been used by MDTA to support new types of transit services designed to meet the needs of the suburban areas.

Choice riders (those who can choose to drive or ride transit) are willing to leave their cars at Metrorail stations and use the heavy rail line into Central Miami, and, if needed, the downtown people mover (Metromover) to their final downtown destinations. The parking garages at the two south ends, suburban Metrorail stations (Dadeland South and Dadeland North) are filled every business day.

In addition, roadways connecting Metrorail and the western suburban areas are crowded at all hours, especially in the peak periods. US 1, the major road connecting Metrorail and the southern suburban areas, is outgrowing in its capacity.

To accommodate this traffic growth, two major transit projects were undertaken: in the late 1980's, Kendall Area Transit (KAT) service was instituted to feed west suburban traffic to and from Metrorail; and in the the late 1990's, the South Miami-Dade Busway was built and new services were implemented to operate thereon.

Kendall Area Transit (KAT)

The Dadeland North Metrorail station was built to cater to vehicular access. An expressway connecting this station to the western suburbs was built, in part, to feed Metrorail. A large, 11 bay bus terminal is also provided at this station. Thus, the physical facilities at this station can accommodate extensive feeder bus and park-and-ride services.

When Metrorail was opened in the mid-1980's, local bus service was realigned to act as feeders to Metrorail. However, the vast majority of riders were either former riders of the express bus service between Southwest Miami-Dade County and Central Miami, or were transit captive riders. There was a need to develop a service that would be attractive to new riders who were accustomed to using their car.

Market research was conducted which showed that the target market could be comfortable using small buses that operate frequently and primarily in their own neighborhood. The run times had to be comparable to automobile run time.

In 1987, using Urban Corridors funding from the State, MDTA started a demonstration project, Kendall Area Transit (KAT), to feed the Metrorail Dadeland North station with three minibus routes. Each route operated on one of the three major east-west arterials in the suburban Kendall area. The buses ran every seven to ten minutes during peak periods. The off-peak feeder service was available on the conventional local bus service.

The buses operated non-stop on the expressway and limited stops (about every half mile) on the arterial streets. Initially, the routes were operated by a private company under contract to MDTA. Each KAT route had an end-of-line park-ride lot on the property of cooperating shopping centers. There were also two other, mid-line, joint-use, park-ride lots.

A major marketing campaign was implemented to initiate the service. Free ride coupons (with free transfers to Metrorail) were mailed to all residents in the target corridor zip codes. The buses were painted bright orange with cat's (KAT's) paws on the buses, bus stops signs, timetable, and other printed material.

The State grant ended in 1991, and MDTA has continued operation of the service using traditional funding sources. Since 1991, the orange colored minibuses were replaced by over-the-road coaches, which were a hit with the passengers. However, the cost of that operation was very high. Under an agreement with the Transport Workers' Union, the service was transferred to MDTA's then new Paratransit Division, a unit that operates minibuses with operators (paratransit driver-attendants) who are paid a lower wage scale. In addition, many of the paratransit driver-attendants are part-timers, which enables more efficient scheduling and operation of the peak period service in terms of operating cost.

The three KAT routes have a combined average daily ridership of about 3000 boardings per day (all data in this paper is based on March 1998 statistics unless otherwise noted). Most passengers are going to or from Metrorail, but there are some using the services for intra-corridor rides. Diversion of riders from cars resulted in a small, but measurable decrease in traffic on North Kendall Drive, the major State Road in the corridor.

The current revenue recovery factor for the KAT routes ranges between 54 and 71%, compared to 50% for all bus routes; and net cost of operation per boarding range from 31 to 63 cents, compared to 77 cents for all bus routes.

The KAT routes successfully attracted suburban, automobile-oriented riders, with their cost statistics better than the the system average. They have been able to attract riders to obviate costly expansion of the parking facilities at the Dadeland North Metrorail station.

South Miami-Dade Busway

The operations of the Paratransit Division have given MDTA the flexibility to try new types of services. The KAT routes are models for routes that were developed for the highly successful South Miami-Dade Busway.

In 1988, the State of Florida bought the railroad right-of-way parallel to US 1 southward from the Dadeland South Metrorail station. It was determined that the most cost-effective use of the right-of-way was as a Busway, an exclusive two lane roadway for buses. The Busway is 13 kilometers (8 miles) long with 15 stations spaced at about every kilometer (1/2 mile). It opened for revenue service on February 3, 1997.

To plan and design the appropriate service to be operated on the Busway, extensive market research was conducted. The market research effort included a random telephone survey of households in the Busway corridor; a survey of automobile drivers on US 1; a survey of passengers on existing bus routes in the corridor; and an analysis of the comments received at public meetings and of letters and phone calls received by public officials and the Transit Agency.

The analyses produced the following major findings:

o About 25% of automobile drivers on US 1 have destinations in the Busway corridor and were potential Busway users.

o Transit riders and potential Busway users were concerned about their complete trip, not just the Busway portion of their trip. For example there are many potential users who work in areas north of the Busway, and there are many transit riders who have destinations in areas south of the Busway.

o Frequency of service and personal security were significant concerns.

o Concerns of automobile drivers about possible traffic delays on US 1 and on cross streets.

o Desires of potential transit riders for convenience while the current transit riders wanted their current travel patterns to be improved.

All of these issues were taken into account in the design of the Busway and the transit service on the Busway. Two existing transit routes were shifted from parallel US 1 to the Busway. Four new routes were designed to serve the Busway as well as neighborhoods which never had transit service or whose transit service needed improvements. The service on routes with expected low ridership were planned to be operated with minibuses as were routes where non-peak period ridership was expected to be low. Using the smaller buses, MDTA was able to provide a high frequency of service at a reasonable cost.

The six routes operating on the Busway provide a combined frequency at Dadeland South of 20 trips per hour in each direction in the peak periods and 6 trips per hour in the midday.

Park-ride lots were provided, one at the mid-point of the Busway and one at the south end of the Busway. Both lots were multi-use lots: the mid-point one is at a County park where golfers used the lot primarily on weekends; and the other at a little used area at a major regional shopping mall. In both cases, the lots are at major intercept points and are owned and would be maintained by entities other than the Transit Agency.

To make the transit service on the Busway easy to use, the standard, uniform, transit fare structure was applied to all Busway service with one exception. That exception was to permit transfers from Busway bus routes to Metrorail at no extra charge. In addition, the Busway park-ride lots provided parking at no charge.

The intermodal terminal at Dadeland South was reconfigured to make the transfer between Metrorail and Busway buses as convenient as possible. The distance between the Metrorail fare gates and the buses is only a few meters.

Stations along the Busway were designed to meet the needs of passengers. Large canopies are designed to minimize weather impacts on passengers; and the stations are open and well lit at night to enhance passenger security.

A major marketing program was developed to promote the Busway. The program had the following goals:

o To convince drivers to use transit as often as possible for all or part of the time for part of their trip;

o To let non-riders know the value of the Busway to the community; and

o To inform then-existing transit riders of the improvements in transit service that would be possible when buses started operating on the Busway.

The MDTA Marketing staff worked with a local television station to develop videos at no cost to MDTA. Several 30-second spots were shown on the cable television network serving the corridor. A four-minute video was shown at dozens of public meetings before civic and business groups, religious organizations, and homeowner associations. Many of the meetings were set in cooperation with the South Miami-Dade Chamber of Commerce. The Chamber made the Busway a major priority in its activities because of the benefits of the Busway to the business community in South Miami-Dade. The public meetings were held in the last six months before the opening of the Busway and during the first six months of operation.

There were two sets of printed materials: 1. Individual route guides developed for the four new routes on the Busway; and 2. a South Miami-Dade Transit Directory was published to look like a telephone directory. The transit directory provided maps and schedules of all routes in South Miami-Dade, all of which either operate on the Busway or feed passenger traffic to the Busway.

A Sunday morning party on the Busway took place the day before the start of revenue service. It was attended by thousands of County residents. Rides on all Busway routes were free for the first two weeks of operation.

Transit ridership on the Busway has exceeded even the highest projections. In general, corridor ridership, which had been hoped to increase 10 or 15% during the first two years of operation, increased about 50% on weekdays and 75% on weekends in the first year of operation. Ridership has been so high, that full-sized buses replaced minibuses in non-peak periods, and, in some cases, articulated buses replaced full-sized buses in peak periods. Ridership continues to increase in the second year of operation.

An extension to the Busway is currently under design. The market research and marketing techniques employed in service development for the initial section of the Busway will be employed for the extension.

Conclusion

By using the knowledge gained from employing conventional business market research techniques to plan and design transit services and facilities, MDTA has successfully implemented suburban transit. MDTA is continuing the application of market research techniques to provide efficient and effective transit service to its customers.

New York City's Socioeconomic Characteristics, Land Use and Travel Patterns

Andrew Bata* and Shoshana Cooper**

Introduction

The transportation system in New York City is shaped primarily by the situation that the density of human activity results in travel levels well in excess of the capacity of the automobile-dominated street and highway network. As a result, rapid transit and commuter railroads form the backbone of the transit system, with buses used for local distribution and miscellaneous tripmaking unserved by the rail systems.

However, the rail system has not grown to accommodate increases and changes in tripmaking patterns, such that both the rail system and the street and highway system are overloaded.

The maximum density is achieved in the Central Business District. Tripmaking is highly polarized into peak travel feeding and leaving the CBD. While population appears to be stable or slow growing, the metropolitan area is expanding; increasing miles of travel and the population appears to be increasing their per capita tripmaking. Employment changes have resulted in a sharper peak, with less shoulder travel, but also more work travel during previously off-peak periods.

The non-CBD areas in New York City and some suburban centers in the New York area also have high densities of activity by American standards. Some geographic boundaries such as rivers create additional travel polarized patterns; these "bottlenecks" may create a more controllable system by limiting overall usage.

* Senior Director, Service Planning, Operations Planning, MTA New York City Transit, 130 Livingston Street 3036, Brooklyn, New York 11201.

** Assistant Director, Operations Analysis, Operations Planning, MTA New York City Transit, 130 Livingston Street 3032F, Brooklyn, New York 11201.

New York City and its Surroundings

New York City is located in the northeast portion of the United States, on its Atlantic Coast at the mouth of the Hudson River. The Northeast corridor is one megalopolis, with large metropolitan areas in Boston, Philadelphia, Baltimore, and Washington all within less than one day's drive from New York City. The corridor also contains many smaller Metropolitan areas such as Providence, Rhode Island and Trenton, New Jersey.

In terms of physical size, New York City occupies only about 0.5 percent of the 122,706 square kilometers (47,377 square miles) that make up New York State's land mass while about 40 percent of New York State's population lives in New York City[1]. Also, New York City has an extremely large economic, political, cultural, and social impact on New York State and the entire region.

To the east of New York State is Connecticut and to the west is New Jersey. The areas in these two states, which are closest to New York City, as well as nearby New York State counties, are frequently described as the Tri-State area. It is difficult to define the exact limits of the Tri-State area, as it seems to be constantly growing as urban sprawl spreads. The distances, which some commuters travel, every day to reach their jobs in New York City is continuing to increase. In fact, there is an increasing number of people who commute over 100 miles each way from communities in the northeastern portion of Pennsylvania, the state to the west of New Jersey.

In the longer version of the 1990 United States Census, some residents were questioned about their average travel time to work. Unfortunately, the tabulation of this data has not kept pace with the growth of the New York Metropolitan area. Of the 50 largest Metropolitan areas in the United States, workers from these three areas are likely to work within New York City:

Travel Time to Work

Metropolitan Statistical Area of Residence	Average Travel Time to Work
New York, New York	26.5 Minutes
Nassau-Suffolk, NY	30.0 Minutes
Newark, NJ	26.2 Minutes

Source: United States Bureau of the Census, 1990 United States Census, http:/www.census.gov

However, this data is somewhat misleading. Anecdotal evidence shows many commuters with commutes of longer than one hour. Furthermore, the commute from Pennsylvania is usually 2 hours or longer.

The size of the New York Metropolitan area and the density of land use within New York City combine to create a "special case." New York City is unique among all other cities in the United States. No other city faces the same issues or generates the same type of attention.

Land Use

New York City is divided into five boroughs: Manhattan, The Bronx, Brooklyn, Queens and Staten Island, each of which is its own county. Manhattan is the central borough of New York City, in almost every sense, and some outsiders even ignore the other four boroughs. Manhattan is an island, largely made of a protrusion of granite, which rises a few hundred feet from sea level[2]. On the west side of Manhattan is the Hudson River, while the Harlem River is on the northeast and the East River on the southeast. In general, Manhattan streets are a grid, with north-south Avenues numbered from First Avenue on the East Side and 12th Avenue on the West Side. The east-west streets are numbered up to 220th Street on the northern tip.

The Bronx is north of Manhattan, past the Harlem River. The Bronx is the only borough that is not on an island and there is no natural geographic boundary between it and Westchester County, to its north. Several bridges and subway lines connect The Bronx to Manhattan. There are also several bridges, which connect The Bronx to Queens, its neighbor to the southeast.

Staten Island is south of Manhattan, with the only direct connection between the two boroughs being the Staten Island Ferry. Staten Island, which is geographically more a part of New Jersey than New York, can be reached from Brooklyn by the Verrazano Narrows Bridge. There are also three bridges, which connect Staten Island to New Jersey. Staten Island has a small, by New York City standards, commuter rail line which mainly brings passengers to and from the Staten Island Ferry. New York City's main rapid transit system operates in every borough, except Staten Island.

Although Brooklyn and Queens are both located on Long Island, most New Yorkers view Long Island as starting at Nassau County, on Queens eastern border. The Newtown Creek separates the northern section of the two boroughs, while there is no natural geographic boundary in the south. Similarly, there is no natural geographic boundary between Queens and Nassau County. From Brooklyn and Queens, there are several bridges and tunnels, which allow vehicular traffic and rapid transit to travel to and from Manhattan.

The size and population density of the five boroughs vary greatly, as demonstrated by the following table:

New York City Land Area, Population and Population Density

Borough	Population	Land Area in Hectares (Acres)	Persons Per Hectare (Acre)
Manhattan	1,487,536	7,355 (18,161)	202.2 (81.9)
The Bronx	1,203,789	10,894 (26,899)	110.5 (44.8)
Brooklyn	2,300,664	18,281 (45,139)	125.9 (51.0)
Queens	1,951,598	28,356 (70,015)	68.8 (27.9)
Staten Island	378,977	15,190 (37,507)	24.9 (10.1)
TOTAL	7,322,564	80,077 (197,722)	91.4 (37.0)

Source: New York City Department of City Planning, Population Division, 1990 Census.

While the population is dispersed throughout the five boroughs, Manhattan is the center of employment. While Manhattan's Central Business District (CBD) is generally defined as the nine square miles south of 60th Street, within the CBD, there are two major concentrations of employment: Lower Manhattan and Midtown. Downtown Brooklyn and Long Island City in Queens form smaller "mini-CBDs" which are also centers of employment. There are many businesses, which generate travel in all of these CBDs as well as in many other locations throughout the five boroughs.

There is no one major manufacturing or industrial area in New York City. Rather there are many smaller areas that are distributed throughout the five boroughs.

New York City has many major, highly respected institutions of higher learning including New York University, Columbia University, and the City College of New York in Manhattan, Fordham University in the Bronx, St. John's University and Queens College in Queens and Brooklyn College in Brooklyn. There are also smaller and/or lesser-known institutions throughout the city. Access to these institutions is quite varied.

Other major trip generators include Yankee Stadium in the Bronx and Shea Stadium in Queens where major league baseball games and other large events are held. Both stadiums are accessible by subway although the proportion of spectators who arrive by public transit varies greatly by day of week and which team is playing. The United States Tennis Center, with its Arthur Ashe and Louis Armstrong Stadiums, attracts large numbers of tennis fans during the U. S. Open, which is held every year in late August. Madison Square Garden, an indoor stadium in Manhattan, also generates many trips, although not on the same scale as the outdoor stadiums. New York also boasts many museums and theaters, which attract a large number of visitors. In addition, there are many major parades and festivals throughout the year, which attract large crowds.

There are two major airports in New York City: JFK is in southern Queens and La Guardia in northern Queens. Neither airport is directly accessible by rail. There are various rail and bus combinations as well as direct bus and van service which provide access to these busy airports, but the complexity of these connections encourages an undesirably high portion of airport access to be made by private vehicles and by taxi. New York is also served by Newark Airport, which is located in nearby Newark, New Jersey. A new rail link will make Newark Airport somewhat more accessible by public transportation, but private vehicles and taxis still compose a major portion of the land entries into the airport.

In Manhattan, there are two major rail terminals: Penn Station which serves the Long Island Railroad, Amtrak and New Jersey Transit, and Grand Central Station which serves Metro North Railroad. In addition, Port Authority Bus Terminal serves many commuter bus lines, both public and private as well as Intra-city bus lines. All three terminals have convenient access to more than one subway line. Since these terminals are all in very congested areas, relatively few customers arrive or leave by private automobile, although each location has an active taxi stand.

Long Island Railroad also has a smaller terminal on Flatbush Avenue in Brooklyn and terminates some trains in Jamaica, Queens. Both locations are well served by the subway. New Jersey Transit also terminates a substantial portion of its service in Hoboken and Newark, New Jersey, where passengers can transfer to PATH, the Port Authority's rapid transit system, to reach Manhattan.

Socioeconomic Elements

In June 1998, New York City celebrated the centennial of the unification of the five boroughs, which form the city as we know it today. In these past one hundred years, the demographic features of the city have changed as different ethnic groups move in and out of the city. When the city was formed, Manhattan was the most populous borough with approximately 1.5 million residents. Brooklyn, formerly an independent city, had slightly fewer than 1 million residents while the other three boroughs were lightly populated. Today, although Manhattan has almost the same population it had one hundred years ago, Brooklyn is now the most populous borough with over 2.3 million and Queens now has close to 2 million residents. The Bronx now has approximately 1.2 million while the population of Staten Island has increased to approximately 375,000[3].

In recent years, as the city's total population increased from 7,306,000 in 1991 to 7,343,000 in 1997 (an increase of about 0.5 percent), more than one million New Yorkers moved out of the city. This exodus was counter-balanced by the arrival of 677,000 immigrants[4]. This population shift is reflected in the changing demographic composition of the City. In 1900, close to 100 percent of the City's population was white. By 1990, whites composed slightly more than 40 percent of the population. Blacks and Hispanics each composed almost 25 percent while the remaining Asians and other non-whites were about 10 percent[5]. The population today is also much older than in 1900, as illustrated by the following chart:

Distribution of New York City Population by Age

	1900	1990
Under 15 Years Old	30.7%	19.4%
15 to 44	53.4%	48.2%
45 to 64	13.1%	19.4%
65 and over	2.8%	13.0%

Source: The New York Times, New York City Centennial Section

At the same time, the number and size of the communities that are considered part of the New York Metropolitan area is growing. This has increased the number of people who are commuting into the city as well as the length of their trip. (This has also increased the frequency of intra-suburb/exurb trips.) The issue of the size of the commuter population is particularly important as New York City continues to recover from the recession of the 1990s during which the City lost more than 300,000 jobs. In 1997, the City's economy grew by 3.1 percent with the number of jobs increasing by almost 54,000, bringing the total employment to over 3,400,000 for the first time since 1990. This job growth was mainly in the construction, trade and service industries while the financial, insurance, and real estate (FIRE) sector lost 9,000 jobs in 1997, with a total job loss in the FIRE sector from 1988 to 1997 of 71,000 jobs[4]. At least some of the job loss in this sector can be attributed to the transfer of work sites to locations outside of the city. Presumably, one of the factors in this transfer is the ease of commuting to new suburban centers. Overall, the average payroll employment (the total number of persons employed in full and part-time non-farm jobs) by industry was as follows:

Average Payroll Employment by Industry

Industry Employment	**1996**	**1997**
Construction	91,200	93,800
Manufacturing	264,500	264,400
Transportation	204,600	206,300
Trade	561,900	579,400
FIRE	472,300	471,400
Services	1,229,000	1,270,700
Mining	300	300
Government	533,800	525,000
TOTAL	**3,347,500**	**3,411,300**

Source: Rent Guidelines Board, Page 11.

The average nominal annual salary in New York City was $46,253 in 1996, an increase of 6.6 percent. However in real terms, as calculated in 1989 dollars, the increase was only 3.6 percent[4].

Travel Patterns

The centralization of employment within Manhattan's central business district creates a great influx of people every weekday morning who arrive using a great number of routes. A large majority of these commuters use public transportation for at least part of this commute. There are many forms of public transportation, including MTA New York City Transit's subway and bus system, Metro North Railroad, the Long Island Railroad, New Jersey Transit's bus and rail system, the Port Authority's rapid transit system PATH and many private bus lines. In 1996, MTA New York City Transit counted 1,689,998 passengers entering the CBD by rail and another 61,699 entering by bus[5]. Since this time, there has been an increase in ridership, mainly due to changes in the fare structure. There are approximately 60,000 people a day who make trips on the Staten Island Ferry[6]. Overall, the method of commutation is detailed in the following chart:

Means of Commutation-Workers 16 Years of Age and Over

	The Bronx	Brooklyn	Manhattan	Queens	Staten Island
Number of Workers 16 years and over	429,777	907,010	754,148	918,06 3	174,090
Car, truck or van	34.4%	31.3%	11.6%	44.2%	64.1%
Drove Alone	24.9%	22.5%	7.8%	33.8%	48.7%
Carpooled	9.5%	8.8%	3.8%	10.5%	15.4%
Public Transportation	56.6%	58.0%	58.4%	47.8%	30.5%
Bus	16.8%	11.8%	14.5%	10.3%	16.3%
Subway	36.9%	44.0%	38.1%	34.6%	3.3%
Railroad	2.0%	1.5%	1.1%	2.3%	1.5%
Taxicab	0.9%	0.6%	4.6%	0.6%	0.4%
Ferryboat	0.0%	0.0%	0.0%	0.0%	9.0%
Walked	7.1%	8.3%	23.0%	6.0%	3.3%
Worked at home	1.3%	1.6%	5.5%	1.5%	1.4%
Other Means	0.7%	0.7%	1.5%	0.6%	0.7%

Source: City of New York, Department of City Planning, 1995 Annual Report on Social Indicators

It should be noted that some commuters use more than one method of commutation, thereby causing a total, which exceeds 100 percent. The variation in the method of commutation, to a large extent can be attributed to the variation in population density, as noted in Section III. Queens and Staten Island, which have the lowest population density, also have the lowest use of public transportation. In Manhattan, on the other hand, with its high density, 23 percent of workers walk to work. The means of commutation and the population density are also reflected in motor vehicle ownership. In 1996, there were 1,715,555 standard passenger vehicles and 1,862,317 total vehicles registered in New York City[7]. In other words, there was approximately one standard vehicle for every 4.3 persons in New York City. Details of vehicle ownership are included in the following chart:

Motor Vehicle Ownership Per Occupied Housing Unit by Borough

Borough	No Vehicles Available	1 Vehicle Available	2 Vehicles Available	3 or More Vehicles Available
The Bronx	61.4%	28.6%	8.1%	1.9%
Brooklyn	56.7%	33.2%	8.4%	1.7%
Manhattan	77.8%	20.2%	1.7%	0.3%
Queens	36.6%	41.4%	17.1%	4.9%
Staten Island	18.2%	37.5%	32.8%	11.5%
NEW YORK CITY	55.9%	31.5%	10.0%	2.6%

Source: City of New York, Department of City Planning, Socioeconomic Projiles: A Portrait of New York City's Community District from the 1980 and 1990 Censuses of Population and Housing, March 1993.

These relatively low vehicle ownership rates reflect the high density as well as the availability of public transportation. Tripmaking in New York City is easier to accomplish without a car because of its density. Anecdotal evidence supports the claim that many households have a vehicle, which is not used daily for commutation, but is used for some personal trips and errands on a regular basis. In addition, many New York City streets cannot accommodate the number of drivers who might otherwise want to use them. This creates congestion, which decreases the ease of using a motor vehicle, and discourages some people from driving. The following chart of travel speeds collected by the New York City Department of Transportation illustrates the increasing severity of this problem:

Manhattan Vehicle Speeds and Volumes

	FY 1991	FY 1992	FY 1993	FY 1994	FY 1995	% Change 1991-1995
Average Vehicle Spd. kph (mph) Avenues	14.6 (9.1)	14.4 (9.0)	14.0 (8.7)	13.1 (8.2)	13.7 (8.5)	-6.5%
Streets	9.7 (6.0)	9.8 (6.1)	9.7 (6.0)	9.0 (5.6)	9.0 (5.6)	-6.6%
Combined	12.1 (7.5)	11.9 (7.4)	11.6 (7.2)	10.9 (6.8)	11.4 (7.1)	-5.3%
Average Daily Vehicle Volumes	1,744,103	1,757,573	1,757,000	1,729,000	1,868,000	7.1%

Source: City of New York, Department of City Planning, 1995 Annual Report on Social Indicators, p. 70.

Although these slow vehicle speeds may discourage some automobile use, they also impede the ability to provide attractive bus service. Furthermore, slow vehicle speeds together with other forms of congestion, increase bus-running times by an average of 23.9 percent for MTA New York City Transit's bus service[8]. In fact, in 1997, it was estimated that congestion increased the cost of New York City Transit bus service in Manhattan by $46 Million dollars annually[9].

This congestion also has a negative impact on New York City's economy as a whole. In particular,

> The dramatic decline of the city's industrial sector can be blamed, in part, on the difficulty of moving goods into, out of, and around New York City. Modern industry's major reliance on trucking translates to a critical dependence on the city's overburdened, outdated highway network[10].

Furthermore, in New York City, and Manhattan in particular, the high cost of real estate (which is related to the high density) has discouraged the development of alleys and loading docks. Deliveries are often made from the curb, which can interfere with bus movement[8]. Or when the curb is occupied, it is not uncommon for deliveries to be made from a double-parked vehicle. This behavior only increases congestion.

The slow vehicle motor vehicle speeds are, to a large extent, a factor of the high number of vehicles crowded into limited roadway space. Another factor is the poor condition of New York City's roadways as noted below.

New York City Roadways by Condition in 1994

Condition	Kilometers (Miles)	Percent
Poor	315.47 (196.07)	28.1%
Fair	558.87 (347.34)	49.8%
Good	247.37 (153.74)	22.0%
Excellent	0.0	0.0%
Total	1121.71 (697.15)	100.0%

Source: City of New York, Department of City Planning, 1995 Annual Report on Social Indicators, p. 69.

These traffic problems are compounded by the geographical separation of the five boroughs, as described in Section III. Intra-borough travel, except between parts of Queens and Brooklyn, requires traveling on a bridge or in a tunnel. Including the bridges that connect different boroughs, there are a total of 2,027 bridges in New York City. Of these bridges, the Queensboro Bridge, which connects Queens and Manhattan, carries the largest amount of daily traffic: On an average weekday in 1997, approximately 158,000 vehicles carrying 400,000 people crossed this Bridge[6].

These bridges are often a source of congestion, particularly near the toll plazas on those bridges, which charge a toll. Although the new electronic toll collection system, EZ Pass, has improved the situation, media traffic reports still have a major focus on the ease of movement on major bridges and tunnels. At times, particularly during rush hour, the congestion on the bridges and tunnels is so severe that it creates queuing on the major highways and city streets, which lead to them. On city streets, this queuing can create spillback, and sometimes gridlock, a term first coined in New York City.

In Manhattan, and in some popular shopping districts in the outer boroughs, free parking can be scarce. At times, drivers looking for parking spots can impede the movement of traffic. More frequently, many drivers park in illegal spots such as bus stops and driveways, or "double-park," meaning to park in the moving lane. There are many parking lots, mainly in Manhattan, which charge relatively high rates. This discourages some people from driving, but the fact that very few lots are empty, particularly on weekdays, implies that the cost of parking does not prevent many drivers from using their automobiles.

On the other hand, traffic speeds in some areas are still too fast, killing and injuring far too many pedestrians. Since July 1996, New York City Department of Transportation has installed 144 speed humps at 72 locations[11]. There are locations, particularly in Manhattan, in which there is pedestrian congestion. That is, pedestrians cannot walk as fast as they would like because of the large number of people crowded into the small area that is allotted to them. This often encourages pedestrians to walk on roads. For this reason, barriers have been placed in some locations in order to prevent pedestrians entering the street or from crossing at locations which are perceived to be unsafe. There are some indications that pedestrian deaths and injuries are becoming less frequent. It is not been proven whether these measures are the main cause of this reduction, or whether the overall reduction in vehicle speeds discussed above has increased pedestrian safety.

Conclusion

The dense land use of New York City is the defining characteristic of travel in New York City. The large number of people who live and work in a relatively small landmass reduces the need to travel and improves the usefulness of walking and public transportation. However, most of these people want and/or need to travel between the areas where they live and to the areas where they work at about the same time. These circumstances, as well as the proximity of many places of employment to each other, increase travel times and decrease the quality and predictability of most people's daily commute.

The difficulty of commuting by private vehicle, as well as the difficulty to find inexpensive parking, encourages the use of public transportation. Many workers perceive public transportation as the only reasonable means to commute on a daily basis. In fact, many consider the subway system to be the lifeline of New York City. Although expanding the existing rail system could increase capacity, it is not realistic to expect that any major new rail lines will be built. However, smaller construction projects may allow for a more efficient use of the existing infrastructure. In particular, there should be a focus on methods to improve the quality of the service by developing innovative ways to remove or minimize the obstacles to service delivery. New technologies can play an important part in developing the most effective use of the current rail system. Furthermore, efforts to maintain and improve the aging transit infrastructure must continue at an appropriate pace that does not prevent the delivery of an acceptable level of service.

In addition, there are policy decisions that can and should be made that can improve the quality of public transit. Since buses can carry more people more efficiently than cars, efforts should be made to give buses priority over other vehicles. This type of transit-first mindset can allow a more cost-effective service that provides a quicker and more attractive service. This will also reduce the number of private vehicles on the road, creating a further reduction in street congestion and an improvement in air quality.

All future construction projects throughout the tri-state area should be carefully examined for their impact on transit. We must be careful not to expand in ways that the existing transit system cannot handle without making the appropriate adjustments. On the other hand, without appropriate growth, the region could stagnate. Every opportunity should be used to develop pedestrian and transit-friendly designs, which discourage the use of private automobiles. Although it may be unrealistic to expect a major re-design of New York City, small additions to projects as they are developed can create improvements in mobility. The cumulative affect of several small projects can be substantial and can create momentum for larger projects. The end result can be not only an improvement public transportation, but also the overall quality of life in New York City and the tri-state region.

Other cities can learn from the New York City experience by creating a managed level density that can support public transportation and discourage automobile use. The new developments should be designed to encourage pedestrian movement and should be planned with public transportation in mind. Then an appropriate level of density will bring a city's destiny of a high quality of life.

Footnotes

1. Microsoft Bookshelf, 1994.

2. The Paperless Guide to New York City, http:/www.mediabridge.com/nyc.

3. The New York Times, New York City Centennial Section, http:/www.nytimes.com/specials/nyc100.

4. The Rent Guidelines Board, 1998 Income & Affordability Study, Pg. 2, 7, 11 and 12.

5. MTA New York City Transit, Operations Planning, Weekday Cordon Court, 1996.

6. New York City Department of Transportation, Frequently Asked Questions.

7. New York State Department of Motor Vehicles data included in New York State Department of Transportation's website.

8. Cooper, Shoshana and Larry Gould (1994), Faster than Walking? Street Congestion & New York City Transit Buses, MTA New York City Transit, September, p. 3 and 7.

9. McKnight, Claire E., et. al. Impact of Congestion on New York Bus Service, University Transportation Research Center (Region II) at City College on New York, April 1997, p. 18.

10. New York City Department of City Planning, New Opportunities for a Changing Economy, January 1993, p. 49.

11. New York City Department of Transportation, Key Accomplishments of the Guiliani Administration for the Department of Transportation, NYCDOT website.

Urban Public Transportation System of New York City and Its Operation and Use by People

Susan A. Carlson[1]

Introduction

The New York transportation system is defined by a little thought about fact: of the five boroughs that make up New York City, only the Bronx is attached to the mainland of the United States. While creating wonderful vistas for the appreciation of Manhattan Island, it has nonetheless made transportation access unusually expensive and complicated. New York City has eleven major bridges and four major auto tunnels reaching within its borders. There are 685 miles of subway and 468 subway stations in New York City. There are more than 3,000 miles of bus routes in the Greater New York Metropolitan Area (GNYMA).

Access to the transit system is further complicated by the fact that the GNYMA constitutes three states: New York, New Jersey and Connecticut and like most metropolitan areas in the world, its borders are constantly being redefined outward in every direction. As a result of geographic realities and a complicated history of ownership, the New York transportation system has evolved into a hodgepodge of separate authorities and agencies governing overlapping and shared routes, often with competing agendas and objectives and a wide range of fare structures, schedules and operations. In the case of rail operations the situation has been further complicated by the agreements of predecessor railroads: Above it all, lurk the ghosts of the nineteenth century railroad barons and Robert Moses, whose visions, dreams, forward thinking and conceits still largely define how we get to, from and through New York.

---

[1]President, Project Planning & Analysis, 56 Juniper Lane, Southport, CT 06490-1062

The recent emphasis of many transportation systems capital projects in the GNYMA has been on the improvement of mobility and customer access. This paper will focus on a description of the current structure and organization of mass transportation in the New York region and discuss one of the more interesting attempts to rationalize and modernize the system, move it beyond its nineteenth century origins and make it more seamless and agreeable for the customer: The effort is work relating to Grand Central Terminal and it actually includes three projects in one: The recently completed renovation of Grand Central Terminal; the expansion of access from the GCT train shed to the north (North-End Access), currently under construction, and the plans to construct a separate tunnel to Grand Central to serve riders from Long Island, utilizing the 63rd Street Tunnel. These projects are particularly challenging because they are being undertaken in the most densely populated and transit dependent city in the United States and involve the most widely used transit station in the country, which is also its premiere urban landmark.

Description of the New York Area Transportation System

As discussed above, the New York Greater Metropolitan area is composed of three states: New York, New Jersey and Connecticut. **More than 40 percent of all commuter rail and rapid transit trips in the U.S. take place in this region.**

Three separate multi-state authorities, discussed in detail below largely serve the NYMA transit needs. They are as follows:

1. The Metropolitan Transportation Authority-a New York State Authority.
2. NJ TRANSIT, an authority of the state of New Jersey.
3. Port Authority of New York and New Jersey, which is a joint New York State and New Jersey authority.

New York State Metropolitan Transportation Authority (MTA)

The MTA subways, buses and railroads carry more than 1.7 billion customers a year, which is equivalent to about one in four mass transit riders in this country. The transportation network is the largest in North America and serves a population of 13.2 million people in the 4,000 square mile area which encompasses New York City; five counties in New York State to the north of New York City; Long Island, and much of Connecticut. The MTA Company is the parent of five separate operating authorities. It is run by a chairman appointed by the governor of New York and has a board with members appointed by the governor and the mayor of New York. The authorities and their operations are discussed below.

New York City Transit (NYCT)

The New York City subway system first opened in 1904, built in response to the blizzard of 1888 which paralyzed above ground New York. Originally a variety of surface and underground private operations, the New York subway and bus system has for many years been under the authority of New York City Transit (NYCT), although many of the rapid lines retain their private distinctions in every day usage within and outside NYCT (for example IRT and IND lines). The rapid system is a combination of tunnels and elevated lines, and operates 24 hours a day, seven days a week. The system operates on a fixed fare so one could theoretically travel the 656 track miles for $1.50 ($1.38 if a $15 MetroCard is purchased). There are 468 subway stations on 25 lines requiring a fleet of almost 6,000 cars.

The NYCT bus system accounts for 80% of all surface mass transportation in New York City. (The balance is largely made up by a fleet of city regulated franchised private operations.) NYCT's bus fleet is composed of 4,000 buses that daily cover 1, 671 route miles on 227 bus routes.

The annual operating budget for NYCT in 1998 is $3.8 billion and it has almost 43,000 employees. Average daily ridership for NYCT is 5.72 million.

Long Island Rail Road (LIRR)

The LIRR is the largest commuter railroad in the country. Interestingly, it still operates under its original name dating back to its charter in 1834. Its major destination is Pennsylvania Station in midtown Manhattan's west side, which is a through station, shared with New Jersey Transit and Amtrak. The LIRR also has terminuses at Flatbush Avenue in Brooklyn and Hunterspoint in western Queens (the latter one subway stop from midtown Manhattan). Most lines pass through Jamaica Station in Queens which is a major transfer point for service east on both diesel and electric lines and where the railroad's headquarters is located.

The rail system operates more than 11 lines, 595 track miles and 134 stations. The fleet is composed of 1,086 rail cars, including 134 new bi-level coaches and 23 "dual mode" locomotives which can operate in both diesel and electric territory. The LIRR has 6,000 employees and daily ridership of almost 270,000.

Metro-North Commuter Railroad (MNR)

MNR is second in commuter railroad size in the U.S. only to its sister agency, the LIRR. The rail lines that make up MNR: the Harlem, Hudson and New Haven Lines are the descendants of the New York Central and the New York, New Haven and Hartford Railroads. MNR became an MTA agency in 1983 after acquiring Conrail's commuter rail services. It operates service in Connecticut through an operating agreement with the state of Connecticut and also operates service west of the Hudson River in New York State through to Hoboken Terminal under an operating agreement with New Jersey Transit. The three main lines converge from the north in the south Bronx and become a four-track railroad. The Park Avenue tunnel begins at 96th Street in Manhattan and proceeds for more than two miles to Grand Central Terminal.

MNR has daily ridership of 223,000 over its six main and branch lines. The lines encompass 758 track miles and 117 stations. Its 1998 operating budget is $771 million and it has 5,443 employees. The fleet includes 850 rail cars, including coaches and electric cars, and diesel and dual mode locomotives.

Triborough Bridge and Tunnel Authority (B&T)

B&T is the revenue engine that drives the MTA trains. Robert Moses founded it in 1933 and its headquarters on Randalls Island, in the shadow of the Triborough Bridge, was also built by Moses. The authority is responsible for all of the seven toll bridges and two toll tunnels within New York City borders. The bridges are the Triborough; Throgs Neck; Bronx-Whitestone; Verrazano Narrows; Henry Hudson; Marine Parkway, and Cross Bay. The two tunnels are the Brooklyn-Battery, linking southern Manhattan and Brooklyn, and the Queens Midtown, which links midtown Manhattan with Queens. (The other bridges within the city of New York are owned and operated by the City of New York, including the Brooklyn, Williamsburgh and Manhattan bridges, which do not have tolls.)

B&T was one of the early agency adopters of E-Z Pass, an electronic toll collection program. The program was completed in 1996 and currently over 50% of customers through the tolls use the payment program.

Long Island Bus (LIB)

The final agency under the MTA umbrella and the smallest is still one of the nations larger bus operations. LIB represents the consolidation of ten private bus routes that links Long Island with New York City. It serves 96 communities within Long Island and Queens, serving five subway stations and seven major shopping malls. LIB has an average weekday ridership of 65,500 along its 53 routes and 684 route miles. It employs about 1,000 and has an operating budget of $77 million.

The Port Authority of New York and New Jersey (PA)

The PA is a bi-state authority whose board is appointed by the governors of New York and New Jersey. The PA is responsible for managing New York and New Jersey's port facilities and areas, including marine ports, JFK, LaGuardia and Newark Airports and all toll bridges and tunnels between New York and New Jersey, including the George

Washington Bridge and the Lincoln and Holland Tunnels. The PA brought E-Z Pass to its facilities in 1997.

The Port Authority Bus Terminal on the west side of midtown Manhattan handles approximately 55 million bus riders a year. Finally, the PATH (Port Authority Trans Hudson) provides rapid transit serves between Manhattan and New Jersey to approximately 66 million riders per year. The service brings riders between Newark and Hoboken and both 34th Street and the World Trade Center, which is also owned by the Port Authority.

NJ TRANSIT

NJ TRANSIT was established in 1979 and soon acquired Conrail's commuter operations in New Jersey, as well as a number of private bus operations. The agency is run by a seven-member board of directors, appointed by the Governor. Total daily bus and rail ridership on the system is 321,000 passengers. The agency covers a service area of 5,325 miles.

Rail System

NJ TRANSIT operates all commuter rail operations within the state of New Jersey and between New Jersey and New York City's Penn Station. The system also connects with the PATH terminal in Hoboken and includes 12 lines in three separate divisions. There are 47 million annual rail passenger trips on 591 daily revenue trains. The system has 542 track miles with stops at 161 stations. NJ TRANSIT uses 104 locomotives and 695 rail cars.

Bus System

NJ TRANSIT operates a fleet of 1,900 buses that travel within New Jersey and to both New York and Philadelphia. The service combines local, and both short and long-distance suburban buses. Service to New York City terminates at the Port Authority Bus Terminal. The bus operation also operates the Newark City subway, a 4.3 miles light rail trolley within Newark which dates from the 1930's and has 11 stations. NJ TRANSIT buses generate 141 million passenger trips each year over 178 routes.

Grand Central Terminal Rehabilitation and Improvements

Background and History

Grand Central Terminal (GCT) is a two-level, terminal station. It is a structural steel building with lightweight latticed columns surrounded by reinforced concrete terra-cotta tiles and ornate decorative stone. The underground train shed is a steel-framed structure with a Manhattan schist foundation.

With a height equivalent to a seven-story building, it stretches between 42nd and 44th Street and from Lexington to Vanderbuilt Avenue, in the center of midtown Manhattan. Its main concourse is approximately 100,000 square feet, with a renaissance sky ceiling and an elegant western staircase made of Tennessee pink marble. The two-level train shed area reaches to 57th Street, while the four track tunnel proceeds over 50 blocks underneath Park Avenue to 96th Street. There are 42 terminal tracks on the upper level and 16 terminal tracks on the lower level. The terminal is used by approximately half a million Metro-North commuters and New York City subway riders day, making GCT the busiest subway station on the system.

The colorful history of Grand Central Terminal has been repeated in many books and periodicals but is worth summarizing. The construction of GCT was the result of the continued growth of New York City in the uptown direction, which saw three New York & Harlem predecessor terminals, the first at Center Street, downtown, the second at 26th Street (Madison Square), and the third, an above ground terminal at the present location of 42nd Street. The third terminal, Grand Central Depot, broke Park Avenue (then Fourth Avenue) in two and made it impossible for traffic to circumnavigate the station and yard, which took up six blocks in the center of the emerging midtown. The growth uptown, related traffic and the emergence of the electric rail car forced the decision to replace the terminal less than 30 years after its construction, with a below ground train shed and platforms.

In 1910 the old terminal was demolished and Reed & Stem and Warren & Wetmore designed the new Grand Central Terminal. The terminal opened in 1913, while the elegant, elevated driveways around the terminal were completed in 1919. GCT was the subject of a famous preservation battle in the 1960's, which spared it the fate of its demolished sister terminal at Pennsylvania Station. After that battle GCT was given landmark status, which did not prevent the terminal from falling into a general state of disrepair over the next thirty years. By the mid-eighties the roof and ceiling were in poor condition and utilities were inadequate and out of date, although the structure was found to be remarkably sound. Also, the original design did not rationalize pedestrian flow north/south and east/west. GCT was constructed with a variety of underground passageways from the trainshed to buildings in the vicinity, but over the years these passageways were closed or eliminated with new building construction in the area. All passenger traffic leaving the terminal is forced to walk south on the platforms to exit the trainshed area. The construction of the terminal had not allowed for the growth of the midtown area north of 42nd Street in the post war area. Over the years, the number of commercial tenants in the building had declined significantly.

Finally, the demands on Penn Station, on the west side of Manhattan, grew in the last 15 years due to ridership growth at LIRR, NJ TRANSIT and AMTRAK and the limited track configuration of the station. Also, approximately half of LIRR commuters' final destination is on the east side of Manhattan. This led to the investigation of alternatives to open up GCT to LIRR riders.

As a result of all of these factors, three projects have been undertaken to restore GCT and improve commuter access. These projects are discussed below.

GCT Rehabilitation

In 1988 a request for proposals was issued by Metro-North Railroad and its parent the MTA, for the design and construction of renovations. Beyer Blinder Belle was selected to lead the architectural rehabilitation and develop a rehabilitation master plan. In the same year retail specialist firm of Williams Jackson Ewing was hired to prepare a retail master plan for GCT. Construction began on the project with the cleaning of the sky ceiling in 1996. The station was officially reopened in October 1998. Project work is expected to cost approximately $200 million at completion. The MTA anticipates that between 2,100 and 2,800 direct construction and construction related jobs are to be created. The MTA also expects that 800 to 900 permanent retail jobs will also be created in the terminal.

The project work includes repairing structural water damage and sealing the structure; replacement of all utilities; increasing the power supply to 13,200V; lead and asbestos abatement, current code compliance, cleaning and repairing the Italian Botticiono marble which covers the lower walls and the pink Tennessee marble floors; restoring and replacing lights and the magnificent chandeliers. Hidden from view until the grand opening in October was a newly constructed eastern staircase to match the existing western staircase. This staircase had been in the original drawings for the terminal and was originally intended to lead to an elevator bank for a building over GCT.

One of the most successful aspects of the construction, was among the least costly and it was done so well that customers came to take it for granted: This was the manner in which pedestrian flows were kept moving throughout the terminal regardless of the construction taking place at the time. Rerouting of passageways had clear signage indicating direction and destinations. Temporary corridors and passageways were assembled with interesting photographic murals papered on the temporary walls. When major changes to routes to the subways began, MTA personnel were on hand to direct traffic.

Commuters walking through the terminal felt an early connection to the project as they monitored the progress of the cleaning and restoration of the sky ceiling, the first and most dramatic step in the renovation. The $4.5 million restoration included cleaning and painting the ceiling as well as replacing the 59, 40 watt light bulbs in the ceiling that represent the stars of a Mediterranean sky, with a fiber optic system. The HVAC system in the attic above the ceiling also was to be replaced with a climate control system for the concourse. The ceiling structure would not permit the ceiling work and the HVAC work in the attic to be done simultaneously using suspended scaffolding, as had been originally planned. As a result, Universal Builders Supply of Mt. Vernon, NY devised an innovative and lightweight, aluminum bridge truss system which moved on tracks and was supported by aluminum scaffolds. The 37 meter-long, nine meter-wide bridge covered a 9.1 meter patch across the ceiling for painting and cleaning. Each night when the terminal was closed, the scaffolding was moved from east to west to its next position.

The GCT retail plan is designed with the intention of to taking advantage of both the huge volume of customers passing through GCT and its upscale commuters. Thankfully, the plan leaves the concourse itself free of retail, allowing maximum passenger flow and allowing the majestic scale of GCT to be the main focus of the concourse. Three restaurants are planned for the main concourse balconies. The lower level is designated as the dining concourse. An indoor market is planned for the Lexington Avenue passageway, featuring a new entrance to the avenue at 43rd Street. Upscale retailers are planned for the space fronting 42nd Street.

In developing its retail plan, the MTA wanted to ensure that pedestrian flows would not be significantly affected by the addition of retail space at the terminal. A detailed study was undertaken that compared the impacts on pedestrian flows of planned retail locations to a no-build alternative. The study found that no significant deterioration would take place and that some of the planned changes in passenger flows would actually benefit the pedestrian. For example, the main corridor from 42nd Street and GCT to the Lexington Avenue station, southeast of the concourse, was widened from nine feet to 34 feet as a result of the restoration.

North End Access at GCT

Opening the north end of GCT, meaning the northern end of its platforms to Park Avenue, has been on the drawing board for 20 years. That is because commuter trains grew in length in the 1980's (a maximum of 10 cars is permitted at Metro-North) and midtown Manhattan, north of Grand Central, grew as a business district. Commuters who work north of GCT and walk from the terminal now must walk as far as five city blocks south from the back of the train to the main concourse of GCT, only to turn around and walk north again. Further, queuing time to leave the platforms, particularly on the lower level, with the bottleneck of stairs, can be substantial on trains of 1,000 passengers or more. If passengers could split and move in two directions, queuing time and overall commute time could be reduced.

The original plan was produced in 1985 by Metro-North with the assistance of Parsons Brinkerhoff. This was followed up by a design by STV. Construction on the project, led by Yonkers Construction, began in 1996 and is scheduled to be complete by 1999. The total project cost, including design, construction and Metro-North in-house labor, is approximately $110 million.

A schematic of the design of North End Access is attached as Exhibit 1. The design is centered on two north south spines, located on the western and eastern sides of the terminal. These spines provide direct access to the GCT concourse and to two cross passageways: The first passageway is located at 45th Street, and provides access to lower level platforms. It is located below the lower level, in the space from an existing baggage tunnel. Passengers exiting trains will walk down to the passageway from platform level. The second cross passageway is located at 47th Street and provides access to the upper level platforms. The passageway is wedged between the upper and lower levels. Passengers exiting trains will also walk down to this passageway.

Street access is provided by escalator at three locations at Park Avenue and one at Madison Avenue. There are also six elevators to be installed, as shown on the schematic drawing. There will be public address, CCTV, vending equipment and passenger information systems provided.

The cross passageways, which were originally conceived as utilitarian tunnels, have evolved into architectural statements in their own rights. The walls are covered in a marble tile, reminiscent of the Botticiono marble of the main part of Grand Central. Tile artwork has been inlaid in the marble walls, with each area of the passageways representing a continent's cultural interpretation of the night sky, again with a tie to the Grand Central concourse ceiling. In fact, the passageways will have starlights in their ceilings, while the main lighting will come from cove lighting, placed at the top of the passageway walls. The ceilings will be made of composite panels, custom colored and cut to fit the various lighting, fire, and PA systems emanating from the ceiling.

The north-south spines have a utilitarian look in their southern portions in the platform areas, where they will remain open to the trainshed, with railings and Terrazzo floors. As they move north there will be a doorway leading to the passageways, where walls are covered with marble tile and with a similar ceiling to the passageways.

The major construction challenge of the project was fitting the 47th Street passageway between the two levels of Grand Central. Overhead train clearances have limited the ceiling height in the passageway to seven feet, six inches. Two 35,000 pound girders were required because lower level tracks prohibited placement of columns. The girders also support two columns, which support the street above.

As the project is completed, excellent signage will be a required finishing touch. The cross passageways are beautifully appointed and inviting but they could be a small, underground maze to the uninitiated.

East Side Access

East Side Access to GCT is now in the planning and Federal Environmental Impact Statement (EIS) preparation stage. The proposed project route is shown as

Exhibit 2. The project will provide rail service into GCT by connecting the Main Line and Port Washington Branch of the LIRR at Harold Interlocking in Queens, to the lower level of the 63rd Street Tunnel, which is currently unused. A tunnel will be constructed from 63rd Street in Manhattan to GCT, where the service will be brought into the western side of GCT through the use of an existing loop track. A ten-track/five island platform facility will be used exclusively by the LIRR. The project will also include a new station in Sunnyside Queens. The 45th Street passageway at GCT, discussed above, will be extended west to include these tracks. Plans estimate that 172,000 riders will use this service each weekday.

Current plans are for the Federal EIS to be completed in 2000. Design will take place 2000 to 2006 and construction will be in the period from 2000 to 2010. Projected costs in 1997 dollars total $4.3 billion, including $2.8 billion for program management, construction and design, $400 million for real estate acquisition and $1.1 billion for rolling stock.

MetroCard: Automating New York City's Public Transit System

Atefeh Riazi[1]

Abstract

On January 6, 1994, New York City Transit introduced Automated Fare Collection to the largest transit system in the world. Known as MetroCard, this system of Automated Fare Collection has helped stabilize ridership and encouraged growth. The significant contributions to ridership growth are, in large measure, attributable to the strategies developed for the marketing and sales of MetroCard. A cornerstone of this initiative has been a multi-faceted out-of-system sales strategy. Out-of-system sales have been used to provide convenient, cost effective, customer-friendly distribution of MetroCards to the public.

Introduction

Four years have passed since New York City Transit (NYCT) began a new and exciting chapter in its own history book. On January 6, 1994, New York City subway customers at the Whitehall and Wall Street Stations walked up to the token booth and purchased, not the traditional round metal token, but a bright yellow-and-blue MetroCard. To gain entrance into the system, they swiped the thin, magnetically encoded farecard through a device on top of the equally new electronic turnstiles. These customers launched Automated Fare Collection (AFC) and, in so doing, brought the world's largest public transportation system into the modern era of transit payment and travel.

---

[1] Vice President & CIO, Technology Division, Department of Operations Support, MTA New York City Transit, 370 Jay Street, Room 1323, Brooklyn, New York 11201

Technical Aspects of Automated Fare Collection

Before proceeding with a discussion about the implementation and marketing of AFC, it is useful to review a few of the technical elements that enabled the successful implementation of AFC.

Turnstiles

The most critical element of an AFC system is the fare media. The NYCT specification calls for a .010" thick polyester credit-card-sized farecard utilizing a high coercivity magnetic stripe capable of holding two tracks of data. Initial requirements specified farecard processing through a swipe read/write unit on 1) the subway turnstile and 2) the Integrated Farebox Unit (IFU) on each bus. However, during the In-Service Qualification Test for the IFU, it quickly became apparent that a swipe read/write unit on a bus was not practical, and the specification was rewritten to have the transporter mechanism which issued the magnetically-encoded transfer to also read and rewrite the farecard.

In the subway, the turnstile was specified to process 30 passengers/minute in both the inward and outward direction. The barrier mechanism that controls the aisle is normally locked in the inward direction and freewheels in the exit direction. Before allowing an entry, a farecard must be swiped and processed by the turnstile electronics. The farecard may be swiped within a speed range of 10 inches per second (ips) to 40ips. At 40 ips, the farecard takes approximately one-quarter second to pass through the reader. During this time the encoded fare information is read, processed, rewritten and finally check-read. If all functions are completed correctly, the barrier arm is released, and the display on the turnstile indicates that it is unlocked and ready for passenger entry. The amount of money deducted and the remaining farecard balance are also shown. If any part of the swipe process does not complete correctly, various error messages -- e.g. "Too slow," "Too fast," "Please Swipe Again," -- are displayed to the customer. Furthermore, the turnstile contains a list of farecards that are not acceptable to the system (the "Negative List") and before releasing the barrier, it's referenced to ensure the farecard just processed is not on it. The Negative List is updated periodically from the central computer (the Area Controller) that controls the entire system.

Integrated Farebox Unit (IFU)

The IFU can process 24 passengers per minute. It contains two main units: a Coin Handling Unit (CHU) and a Ticket Processing Unit (TPU). The CHU accepts the coins and tokens through an entry bezel that permits multiple coins to be deposited simultaneously. The inserted coins are singulated and passed through a verifier where they are registered; an incrementing passenger and bus operator display is updated as the coins are processed. The CHU processes 10 coins per second.

The TPU has two purposes. It can issue a magnetically encoded paper transfer to passengers paying with coin or a token, or it can serve as the reader/writer for farecards and encoded paper transfers.

The TPU contains two rolls of 500 unencoded transfers. At time of issue, a transfer is separated from its roll, printed with the route number and direction, bus number, date and time and then encoded with all the information required for processing in another farebox on the second leg of the journey. When reading/writing farecards, the TPU first processes then returns them to passengers through the entry bezel. Encoded paper transfers are checked for validity. When verified, they're magnetically canceled and sent by the IFU to a capture bin on the operator's side of the unit.

Computer Systems

Turnstiles and other station-level equipment are connected to a local Station Computer (SC) in a communications network that links the system's 469 stations and 19 bus Depot Computers (DC) to the Area Controller (AC). The AC itself is a mainframe using twin IBM ES9000/620 and 146 Gbytes of disc storage. Housed in the central data processing facility in Manhattan, the AC provides the centralized reporting and control functions that support the live system. It captures data on sales transactions and usage, and disseminates fare-table data and Negative List information. Should a communications failure occur, all station- and depot-level devices operate in an autonomous, stand-alone mode. Data is retained in the end device until 1) communications are restored and 2) the AC confirms it has been securely stored.

Implementation of Automated Fare Collection in New York City

Implementation occurred in two phases. The first "core" 69 stations, located primarily in Manhattan's midtown and central business district, along with representative stations from each of the other four boroughs, were implemented between January and April of 1994. An evaluation period of approximately one year was built into the schedule to address unforeseen problems. Fortunately, no such design issues surfaced, and Phase 2 started on schedule in July 1995.

The last station installation was originally scheduled for December 1997. However, after approximately 150 stations had been equipped, there was such strong favorable public reaction that full installation was accelerated to May 1997.

As work in the subway headed towards completion, parallel conversions put the new IFU on Transit's entire bus fleet. This involved converting all 3,800 buses and installing a Depot Computer and probing equipment at the fueling lanes of 19 depots. To bring AFC to the area's complete public transit network, similar fareboxes were put on 1,200 buses operated by seven private companies under a franchise agreement with the New York City Department of Transportation (NYCDOT), and on 354 buses operated by MTA Long Island Bus (MTA-LIB).

When the last station was converted to AFC, and the use of MetroCard became possible throughout the entire system, it was then feasible to introduce fare initiatives. To support a seamless transportation network, NYCT introduced. the Unified Ticket Software on July 4, 1997. This permitted intermodal transfers through the MetroCard. Additionally, new fare policies were also implemented to increase customer convenience. January 1998 witnessed the introduction of volume discounts, one ride free for every 15 purchased. Volume discounts in the form of monthly and weekly passes were implemented in July 1998. The next milestone is the release of the One-Day pass, which is expected in early 1999.

MetroCard Benefits

Today, the MetroCard has significantly contributed to growth in ridership. As indicated on Chart 1, between the 1st Quarter 1996 and the 1st Quarter 1998, over 500,000 more rides were made daily on NYCT services (bus - 322,000; subway - 192,000).

MetroCard has greatly reduced fare evasion. The rate shrunk from 5.91% of total ridership in 1991 to 1.08% in 1997, an 82% reduction. Due to structural

Chart 1: Ridership Growth
1st Qtr 96 -- 1st Qtr 98

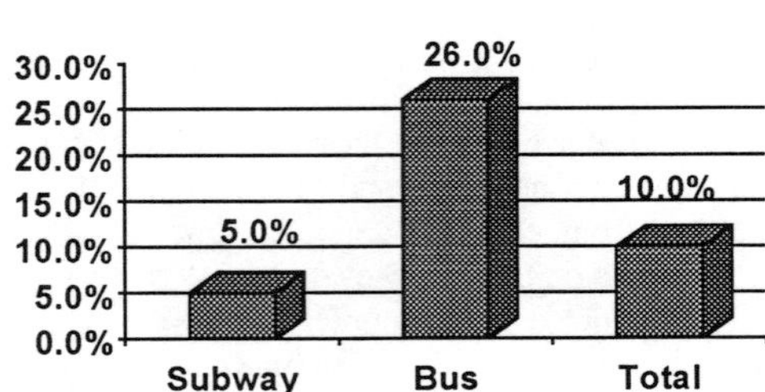

modifications, NYCT realized fare abuse savings amounting to $64 million--$41 million on subways due to new turnstiles and other fortifications at stations and $23 million on buses with the introduction of magnetic transfers. As indicated by Chart 2, Equipment reliability has also increased:

Chart 2: Equipment Maintenance

	Pre-AFC	1990 (Projection)	1998 (Actual)
• Reliability (MCBF)	<30,000	>120,000	>1,500,000
• Availability	90%	99.7%	99%
• Maintainability (MTTR)	7.5 hrs	0.5 hr	0.5 hr

Benefits have not only accrued to NYC Transit: because of fare policy enhancements, customers now receive:

- free intermodal transfers
- free transfers between NYCDOT Private Buses and LI Bus
- Bonus program (11 rides for the price of 10)
- Express Bus fare rollback (from $4.00 to $3.00 fare)
- Unlimited Ride Passes (Monthly and Weekly Time-based MetroCards)
- Free Staten Island Ferry crossings

MetroCard Marketing

Overview of MetroCard Sales

The role of the MetroCard Sales unit is to entice customers to purchase MetroCards through out-of-system sales and distribution channels. Such convenient, cost-effective and revenue-enhancing channels provide an alternative to the in-system sale of fare media. MetroCard Sales currently achieves this objective through the following established programs: Retail Sales, Employer Sales, Agency Sales, Mobile Sales, Subscription Sales, and Corporate Sales. The following synopsis offers a description of each of these programs and the enhancements they have provided to MetroCard sales:

- The *Retail Sales* program, through its growing network of retail MetroCard sales merchants located throughout the New York City metropolitan area, has supported Automated Fare Collection (AFC) since the start of the AFC rollout in 1994. Customers may purchase $6, $15, $30, and 7-Day Unlimited Ride MetroCards from participating merchants.

- *Employer Sales*, in association with TransitChek, coordinates the employer-based transit subsidy program. MetroCards are distributed to employees of participating companies on a pre-tax basis. Employees may receive 7-Day, 30-Day, $15, $21, $30, or $35 MetroCards.

- *Agency Sales* handles MetroCard sales by MetroNorth Railroad, Long Island Rail Road, Long Island Bus, and the NYC's seven private franchised bus lines. These agencies all offer Pay-per-Ride MetroCard products. In addition, the commuter railroads sell joint railroad ticket-MetroCards, which can include a 30-Day Unlimited-Ride MetroCard option, and Long Island Bus sells the 7-Day Unlimited-Ride MetroCard.

- *Mobile Sales* serves the general public, reduced-fare and bus-only customers with its two buses, which visit senior centers and public sites and also make scheduled stops at selected fixed locations. These buses not only sell MetroCards, but also add value and handle applications for Reduced-Fare MetroCards.

- *Subscription Sales* currently serves reduced-fare customers through a mail-and-ride program in which participating customers pay by mail for the trips they have taken in the preceding month.

- The *Corporate Sales* program builds strategic alliances for card distribution and cost recovery, benefiting MetroCard Sales as a whole and enhancing the Retail Sales program in particular. The Corporate Sales program works closely with the MetroCard Product Development unit to develop MetroCard products and card holders which are value-added promotions for MetroCard customers purchasing their fare through out-of-system channels.

Accomplishments in 1998

MetroCard Sales has successfully faced many challenges in 1998, including preparing for the introduction of the 10% bonus program, the rollback of express bus fares, and the introduction of time-based passes. Each introduction of new fare policies and farecard products entails change-out or build-up of card inventories, information campaigns for merchants and customers and development of back-office policies and procedures. Furthermore, each MetroCard Sales program has been actively expanding or developing new program-specific initiatives.

1998 accomplishment highlights, by program, include:

- *Retail Sales* has been streamlining its back office fulfillment operations while increasing sales and expanding the geographic coverage of the merchant network. In particular, the number of retail chains in the merchant network has grown in 1998. Major chains added to date include Emigrant Savings Bank, Walgreen's, New York City College Bookstores, Amerada Hess Corporation, Lori's Hallmark Shops, Gateway Newsstand, D'Agostino's Supermarkets, Waldbaum's Supermarkets, and Edwards Super Food Store. Retail Sales also started an ATM sales pilot of Triplex MetroCards this year, which has been successful.

- *Employer Sales* has increased its sales and expanded its fare media product offerings.

- *Agency Sales* has also increased its sales and expanded its fare media product offerings.

- *Mobile Sales* has introduced a monthly fixed-stop service for its two buses and augmented its services with the introduction of three vans in October 1998.

- *Subscription Sales* has completed its one-year pilot with reduced-fare customers. Participating customers were happy with the program and increased their use of public transit as a result of the post-payment program.

- *Corporate Sales* has had a number of highly successful promotions to date: the "MetroCard Rewards" Broadway discount promotion, the Emigrant Bank "City of Immigrants" MetroCard series, and the JVC Jazz "Take the A Train" MetroCard series. In addition, the pilot of hard plastic MetroCard holders, designed to protect the MetroCard and keep it handy, has proved popular. Further research and development on cardholder products are underway.

Five-Year Goals

In the upcoming five years, MetroCard Sales expects its sales to continue to grow as it maintains and expands its programs. Its Retail Sales program will continue to make MetroCard purchases convenient through its growing network of merchants. This program will also augment sales through the introduction of retail vending, currently under development, while at the same time becoming more cost-efficient by streamlining back office operations. The Corporate Sales program's initiatives, including the continued development of promotions as well as the development of new products and sales and distribution alliances, will continue to support Retail Sales as well as introduce new out-of-system sales channels. The Employer Sales and Agency Sales programs will continue to promote and facilitate MetroCard usage in the metropolitan area. MetroCard Sales will continue to target the needs for more accessible MetroCard sales through the Mobile Sales and Subscription Sales programs. Overall, MetroCard Sales will continue to balance customer service with cost containment.

In the 1996 Strategic Business Plan, MetroCard Sales targeted year-end goals for out-of-system sales. With the advent of time-based passes and the exclusion of monthly unlimited-ride passes from retail sale, the market share targets have been revised, as shown below:

Chart 3: Strategic Business Plan Out-of-System Sales Goals

Year	1996 SBP Year-end Goal	Revised Year-end Goal
1998	9%	8.6%
1999	12%	10.0%
2000	15%	12.5%

However, as the MetroCard matures, it may be possible to reach the original 15% goal by the end of five years through the further development and integration of new distribution channels and partnership activities, which would increase sales and cost-recovery savings associated with card design, production, and distribution. A projection based on this assumption is presented in Chart 4.

Chart 4: Projected MetroCard Sales—1999 through 2003

Year-end:	1999	2000	2001	2002	2003
Without 30-Day Unlimited Ride Card Sales	10.0%	12.5%	13.5%	14.0%	14.5%
number of merchants	3,900	4,000	4,150	4,200	4,200
With Full Fare Subscription, AVT, AVM	5.0%	10.5%	17.5%	18.0%	18.5%
Total	15.0%	22.5%	30.0%	32.0%	33.0%

Goals and Strategies by Program

Outlined below, by program, are the goals and strategies NYC Transit plans with respect to MetroCard sales.

Retail Sales

Goals: To continue to 1) build sales; 2) increase the quantity and quality of merchants in the MetroCard retail sales network; 3) introduce automated sales and add-value capability, providing convenient options for customers and efficiencies for merchants and NYCT; 4) increase the efficiency and security of order-taking and fulfillment processes.

Implementation strategies: Retail Sales is currently the mainstay of the MetroCard Sales programs. In the next five years, Retail Sales plans to continue to expand its network of out-of-system MetroCard merchants beyond the current base of about 3,600 merchants in the New York City metropolitan area, with an emphasis on quality retail chain establishments. Maintaining a reasonable level of commissions will be important for recruiting and keeping merchants, because merchants are reluctant to take on the financial risk of pre-payment.

However, it is expected that the retail network will reach a saturation point when it reaches 4,300 merchants. Retail Sales expects to gain further penetration by emphasizing vending through out-of-system AVMs and AVTs (phase 1 covers 55 AVMs). A number of major chains have already expressed interest in vending through machines and in providing add-value capability to their customers. Vending machines provide added security against retail shrinkage and, with add-value capability, they lower NYCT's card production and distribution costs.

Furthermore, Retail Sales will continue to explore other retail and vending partnerships. For example, this year Retail Sales has begun a pilot with Republic Bank to vend MetroCards through the bank ATMs. Plans are underway for expansion with other banks.

While maintaining merchant recruitment, Retail Sales plans to outsource fulfillment activities, starting in 1999. This will lower the costs of growth and increase efficiency.

Employer Sales

Goals: To increase TransitChek MetroCard sales.

Implementation strategies: Employer Sales is the most cost-efficient MetroCard Sales programs. It expects sales of TransitChek MetroCards to increase in the next few years as a result of changes in legislation in 1998. The new legislation gives employers the option of providing the tax-free transit benefit to their employees as part of their total salaries instead of on top of their salaries, thereby reducing the employers' payroll taxes and making the program more attractive to employers.

Agency Sales

Goals: To promote and facilitate MetroCard usage among customers using transit services which link with NYCT's subways and buses, to help create a seamless transportation network in the metropolitan area.

Implementation strategies: Agency Sales plans to continue to expand the convenience of MetroCard purchases for the customers of agencies whose services connect with NYCT. With the commuter railroads, there are plans to install new ticket vending machines which will dispense both MetroCards and zoned railroad tickets and which will also accept credit cards. Long Island Bus is scheduled to receive 7 MetroCard Automated Vending Machines. The city's seven private franchised bus lines currently have a mail sales program which includes the sale of $30 Pay-per-Ride MetroCards and $120 Express Bus Plus Unlimited Ride MetroCards.

This MetroCard mail sales program is in its infancy, and it is expected to grow. It may possibly be combined with the Subscription Sales programs. Extension of MetroCard sales to other transit properties is also under consideration.

Mobile Sales

Goals: To expand MetroCard sales, service, and outreach to reduced-fare and bus-only customers and provide added convenience for the general public.

Implementation strategies: Mobile Sales will maintain its buses, which began service in September 1996, and it has added 3 vans to its service fleet. The vans have debit/credit capability, in order to expand service to full fare customers in addition to the initially targeted populations. Mobile Sales will also explore the viability of mobile AVMs, similar to the mobile ATMs used by banks at special events targeted to the general public.

Subscription Sales

Goals: To expand the pilot reduced-fare program to all reduced-fare customers.

Implementation strategies: Subscription Sales is currently preparing an RFP for back-office services, in preparation for expanding the program beyond the current 922 active pilot customers to the full reduced-fare population. In the evaluation of the pilot, it was found that customers' transit use increased by about 50% after they joined the Mail&Ride program, taking advantage of the ease and convenience of this new fare payment option. These results suggest that expansion of the program will generate new trips, and consequently new revenue, in the next five years.

In addition, MetroCard Sales has proposed a Subscription Sales program for full-fare customers, pending approval. Subscription Sales is one of the least expensive of MetroCard distribution methods. Expansion of this program would make the MetroCard available in a convenient manner to all bus and subway customers, and would be expected to generate new trips and new revenue, as has been demonstrated by the EZPass program and the Reduced-Fare Subscription Sales pilot. It is projected that a Subscription program for Full Fare customers to purchase all Full Fare MetroCard products could account for 15% of all MetroCard sales within two years of start-up.

Corporate Sales

Goals: To increase awareness of MetroCard products and promotions, to generate revenue through expired card value, and to offer promotions which both pay for themselves and add value for transit customers.

Implementation strategies: Corporate Sales will continue to focus on corporate sponsorship; alliances for promotion, distribution, and sales; and the development of new products and services. The promotional MetroCards, developed by MetroCard Product Development for sponsors found by Corporate Sales, have proven popular with the public. These commemorative MetroCards draw customers to the Retail Sales merchant network and bring in revenue from both sponsors and unused fares on pre-encoded cards. There are plans to issue more of these MetroCards in the future, pending sponsorship. Corporate Sales is preparing to launch a joint MetroCard-Phone Card pilot, which will use the partners' distribution network as well as NYCT's own distribution network. Corporate Sales is also developing an Internet sales pilot, whereby a partner will take orders and deliver MetroCards to individuals. This pilot has long-term, ongoing institutional expansion possibilities, including partnering with travel agencies which are already conducting some business over the Internet. Overall, partnerships with big partners are critical to the extensive expansion of Corporate Sales programs. Corporate Sales is researching other opportunities for the future, including customer-affinity programs and the development of a multi-application MetroCard.

Conclusion

To support the growing popularity of the MetroCard, the array of fare products, and rising ridership, MetroCard Sales will continue to develop existing sales channels and introduce new programs. Together these provide a multi-faceted out-of-system sales strategy for convenient, cost-effective, customer-friendly distribution of MetroCards to transit riders. Other cities considering the implementation or expansion of fare media options should pay close attention to developing a sales network which employees these strategies. In addition to assisting ridership stability, these strategies have significantly contributed to ridership growth at NYC Transit.

Socio-Economic Characteristics, Land Use and Travel Patterns

Hiroshi Matsumura[1] and Hitoshi Kawata[2]

Introduction

Osaka City is located in the central part of Japan. The city is bordered to the west by Osaka Bay and to the south and the north by Yamato and Kanzaki Rivers, respectively. The city serves as business center for the western part of the country as well as Osaka metropolitan area, which includes Kyoto, Kobe and Nara, covering the area within a radius of 50 km from the city. The city now covers an area of 220.66 km^2, including newly reclaimed off shore lands. Most of the urban area is flat land, some 3 m above sea level.

Socioeconomic Aspects

Population: With the city's highest population of 3.16 million in 1965, it began to take a downward turn because of the outflow of population to the residential suburbs. In 1995, its nighttime population was 2.60 million, the third highest in Japan. The population of the Osaka Metropolitan Area is 17.1 million. Osaka attracts daily 1.5 million people from outside and has 290,000 citizens working outside, having a daytime population of 3.8 million. This large net inflow causes a variety of transportation problems as described later.

Age Distribution: Classified by age group, the population under 15 years of age, between ages 15 and 64, and 65 years and older are 352 thousand (13.5%), 1,879 thousand (72.2%) and 366 thousand (14.1%), respectively. Compared with the 1985 census results, these groups decreased by 26.2% and 0.04% and increased by

[1] Managing Director, Osaka City Foundation for Urban Technology, 3-5-22, Kitahama, Chuo-Ku, Osaka 541-0041, Japan

[2] Assistant Manager, Transportation Policy and Kansai International Airport Project Department, Planning Division, Planning and Coordination Bureau, City of Osaka, 1-3-20, Nakanoshima, Kita-ku, Osaka 530-8201, Japan

35.1%, respectively. Decrease in the younger generation is caused by an outflow to the suburbs when couples marry and decline in birth rate. The advanced age group is likely to further increase in the future.

Economic Base and Employment by Industry Type: Osaka City has a high concentration of economic activities, second only to Tokyo. The gross regional products of Osaka and its metropolitan area in 1993 were US $136 billion (20.9 trillion yen) and US $464 billion (69.6 trillion yen), respectively.

Daytime workers in the city in 1995 amounted to 2.472 million persons, an increase of 140,000 persons (6.0%) from 1985. As for industrial structure, the tertiary industry constitutes the greatest number of workers at 1.753 million (71% of the total), secondary industry of 699,000 (28%), and primary industry of merely 1,200 (0.04%).

Car Ownership: The number of automobiles registered in Osaka City in 1993 totaled 933,366. This number has continued to grow. By vehicle type, passenger cars account for 50% and continue to increase, while trucks are decreasing.

Furthermore, the rate of growth in drivers licenses is rising because of the increase in the issue of licenses to the aged and to women. This would accelerate the problem in the future, considering from the previous survey results.

Land Use and Locations of Major Facilities Affecting Travel Behavior

Land Use and Distribution of Employment: The central area is surrounded by Japan Railways' (JR) Loop line. The area is the heart of business and commercial activities, establishing high-dense business center. It occupies six principal wards, merely 17% of the city area, but has 41% of the 3.8 million daytime population of the city.

The west block is a mix of commercial and residential uses. It reaches Osaka Bay, where conventional port facilities and factories of heavy industry have been converted to the new development for commercial, amusement, and housing. Artificial islands are expected to be developed into new business centers to balance with the existing central area. Primarily residential areas exist in the east and the south and there is a mix of residential, business, and commercial uses in the north.

Major Travel Generators: Major travel demand for business and commercial activities is generated by the Central Business District (CBD) within the JR Loop line, particularly Umeda, Namba and Tennoji – the three largest business centers of the city - where gigantic railway terminals are located. Strong business and commercial activities are concentrated in the central area from north to south. The most symbolic Midosuji-street connects these districts. Now, many development projects are underway or planned, to avoid excessive concentration in the CBD and to facilitate more balanced development all over the city by reinforcing east-west urban axis, utilizing new reclaimed lands in the Bay area.

Location of Major Transportation Terminals

Railway Terminals: Railway terminals are located on the radial intercity railway lines, operated by the private companies, to connect with the intracity municipal subway. The three largest railway terminals within the city are Umeda, Namba, and Tennoji. In Umeda terminal, six private and three subway lines connect serving 1.572 passengers trips and 780,000 transfers daily. In Namba terminal, four private and two subway lines connect serving 743,000 passengers and 422,000 transfers daily. In Tennoji terminal, two private and two subway lines connect serving 566,000 passengers and 356,000 transfers daily.

Underground shopping malls, large department stores, theaters and high-rise buildings for business are jointly developed within the terminals forming enormous business and commercial concentrations. There are also several medium-sized railway terminals; for instance, Shin-Osaka terminal serves for Bullet train, JR intercity line, and municipal subway.

Bus Terminals: The intracity bus system, which is operated mostly by the Municipal Transportation Bureau, comprises of trunk and branch lines. Bus terminals function as transfer points among these two types of bus lines and railways. The Transportation Bureau provides a sophisticated Bus-Rail System in which bus departures are adjusted to subway trains' arrivals through computerized information interaction, along with providing schedules of connecting rail and bus services to the passengers.

Airports: There are two airports in the vicinity of the city. Kansai International Airport (KIX), Japan's first 24-hour airport, is located in Osaka Bay, 35 kilometers south of the city center. Osaka International Airport is located north of the city, now serving only domestic flights.

Two private railway lines serve the KIX from the city with a travel time of 29 minutes from the city center. Limousine services provide transportation to KIX from several spots in the city, the largest of which is the Osaka City Air Terminal near Namba, where passengers for international flights are able to check-in before departing for the airport. Mainly limos and monorail service connected to the municipal subway Line No. 1, Midosuji line, serve Osaka International Airport.

Major Transportation Infrastructure and Network

Roads: The city has an elevated expressway system and ordinary surface road network.

Expressways: Hanshin Expressways form radial and circular networks. Together with the national highway, they mitigate congestion of surface roads as well as facilitating smooth traffic flow for long travel. Currently, there are four expressways extending 35.4 km within the city. A new addition under construction will increase the system to a total of 98.24 km.

To counter the environmental problems such as noise, vibration and obstruction of sunlight, particularly in residential areas, "environmental belts," strips of land about 10 to 20 m wide separate the highways from residential areas.

Ordinary Roads: Ordinary roads, comprising arterial roads, feeder roads and special streets, form a network of mostly a grid pattern that runs east to west and north to south.

Arterial roads consist of four lanes or more. Running through the heart of the CBD from north to south is 44 m-wide Midosuji Street, which connects Umeda and Namba, districts. Four adjacent streets in the north-south direction, including Midosuji, are one-way alternate streets. Arterial roads are situated at intervals of 500m in the city center and about 1 km in the outskirts.

Feeder roads, with two to four lanes, serve primarily the residential areas. In addition, special streets exist for non-automobile traffic, such as pedestrians, bicycles, automated guideway transit, and underground pedestrian walkways.

Public Transit

Railways: Railway network in the city consists of JR, five major private railways with over ten lines and seven municipal subway lines. Total length of the network is 245 km within the city and 1,433 km in the Osaka metropolitan area. The JR line forms a radial and circular network, with the Loop Line located at its center, linking the city with surrounding areas. Private railways also provide intercity services from the terminals.

Seven subway lines form a grid network within the JR Loop line, stretching out diagonally to the suburbs. The total length of municipal subway is 115.6 km with 111 stations. As for its route alignment, each line is allotted at an interval of about 1 km within the JR Loop line and 1.5 km in the suburbs to cover all the city combined with private railways. To minimize the need for transfers, three reciprocal through services between municipal subways and private railways are being provided.

Regarding ridership, the railways in total carry 14.2 million passengers and 8.6 million passengers daily in the Osaka metropolitan area and in the city, respectively. They account for 79.5% and 90.4% of the total transit trips in the metropolitan area and city, respectively. The municipal subway serves 2.65 million passengers daily. It should be noted that Midosuji Line (Line No. 1) alone carries nearly a half of the total subway passengers.

To reduce construction cost, a medium capacity subway was developed and first introduced to the subway Line No. 7 in 1990. The vehicles are driven by linear induction motor, which enables low vehicle floor. The system meets the traffic demand, which ranges from 20,000 to 30,000 people/hr, which is the capacity between that of bus (max 3,000 to 6,000 people/hr) and conventional subway (max. 30,000 to 50,000 people/hr). With strict budget restrictions, this system is expected to be used for future railway projects in Osaka.

Automated Guideway Transit (AGT) system, called New Tram, has been serving the newly developed waterfront area, with 6.6 km of network carrying 60,000 passengers daily.

Bus System: Currently, the Transportation Bureau has 107 service routes with 457.6 km of route length. A fleet of 930 buses provides intracity service for 320,000 passengers daily.

The bus network is designed based on the so-called Zone Bus system, which comprises trunk lines and branch lines. Trunk lines serve the areas not easily accessible to railway lines as a primary public transit. Branch lines mostly serve residential areas as a feeder service to railway stations or trunk lines. There is a flat fare of $1.40 (200 yen). Transfers between trunk and branch lines are free for mitigating inconvenience in transfer.

To secure bus punctuality, in cooperation with the traffic police, the city applies many traffic control systems. Bus exclusive lanes and priority lanes, paved in yellow or brown, give a right of way to public buses in connection with traffic signal control which shortens duration of red signal when detecting buses approaching to the intersections. Bus location system, in addition, not only shows passengers the location of buses but also transmit data on locations of buses ahead and behind to avoid a series of buses from being jammed up.

Administration and Finance for Public Transit

Railway Projects: Railway sectors are classified into three types depending upon the extent of local government's involvement. There are public sector (Municipal Transportation Bureau); tertiary or semi-public sector which is organized and invested jointly by private companies and local governments; and a purely private sector.

In addition, there are three types of financing in terms of who constructs and operates the railway. What is called primary enterprise in Japan means that the company for itself raises its own funds for railways, and builds, owns and operates it. Secondary enterprise owns vehicles and operates on railway infrastructure leased from its owner in return for lease payment. Lastly, tertiary enterprise is engaged in funding, building, owning and leasing railways, but is not involved in operation.

Tertiary enterprise is derived from the growing needs for extending private railways into the city or promoting new development, like artificial land or new town projects, that are expected to give benefits to both public and private sectors.

As for the subsidy system, Ministry of Transport specifies the urban railway projects to which types of sector they belong. Subsidy for the tertiary sector is proportional to the share of capital investment that local governments make and coincides with subsidy rate for municipal subway when 100% of capital investments of tertiary sector is made by local government.

Bus: For financial assistance, a subsidy is granted for a small portion of capital investments such as improvement of bus stops, procurement of low-pollution vehicles but nothing for operation, which is the most expensive portion.

Travel Pattern

Following is a summary of the Person Trip Survey in 1990, which describe all the travel made from a 5% sample surveyed excluding movement of trucks.

Total Volume of Traffic: 43.3 million trips per day are made in the Osaka metropolitan area; out of which, 11.3 million (or 26.2%) trips originate or terminate in Osaka City. While residents of Osaka City make 81% of the intracity trips, non-residents make 80% of in-flowing and out-flowing trips. Regarding geographical distribution of Osaka-based trips, 60.2% of them are intracity trips and 17.8 % from/to the north, 13.2 % from/to the east and 8.8 % from/to the south. Excluding bicycle and walking, nearly two-thirds of them are made between the city and outside of the city.

Trip Purposes: Both trips to work and for business share nearly 20% of total trips. In the CBD, business trips dominate, capturing around 25%. It is notable that 36% of the business trips made in the Osaka metropolitan area are Osaka-based. It should also be noted that trips attracted by the CBD increased by 12.0% although the total of Osaka-based trips rose by 1.3% since 1980 to 1990.

So, in planning the transportation system, it is important to consider how to handle massive flow into/out of the city, especially to the CBD for commuting and business trips.

Trips by Mode and Their Spatial Distribution: The railway accounts for 40.1%, the largest share. Over 70% of inflow and outflow trips are made by rail and less than one quarter by auto because of the availability of dense rail network and frequent service both by private operator and the Transportation Bureau. Excluding walking and bicycling, mainly used for a short-distance trip, the railway and auto usages in the city are 52.8% and 40.6%, respectively.

By trip purpose, nearly 60% of the commuting trips are made by rail. 84.4% of the in-flowing commuting trips are made by rail. Regarding business trips, auto use (38.4%) dominates rail use (10.4%). Excluding walking and bicycling, auto usage rate reaches as high as 63%.

In addition to the above person trips survey, which deals only with passenger cars, trips by all autompbiles, including vans and trucks, are surveyed by the Automobile Traffic Volume Survey. Out of the total of 2.85 million auto trips observed in 1996, 1.32 million trips are out-flowing/in-flowing traffic and 1.53 million trips are intracity.

Commuting trips and business trips by automobiles, excluding freight transportation, account for 11.3% and 16.5%, respectively. These are the main targets, the city should try to convert to public transport use. Particularly, it is important to suppress inflow and outflow for commuting and intracity business trips, which share a large portion.

Bus trips account for only 1.7% in 1990, down further from 2.4% in 1980. Since the Transportation Bureau operates both buses and subways, the bus network has been reduced from a managerial point of view as new subway lines have opened.

Trip Distribution by Hour: Peak hour for commuting to work is from 8 a.m. to 9 a.m. and business trips are distributed uniformly throughout the daytime. Railway operators should make a large investment in new construction or in improvements to handle the extremely high concentration of passengers.

Travel Problems Associated with Auto Traffic

Traffic Congestion: With a high rate of auto use for business in the CBD, road congestion is observed in such roads that link the city center and the satellite cities. Accordingly, the traffic congestion reduces the bus-scheduled speed down to as low as 12.8 km/hr, as well as 20 to 25 km/hr for travel speed by auto within the city.

Continuous Increase in Auto Traffic: The inflow/outflow and the total kilometers run by auto have been rising. Also increasing is the number of registered cars. Since the high price of land and spatial limitation has slowed surface road network expansion, the present increase in auto traffic had to be served by expanding elevated expressways. For the future, however, public transportation is and should be expected to serve the growing needs for mobility since expressways are more difficult to build considering strong environmental concerns of residents, cost constraints and transport efficiency.

Illegal Street Parking: Lowered travel speed by car is partly caused by notorious and illegal street parking. In 1996, the number of illegally parked cars on streets was 123,000 vehicles, which translates into 3.1 vehicles for every 100m of all the streets.

Travel Problems Associated with Public Transit

Public Transport for the Areas that are not accessible to Railway Stations: In the city's master plan, the entire city is to be served by public transit system consisting of railway or subway, whose station covers the area within a radius of 500m, and a bus system, which would have bus stops every 350 m. It is, however, true that some areas are served only by bus, which is often hampered by traffic jams, despite bus priority measures applied.

Train Congestion: Massive in-flow to CBD from 8 a.m. to 9 a.m. by train causes serious congestion, with vehicle occupancy ratios as high as 170 to 200%. The goal is to decrease the ratio to as low as 150% by a longer train formation, frequent services or dispersing passengers by new line construction. The biggest problem is that revenue increase will not compensate for the expenditure required for such investment above, especially in the case of private carriers, which are hardly subsidized in the nationally controlled subsidy system.

Transfers: Although congestion at the major terminals is serious, there is very little room to widen or expand passenger corridors in the terminals to restore a smooth passenger flow. Also, the need for negotiating levels during transfers in the terminals discourages potential users from choosing railway system.

Construction Cost of the Railways and Financing: Since the entire city has already been highly developed, the only available rail-type transit is underground subways. However, the cost of subway construction has risen rapidly, reaching approximately $200 million (30 billion yen) per km in 1999 for a conventional subway. It is strongly suggested that a small scale of the new transit system on the surface should be studied and applied appropriately to the areas which will not generate as much demand as conventional rail system would fit.

Despite strong needs for rail-type public transit, raising new construction capital is not easy, even for the public sector, because it suffers from accumulated debts and the national subsidy only accounts 25% of the total cost.

A lot of discussions are expected regarding subsidy rate, types of sectors entitled for subsidy, expansion of financial sources, transfer of earmarked taxes etc., for financing rail systems. A railway-type project is very crucial now for improving overall urban transportation situation.

The bus problem is related to many factors. Since the late 1960s, when its ridership peaked at 1.2 million trips per day, the expansion of the city's subway system and motorization diminished the bus ridership, which lead to route reduction. This accelerated automobile use, formed a vicious cycle of aggravated road congestion, worsened on-time performance of buses, and further reduction of bus ridership.

Policy for Solving the Problems and Future Perspectives

In view of the current trend in auto traffic, it is predicted that demand for automobiles will grow unlimitedly due to the multiplying effect of further motorization by the elderly and women. Both of these groups are expected to be employed more and have a higher rate of owning a drivers license.

It is highly possible that the road supply would not meet the demand. Drastic measures are necessary to improve public transit system, because constructing or widening roads and its resulting administration sometimes cost, - US $200 to $280 million per km (30 to 40 billion yen) in an area that is already developed - even more than that of new subway line.

In addition, the city plans to reduce NO_2 emission by 31% in 2000 compared with that of 1992. Reduction of CO_2 is also needed from the viewpoint of global warming.

Osaka will be engaged in emphasizing and refurbishing the public transit system for years to come. First, the railway network is to be steadily expanded by 83 km in keeping with regional development as a critical policy in order to both serve the areas not easily accessible by rail and to mitigate congestion of the existing lines.

Second, for reducing the inconvenience of transferring, reciprocal through services planned to be proceeded by connecting the existing lines of different operators. Stored-fare system is working out quite well and has been positively accepted by many passengers since the system is in common with many railway and bus companies, so that one universal ticket is accepted.

Third, for reducing construction cost, mini-subway and AGT are possibilities. Another current option is to electrify existing freight rail and convert it to passengers' use. In cooperation with strict controls, LRT could be possible in the future even in 25m wide narrow streets.

As for financing, it should be noted that several new railways were constructed jointly with private redevelopment projects, so that development benefits were transferred to the rail projects, reducing their project costs. This could be applied to railway projects closely related with a new developments.

Regarding auto traffic control for smoothing surface transit's flow, bus priority, bus exclusive lanes, bus priority traffic signals and the new bus system with protruded bus stops to vehicle lane for running free from illegal parking are to be expanded.

To increase bus ridership, several diverse bus services are to be provided. Specifically, for wheelchair-bound passengers and/or the elderly, buses with wheelchair lifts were first introduced in 1991, non-step bus was introduced in 1996, whose floor is low so that passengers do not need to step up when boarding.

From environmental concern, low-pollution buses have steadily been introduced into the system. Examples are hybrid buses, natural gas buses, and so-called minimal idling buses, which stop engines at red signals. In addition, a discount fare is applied on Fridays and on the 20th of each month with one-day tickets, which sells at US $4 (600yen) instead of regularly US $5.80 (850yen), for all the municipal subway, AGT and bus to reduce auto traffic.

Lastly, it is very important to share the experiences with the public transportation oriented policies of many cities. It is expected that the experiences of Osaka City would help other cities in their transportation system planning.

References

1. "Results of Person Trip Survey 1990 in Keihanshin Metropolitan Area, - Traffic Flow in Osaka City -," Planning Bureau of Osaka City Government, 1992.

2. "Current Statistics of Osaka City," Osaka City Government, 1996.

3. "City Planning in Osaka City," Planning Bureau of Osaka City Government, 1997.
4. "Result of Road Traffic Census in 1994."

5. "Subway/NewTram/Bus," Osaka Municipal Transportation Bureau, 1998.

6. "Almanac of Urban Transport," Research Institute of Transport and Economics, 1997.

Existing Urban Public Transportation System and its Operations and Usage

Yasumasa Hayashi[1]

Introduction

There are many types of railways in the Osaka City metropolitan area. They are coping not only with the inter-urban transportation but also urban transportation in the metropolitan area.

The JR (Japan Railway Company, which used to be Japanese National Railways) Shinkansen, is in the north of Osaka City and runs east to west along the national corridor. It connects Osaka with Tokyo and other major cities like Nagoya, Hiroshima, and Fukuoka.

The Osaka Metropolitan area includes the major Kansai cities Kyoto, Kobe, Nara, Wakayama, and Osaka. Regarding inter-city transportation within the metropolitan area, the transit system is characterized by the fact that although Osaka is at the center, Kyoto and Kobe both have individual economic foundations and the flow of people is not solely in one direction. In this area, private railways have shown strong development for a long time.

Inter-city railways within the metropolitan area are arranged radially centering on Osaka City and linking Kyoto, Kobe, Nara, and Wakayama. The railway network is formulated along the radial corridor to meet the transport demand, and especially on both the Osaka-Kyoto corridor and the Osaka-Kobe corridor. There are three railway lines respectively to serve high-density transport demand. These railways are mostly operated by private companies such as JR and five major companies, which are the Hankyu, Hanshin, Kintetsu, Nankai, and Keihan Corporations.

---

[1] Director of Planning Division, Transportation Bureau, City of Osaka, 1-12-52, Kujo-Minami, Nishi-Ku, Osaka 550-0025, Japan

Inside Osaka City, the Municipal Transportation Bureau is taking charge of transport demand and it has succeeded, since the Osaka municipal streetcar started its operations in 1903. The JR Osaka Loop Line has been in operation since 1961. Recently, some private railways have started operations beyond the JR Osaka Loop Line. The area inside the Loop Line, which is recognized as the CBD (Central Business District), consists of shopping spots like Umeda, Namba, and Tennoji, and business areas like Yodoyabashi and Hommachi. Almost all the people from the suburban areas use inter-city railway and transfer to the subway system to get to their destinations in Osaka City.

The subway network, operated by the Transportation Bureau, has been expanded since its commencement and it forms a grid network inside the Loop Line and radial network outside of it (Figure 1).

Bus

The bus system generally has two major functions: the first as feeder system to railway stations and the second as trunk system for the areas not served by rail service.

In the Osaka metropolitan area, there are many bus operators, such as the municipal bureau and private companies, which also operate the railway lines. Inside Osaka City, mainly the Osaka Municipal Transportation Bureau operates the bus system. It has 107 routes with total length of 440.2 route km and carries 320,000 passengers daily.

Other systems

In Osaka City, Japan's first municipal streetcar was commenced in 1903. After the 1960's motorization, the tramway system could not keep up either speed or punctuality and finally, it was abandoned in 1969. Now, Hankai Electric Railway operates two lines of streetcars. They are 18.7 km in length and connect Osaka City with Sakai City in the south of Osaka.

The New Tram system, the Japan's first developed AGT (Automated Guideway Transit), was introduced as a brand-new transit system suitable for a newly reclaimed town, 'Nanko Port Town.' This system consists of 4-car trains with rubber tires traveling on a guideway of concrete track. It has a capacity of about 5,000-passenger per hour. It is usually used for commuter transportation.

On the outside of Osaka City, there is a monorail system that runs in a circular direction and connects radial railway stations in the satellite cities, Osaka Airport, and Expo Memorial Park. An extension to a major hospital was completed in 1998, and another extension project to the International Culture and Park City is under construction.

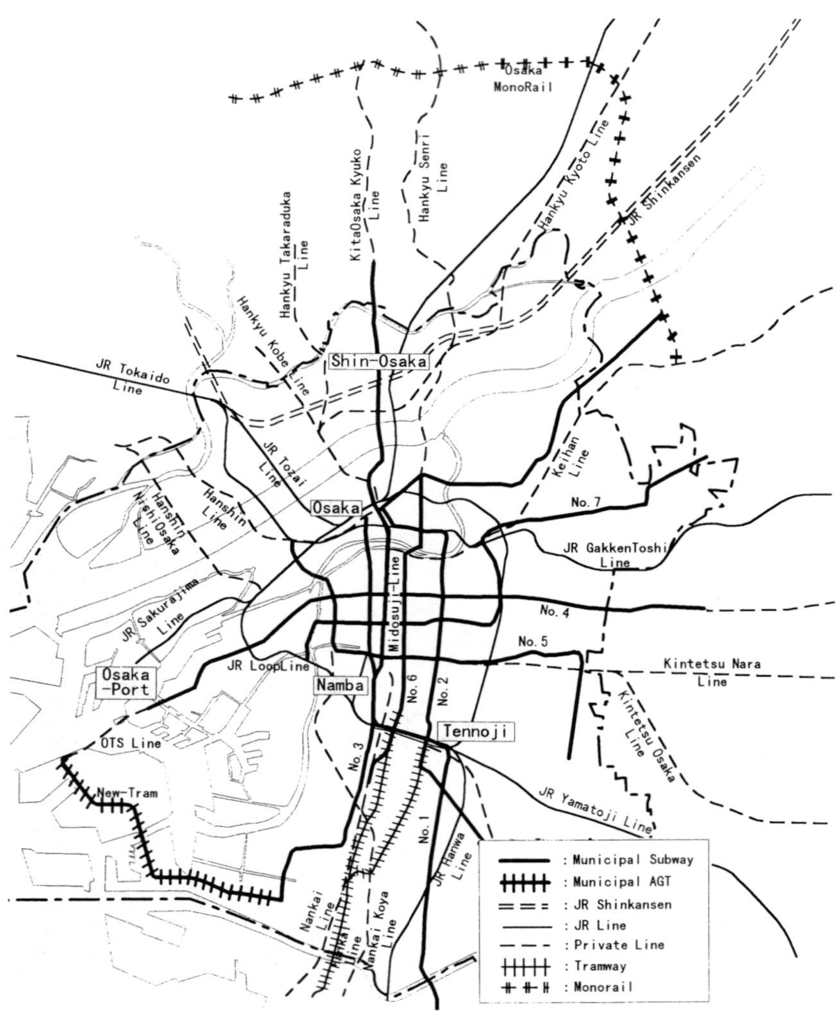

Figure 1: Railway Network in Osaka City

Organizations of Carriers

Rail Carriers are categorized into three groups. The first group is the public sector, such as the Transportation Bureau of Osaka City Government; the second is the private sector consisting of the five major railway companies; and the third is so-called "tertiary sector" carriers, which are financed by both local government and private companies.

While these carriers provide the railway service, another service called "mutual through operation" is provided on some lines. This service is provided on more than two connecting lines of different carriers and the carriers operate their trains beyond the connecting point. In Osaka City, there are three mutual through services between municipal subway, and three private companies.

Bus carriers, that have routes in Osaka City, are the Transportation Bureau of Osaka City Government, which provides municipal bus service, and eight private bus companies. Most of the bus routes in Osaka City belong to the Osaka municipal bus service, while private bus routes mainly connect major terminals in the city to suburban cities, Osaka Airport, or Kansai International Airport.

Subway and New Tram

General Outline: The history of the Osaka municipal subway begins from the specific transport demand in Osaka City. Since the first subway Line No.1 was commenced, the north-south corridor from Umeda through Namba to Tennoji had been the most congested. Due to this demand, the subway network was expanded to mitigate the congestion of Line No.1.

For the opening of Expo '70, the north-to-south lines No.2, No.3, and No.6 and east-to-west lines No.4 and No.5 were constructed to make a grid network in the city center with total subway length of 64.2 km. After Expo '70, the subway network expanded to the satellite cities to meet the increasing transport demand caused by the increasing suburban population. The total length of the subway network was 94.1 km in 1985.

Meanwhile, New Tram system was introduced in 1981 as a main transit with adequate capacity to meet the demand and causing less noise and vibration.

Japan's first linear induction motor driven subway system was introduced to the Line No. 7 in 1990. This line was extended westbound to the city center of Shinsaibashi and Osaka Dome to meet the demand between the city center and the suburban area as well as to connect with the other subway lines. It was extended eastbound to Kadoma-Minami to cope with the demand to Kadoma-Minami development and Namihaya Dome, which is a large-scale swimming pool.

Now the Osaka municipal subway comprises seven subway lines and one New Tramline and the total length of the subway is 115.6 km and specifically, New Tram is 6.6 km in length.

Fare Structure: Fares are collected using tickets, commutation and school passes, and prepaid cards. These are all magnetic cards and strictly checked at both in and out wickets. If the amount of fare is not enough, passengers must pay the amount before going out of the wicket.

The sectional fare system is adopted for the subway (including New Tram), and the flat fare system is adopted for the bus. $1.40 (200 yen) is needed for the first ride on the subway, which is the same as the bus fare. In transferring from subway to bus, $0.70 (100 yen) is discounted (equal to the half of bus fare). Special discount tickets like a coupon ticket, a Day-Long ticket, "No My-car Day" Pass, and bus daytime ticket, are sold to enhance the use of public transportation. Bus daytime ticket which has 25% discount ticket is available just for bus travel during off-peak period of 10:00 a.m. to 4:00 p.m. A Day-Long ticket, which costs $6.00 (850 yen), can be used for travel using all subway lines, New Tram and bus routes, for all day long. And "No My-car Day" ticket costs $4.20 (600 yen) and is available on "No My-car Day" and Fridays in every month. "No My-car Day" is 20th in every month and on that day the use of public transportation is promoted instead of the use of cars by Osaka City and Osaka prefecture.

Moreover, the special fare system for the aged and disabled passengers is introduced. There is no charge for those over 70 years old or for extremely disabled passengers. Mildly disabled passengers are offered a 50% discount. These special fare systems are supported and subsidized by the Welfare Bureau of Osaka City.

Condition of Management: Ninety percent of revenue is collected from the fare box and 10% from other sources, such as advertising. Expenditures are 42% for personnel expenses, 22% for interest, and 15% for other expenses, such as electric charges. The operating revenue of subway and New Tram services showed deficit figures of $173 million (24.7 billion yen) in 1996 fiscal year (Table 1). This deficit is attributable to the increased costs of capital associated with investments, such as facility improvements and New Line project, and the high operation and maintenance costs, of which most part is personnel expenses.

In order to recover financially and to continue to provide stable transportation services on a longer-term basis, it is urgently necessary to restore sound management. In this respect, since 1995, the city has formulated and enforced the "Better Subway Service Plan". The plan contains measures to streamline overall operation and maintenance and to increase revenues and public subsidies. The plan also includes curtailment of 2,000 personnel with the introduction of automatic wicket facilities and streamlining the business structure.

Table 1: Financial Record of Subway Operations (Unit: US $ in millions)

		1975	1980	1985	1990	1994	1995	1996
Income	Fare income	350	630	958	1,218	1,306	1,321	1,315
	Miscellaneous income	22	29	38	65	78	80	81
	Non-operating revenue	115	207	149	208	47	49	54
	Total	487	866	1,145	1,491	1,431	1,450	1,450
Expenditure	Labor costs	254	434	531	658	728	696	702
	Energy Costs	21	46	60	56	56	56	57
	Repair expenses	16	29	33	49	65	62	69
	Depreciation expenses	63	99	146	267	322	319	362
	Interest payment	164	238	345	412	386	375	363
	Other expenses	26	38	59	98	131	141	123
	Total	544	884	1,174	1,540	1,688	1,649	1,676
Ordinary	Profit and Loss	-57	-18	-29	-49	-257	-199	-226
Extraordinary	Profit	0	1	11	110	23	0	11

Now, the cost of new line project is subsidized by National Government for 25% and by local governments for 50%.

Safety System: Safety is the most important issue. Every line has CTC (Centralized Train Control) system in its own control center for safe and punctual train operation. The operation is totally supervised by PTC (Programmed Train Control) system and ARC (Automatic Route Choice) system in normal operation. It is carefully controlled by ATC (Automatic Train Control) in an emergency. The electric power control center controls all the electrical facilities, like electric power, station lights, and fans for ventilation. An electric generator is provided to supply power in case of power failure.

Service Improvements in Rail System

Reinforcement of Conveyance Capacity: Along Line No. 1, where the peak-hour congestion is the most severe, the city has taken a lot of measures to increase the conveyance capacity for the target of 10-car train service. Recent improvements are platform and switch-back line extensions, improvements of signaling system, and TV monitors for conductors to watch the passengers on platforms for 10-car train service. In 1987, 9-car train service started, and after that the plate for the gap between the platform and cars and the emergency stop button were equipped, and the TV monochrome monitors were changed to color. Ten-car train service started in 1996. At the same time, Lines No. 2, No. 3, and No. 6 increased their capacity during peak hours to reduce the congestion ratio to under 150%.

Improvements of Facilities for the Aged and Disabled: Station facilities and cars have been improved so that they are easier for the elderly and the physically handicapped to use.

In order to install escalators and elevators, not only in the stations of newly- built subway lines but also in the old stations, the city has formulated the "E-E-Machi Plan" (Good Town Plan with Escalators and Elevators). The plan is targeting installation of escalators or elevators in all stations by 2003; this would enable passengers who cannot use the stairs.

In order to provide barrier-free facilities in trains as well as in stations, amenities are being provided in cars and stations for wheelchairs.

Improvements for Amenities: In order to enhance comfort in the subway stations, an attempt is being made to refresh and renew the stations, which were built more than 50 years ago. Installation of air conditioning systems in all cars was completed in 1995 to protect the passengers from the hot and humid climate in summers. The installation of air conditioning systems in all stations will be completed by 2001.

Improvements in Ticketing System: Starting in 1996, a new ticket system called "Through Kansai" was implemented. The same prepaid card can be used for subway, city bus, New Tram, and 4 private railways (Hankyu, Hanshin, Nose and Kita-Osaka Rapid Railway). In 1997, the Hankyu Bus and OTS (Osaka-ko Transport System), a corporation responsible for the Nanko-Minatoku connective line, has joined the "Through Kansai" network. It is expected that 22 transport companies operating railway and bus services will join the network in 1999, and the card for "Through Kansai" will be effective for over 700 km of railroad service and 2,600 km of bus service.

New Information Service System: In addition to the information centers at major subway stations, a new center named "Access Guide" with staff who can speak English is available. An "Inquiry Center by Phone" is also provided. New information service system is also available through a web site on the Internet since 1997. Computer terminals are installed at major subway stations and bus terminals to display public transportation information, such as fares and times as well as the events taking place along the rail lines.

Bus System

General Outline of City Bus: Since the first publicly managed bus service was launched in 1927, city bus has been taking a great role as a familiar means of transportation supporting the citizens' daily life. The average daily bus passenger volume was approximately 1.2 million in 1964, but due to the drastic increase in automobile traffic and subway network development, the bus ridership dropped to as low as 320,000 in 1987. There are now 105 routes and the total route length is 447.6 km.

Bus Network: Now, inside Osaka City, subway provides trunk transportation and bus service is taking the important role of providing trunk transportation in the communities, which are not covered by subway lines.

In Osaka City, the "Zone Bus System" has been introduced. Under this system, complicated conventional bus routes have been consolidated into one main route and several local routes support the main route to enhance bus service efficiency. This system allows passengers a combined use of a main-route bus that connects a terminal or railway station to each local area, and a local-route bus that serves each area. This system contributes to bus service punctuality and efficient vehicle use, while expanding the service area without an increase in fare.

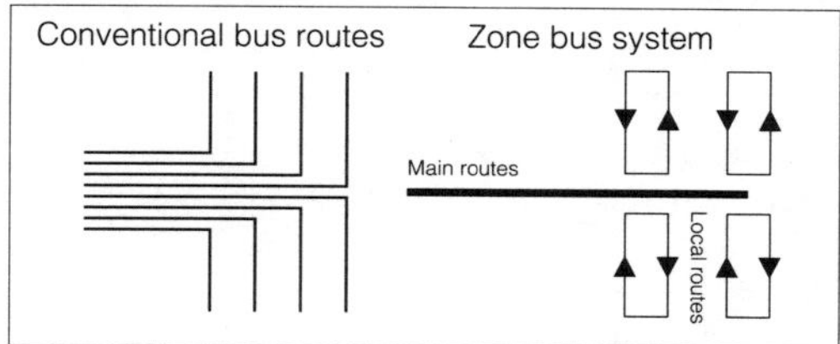

The City has 12 sightseeing bus tours in operation covering scenic spots, such as Osaka Castle and Kaiyukan Aquarium, in the hope of enhancing Osaka's tourism. Since 1996, a limousine bus service between OCAT (Osaka City Air Terminal) and the Kansai International Airport has been in operation. Two privately operated bus companies provide the service.

Condition of Management: Sixty percent of revenue is from fare boxes, and 70% of expenditures is personnel cost. The remainder of expenses includes interest, depreciation, and other costs. The managerial balance of the bus services was in the red to the amount of about $8.4 million (1,200 million yen) for the 1996 settlement of accounts. The deficit is due to many bus routes with low ridership, shift of passengers to subway lines after the development of new subway lines, and high personnel expenses (Table 2).

Service Improvements in Bus System

Vehicle Improvement: Anticipating the arrival of an aging society and encouraging social participation of the physically handicapped, buses with wheelchair lifts were introduced in 1991. "Non-step" bus, in which the floor is as low as 34 cm above ground level and with retractable slope equipped for the wheelchair users, is introduced in 1996. The floor level on these buses is 50 cm lower than the conventional one.

Considering the deteriorating environment, buses with low adverse impact on the environment, such as hybrid buses and buses using natural gas, were introduced. A hybrid bus requires less load on its engine when starting and accelerating, and thus emits less nitrates and black smoke. Incorporating a generator-motor into its engine, an electric hybrid bus stores braking energy that is generated during deceleration as electricity, and applies the electricity to gain drive power for starts and acceleration, thereby reducing the engine load. In Osaka City, the hybrid bus was launched in 1991.

Table 2: Financial Record of Bus Operation (Unit: US $ in millions)

		1975	1980	1985	1990	1994	1995	1996
Income	Fare income	97	117	132	138	157	159	156
	Miscellaneous Income	12	29	44	47	54	57	60
	Non-operating Revenue	50	69	45	48	45	44	44
	Total	159	215	221	233	256	260	260
Expenditure	Labor Costs	195	175	169	178	189	186	187
	Energy Costs	8	12	10	7	8	8	9
	Repair Expenses	4	4	5	5	8	10	10
	Depreciation Expenses	11	10	16	19	19	25	29
	Interest Payment	29	13	8	7	7	9	9
	Other Expenses	13	15	18	19	27	27	27
	Total	260	229	226	235	258	265	271
Ordinary	Profit and Loss	-101	-14	-5	-2	-2	-5	-11
Extraordinary	Profit	0	49	7	18	3	0	0
Extraordinary	Loss	0	6	0	0	0	0	0
Net Profit and	Loss	-101	29	2	16	1	-5	-11

A natural gas bus was introduced in 1995 and it emits neither sulfates nor black smoke and reduces nitrates by 60% to 70% in comparison with conventional diesel-fueled vehicles.

Bus Terminal Development: In transferring between subway and bus, users like convenience as well as discount service. By the end of 1996, 10 bus terminals had been established and these terminals are equipped with fully air-conditioned waiting rooms. Some of them also have bus information panel in front of the wicket of the subway station. During non-peak times, sometimes bus departure time can be adjusted to coincide with the subway.

Bus Location System: The Bus Location System, which notifies passengers of the approach of buses and tries to relieve customer frustration during waiting time, was introduced in 1981. The information is provided by the bus control center, which collects bus location data through antennas both on-board and on the road .

Bus Priority Policy and "Urban New Bus System": In an effort to increase the punctuality of bus service, the city has introduced a number of measures jointly with the Osaka prefecture police and other authorities. These measures include bus exclusive lanes, roads only for buses, and bus priority lanes with bus priority traffic lights.

The "Urban New Bus System" was introduced in 1986. It is comprised of various measures including:

(1) The introduction of bus priority lanes and bus priority traffic lights
(2) The introduction of fully air-conditioned vehicles with wider doors and lower floors to better satisfy urban transportation needs
(3) The installation of the bus location system
(4) The provision of shelters for bus stops

So far this system has been introduced to two routes. One of the two is the route between Sakuragawa Station and Tsurumachi-4-chome. The system is quite successful through utilizing bus exclusive lanes for 24 hours and "bus terraces", with which buses are not delayed by the cars parked on the street.

Use by People

Presently the subway carries 2.7 million passengers daily, New Tram carries 68,000, and buses carry 320,000 per day. The ridership of each Subway line and New Tram are indicated in Table 3. The congestion during morning peak hours is extremely severe and the ratio is almost over 100%.

Table 3: Outline of Subway Lines

Item/Category	Line Name							Total
	No.1	No.2	No.3	No.4	No.5	No.6	No.7	--------
Origin	Esaka	Dainichi	Nsihi-Umeda	Osakako	Noda-Hanshin	Tenjinbashisuji-6-chome	Taisho	--------
Destination	Nakamozu	Yao-Minami	Suminoe - Koen	Nagata	Minami Tatsumi	Tengacha-ya	Kadoma - Minami	--------
Operating Distance	24.5km	28.1km	11.4km	15.5km	12.6km	8.5km	15.0km	115.6km
No. of Stations	20	26	11	13	14	10	17	99
No. of Vehicles	400	246	132	108	68	136	100	1190
Daily Average								
Vehicle Running Mileage (km)	107,541	57,214	25,969	30,735	16,653	29,695	9,502	277,309
Passenger Volume	1,394,487	577,347	352,620	291,389	224,155	365,849	61,969	2,646,421
Running Intervals (min.)								
Morning Rush Hour	2	2.5	2.5	4	3.75 – 4	2.75 – 3.5	3	--------
Afternoon	4	5	5 – 6	7	7	5	7	--------
Evening Rush Hour	2.5	3.5	3.5	5	4.75	3.5	3.5	--------

All lines are congested during peak-time, especially the ratio of the Line No. 1 is over 150%. Regarding the congestion ratio, the Report No. 13 of the Council for Transport Policy issued in 1992 suggested that the ratio is to be less than 150%, and the city is targeting for the condition. The city has made every effort to decrease the congestion of the Line No. 1 by shortening the interval (headway) to 2 minutes and operating 10-car trains with longer platforms. Recently the city is encouraging staggered office hours and flex time system to the companies along the Line No. 1.

Conclusions

In Osaka City, a new urban center is being developed in the waterfront area to decentralize the excessively concentrated business activities. In order to connect the newly developed center and the existing urban center, two subway lines are being planned. Future railway plans are in Osaka City's Master Plan, which comes from the Report No. 10 of the Council for Transport Policy issued in 1989.

One of the two subway new lines is now planned along the north-to-south corridor in the east area of Osaka City. This area is highly populated and the development of public and private housing is in process. Now, this area has minimal railway service. The construction of the new line, the subway Line No.8, is going to start in 1999 and should be completed by 2006.

Another new subway line, which is called "Hokko Techno-port Line", connects a reclaimed island with the city center. It provides the important rail access to "Maishima Island" and "Yumeshima Island", the sites of the Olympic Games, which Osaka City is hosting in 2008. This line will connect with the subway Line No.4 and OTS Line, and the mutual through service will be provided.

In order to have a city with amenities and comfort for the coming new century, and if the environment of the earth and energy efficiency are considered, it is absolutely necessary for Osaka City to control the use of private automobile, to shift automobile trips to public transit, and to keep an adequate balance between automobiles and mass transit. The subway has the least estimated damage to the environment with very little noise and vibration and needs less energy per person transported. It is indispensable to develop the public transportation network, which consists of subway, bus, and other public transit like LRT (Light Rail Transit), or APM (Automated People Mover). A large amount of money is needed for the subway construction. It is a very important issue to reduce the construction cost of the subway.

References

"Subway Construction in Osaka City (Historical Presentation Covering the Last Fifteen Years)," Osaka Municipal Transportation Bureau, 1985.

"Subway/NewTram/Bus," Osaka Municipal Transportation Bureau, 1998.

"Municipal Transportation in Osaka," Osaka Municipal Transportation Bureau, 1997.

"Outline of Railways," supervised by Railway Bureau of Ministry of Transport, 1997.

Technology Innovation in Urban Transit System and its Project

Development of the Small Size Subway System with Linear Induction Motor

Toshimitsu Uebayashi[1]

Abstract

In the 1960's, Osaka City started the investigation for miniaturizing the subway for the purpose of developing an intermediate capacity transit system along the Line No. 1, which was most crowded. Subway construction cost was rising. In 1977, the construction cost of subway tunnel was more than US$70 million per km. In those days, the subway network in Osaka City was far from satisfactory and expanding the subway network became difficult due to the high cost of construction. With this background, the city investigated to miniaturize the subway to construct it at lower cost and to utilize the underground space effectively in the metropolitan area. The city pursued the most minimized subway with all facilities minimized technically. This paper covers the development of the most miniaturized and economical subway system using the Linear Induction Motor (LIM). This small-sized subway could be built within 80% of the conventional subway system cost with the same passenger capacity.

History of Investigation to Develop a Small Size Subway

In the 1960's, Osaka City started the investigation to miniaturize the subway associated with the double-track Line No. 1, which had been the most crowded. In 1971, Report No. 13 of the Council for Urban Transport Policy suggested the introduction of an intermediate capacity transit system along Line No.1 to mitigate the congestion between Umeda and Namba. In accordance with this report, the City established a "New Transit Research Committee." The

[1] Staff Officer, Planning Department, General Affairs Division, Transportation Bureau, City of Osaka 1-12-52, Kujo-Minami, Nishi-ku, Osaka 550-0025, Japan

Committee has researched four systems such as subway with LIM (Linear Induction Motor) cars and AGT (Automated Guideway Transit) and concluded the research in 1972. As per the conclusions of the research, AGT system was put into practice under the name "New Tram." The New Tram project was implemented as a main transit system connecting the new reclaimed island Nanko Port Town with the city center.

Subway construction costs were gradually rising and by 1977, the construction cost for 1 km of tunnel was more than US $70 million (10 billion yen). However, in those days, the subway network in Osaka City was not satisfactory because there were many areas for which the railway service was not available. In order to expand the subway network in the corridors with travel volumes of less than 50,000 persons per hour, it became difficult to construct subways at lower cost.

There was an easy way to construct elevated transit at a lower cost, but in downtown Osaka City, there was no space left for elevated transit. Instead of AGT and monorail, a miniaturized and economical subway was needed that enables the City to utilize the underground space effectively. For the investigation, the City has established a committee within the Bureau to pursue the most minimized subway with all facilities minimized technically.

The basic concept to miniaturize the subway was mainly to minimize the tunnel excavation cost, which accounts for 67% of the total cost. The City first targeted the excavation area down to half of the conventional subway excavation area and researched from various points like vehicles, electric facilities and track structures (Figure 1).

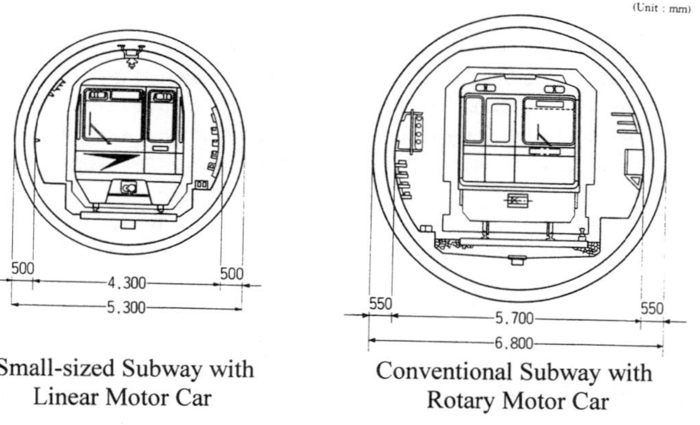

Figure 1: Comparison of Shield Tunnel Cross Section

Finally in 1981, the City completed the basic specifications for a technically minimized subway. The research results compared with conventional subway and New Tram is shown in Table 1. It shows that New Tram is superior to others in curve radius and gradient due to the rubber tires and the steerable truck, but the small-size subway tunnel was smaller than the New Tram's because the lower floor could be applicable and the room for a guide rail was unnecessary. The cost of this small-size subway is 67 % of the conventional subway. If the capacity would be equal, the construction cost of conventional subway would be 80% (Table 2).

The possibility of further miniaturization by introducing a linear induction motor was also investigated. It was recognized that the three-phase induction motor or the Linear Induction Motor (LIM) is most advantageous to make the floor level lower. But it has been found that the three-phase induction motor is difficult to control speed and its development could not be realized easily. The development of a VVVF (Valuable Voltage Valuable Frequency) inverter, which could be equipped to the vehicle, was indispensable to use this LIM on the vehicle. So, the City has developed Japan's first VVVF test vehicles and tested on Line No.4 from 1981 to 1982. The test results were utilized in "The research of small size subway vehicles with linear induction motor" performed by the JREA (Japan Railway Engineering Association). After that, the 12-m test vehicle (LIM-2) was manufactured and it was tested under "The Practical Experiment of Linear Induction Motor Subway" by the Ministry of Transport and Japan Subway Association for one year on the test line built in Osaka Nanko. The results of the experiment gave the City the opportunity to introduce the linear induction motor car on Line No.7.

Plan of New Subway Line No. 7

Osaka City celebrated the centennial anniversary of municipality in 1989. The "Flower Exposition," a part of the centennial ceremony, was designated as "International Garden and Greenery Exposition." The site selected for the exposition was Tsurumi-Ryokuchi, located in the east area of Osaka City, which consists of an artificial hill made with soil from subway excavation and garbage.

In those days, there was only bus service as a transport between the city center and the east area. Another transportation system was needed as an access transport to the Expo site. The demand for access to the Expo site was estimated at 220,000 people for an average day. A subway to meet this demand was recommended for the success of Expo. The Keihan Railway Line and JR Gakken-Toshi Line were serving the eastern part of Osaka City. There was significant transportation demand between this area and the city center. The trains of these railways were highly crowded during peak hours. The main road, which runs east to west in the area, is chronically congested because it connects to the national road Route 1. Both municipal and private buses on the main road were often delayed due to the traffic congestion. The improvement of transportation

Table 1: Comparison between subway, small size subway, and New Tram

Item	Part	Unit	Conventional Subway	Small-Size Subway	New Tram
Gauge		(m)	1.435	1.067 or 1.435	1.600
Electric Power		(V)	750 dc or 1,500 dc	750 dc or 1,500 dc	600 ac 3-phase
Maximum gradient	Main	(%)	3.5	3.5	3.5
	Siding	(%)	4.5	4.5	9.0
Minimum radius	Main	(m)	120	120	70
	Siding	(m)	55	55	30
	Station	(m)	330	200	300
Maximum cunt		(mm)	150	150	160
Car height		(m)	3.745	2.950	3.150
Car width		(m)	2.800	2.450	2.290
Car length		(m)	18.000	12.000	7.600
Ceiling height		(m)	2.080-2.265	2.020	2.055
Floor height		(m)	1.190	0.850	1.050
Wheel diameter		(mm)	860	660	1012
Car capacity	Leading	(person)	130	65	72
	Middle	(person)	140	74	75
Average speed		(km/h)	40	40	30
Minimum headway		(minute)	2	2	2

Table 2: Comparison of construction cost **(Unit:%)**

Item	Small Size Subway	Conventional Subway	Conventional Subway
Capacity (Passenger/hr)	20,000	20,000	50,000
Operation (No. of Cars) (Interval)	6-Car 2.5 minutes	4-Car 3 minutes	8-Cars 2.5 minutes
Cost (in % of total)			
Tunnels and stations	36.0	49.4	55.9
Electric facilities	6.8	7.4	8.5
Depot and inspection shop	4.8	4.2	7.0
Rolling stocks	8.5	5.5	12.2
Interest and office expense	10.9	13.8	16.4
Total	67.0	80.3	100.0

service was earnestly needed under such conditions.

The City decided that the subway route, recommended in Report No. 13 of the Council for Urban Transport Policy issued in 1971, to be ready for the Expo. The City also decided to proceed with the total plan consisting of Line No. 7, which connects the Tsurumi-Ryokuchi Expo site to Kyobashi and further to the city center along with the associated road improvements. The City initiated the segment from Kyobashi to Tsurumi-Ryokuchi, which is needed for the Expo's access.

Practical Utilization of Linear Induction Motor Cars

The completion of the construction of Line No. 7 before April 1990, the opening of the Expo, was the highest priority. In order to manufacture the cars one and half years before the opening, the city tested the LIM driven car to decide whether they should be introduced or not. The Japan Subway Association conducted the tests for 6 months. To investigate safety and reliability, the City manufactured 15-m LIM cars, which have more capacity than the LIM-2, and also rotary motor cars for comparison.

The testing was accomplished to understand in detail regarding vehicles, trucks, signals, communications, electric power, and impacts on the environment and the passengers.

Introduction of Linear Induction Motor (LIM) Driven Cars to Line No. 7

The introduction of LIM driven cars to Line No. 7 was finally decided after the recognition of the safety and reliability through running tests and careful investigation. Regarding the subway system with LIM cars, various points were checked carefully and in detail as documented below:

Equipping: A method of equipping LIM to the frame of a truck was selected as the best way to reduce vibration and strain to the motor, to use simple structure, and to adjust the gap easily.

Steering: The self-steering system was chosen because the structure of the truck was simple and the stability of running both in straight and curved paths was assured.

Reaction Plate: The reaction plate is comprised of one unit of aluminum and steel plates. The reaction plate is available for three different types: flat plate type, cap type, and end bar.

Gap Between the Linear Motor and the Reaction Plate: Considering the frame system of equipping the LIM deflection and wearing depth of rail, and the error of installation, the standard gap between the linear motor and the reaction plate was decided to be 12 mm.

Osaka City made the decision of using the LIM partly because the system can realize the low-cost subway by the reduction of excavation cost. It also can be applied to steep slopes and sharp curves to overcome when this line was extended to the city center. Thus, the LIM subway system was considered as a brand new type of subway available to smaller cities.

Subway Line No. 7

Osaka Subway Line No. 7 started its operations in March 1990 as the Japan's first LIM subway, which is 5.2 km long and has five stations from Kyobashi to Tsurumi-Ryokuchi. During the Expo 90, 4-car trains were operated at 2 minutes and 30 second intervals during peak hours. This line carried 8 million passengers, i.e. 35% of 23 million visitors to the Expo.

The number of passengers carried by Line No. 7 decreased after the Expo because of the inconvenience caused by the disconnection with Line No. 1 to No. 6. So, Line No. 7 has to be connected with other subway lines by extending it to the city center as soon as possible, and at the same time the subway network should be expanded owing to this extension.

From 1990 to 1991, there were a lot of development projects along the planned route of Line No. 7. Osaka Business Park (OBP) had been developed as the new sub center with more than ten high-rise buildings. The Nagahori Dori (Street) Improvement Plan is to reconstruct the underground transportation network along the Nagahori Dori, which consists of subway station, underground passage, underground shopping mall, and underground car parking lots. Osaka Dome is a multi-purpose dome for baseball games, concerts, events, and so on, and the area including Osaka Dome was called Iwasakibashi and was going to be redeveloped as business and leisure complex. Namihaya Dome is a large-scale swimming pool and was decided to be the main site of the National Athletic Meet "Namihaya Kokutai" in 1997. These projects were in process, and to meet the transport demand to these projects, the extension project of Line No. 7 had to begin.

In 1991, the construction of 5.7 km from Kyobashi through OBP to Shinsaibashi, the downtown adjacent to Nagahori Dori, was started. In 1993, the construction began for 2.9 km subway service from Shinsaibashi through Osaka Dome to Taisho. In 1994, the construction of 1.3 km from Tsurumi-Ryokuchi to Kadoma-Minami, near the Namihaya Dome, was started.

In the alignment of Line No. 7 extension, the linear motor subway's ability to negotiate steep slopes and sharp curves is fully utilized. Near the Tamatsukuri station, a 102m curve radius was adopted and the tunnel was constructed more economically under the road without invading private land.

And in six parts of the tunnel, 4% to 5% steep slopes were used to avoid the river, the service tube, or to put the shallower stations.

As another topic, in excavating for Osaka Business Park station, the world's first Multi-Face shield machine with three cutter faces was developed and used to construct the station under a building and a sewage tube.

Regarding the operation of the system, two men (driver and conductor) initially operated the trains, but the facilities like the island platform and the driving equipment were designed for one-man operation. In 1996, one-man operation was started with additional facilities such as TV monitors and rear view mirrors, which enhance safety with one-man operation.

Expectation for the Linear Motor Subway in the Future

Japan's first linear motor driven subway Line No. 7 was extended and its total length is now 15.0 km, which is fourth in the length of the Osaka subway lines. It has been more than 30 years since the City started the investigation of a small size subway. The practical use of a linear motor subway is an informative development as well as epoch-making. This new subway system would be available not only as sub-trunk lines with less transport demand than the main line in metropolitan cities but also as trunk lines in minor cities.

Osaka City's total subway network length is currently 115.6 km. The City plans to expand the network to improve mobility. The City is now planning Line No. 8, which runs north to south in the east area of Osaka City using the same type of linear subway system of Line No. 7.

In metropolitan cities, when we pursue the city life with good mobility and better amenities for well-being, the urban infrastructure is indispensable. In the 21st century, when there is high movement of people, goods, and information, the means of communication are more and more important. The role of transportation system is crucial. In that sense, a comprehensive transportation policy is absolutely necessary to achieve adequate balance between transit and automobile. The automobile travel should be minimized to utilize the limited urban space effectively, to minimize the damage to the environment, and to conserve the petroleum resources in the earth. For this reason, the City wants to build more subway lines. In order to meet the objective, continuous endeavor to design a less costly subway and to perform more effective operation and maintenance are necessary.

References

1. Tadayoshi Yamanaka, Hisao Sawada, "Linear Motor Driven Car of Tsurumi-ryokuchi Line, Subway," Osaka and its Technology No.19, Osaka Municipal Government, 1990.

2. "Subway Construction in Osaka City (Historical Presentation Covering the Last Fifteen Years)," Osaka Municipal Transportation Bureau, 1985.

3. "Subway Construction in Osaka City from 1986 to 1997," Osaka Municipal Transportation Bureau, 1998.

4. "Subway/NewTram/Bus," Osaka Municipal Transportation Bureau, 1998.

5. "Municipal Transportation in Osaka," Osaka Municipal Transportation Bureau, 1997.

6. "Construction Project of the Linear Motor Subway - Osaka Subway Line No.7 from Kyobashi to Tsurumiryokuchi - ", Osaka Municipal Transportation Bureau, 1993.

Socioeconomic Characteristics, Land Uses and Travel Patterns in Pusan

Hang Mook Yoon[1], Sang-Soo Lee[2], Yong Eun Shin[3]

Introduction

Located at the southeastern tip of the Korean Peninsula, Pusan Metropolitan City has developed into a major port for international trade, connecting the land and sea, as well as one of major centers of the Korean economy, since its opening as international port in 1876. It is currently the 5th largest international port in terms of container handling volume, and it is the logistics center of southeastern coastal industrial belt as well as the northeast Asian distribution Center. It is also known as a resort city, offering outstanding sightseeing opportunities during the summertime.

Pusan is the second largest city in Korea and occupies an administrative area of 749.4 square kilometers with a population of over 3.8 million in 1997. It was named Pusan City in 1949 and became Pusan Metropolitan City in 1995.

Like other industrialized cities, Pusan has experienced a rapid urbanization along with rapid motorization process over the last few decades. In particular, the number of cars has increased by an average of 13.5% per year over the last 5 years. The result is obvious: the city is experiencing severe congestion and air pollution. This situation is likely to exacerbate in the near future as travel demand increases. As a response, the government has developed an extensive package of policies to cope with the problems. Under the assumption that the rapid motorization will

[1] Assistant Prof. Dept. of Urban Engineering, Dongeui University, 24 Gaya-Dong, Pusanjin-Ku, Pusan, Korea

[2] President, Transportation & Environment Research Institute, Pusan Branch, 4Fl. Chonji Bldg. 607-1, Chachon 1-Dong, Dong-Ku, Pusan Korea

[3] Assistant Prof. Dept. of Urban Engineering, Dongeui University, 24 Gaya-Dong, Pusanjin-Ku, Pusan, Korea

continue, the package includes a combination of public transportation system improvement plans and large-scale capital investments in rail rapid transit systems and urban expressways.

The objective of this paper is to understand the nature of Pusan City in terms of the need for public transportation. This paper consists of three parts. The first part will discuss the socioeconomic and land use elements that affect the transportation environment. In the second part, travel patterns including traffic volume and modes of travel will be presented. Finally, the policy direction for public transportation system will be discussed.

Socioeconomic and Land Use Characteristics

Population

Like other cities in industrialized nations, Pusan City has seen a rapid urbanization process. Its population has grown almost tenfold to over 3.8 million since 1945, when the population of Pusan was estimated at slightly over 360,000 people. There were two important factors for population growth: the Korean War in the early 1950's and economic growth policies implemented by Central Government in 1960-70's.

As Table 1 shows, the population in Pusan has increased consistently, reaching over 1 million in 1956 as a result of incoming refugees due to the Korea War. With a rapid influx of large population into Pusan from rural areas, the population was over 2 million in 1972, over 3 million in 1979 and near 4 million in 1994. From 1965 to 1975, the average growth rate peaked at 6.1% per year. Since then the rate has slowly decreased. In Figure 1, one can notice that the growth of population has stabilized or even decreased over the recent years. Despite this trend, a slight increase of the population is expected in the near future, mainly due to numerous recent large-scale development projects within the city boundary. The population for year 2001 is thus forecasted to be over 4 million.

Employment

With the growth of economy, the employment has increased as well. Table 1 indicates the constant growth of employment in Pusan. In 1996, the total employment reached close to 1.7 million. The employment has increased at an average annual rate of 4.53% over the last decade.

Though Pusan is the second largest city in size, its industrial composition centers mostly on the light industry and tertiary economic activities, generating only one-tenth of the national GNP. This includes textile, shoes, constructing-machine and service and infrastructure industries. These activities had, however, shown a decreasing share since 1991. It is expected to increase gradually due to the city's investment in the development of industrial sites. It is also expected that the total employment will reach over 2.1 million by year 2001. The share of the service industry is expected to increase to 1.63 million (share of 75.98%) and that of the secondary industry to 0.5 million (22.48%).

Table 1. Trend of population and employment

Year	Population
1936	206,386
1946	362,920
1956	1,002,391
1966	1,426,019
1976	2,573,713
1986	3,578,844
1996	3,878,918

Year	Employment
1986	1,124,000
1988	1,207,000
1990	1,525,000
1992	1,597,000
1994	1,627,000
1996	1,671,000

Figure 1. Population and Employment: 1986-1996

Number of Vehicles Registered

Along with the growth of economy, the number of vehicles registered in the city has been increasing very sharply. In 1986, the total number was only about 124,000. The number has increased by an average annual rate of 15% over the last 10 years, reaching 671,958 in 1996. As seen in Table 2, the number of passenger cars has increased at a higher rate than those of other types of vehicles. The proportion of cars has increased from 58.2% to 68.6% during the last 5 years. It is projected by the city that the number of vehicles registered will be 1 million by year 2002.

Table 2. Number of Vehicles Registered

Year	Passenger Cars	Trucks	Buses	Others	Total
1986	59,099	44,526	13,941	6,894	124,460
1987	75,444	51,332	18,535	7,587	152,898
1988	97,022	58,560	23,530	8,497	187,609
1989	130,501	73,916	28,398	2,121	234,936
1990	167,164	84,016	33,619	2,259	287,058
1991	213,485	94,103	36,462	2,387	346,437
1992	259,489	104,388	40,187	2,692	406,756
1993	309,488	114,944	42,384	2,806	469,622
1994	361,209	125,313	45,305	3,281	535,108
1995	417,752	135,001	47,318	4,007	604,078
1996	476,275	141,138	50,247	4,298	671,958

The Geographical Features

Two geographical features characterize Pusan: its location at seaside and its mountainous terrain. Pusan has been an important port and a logistic center. Both land and sea generate heavy truck traffic within the city area. The mountainous terrain limits the usable land for development as well as for transportation facilities (Figure 2).

The area of Pusan includes a small peninsula, several small islands and many coastal areas. Broadly, the Nakdong River divides the Pusan region into two areas: the east hill area and the west plain area. As urban areas have been developed along the west-side coastline, major roads of the city follow this coastline from north to south. Such form of streets restricts the functioning of Pusan serving as a center of metropolis for surrounding areas. Though the west plain area mostly consists of agricultural farmland, this area has recently been rapidly urbanized through many apartment development projects. The area of urbanized region in Pusan is only 167.75 km^2 (19.7%) and the area of greenery region and the surface of the sea is 682.04 km^2 (80.3%) out of all the urban planning area of 849.75 km^2.

Figure 2. Topographical Map of Pusan

Travel Patterns

Road Inventory

The road network consists basically of ring and radial roads as displayed in Figure 3 which shows major existing surface streets and expressways as well as planned roads. The Bunyoung-Ro and Dongseo freeways (shown as two crossed solid lines within the inner beltway with no numbers) are the backbone of the Pusan freeway system. The total length of these freeways is only 47.7 km. The major arterials in Pusan are Jungang-Ro, which connects northern and southern areas and Sooyoung-Ro, Chungryul-Ro & Gaya-Ro, which connect eastern and western areas. These roads are not shown in Figure 3.

The length of the road system in 1984 was 2,453 km and has increased substantially during the last 15 years. The total length of roadways in Pusan was 3,042 km in 1994 (See Table 3). However, road provisions in Pusan are well below that of Western motorized cities. The area devoted to the roads (road rate) in Pusan is only around 19.7% of total area. Due to the lack of capacity, major roads are heavily congested throughout the day. The length of paved roads has increased as well, increased from 918 km in 1984 to 2,179 km in 1997, as a result of constant investments on road facilities.

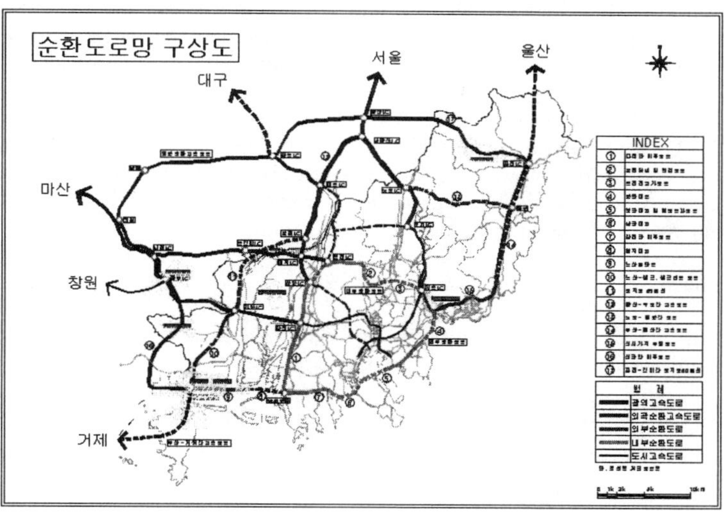

Figure 3. The Road Network in Pusan: Existing and Planned

Table 3. Road Inventory

Year	Length(km)	Road rate(%)	Paved(km)
84	2,453	11.01	918
92	2,647	13.80	1,337
93	2,786	14.40	1,797
94	2,786	14.80	1,838
97	3,042	16.40	2,179

Source: Statistical Year Book of PMCG, 1984-1994

Modes of Travel

Pusan offers a variety of transportation modes. As seen in Table 1, which showed the basic information on each mode, the trend of daily trips has been in line with the rapidly increasing trend of vehicle ownership in Pusan. In Table 4, one can also notice that use of passenger cars has increased rapidly and this is expected to continue to increase in the future.

Metro Line No. 1 extends for a length of 32.5 km and connects most of the urban centers from the north end of the developed area within the city. Line 1 was completed in 1994, and the first phase of Line 2 (Hopo-Soemyon Line) is expected to be completed by March 1999. The number of average daily passengers for Line 1 was 548,000 in 1995, which amounts to over 9% of total daily passenger trips.

Bus is the dominant transport mode in Pusan, carrying the most passengers. In 1997, 42 bus companies operated 232 routes with 2,797 buses. 44.7 % of the total bus routes serve the downtown area of the city. Buses compete with the Metro rather than providing feeder service. Bus patronage is expected to decrease because of both worsening road traffic congestion and increasing Metro services.

Mode Share

Table 4 shows the growth of average daily passengers and freight traffic by different modes, including existing and expected ones.

Table 4. Travel Demand by Mode, 1990-2011

(in thousands)

Class		1990	1997	2001	2011	Average annual rate of increase(%)	
						90-97	97-11
Container freight*		2,273	4,129	6,125	8,677	8.9	5.45
Passenger trips		6,278	6,327	7,163	8,505	0.11	2.18
Mode Trip	Bus (%)	2,941 (46.8)	2,224 (35.2)	2,034 (28.4)	2,220 (26.1)	-3.91	-0.01
	Passenger Car (%)	758 (12.1)	1,357 (21.4)	1,175 (16.4)	1,318 (15.5)	8.68	-0.21
	Taxi (%)	1,358 (21.7)	1080 (17.1)	774 (10.8)	757 (8.9)	-3.22	-2.51
	Metro (%)	482 (7.6)	623 (9.8)	2,149 (30.0)	2,977 (35.0)	3.73	11.82
	Others (%)	739 (11.8)	1043 (16.5)	1,031 (14.4)	1,233 (14.5)	5.05	1.2

* T.E.U (Ton Equivalent Unit)

Between 1990 and 1997, the freight traffic tripled with an annual average rate of 8.9%, increasing more rapidly than the passenger traffic. It is expected that the trend will continue in the future. In 1997, 35.2% and 9.8% of total daily person trips are made by bus and metro, respectively, while private passenger cars carried 21.4% and taxis carried 17.1% of the total passenger trips. The patronage of the metro is expected to increase when the metro Line 2 begins its operation in 1999; then bus share is expected to decrease by 3.0% annually. The share of auto has increased sharply; yet its percentage share in the future levels off though it increases.

Traffic Volumes and Congestion

The Traffic Volume Survey of 1995 conducted by Pusan City indicates that traffic volume is concentrated along a few major arterial roads. They are Jungang-Ro, Gaya-Ro, and Sooyoung-Ro, all of which are major arterials within the city area. This is an inevitable consequence due to the geographical characteristics of Pusan's road network, i.e., lack of by-pass roads and a secondary road network. As shown in Table 5, most of the arterial roads carried similar amounts of traffic for the last three years. This can be interpreted to mean that all the roads are already saturated. Vehicle speed on major arterials during the peak periods averages 27.6 km/h and 10-20 km/h in the CBD.

Table 5. Traffic Volume of Major Arterials

(Unit: pcu/day)

Name	Direction	No. of Lane	1993	1994	1995
Jungang-Ro	S->N	3	20,536	30,043	28,619
	N->S	3	28,966	27,199	26,489
Sooyoung-Ro	W->E	3	27,206	30,246	23,960
	E->W	3	20,007	30,540	24,656
Gaya-Ro	E->W	3	19,225	19.008	16,028
	W->E	3	16,410	18,306	17,010
Chungryul-Ro	W->E	4	27,457	31,220	20,370
	E->W	4	29,301	31,220	24,798

Source: '95 Road Traffic Volume Survey, 1995, PMCG

Conclusions

Pusan is the second largest city in Korea with a total population of approximately 3.9 million. Despite the geographical limitation of the city, the number of cars has been increasing. Although the government has relied rather heavily on highway expansion as its policy for relieving traffic congestion, its supply of transportation infrastructure has been continuously outstripped by a tremendous increase in travel demand.

As a solution to the problem, the Pusan Metropolitan City Government has made "promote of competitiveness of public transit system" as the first priority of transportation policy. Its basic goal is to divert auto and taxi users to the transit system by improving the quality of transit services, while discouraging the use of autos. As one step to achieve this goal, the city plans to integrate the components of the public transportation system, i.e., metro and buses, which are currently managed and operated separately, into one organized system. It is essential for a public transportation system to coordinate services of all modes of transportation to attract passengers to the transit system. To do this, much work has to be done in Pusan.

Most urban developments in Pusan are concentrated along the north-south and the east-west corridors, which therefore can be served by a relatively simple public transportation network. Metro Line 1 already passes through the north-south corridor. The Line 2 is being built through the east-west corridor. The existing bus routes should be restructured to feed metro stations. This simple metro/bus route network will be able to cover the whole area where Pusan citizens reside, and also it will be able to make public transport system accessible within 300 meters of walking distance.

Pusan City is also planning to introduce a combined fare system for metro and bus systems. As a first step, a discounted fare should be applied to the passengers transferring between bus and metro. Later, the fare levels should be determined by travel time spent or distance traveled regardless of mode of transit used. This scheme should be expedited by the adaptation of 'Hanaro Traffic Card' as a transit fare collection system.

References

1. Pusan Urban Transportation Management Project, Pusan Metropolitan City 1996.

2. Urban Planning of Pusan, Pusan Metropolitan City, 1997.

3. The Guide to Foreign Investment in Pusan, Pusan Metropolitan City, 1996.

4. Road Traffic Volume Survey, Pusan Metropolitan City, 12. 1995.

5. Statistical Year Book, Pusan Metropolitan City, 1984-1994.

Urban Public Transportation in Pusan and its Operations and Use by People

Yong Eun Shin[1], Chi-Gook Choi[2], Ik-Doo Choi[3] and Hang Mook Yoon[4]

Introduction

Pusan offers a wide variety of modes of public transport with varying degrees of services and fare levels. The main ones are buses and metro, which together carry 45% of total daily passenger trips. In addition, there are taxis that carry about one million person trips daily.

This paper describes the public transport system in Pusan, discussing the system characteristics, operations, and usage by people. Focus is given on three main modes of transit: bus, metro and taxi. The current situation will be first discussed, followed by a description of each mode. It will also discuss current problems of transit system in Pusan and future plans to deal with the existing problems.

Overview of Public Transport System in Pusan

The first fixed-route transit line in Pusan started with a streetcar line (10.9km) in 1910. By 1935, the network had grown to 21.6 km with a total of 6 streetcar lines, carrying about 80,000 person-trips each day. However, all streetcar

---

[1]Assistant Professor, Dept. of Urban Engineering, Dongeui University, 24 Gaya-Dong, Pusanjin-Ku, Pusan, Korea

[2]Chief Research Fellow, The institute of Policy Development, Pusan Metropolitan City, 1000, Yonsan 5-Dong, Yonje-Gu, Pusan, Korea

[3]Director, Transport Planning Division, Pusan Metropolitan City, 1000, Yonsan 5-Dong, Yonje-Gu, Pusan, Korea

[4]Assistant professor, Dept. of Urban Engineering, Dongeui University, 24 Gaya-Dong, Pusanjin-Ku, Pusan, Korea

lines were closed in 1968 to augment the road capacity. Since then, buses have been the main mode of public transportation system in Pusan, carrying a majority of transit trips. Taxi services began to operate in 1919 and by 1956, 680 registered taxis operated. The Metro Line began to operate in 1985 with the opening of a section of Line 1. In spite of increasing auto ownership, these three modes have been crucial means to move people around the city area. This can be seen in Table 1, which presents the trends of modal shares in Pusan.

Table 1: Daily Passengers by Mode, 1980-1997

(in thousands)

Modes	1980	1985	1990	1993	1995	1996	1997
Bus	2,417 (71.4%)	3,040 (57.0%)	2,941 (46.8%)	2,862 (43.0%)	2,515 (37.6%)	2,606 (39.3%)	2,224 (35.2%)
Metro*	-	213 (4.0%)	482 (7.6%)	515 (7.7%)	586 (8.7%)	597 (9.8%)	623 (9.8%)
Taxi	608 (18.0%)	997 (18.7%)	1,358 (21.6%)	1,261 (18.9%)	1,232 (18.4%)	1,273 (20.8%)	1,080 (17.4%)
Auto	181 (5.3%)	579 (10.9%)	758 (12.1%)	1,189 (17.9%)	1,428 (21.4%)	1,184 (19.3%)	1,357 (21,4%)
Total	3,386 (100%)	5,333 (100%)	6,278 (100%)	6,659 (100%)	6,685 (100%)	6,124 (100%)	6,327 (100%)

Notes: * Opened for service on July 19, 1985.
Source: Pusan Metropolitan City White Book, 1998, Bus Service Improvement Plan, 1997

In 1997, bus and metro made almost 2.9 million person trips each day. If taxi trips are added, the number reaches nearly 4 million, which represents about 70 % of total daily person-trips. Currently, two more metro lines are under construction. The opening of these lines in the near future is expected to increase the combined bus and metro transit share by close to 60%. There are also ferries and commuter rails in Pusan, but as they carry only a small fraction of passengers, they are not discussed here.

Buses

Buses have been the dominant transit mode in Pusan, carrying the largest portion of transit users and providing a comprehensive transit network over the entire city area. In 1997, buses carried over 2.2 million passengers each day, which accounts for over 35% of the total daily passengers. They are owned and operated entirely by private companies, and they offer scheduled fixed-route services at relatively cheap fares and provide long-haul-services on major trunk routes as well as short-haul services (e.g. local and feeder) to major transit stations and major shopping centers.

Operation

Buses in Pusan are classified into four categories: regular buses, all-seated buses, feeder minibuses, and express buses. The all-seated buses are designed to offer a seat for every passenger without standees (45 seats/bus compared to 20 seats/bus for regular buses). Minibuses provide feeder services connecting major stations and dense residential areas. Special buses connect large residential complexes with the Central Business District (CBD). Table 2 summarizes some of the important operational characteristics of buses in Pusan.

Table 2: Bus System in Pusan, 1997

Type of bus	No. of Buses	Riders (prs/day) ('000)	No. of Routes	Daily Vehicle Runs	Avg. Headway (min.)	Avg. route distance (one-way km)
Regular Bus	2,074	1,913	170	32,607	10.7	18.9
All-seated Bus	723	311	62	9,217	10.9	24.9
Feeder Minibus	313	176	115	11,237	-	6.6
Special Bus	20	-	2	76	10-15	-
Total	3,060	2,400	349	53,207	-	-

Source: Bus Service Improvement Plan, 1997

At the end of 1997, there were 42 bus companies and 79 minibus companies operating 3,060 buses running on 349 routes. As seen, the regular buses are most extensive and carry the most passengers, running on 170 routes with 2,074 buses. The buses average route length is 18.9 km and average cycle time (operating and terminal times together), is 160 minutes. There are 723 all-seated buses running on 62 routes with the longest average route length 24.9km. These two types of buses provided long-haul service. For a short-haul transport, there are 115 routes for minibuses with a fleet of 313. Their average one-way route length is only 6.6 km. There are 20 express buses running on 2 routes, each of which has 5 stops between terminals.

The headway varies with the route and type of bus. The headway is mostly constant throughout the entire day. All-seated buses operate 27 night-owl services on 5 routes, starting from about 11:00 PM to 5:00 AM next day. The headway of this night-owl service varies from 5 to 40 minutes, depending on the travel demand.

Fare System

Bus passenger can pay the fare by cash, bus tokens, or Hanaro Traffic Card. On any given route the fare is a flat rate, regardless of the distance passengers travel on the bus line. However, the fare levels vary with type of bus as well as with passenger's age, as seen in Table 3. Night-owl services charge a higher fare (1,300won, $1.10) compared to the base fare of it. To stimulate the use of Hanaro Traffic Card, 5-8% discounted fare is given to the card users. The all-seated buses and express buses have fixed fares for all passengers.

Table 3: Bus Fare System (as of January '98) (Unit: $: 1$=1,200 Won)

Type	Regular bus			All-seated Bus	Feeder Minibus			Express bus
	Adult	Student >10 yrs	Student <10 yrs		Adult	Student > 10yrs	Student <10 yrs	
Base fare (cash)	0.43	0.29	0.17	0.83	0.38	0.24	0.15	1
Discounted	0.42	0.28	0.17	0.81	0.37	0.23	0.15	0.92

Source: Bus Service Improvement Plan, 1997

Facilities

To enhance bus service through reduced operating time, a bus-only lane, for a length of 1-km, was introduced in 1987. The length of the bus-only lane has increased continuously and in 1997, the bus only lanes were in 16 sections of major corridors with a total length of 60.1 km. The length of the 24-hour bus lane is 1.7km in 2 sections, while the length under directional time control is 58.4 km in 19 sections. The benefits resulting from the adoption of bus-only lane are reduced travel time for passengers and operational efficiency for operators. The survey shows the bus-only lane substantially increased the speed of bus travel (Table 4) compared to the general-purpose lane.

Table 4: Bus Operating Speeds, 1996 - 1997

		1996	1997
Op. Speed (Km/hr)	Bus-only lane	19.38	30.44
	General purpose lane	22.24	23.57

Source: Pusan Transportation: 100 years, 1999.

While all buses are equipped with heating and automatic sliding doors, only about 30% of the buses are equipped with air-conditioning. Currently, no buses are equipped with a lift for the handicapped.

Organization and Management

As indicated earlier, all types of buses in Pusan are privately owned and operated by 42 bus companies and 79 feeder minibus companies in 1997. Among them, the number of bus companies that own more than 100 buses are only 8, while 9 companies operate fewer than 40 buses. On average, each company owns 66.6 buses. The bus industry, in general, consists of a small-scale and financially very weak business. About 8,000 employees work in the bus industry, including 6,250 drivers and 540 mechanics.

Financial Aspects

The bus system is operated with a limited government subsidy, with the exemption of toll fees. The capital and operating costs of the bus industry depend entirely on the fare-box revenue. Its total asset is $9.3 million (11.2 billion won), while its total debt reaches $122 million (146 billion won) and on average $2.8 million (3.4 billion won) per company. The debt, overwhelmingly exceeding the asset, indicates the financial weakness of bus industry in Pusan. According to the City of Pusan (see Table 5), a regular bus is running daily with a deficit of about $21 per day, while an all-seated bus is making a profit of $6, not including their capital costs.

System Usage

Pusan City conducts the survey of passenger counts annually. Figure 1 shows the results of the surveys between 1985 and 1997.

The Figure shows that the ridership peaked in 1989 and has decreased constantly, reaching the lowest bus transit share in 1997. In spite of the increased number of buses, the buses have lost a substantial portion of passengers due to the increased modal shares of auto and metro. This situation will be expected to exacerbate with the opening of two new metro lines in near future.

Metro

To cope with the increasing road traffic, Pusan started to construct the metro system in 1981. Compared with the other cities of similar size, Pusan was a late starter in this system. Pusan now has one metro line connecting the east and north sides of Pusan, running in the main traffic corridor. Two more lines are under construction.

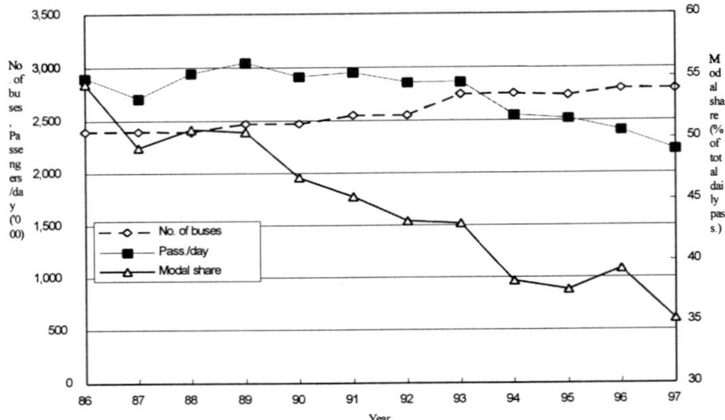

Figure 1. Trends of Bus System 1986-1997

Metro Line 1 started its full operation in 1994 after 14-years of 4-staged construction. With the completion of the first-stage, Pusan metro started its partial operation in 1985. Line 1 has 34 stations and operates on underground and elevated tracks with a combined length of 34.1 kilometers. Lines 2 and 3 will be completed in year 2001, and after completion, the metro network will cover most of the city areas with a total length of 101.1 kilometers and 100 stations. Pusan plans to have its fourth metro line in the future. Line 4 is at an early phase of planning so its alignment and detailed information are not yet available. Table 5 summarizes the information on the metro system in Pusan, including the future two lines.

The amount of construction cost for Line 1 was approximately $813 million (975 billion won), which was shared at 16.3% by the Central government, 10.6% by Pusan City, and 73.1% through loans and bonds. Lines 2 and 3 will be funded by the Central Gov't and the Pusan City, sharing equally.

Operation

The headway of the Metro Line 1 is between 3.0 and 3.5 minutes during peak periods. During the off-peak periods, it increases to 5.5 minutes. On an average working day, a total of 408 trains run in both directions, while 350 trains run during weekend days and holidays. Its one-way operation time is 61.3 minutes and average operating speed is about 34.1km/h, which is much faster than that of road traffic. Line 1 operates for 19 hours 10 minutes a day, starting at 05:20 AM and ending at 00:30 AM the next day.

PUSAN METRO SYSTEM MAP

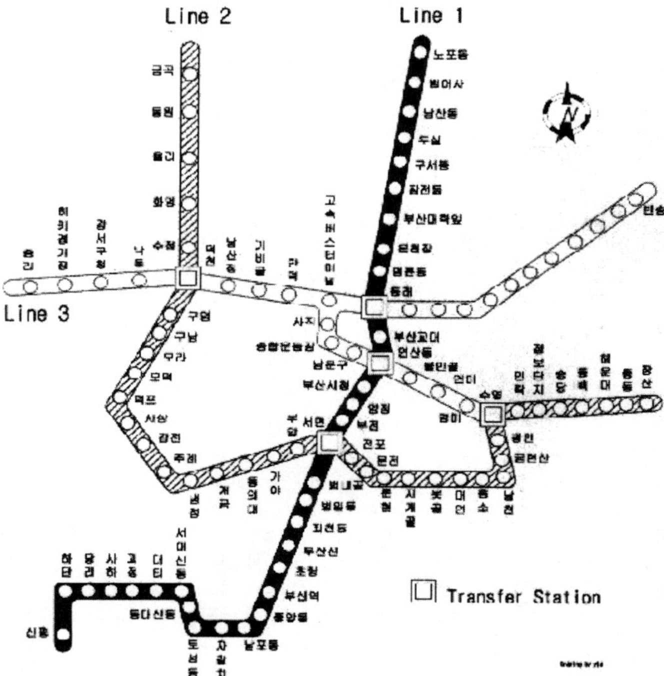

Figure 2. Schematic Network of Pusan Metro System

Fare System

Metro employs a distance-based fare system with varying fare levels by age (regular and student). The metro riders can pay the fare by cash, metro tickets, or Hanaro Traffic Card. Base fares for adults and students with less than 10-km trips are $0.39 (450 won) and $0.19 (230 won) respectively and with a longer than 10-km trip fares are $0.42 (500won) and $0.21 (250 won). With the use of the Hanaro Traffic Card, those fares are discounted by 15%.

Table 5: Metro System in Pusan

	Line 1	Line 2*	Line 3*
Length (km)	32.5	39.1	29.5
N. of stations	34	39	27
Const. period Stage:	1981 – 1994 1: 16.2-km, '81-'85 2: 5.4-km '83-'87 3: 4.5-km '84-'90 4: 6.4-km '90-'94	1991-2001 1: 22.4-km, '91-'98 2: 16.7-km, '94-'01 3: 11.1-km, '98-'01	1997-2001 1: 18.3-km, '96-'01 2: 11.2-km, '96-'01
Costs US $ (won) Share	$ 813 M.(975.1 B.) Central Gov't: 16.3% City Gov't: 10.6% Borrowings: 13.1%	$2.11 B. (2530.7 B.) Central: 50%, City : 50%	$1.17 B. (1401.5 B.) Central: 50%, City:50%
Rolling stock	360	-	-

Notes: *Expected figures; Source: Annual Report on Operation, Pusan Urban Transit Authority (PUTA), 1998.

Facilities

The Metro system is equipped with numerous state-of-the-art features, such as a full weather control system, Automatic Train Protection (ATP), Automatic Train Operation (ATO), and Automatic Train Supervision (ATS). The system has one-man train operation. Many stations are equipped with facilities for the handicapped; 31 stations have a total of 110 wheelchair lifts; all stations have wheelchair ramps; 12 stations have bathrooms for use of the handicapped. There are 9 stations with 38 escalators. Twelve stations have park-and-ride facilities with a total capacity of 3,044 spaces. There are bike-rack facilities at 12 stations with a total capacity of 710 bicycles.

Organization and Management

In order to properly operate and construct the Metro system in Pusan, Central Government established the Pusan Urban Transit Authority (PUTA) in 1988. Currently the PUTA is fully in charge of operation of Metro Line 1 and the construction of metro Lines 2 and 3. The total staff is 2,566, comprised of 6 directors, 440 other staff, and 770 field workers. In addition, there are 34 transit police officers. The general manager of PUTA is appointed by the Central Government, and the PUTA consists of an auditor, 14 departments, 35 sections and 9 field offices.

Financial Aspects

The Central Government subsidizes the operating expenses of PUTA, while both the Central and City Governments subsidize the capital program. PUTA suffers from a huge deficit. As the subsidies are not sufficient, PUTA has to borrow from banks for daily operations as well as the construction of Metro Lines. The operating ratio of Metro Line 1 is approximately 0.21 in 1997, meaning that the fare-box revenue covers only 21% of costs of operation (Table 6).

Table 6: Revenue and Expenditure Budget of PUTA 1997

($ million – won million)

Revenue		Expenditure	
Total	$773 (927.1 bil)	Total	$773 (927.1 bil)
Operation (Line 1):	$510 (612.2 bil.)	Operation (Line 1)	$510 (612.2 bil.)
Fare box	$106 (127.7 bil.)	Operating cost	$151(181.2 bil.)
Central Gov't	$111 (133.5 bil.)	Interest & principal	$355 (425.5 bil.)
Borrowings	$293 (351.0 bil.)	Other	$4.5 (5.5 bil.)
Capital	$262 (314.9 bil.)	Capital:	$262 (314.9 bil.)
Central Gov't	$124 (148.6 bil.)	Line 2:	$172 (206.4 bil.)
City	$126 (150.9 bil.)	Line 3:	$77.6 (93.1 bil.)
Borrowings	$3.7(4.4 bil.)	Other	$9.2 (11.0 bil.)
Other	$9.2 (11.0 bil.)		

Source: Annual Report on Operation, Pusan Urban Transit Authority, 1998

With the fare box revenue, the PUTA can pay only a fraction of the interest and principal of the loans, which amounts to about 70% of the total operating expenses. At the end of 1997, total asset of PUTA amounts to $1.5 billion, while the total debt is about $2.3 billion, which indicates the financial instability of the PUTA.

The unpredictability and uncertainty of capital funding for future Metro Lines distort Pusan's public transit future. The funding situation for construction of Lines 2 and 3 is especially disturbing. The first section of Line 2 is expected to open in March 1999, yet the second section of Line 2 is continuously delayed due to the lack of capital. Moreover, the construction of Line 3 has no guaranteed fund yet.

System Usage

In 1997, Metro Line 1 carried 598 thousands passengers each day, which accounts for 9.8% of total daily person trips. Since the opening of the Metro Line 1, the ridership has continuously increased, as shown in Figure 3.

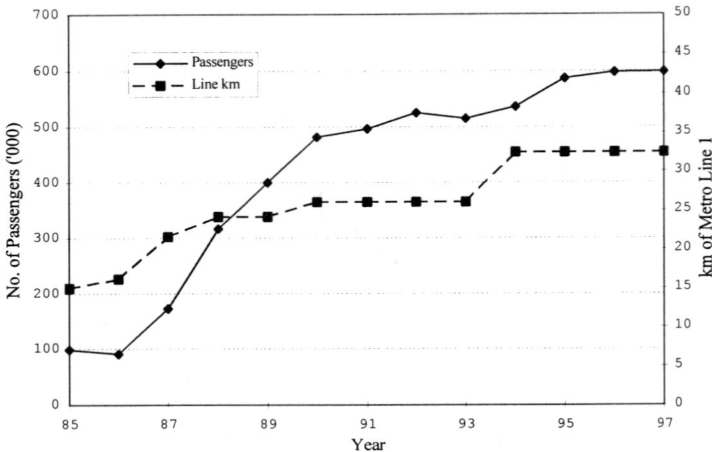

Figure 3. Metro Line 1 1985-1997: daily riders and lengths of Metro Line

When the first section of Line 1 was opened in 1985, only 80,000 passengers rode the Metro each day. But the number has rapidly increased until the final stage of construction. After its completion in 1994, the growth has stabilized. Figure 4 shows the detailed trend of Metro daily passengers from 1985 to 1997. It is expected that 30% of the total daily person trips will be made by Metro when Lines 2 and 3 open.

Taxi

Since taxis provide a semi-paratransit service rather than a regular fixed-route service, taxis do not belong to a typical public transit mode. Yet taxis in Pusan play an important role in carrying passengers by providing a personalized door-to-door, convenient and fast service. A brief description of characteristics of operation and usage of taxi in Pusan follows.

In 1997, taxis carried over 1.08 million persons each day and the fleet comprises 23,272 registered vehicles. In terms of types of ownership, taxis in Pusan are divided into 2 groups: taxis owned by taxi companies, and taxis owned by individuals. There are total of 106 taxi companies operating 11,251 taxis and 12,297 taxis owned by individuals. They can also be classified into regular taxis and special taxis. A special taxi equipped with a cellular phone charges a much higher fare for providing an on-call service. The target passengers of special taxis usually include foreigners and tourists, though they are also open to the general public.

The taxi industry is strictly regulated by the Government, such as licensing of drivers and entry of new vehicles, drivers' training, days of operation and fare levels. Taxi fares differ considerably between regular and special taxis. For example, for a regular taxi, the fare starts with $1.10 (1,300 won) with a base distance of first 2 kilometers. After the first 2 km, the fare is computed by a combination of time spent and distance traveled (Table 7). There is an additional charge for the night-owl services from midnight to 4:00 AM.

Table 7: Taxi Fares: Special Taxi and Regular Taxi, 1997

(Unit: $)

	Base fare	After 2 km	
	(First 2 km)	Distance fare	Time fare
Regular taxi	1.08	0.08/247 m	0.08/51 sec.
Special taxi	1.00	0.16/250 m	0.16/60 sec.

Figure 4 shows the number of daily passengers carried by taxi from 1986 to 1997. The ridership reached its peak in 1993, after that it has declined, while the number of registered taxis has continually increased. In particular, there was a sharp decline in 1997. This is due to the recent economic crisis in Korea which made people avoid taxis, the high fare transport mode.

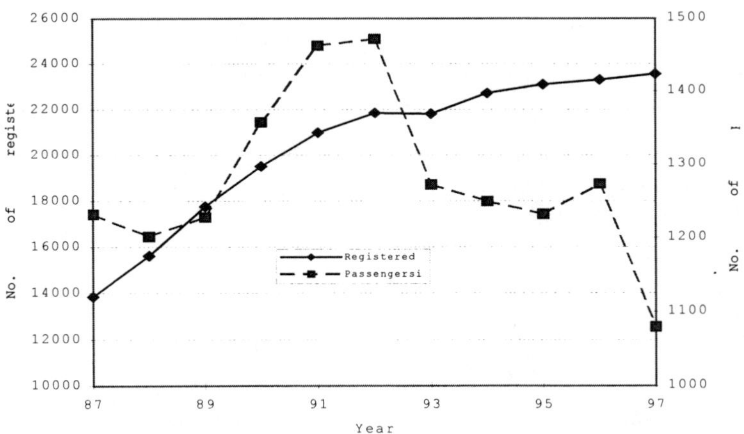

Figure 4. Taxi 1987-1997: daily passengers and registered taxis

Problems and Future Plans

In a city like Pusan, where there is no prospect of expanding the road facilities, public transport must play an important role in moving people around and carrying a large portion of passenger travel. There are, however, numerous problems that need to be fixed or improved. The major ones can be summarized as follows:

1. Uncoordinated services among different modes.
2. Lack of rail-based transit systems
3. Decreasing bus transit ridership
4. Uncertain financial resources for the capital program of rail system

According to the scenario proposed for the Year 2011 Plan, the number of total person trips in the city will increase to 15 million trips a day by the year 2011. Pusan City has proposed a package of a general transport plan for year 2011 to cope with the problem. The package implies that by year 2002, the modal share of buses and metro will be about 60% and will increase to more than 70% by year 2011. In the short term, the city will focus on improvement of the quality of existing transit services: widening the application areas of Hanaro Traffic Card, rationalizing the bus routes by integrating them with the metro lines, speeding up the bus services by introducing more bus-only lanes.

In the long-term, the city plans to expand rail-based transit systems by introducing light-rail lines and commuter rail lines. By year 2011, Pusan will have 5 light-rail lines with a total length of 89.5 km and 3 commuter rail lines. A federation of bus companies is planned to deal with their financial dilemma.

It is obvious that Pusan's density of population, employment distribution, as well as its geographical limitation require an efficient form of public transport system. The policy package for year 2011 is formulated to establish such an efficient transport system.

References

1. Pusan Urban Transportation Management Project, Transportation and Environment Research Institute Ltd., A Report to Pusan Metropolitan City, 1996.
2. Pusan Metropolitan City White Book, Pusan Metropolitan City, 1996, 1997, 1998.
3. Bus Service Improvement Plan, Pusan Metropolitan City, 1997.
4. Annual Report on Operation, Pusan Urban Transit Authority, 1998.
5. Year 2011 Plan for Pusan Metropolitan City, Pusan Metropolitan City, 1999
6. Pusan Transportation: 100 years, Pusan Metropolitan City, 1999.

Innovative Electronic Fare Collection: Hanaro Traffic Cards

Yong Eun Shin[1], Hang Mook Yoon[2], Chi-Gook Choi[3]

Abstract

Pusan adopted a 'Hanaro Traffic Card' system for its transit system, using contactless smart card technology. It has many notable and unique features. It is a contactless, prepaid and rechargeable card system that is designed to work for multiple modes of transit operated by multiple operators. The City Government spent almost 3 years to initiate and develop the system through the coordinated efforts of public and private sectors. Currently, the system is installed on most transit modes in Pusan – bus, metro, taxi, and minibus. The system is generating numerous benefits for transit users as well as operators. The purpose of this paper is to describe the Hanaro Traffic Card system and to discuss the process and background work of system development, the system features and resulting benefits to all related parties, including users, operators and the city.

Introduction

Fare collection method is one of the important elements that affect transit operations. It directly determines quality of service for the users and operating costs for the operators. In recent years, a number of cities across the world have pursued advanced electronic technology in fare collection for their transit systems in search for greater efficiency and operational effectiveness as well as for attraction of more transit riders by offering enhanced transit services.

---

[1]Assistant Professor, Dept. of Urban Engineering, Dongeui University, 24 Gaya-Dong, Pusanjin-Ku, Pusan, Korea

[2] Assistant Professor , Dept. of Urban Engineering, Dongeui University. 24 Gaya-Dong, Pusanjin-Ku, Pusan, Korea

[3] Chief Research Fellow, The institute of Policy Development, Pusan Metropolitan City, 1000, Yonsan 5-Dong, Yonje-Gu, Pusan, Korea

In February 1998, Pusan City adopted a Hanaro Traffic Card (HTC) system, employing a smart card technology, to achieve the above purposes. The system is noteworthy in that it is the first integrated fare payment system in the world. It applies smart card technology to multiple transit modes operated by multiple service providers. The card also applies to other transport facilities, such as toll fees and parking charges. The system was developed and is operated through the coordinated efforts of the public and private sectors, overcoming obstacles and conflicts among the organizations involved in the process of system adoption and operation. The HTC system has been a great success. As of July 1998, less than 6 months of its full operation, a total of 1.6 million cards have been sold, and 55% of subway travelers and more than 45% of bus riders are using the cards.

This paper describes the Hanaro Traffic Card System in Pusan. It discusses the process and background work of system development and the operation of the Hanaro Traffic Card system. It includes the description of system structure, present usage, and the costs and benefits resulted from the system operation.

Introduction of Hanaro Traffic Card System

Note that the term, Hanaro, means "by only one." As this word indicates, the Hanaro Traffic Card (HTC) system was originally developed as a multi-purpose card system to be used not just for travel related fare payments but also for payments of other merchandises and services. The Hanaro Traffic Card is a dual-interface (both contact and contactless), pre-paid, value-stored and rechargeable card embedded with an Integrated Circuit (I.C.) chip, employing Radio Frequency (RF) technology. The contactless feature allows it to remain inside the purse or wallet during the transaction with a maximum working distance of 10cms between the card reader and card.

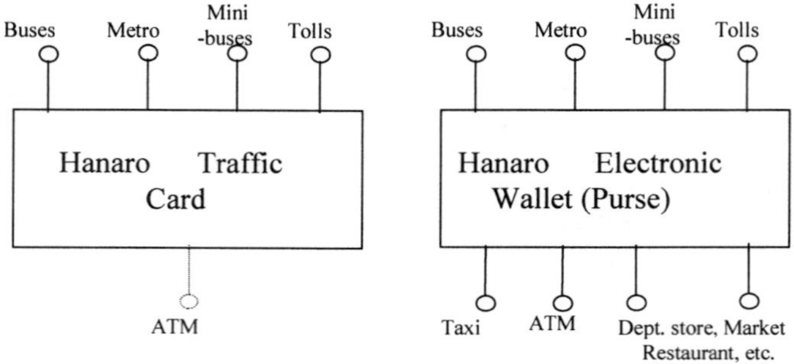

Figure 1. Uses of Hanaro Traffic Card and Hanaro Electronic Wallet

The HTC system provides two types of cards: Hanaro Transit Card and Hanaro Electronic Wallet (or Purse). As seen in Figure 1, both cards can be used for tolls, buses, feeder minibuses and metro, while the Wallet is a credit card that can be used for purchasing merchandise as well as for paying taxi fares. Note that the Traffic Card can be used for an Automated Teller Machine (ATM) of a bank only if the user makes a contract with the bank. There are three different colored transit cards by coded fare level: yellow card for adults, blue for college students and green for 7th to 12th grades. Buses display signs indicated 'card reader installed' on the front and rear sides. Inside the buses there are instructions for the use of Hanaro Card and each recharge terminal also displays a sign indicating the 'Hanaro Traffic Card Recharge Center.'

Though the city formulated the basic policies and plans along with numerous regulatory measures that would support the system development and implementation, a number of organizations having different interests were involved in developing the system. Introducing an integrated payment system like the HTC system was indeed a complex undertaking with a variety of public and private organizations involved. As seen in Table 1, there were numerous participants with specific responsibilities in the system development and operation.

Process of System Development

The Hanaro Traffic Card system was first conceived and initiated by the City Government in July 1995, as a means of achieving city's long-term goals that will direct the City towards the 21st-century. It took over two and half years of rigorous work by the participants in the system until it was fully implemented. Table 2 summarizes some of the important events chronologically that happened during the system development.

To implement the plan efficiently and compromise the conflicting interests of different organizations, the Hanaro Traffic Card Executive Council (HTCEC), which includes all relevant parties, was established. This was a key organization in setting up the detailed plan for the system implementation, including organizational, financial and technical plans for the full operation of the HTC system.

Before its full operation, the system was first tested in a trial from April to September 1996 on one bus line (16 buses), metro (3 stations) and taxies (10). Then an 8-month full test (installing the system on 517 buses, all metro stations and 5,500 taxis) was performed. During this period, numerous technical problems were found and then fixed. After the successful completion of this test, the system started a partial operation on September 1997. The system began full operation on February 1998, including all buses. The system was later expanded to all minibuses and tollgates on 3 urban expressways. The full operation of tolls and feeder minibuses was started in August 1998.

Table 1. Participants, Organizations, and Responsibilities

System participants	Organizations	Major responsibilities
Government	Pusan Metropolitan City Government	♦ Develop policies ♦ Set schedule of system development ♦ Support system development ♦ Compromise the conflicts among the parties
Operators	-Bus Union, -Taxi Union -Pusan Urban Transit Authority (PUTA)	♦ Install card readers; ♦ Collect and transmit transaction data
Management bank	Former Dong-Nam Bank (Korea Housing Bank and Pusan Bank)	♦ Loan service, Pay card issuance cost ♦ Financial transaction, Card issuance, ♦ Recharge system, On-line system
Seconday management organization	Korea Information and Communication Co. LTD. (KICC)	♦ Design system and specifications; ♦ Operate the enitre network system; Design Value-Added Network (VAN) system.
Private enterprises	Kyungduk Co. LTD. Korea Data Comm. Co. LTD.	♦ Develop card readers; ♦ Establish system network; ♦ Install terminals.

Source: Adoption of Hanaro Traffic Card System, Pusan Metropolitan City, 1997

Issues and Obstacles

There was a great deal of concerns for the city on adopting such a new technology as electronic fare media because there was no existing system that could be used as a model for the development of the Hanaro Traffic Card system. In addition, there were issues such as riders' acceptance, financial management among various organizations, and expenses associated with the system. It is no surprise that, in the process of introducing the HTC system, numerous obstacles, to be resolved before the system adoption, were encountered.

Table 2. Diary of Hanaro Traffic Card System Development

Date	Major events	Contents
July 1995	Devised a plan for system development	Base plan/ Appointed a managing organization: Dong-Nam Bank
Sept 1995	Organized Hanaro Traffic Card Executive Council (HTEC)	Detailed system plan
March 1996	Selected manufacturers of card readers and system developer	
Aug 1996	Held Hanaro Steering Council meeting	Concluded an Agreement of Hanaro Traffic Card Operation
Dec 1996 July 1997	Full Test	All metro stations (581 terminals) / Bus (517), Taxi (5,500)
Sept 1997	Partial operation	All metro station (581 terminals) / Bus (517), Taxi (5,500)
Feb 1998	Full operation	All metro station / All buses (3,003) Taxi (5,500)
Aug 1998	Full operation:	All minibuses (341) Urban expressways (28 gates)

Source: Development of Hanaro Traffic Card System, Pusan Metropolitan City, 1998

The major opposition against the HTC system came from transit companies. The concern about exposure of their financial transactions was the main reason. As a voluntary financial and transaction report had been a custom for years, companies were reluctant to expose their financial status. The bus companies, in particular, were strong opponents of the Hanaro Project. The city played a major role, resolving this impasse through strong regulatory supports and continuous persuasion, and emphasizing the system's benefits to the companies. In addition, three issues attracted particular attention at the initial stage of the project.

- Who is going to pay for the cost of the system? How will that cost be shared?
- Is the technology reliable and workable?
- Are transit riders going to use the card and to continue to use it?

The city was a major force in resolving the first issue. The city made sure that there would be regulatory supports for service providers, which would allow them to raise all fares to cover the costs expected to be incurred from the system operation. The total for system development and installation cost was approximately \$26.5 million (31.8 billion won). The money was mainly paid by private sectors: 5% by the city, 11% by the Pusan Urban Transit Authority (PUTA), and 84% by the private sectors (participating bank and transit operators).

There should be absolute user faith on the new system developed and the technology for the success of the system. Several tests of the system were thus performed in order to ensure the system reliability. As to the reliability of network and its costs, former Dong-Nam Bank convinced the other parties that their existing banking network could be readily modified to accommodate the entire Hanaro network system.

Riders' acceptance of the system was a questionable issue since no system was in use in other cities. No one could anticipate how the users would react to this system. The operation of the system so far shows that this was an unfounded fear. According to the survey performed by the City on April 1998, 76.5% of citizens are satisfied with the system and 85.8% of them consider the system a significant improvement tool for transit system.

Description of System Structure

The HTC system is working with a network of computer terminals that are connected to the mainframe computer located at the Hanaro Traffic Information Center (HTIC). Broadly speaking, the HTC system consists of following five sub-systems: 1) card issuance system; 2) recharge system; 3) card payment and reimbursement system; 4) finance management system; and 5) finally Value-Added-Network (VAN) system composed of a network of computer system which connects all of the above systems together, including distribution system, Data and Information System.

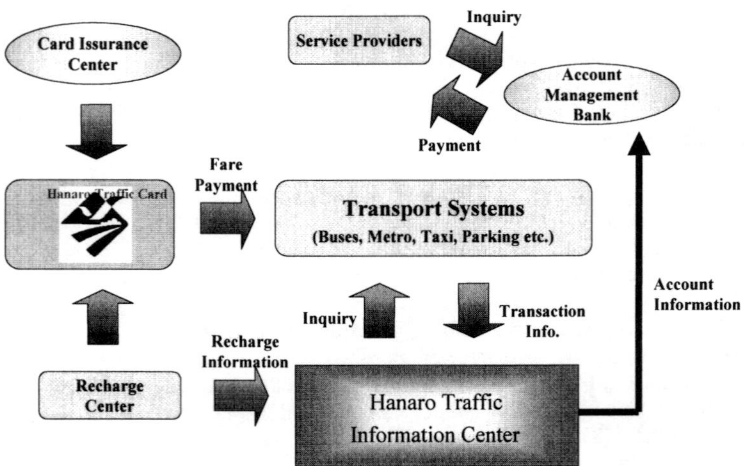

Figure 1. Schematic diagram of HTC system

Figure 1 schematically shows how all of above sub-systems work as one system. To readily understand the system shown in the figure, let us consider the case of bus rider and bus company. Bus riders purchase the card (card cost: 2,000 won - $1.67) from card issuance center, such as kiosks, banks or general stores, and store a money value up to 70,000won ($58.30) for regular and 30,000won ($25) for student riders. Card readers installed on buses deduct appropriate fare values from the cards each time they are used. Information stored on the card reader is transferred to the terminal at bus company, and then the company sends it to the mainframe computer at the Hanaro Traffic Information Center (HTIC). At the same time, the recharge information stored on the terminal at recharge center is also sent to the HTIC. HTIC then sends all the transaction information to the management bank, which then pays the bus company the amount of money corresponding with the amount of services provided by the company. HTIC handles not only all transaction data, but also its mainframe computer generates the traffic information associated with the transit riders, while underlying financial and operational data are reported to the operators.

Usage of Hanaro Traffic Card

As of July 1998, the card readers are installed on the entire bus fleet (3,003) and feeder minibuses (348), all metro fare gates (50), and more than 8,000 taxies that are owned by individual drivers. There are 949 card selling and recharge terminals installed at kiosks (former token selling places), general stores, and management bank (Dong-Nam Bank). This number is increasing day by day as a result of increasing card demand.

The citizens' reaction to the system is very positive, and thus the uses of Hanaro Cards are also increasing day by day. Figures 2 and 3 present the trends of the number of cards sold and the numbers of metro and bus riders for the first 8 months of system operation.

It can be seen from the figures that since the beginning of system operation, the numbers of cards sold and card users are constantly increasing. However, the use of cards for metro and buses were either leveling off or slightly decreased after June '98. This is attributed to the 10%reduction in total passenger trips in Pusan as a result of the recent economic crisis.

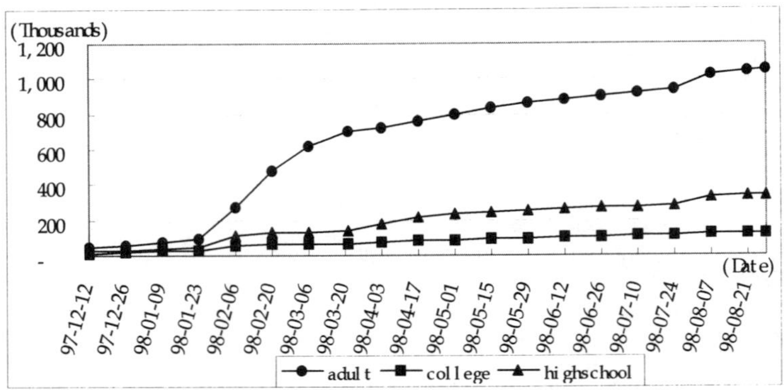

Figure 2. Number of cards sold: Dec. 1997 - Aug. 1998

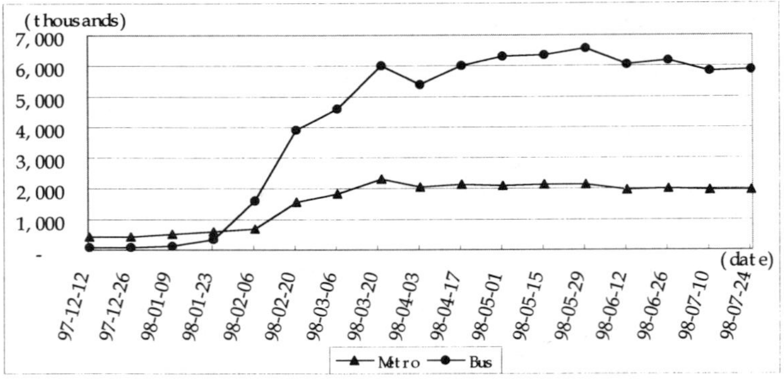

Figure 3. Uses of Hanaro Cards: metro and buses (Dec. '97 - Aug. '98)

As of September 1998, about 45% of bus riders, 56% of metro riders and 28% of minibus riders are card-users, while the card use for taxi trips is very limited (Table 3). Taxi drivers are concerned about the low usage problem, since they also paid for the cost of terminal installation.

Table 3. Hanaro Card Usage - Sep. 1998

Transport System	Daily riders using Hanaro Card (A)	Total Riders (B)	% A/B	Operation
Bus	1,000,000	2,220,000	45.0	Full
Metro	350,000	620,000	56.5	Full
Taxi	30	1,080,000	-	Partial

Source: Development of Hanaro Traffic Card System and its Effectiveness, Pusan Metropolitan City, 1998

The Costs and Benefits

To develop and operate a complex integrated fare system like the Hanaro Traffic Card, the city should invest a great deal of money to receive the great deal of benefits associated with it. The experience of the Hanaro system operation clearly demonstrates this. Indeed, there are numerous invaluable monetary as well as non-monetary benefits to all related parties. Table 4 presents major costs and benefits of the Hanaro Project.

Table 4. Costs and Benefits of Hanaro Traffic Card System

Participants	Costs	Benefits
Pusan City Gov't	·Subsidy for the cost of card purchase: $ .75/ea.	·Enhanced Policy formulation processes ·Enhanced fare levels ·Reduced traffic jam at toll gates ·Survey cost reduction : $87,000/yr
User	·Card cost : $1.67/ea.	·Fare discount: $7.70/persons/yr
Bus Companies	·Card reader : $800/veh ·Transmit System : $6,000/ea.	·Reduced Manpower (cash, token, ticket handling) $812,000/yr ·Token vending machine : $130,000/ea ·Token and ticket products cost : $60,000/yr ·Reduced frauds.
Taxi Companies	·Card reader & Cellular phone : $500/veh	·Cash handling manpower : $95,500-191,000/yr ·Profit for phone device : $5,600,000/yr
Pusan Urban Transit Authority	·Metro Hanaro System Installation - Line 1 : $2,464,000 - Line 2 : $1,087,000 - Line 3 : $797,000	·Ticket production cost : $303,000/yr ·Manpower reduction - Line 1 : $1,480,000/yr - Line 2 : $3,220,000/yr - Line 3 : $1,970,000/yr ·Ticket vending machine : $372,000/ea.

Source: Development of Hanaro Traffic Card System and its Effectiveness, Pusan Metropolitan City, 1998

Table 4 indicates that there are numerous benefits that may exceed the costs. For the City Government, the data and information generated by the Hanaro system eliminated the costs of an annual survey. The better quality and more accurate information also enhanced the decision making process, which enabled decision-makers to formulate more rational policies.

Before the adoption of the system, the fare payment methods for transit system varied with modes. Bus riders can pay fares by cash, bus tokens, or bus tickets, while metro riders pay cash at booths to get train tickets, or by train pass (monthly, weekly). As bus drivers do not carry change, bus riders should carry exact change, tokens or tickets. It is thus not difficult to conceive that for the passengers, who regularly use the card for their transit rides, the biggest benefit is its convenience of fare payment. In other words, transit became more convenient through the following service improvements:

- Elimination of the burden of carrying cash and exact change and buying bus tokens and metro tickets;
- The same card can be kept and reused;
- Faster boarding; and
- Multiple fares can be paid on the card, such as for friends travelling together;

Besides, the cardholder is entitled to get a 5% discount on a regular fare when they use the Hanaro Card for bus and metro. This is a direct monetary benefit to the transit riders.

For the transit operators, the single most important benefit is that they can reduce the costs of coin handling through the reduction of manpower and can increase efficiency of management of bookkeeping and financial transaction. In addition, tokens and tickets production cost has been substantially reduced and ticket vending machine is becoming slowly obsolete. Further, since all fares are pre-paid, the interest accrued from the prepaid money is additional income to the operators. Furthermore, the information collected from the system allows more rational analysis and evaluation of the service planning and fleet management. Smart cards are also a means of limiting fraud. Being difficult or impossible to duplicate, unlike paper tickets, the system can thus eliminate fraudulent travel.

The Problem and the Future

So far the system has performed better than expected, causing no major technical problems and making almost every organization involved satisfied with the system. The citizens' reaction to the system is in general positive. Some areas that needs to be corrected and improved for the further utilization of the system potential, are summarized below:

- Absence of a long-term plan integrated with Hanaro System;
- Shortage of card selling and recharge centers
- Very limited usage for taxis;
- No discounted fares for the intermodal transfer passengers;
- Potential problem of dealing with unusable cards;
- Need for integrated card with regional transit system.

Pusan City attempts to resolve these problems by improving the system and widening the application areas of the system. Some major ones are: 1) the Year 2011 Plan includes the Hanaro system as one of the major components that will be integrated with a city-wide ITS plan; 2) Pusan City plans to introduce discounted fares for intermodal transfer trips among different modes; 3) a combo-card which will combine both electronic wallet and transit card into one card will be introduced; 4) more recharge centers will be added; 5) finally, city government plans to widen the application area of the Hanaro System to transport facilities, such as toll, public parking facilities and other transit modes, such as ferries and commuter train.

As shown in Figure 4, in the future there will be a combo-Hanaro Card that integrates the Hanaro Traffic Card and the Hanaro Electronic Wallet with many more application areas.

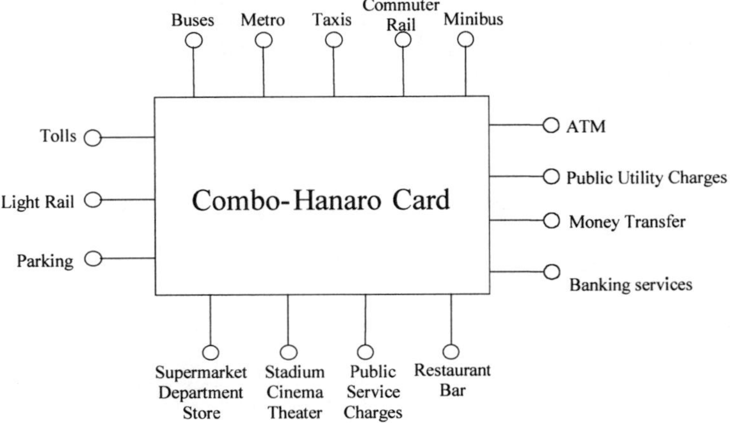

Figure 4. Future Combo-Hanaro Card

Conclusions

The success of the Pusan Hanaro Card system is a result of the strong efforts and resolution to implement the system not just by City Government but also by the private parties including service operators. As indicated, there were obstacles and oppositions against the system. Coordinated efforts of various organizations overcame all these obstacles. Keep in mind that the system is still in its initial stage, and the future of this system is promising in many fields of transportation. Finally, the experience of this system would be very helpful to many cities that plan to introduce an integrated fare system, using a smart card technology. Pusan City showed that it is possible and is eventually very helpful to all the related parties.

Reference

1. Impact Analysis of "HANARO" Card, Transport Policy Study, Pusan Metropolitan City, 1996.

2. Adoption of Hanaro Traffic Card System, Pusan Metropolitan City, 1997.

3. Development of Hanaro Traffic Card System, Pusan Metropolitan City, 1998.

4. Year 2011 Plan for Pusan Metropolitan City, Pusan Metropolitan City, 1999.

5. Transportation Base Plan for Pusan Metropolitan City, Preliminary Report, Pusan Metropolitan City, 1998.

LEADERSHIP AND PROFESSIONAL DEVELOPMENT FOR PUBLIC TRANSPORTATION

Benjamín Colucci, M.ASCE[1]
Antonio González-Quevedo, M. ASCE[2]
Lydia Elena Mercado-Sherman[3]
Kenneth Kruckemeyer, M.ASCE[4]
Nigel H.M. Wilson[5]

ABSTRACT

Since 1994 the Puerto Rico Department of Transportation and Public Works has collaborated with the University of Puerto Rico (UPR) and the Massachusetts Institute of Technology (MIT) in an innovative leadership and professional development program. The goal of the program is to develop the technical professional human resources required for planning, designing, building, operating and maintaining an urban rail system. Graduate and undergraduate students from UPR & MIT are selected to participate in two special short courses, work internships, a visit to a transit system and assisted research. Upon graduation the students are given priority for available positions with project consultants and contractors.

BACKGROUND

A rail transit project improves accessibility and mobility of a metropolitan region and it can also be a unique opportunity to develop the next generation of rail transit professionals. Puerto Rico's Tren Urbano Project has strategically taken advantage of the environmental, design and construction phase to involve university students in research, professional development and educational activities (Pesquera, 1999). The

---

[1] Civil Engineering Professor at the University of Puerto Rico and Co-Director of the Transportation Technology Transfer Center, University of Puerto Rico, P.O. Box 9041, Mayaguez, PR 00681.
[2] Civil Engineering Professor at the University of Puerto Rico and Director of the Civil Infrastructure Research Center, University of Puerto Rico, P.O. Box 9041, Mayaguez, PR 00681.
[3] Technology Transfer Manager, Tren Urbano Office, Frederic R. Harris, Inc., 398 Jesús T. Piñero Avenue, San Juan, PR 00918
[4] Tren Urbano Design Consultant and Research Associate at Massachusetts Institute of Technology, Room 1-232, Cambridge, MA 02139
[5] Professor of Civil and Environmental Engineering at Massachusetts Institute of Technology, Room 1-240, Cambridge, MA 02139

Tren Urbano UPR-MIT Professional Development Program (referred to as the UPR-MIT Program) is an interdisciplinary, bilingual, multi-cultural, multi-campus program sponsored by the Puerto Rico Department of Transportation and Public Works (DTPW), in conjunction with the University of Puerto Rico (UPR) and the Massachusetts Institute of Technology (MIT).

In 1993 as the Tren Urbano Project was being shaped, the Secretary of DTPW, Dr. Carlos I. Pesquera, realized that the long-term success of the rapid transit system in San Juan would depend on the development of highly skilled rapid transit professionals and technicians. Dr. Pesquera along with the Highway and Transportation Authority (HTA) Executive Director, Dr. Sergio L. González, both former university professors, understood the relationship between academic formation, research and professional practice. At the suggestion, of DTPW consultant and former Massachusetts Secretary of Transportation, Frederick P. Salvucci, DTPW/HTA decided to integrate a technology-sharing component as part of the overall Tren Urbano Project. The UPR-MIT Program embodies the high value placed by the project on five primary areas: real-world learning experience, applied research, multidisciplinary perspective, leadership development and cross-cultural interaction.

The primary objective of the Tren Urbano UPR-MIT Professional Development Program is to develop local professional leaders in transit system planning, design, construction and operation. The secondary objectives are (1) to strengthen the educational and research programs in key infrastructure-related disciplines at the University of Puerto Rico; (2) to establish a model for cross-disciplinary cooperation among UPR faculty in engineering, architecture, and urban planning, working together with experts from government and industry; and (3) to develop a collaborative relationship between the University of Puerto Rico and the Massachusetts Institute of Technology.

KEY PARTNERS

The UPR-MIT Program is sponsored by the Department of Transportation and Public Works (DTPW) of the Government of Puerto Rico, and the Puerto Rico Highway and Transportation Authority (PRHTA) through the collaboration of three entities: 1) The Tren Urbano Project, 2) The University of Puerto Rico and, 3) The Massachusetts Institute of Technology. Each partner has a specific role and function with specific objectives and benefits.

KEY PARTNERS

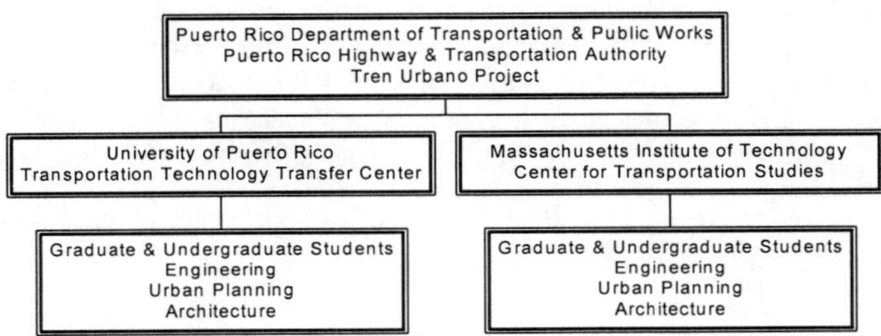

The Tren Urbano Project

The Tren Urbano Project is a living laboratory for the formation of young professionals in rail transit. The project itself is the context and object of study and research. The project also provides internships for engineering, architecture and urban planning students.

From the project side, the Tren Urbano Office manages a subcontract with each university, coordinates program activities with the universities, plans and implements the work internships, serves as liaison for student inquiries, and evaluates the effectiveness of the program.

The Universities

The two universities -- UPR and MIT -- recruit and select the students; plan and implement two short courses; guide student research; provide faculty advisors to students; provide undergraduate and graduate research assistantships; plan and implement site visits to operating transit systems; and evaluate the effectiveness of the program.

The University of Puerto Rico (UPR)

The Civil Engineering Department at UPR through the Transportation Technology Transfer Center coordinates, plans and executes all program-related activities (since January 1999) and through the Civil Infrastructure Research Center prior to 1999. Three schools participate in the program: 1) the School of Engineering located at the Mayagüez Campus, 2) the Graduate School of Planning and, 3) the School of Architecture, both located at the Río Piedras Campus. Designated professors from the three schools advise the students on their research projects and also participate in all technical, social and cultural activities. Faculty participation provides academic rigor to the activities. UPR students are from undergraduate and graduate levels.

The Massachusetts Institute of Technology (MIT)

The MIT component of the collaboration is based in the Center for Transportation Studies, and many of the faculty and students in the program are a part of this Center. Students and Faculty from the Departments of Civil and Environmental Engineering, Urban Studies and Planning, and Architecture are also an integral part of the program. Most MIT students are in the master's program, and their research, financially supported by Tren Urbano Project, is typically incorporated into their Master's Thesis. Each year, two to five undergraduates also participate in the program typically working closely with specific graduate students.

PROGRAM ELEMENTS

The program consists of six (6) key program elements which are focused on giving students a well-rounded academic and practical professional formation, as depicted below and described in the sections that follow.

Program Element 1	**MIT Summer Short course on Public Transportation in Boston** *(June or July)* Introduction to Tren Urbano Project Introduction to Public Transportation History and development of Boston's transit system Field trips to MBTA, Central Artery Project, and others Best & Worst of Boston Team Project MIT student research presentations Recreational / Cultural Activities Cultural immersion in transit-oriented city *(for UPR students)*
Program Element 2	**UPR Short Course on Tren Urbano & Transportation in Puerto Rico** *(January)* Update on Tren Urbano Progress International Transit Systems Existing Public Transit in SJMA Student Research Project Poster Exhibit One-on-one meetings with Tren Urbano staff Field trips to view construction progress Helicopter Tour of Alignment Cultural immersion *(for MIT students)* Recreational / Cultural Activities
Program Element 3	**Student Research Project** *(during 1 or 2 years)* Research topic selected based on project need and student interest
Program Element 4	**Professional Practicum / Internship** *(Summer)* Internships with Tren Urbano consultants and contractors

	Placements made according to available positions & student interest Mid-point and final evaluations
Program Element 5	**Site Visit to Operating Transit System** *(Spring break)* Study and observe operating rail transit system Compare / contrast with Tren Urbano Dialogue with professionals
Program Element 6	**Possible employment opportunity with contractor or consultant** Work internship frequently leads to employment

In response to special opportunities, several offshoot activities have been implemented that enrich the experience for the students and increase program exposure. UPR students participate in various forums where they present the results of their research projects. During the past two years, UPR students, in collaboration with the Tren Urbano Office, have also organized a Tren Urbano Fair at the UPR Mayagüez campus to provide general information about the project to the university community. MIT students have presented their research findings at the TRB conference in Washington, DC.

The MIT Summer Short Course on Public Transportation in Boston

The summer course at MIT is an opportunity for the students to learn about the evolution of the public transportation system in Boston, and to see how this system has shaped the development of the Boston metropolitan area. For the UPR students this course comes at the beginning of their participation in the program, while for most MIT students it is nearer the mid-point of their Tren Urbano studies and research.

The UPR students begin the summer course in San Juan, where they are introduced to the Tren Urbano Project through two intensive days of presentations. The urban development of the San Juan Metropolitan Area (SJMA), the changes in transportation needs and facilities, particularly the post-World War II, wide-spread suburbanization and exponential growth in the use of the automobile, set the stage for studying the strategy for the design, construction and operation of the Tren Urbano Rail System. The long-term considerations of system expansion, island-wide transportation, joint development and land use are also discussed.

The UPR students then travel to Boston to meet their MIT colleagues, and to spend a week studying and experiencing a city that has a highly developed public transportation network, and that has had a rapid transit system for nearly a century.

The course of study at MIT includes field trips to study transit design, construction, operations and maintenance first hand. Students supplement the formal field trips with personal exploration of the city by transit as they travel by transit to several events that include a harbor cruise and a Red Sox baseball game.

Group 4 – 1994 UPR students at MIT

A course project is assigned that mixes students from both institutions and across disciplines. The project requires keen observation of the best and worst aspects of the Boston transit system, and it asks the teams to make recommendations for achieving the best system in San Juan. These projects are presented by the teams on the final day of the course.

The instruction at MIT presents both details and broader considerations of public transportation. The sequence of history, design, construction, operations and maintenance is followed. The presentations are made by a variety of MIT professors from Civil Engineering, the Department of Urban Studies and Planning, and the Center for Transportation Studies. Local professionals, transportation officials and community activists all add to the breadth of experience that is conveyed in the course.

Case studies of specific transit lines are used to organize the presentation of course material, and to focus the field trips. For example, the Boston's Southwest Corridor Orange Line is presented to illustrate a design process with active community participation, and how the resulting line has achieved public support, personal security and even substantial volunteer maintenance of the parkland that was built as part of the transit project.

The specifics of Boston transportation projects are also placed in the context of land-use and development; local, state and federal policies and politics; economics, and environmental considerations.

The interaction between the University of Puerto Rico and MIT students takes place on many levels. Non-academic contact at the dinners, evening events, and a weekend excursion to Vermont provide an opportunity to learn about another culture, and forms a basis for learning about professional work in different environments.

In summary, the summer course at MIT is intended to provide an intense introduction to the central issues of public transportation through expert presentations and the direct experiences of the students. It also serves to spark students' curiosity about particular topics that will become the focus of their research for the coming year.

Over the five years, the summer course has been revised and expanded, in response to the suggestions of students and faculty, and to synchronize with the progress on the Tren Urbano project itself.

The UPR Short Course on Tren Urbano and Transportation in San Juan Metropolitan Region

The January course in San Juan is structured to acquaint MIT students with the San Juan Metropolitan Region (SJMR), to learn first-hand about the transportation challenges and opportunities of the region, and become acquainted with the strategic role of Tren Urbano. During the short course, the MIT and UPR students are exposed to a number of distinguished professionals who give talks and lead discussions on topics ranging from aspects of Puerto Rican culture to specific technical details about the Tren Urbano project.

Group 3 – 1996 UPR & MIT students after helicopter ride of Tren Urbano Alignment

Representatives from the government of Puerto Rico, as high as the Secretary of Transportation and Public Works, also make presentations about critical issues and share the challenges and vision for public transportation in San Juan.

The San Juan course is a blend of learning experiences – presentations, panels, field trips, individual meetings with staff, exploring the San Juan area by public transportation, and a helicopter ride to view the alignment. The subjects covered vary slightly every year in response to the stage of the project. The topics presented over the last few years have included: transit systems in South America, construction progress, bus and público service in the San Juan metropolitan area, intermodal integration, operations & maintenance, systems, financing, marketing and community relations. The program is enriched by sociocultural activities to foster interaction between MIT and UPR students as well as increase cross-cultural understanding. Each participants receives a binder with handouts of each presentation which serves as the proceedings of the event.

The students present the status of their respective research projects and receive feedback from the other students, professors and other participants on how to proceed with their research work. During the first four years, student research projects were presented orally aided by multi-media presentations, but beginning in 1999 students prepared posters that were on exhibit at the Tren Urbano Office. The poster exhibit became part of the program and each day one hour was dedicated to reviewing several posters and providing feedback on the research conducted to date with suggestions for further consideration.

The Research Experience

The student research projects are intended to be beneficial to the student and the Tren Urbano project. The idea is that as the Tren Urbano project progresses the focus of the research agenda also shifts to keep it in tune with the project priorities, while the student researches a topic that is relevant to his/her academic interests. There is a balance to be maintained in the research program. On one hand, it is essential that the research be highly relevant to the current and emerging decision-making domain for Tren Urbano. On the other hand, it is essential to have a clear distinction between the applied research appropriate to a university and the consulting work necessary to support the Tren Urbano project. Typically, graduate students will be involved in a single research project that takes at least nine months to complete and there is a strong learning component to it.

In the fall, students at each university select research topics, under the guidance of one of the professors involved in the program and in consultation with the project Most of UPR engineering students are undergraduates, while planning and architecture students are mostly graduates. The graduate students are usually involved with the program for two years and their research becomes the masters' theses. Undergraduate students conduct research for six credit hours (3 per semester) as part of the undergraduate research course. They must submit a report of their research to fulfill a requirement of the program.

While the majority of MIT students involved in Tren Urbano are graduate students doing thesis research, a substantial number of undergraduates are also involved, typically working closely with a graduate student under the direction of a faculty member. This has been found to be a very effective way to interest MIT undergraduates in particular areas of study and potential careers.

The MIT research program has focused principally on transportation systems, construction management and urban planning, with a strengthening emphasis on the procurement process and construction management as the project has now moved into the construction phase. Transportation has been a strong area from the outset both in terms of aspects of the rail system itself and particularly the improvement of the existing bus and publico systems and their effective integration with Tren Urbano. An important emerging area is the analysis of alternative extension strategies and priorities.

Since 1994, one hundred and twenty six (126) research projects have been completed or are in process. The range of subjects covered include airport access, station design, construction management, controls, electrical engineering, environmental impacts, fare collection and policy, finance, geotechnical engineering, intermodal integration, joint development, economic development, land use and zoning, operations and maintenance, parking, project management, rail transit, seaport planning, traffic engineering, transportation planning, urban design and vehicles.

Professional Practicum -- The Work Internship

During the second summer in the program, students have the opportunity to do a ten-week, full-time internship with Tren Urbano contractors and consultants. The internship experience is coordinated by the Tren Urbano Office through its technology transfer department. The internship has the two main objectives: (1) To provide students a meaningful and professionally-relevant work experience, and (2) To provide a forum for professional development.

The internships are mostly in areas such as quality control, safety, technical services, field engineering, geotechnical, traffic engineering, transportation planning, architecture, urban design, graphics, interface coordination, vehicles, systems, scheduling and project management. The great majority of the internships are in San Juan but during the past three years there have been interns in Boston, Sacramento, and Minneapolis. Each Tren Urbano contract is required by contract to provide work internships to students.

UPR students are given priority for summer work internships. If there are MIT, non-Puerto Rican students with Spanish language skills whose participation in a work internship would benefit the project and there are available placements, then every effort is made to assign those students to summer internship positions.

The success of the internship experience begins with the preparation prior to the arrival of the interns. Typically, this process begins in January/February when students express their desire to participate in the internship by completing an internship application. Simultaneously, contractors and consultants are required to develop job descriptions for the internships. These descriptions and the intern applications are the background documents for the Interview Session coordinated by the Tren Urbano Office. Students and supervisors attend the interview session and after mutual interviews express their preferences. Based on the expressed preferences, students are assigned to internship placements.

Students spend ten weeks at their work sites. They also participate in four bi-weekly lunch-time meetings. Two of these meetings are designed to enhance professional development and the other two are evaluation sessions. Evaluations are both written and oral. The summer internship culminates with a recognition evening for interns and supervisors. Internships frequently lead to jobs, particularly when the student graduates in June, does the internship in June & July and is hired in August. After the internship students also return to the university to complete academic work, while others go on to pursue graduate studies.

Tour of Operating Transit System

UPR Group 4 (1997) students and faculty during visit of Medellín Metro

During Spring Break (in Puerto Rico this is celebrated during Holy Week), the UPR students and professors visit another city to study first-hand another rail based transit system. This is a formal two to three day activity including lectures, panels and visits to stations, the control center and maintenance facilities. The students have a strong background at this time so they absorb all the information provided them and contrast and compare the projected San Juan system and the Boston system. The systems visited since 1994 are Caracas, Venezuela; Miami, Florida; New York City and Medellín, Colombia.

Work Opportunity with Tren Urbano Consultant or Contractor

After students complete the internship period and their academic studies, every effort is made to place them in an entry-level professional job with one of the companies working on the Tren Urbano project. Placement with these companies varies upon the availability of positions and the volume of Tren Urbano work.

Program Funding

The PRHTA has invested in excess of $9 million in the technology-sharing program and the University of Puerto Rico invested $330,000. The PRHTA was also successful in securing a grant of $750,000 from the Puerto Rico Economic Development Administration to offset the cost of the program.

The cost elements that have been funded by the project for the universities include research assistantships, short course expenses, travel and lodging, faculty advisor stipends, administrative support and overhead. Tren Urbano Office and contractor cost elements include administrative support, travel & lodging expenses, internship stipends and overhead.

Program Results

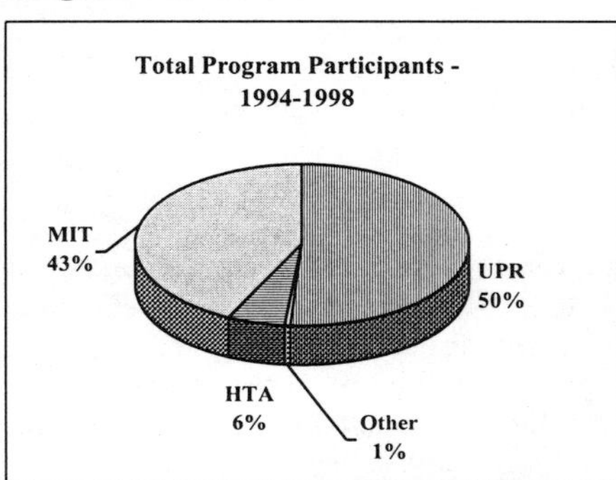

Since its inception in 1994 there have been 145 program participants (74 from UPR, 62 from MIT, 8 from PRHTA, and 1 from the University of Texas). Estimates to 2001 indicate that over 200 students will participate in the program. Twenty (20) students from UPR have been hired by Tren Urbano contractors and consultants. Some have already assumed leadership positions in their designated positions. Some MIT students have also found employment with stateside companies involved in Tren Urbano work. For the PRHTA engineers who have participated in the program, the experience has provided them with a broad knowledge of public transit and enhanced their leadership skills.

Beyond the statistical results, at the UPR, the program has been a catalyst for increased student research exposure. Students share their research results at a variety of academic forums: the Junior Technical Meeting (National Science Foundation), at COINAR (Congreso de Ingeniería, Agrimensura, y Ramas a Fines), at FoPER (Forum to Promote Engineering Research), EPSCoR (Experimental Project to Stimulate Competitive Research).

The investment in the UPR-MIT Program, in addition to the many academic advantages it brings to the UPR, results in two valuable long-term benefits for Puerto Rico: (1) strengthened teaching and research programs in the College of Engineering, the School of Architecture, and the Graduate School of Planning, and (2) significantly increased public transit professional expertise.

For MIT, the academic advantages reside in the real-world experience, applied research, interdisciplinary perspective and cross-cultural interaction. The Tren Urbano project provides an extraordinarily stimulating practical application of the theories presented in the Institute's coursework. The summer program in Boston, the winter program in San Juan, and student research projects all give students intimate knowledge of the specifics and broader aspects of real transit projects.

The Tren Urbano project also provides a parallel in the academic environment of the collaboration between disciplines that is required in the creation of successful transportation projects. This collaboration, emphasized in the winter and summer courses, is a feature of the presentations and discussions at the weekly student/faculty meetings during the year and in the organization and supervision of research projects.

For MIT students, the program provides an unusual opportunity for studying with students and professors from Puerto Rico, and for learning about a transit project that must be successful in a non-anglo environment. Thereby developing the cultural professional sensitivity and skills required for work in the 21st century.

Ingredients for Success

The importance of using real projects as a centerpiece for the academic pursuits of teaching and research seems readily apparent to those intimately involved in the UPR-MIT Program. The students who have participated in the program, both from MIT and UPR, have already found immediate use of their experiences in employment and further academic studies. More importantly, they are primed with skills and knowledge for a lifetime of multi-disciplinary, collaborative work.

The program has been successful because there has been strong commitment to the program from the very top of the agencies responsible for Tren Urbano - DTPW and PRHTA. There is an appreciation that the participating universities can indeed provide value to the project itself. There is mutual trust, respect and openness among all parties. The Tren Urbano project has sufficient size and innovative content to justify a program with extensive university involvement.

The Tren Urbano UPR-MIT Professional Development Program has demonstrated that the development of a rail transit system can be a significant opportunity and learning laboratory for developing the future professional leaders who will shoulder the responsibility for excellence and efficiency in urban public transportation well into the 21st century. The funds invested in developing the professional leadership capability in rail transit are already paying dividends. The program is living proof that transportation agencies, universities and private companies can work successfully in the formation of professional leaders for the transit industry.

REFERENCE

Pesquera, Carlos I. “Tren Urbano and Related Transit System Improvements in San Juan, Puerto Rico”, paper presented at the ASCE First International Conference on Urban Public Transportation Systems, Miami, Florida, March 21-25, 1999.

Socioeconomic characteristics, land use and travel patterns of São Paulo, Brazil.

Eduardo A Vasconcellos[1]

Abstract

The São Paulo metropolitan region has been experiencing intense growth in the last decades: the population doubled and the motorized vehicle fleet was multiplied by six from 1970 to 1996. The region has consolidated itself as the most important economic and political region in the country, with the tertiary sector corresponding to 70% of jobs. Average income, although high for Brazilian standards, is relatively low, with 58% of people earning less than U$ 500 a month (and highly skewed towards the upper strata).The region has also experienced increasing transportation problems, that are unevenly distributed among social groups and classes, considering those with and without access to private transportation. Most of the problems are related to the sharp increased in the use of private transportation and the corresponding decrease in the use of public transportation: the percentage of congested roads in the afternoon peak is currently 80% and average bus and auto speeds are very low (12 km/h and 17 km/h respectively). Severe congestion is causing a waste of 300 million passenger-hours per year in the city (for bus and auto trips), and atmospheric pollutant concentration is inadequate in 10% of the days. In addition, the city of São Paulo presents some of the highest traffic accident figures among large cities in the developing world, with 60,000 injured people and 2,300 fatalities in 1995.

Current problems are challenging the region's economic efficiency and its position as a candidate world city and have also been promoting an intense debate on alternative transportation solutions, that include the coordination of urban, transport and traffic problems at the metropolitan scale, the provision of high quality public transportation and the restraint to the use of private transportation.

[1] Associate Director, Associação Nacional de Transportes Públicos – ANTP, Rua Augusta 1626, 01304-902, São Paulo, Brazil; fax (55) 11 253 8095; e-mail : eduardo@antp.org.br.

Introduction

The São Paulo metropolitan region is the largest in Brazil and in South America as well, with an area of 8,051 km^2 and an estimated population of 16.6 million in 1996. It is formed by 39 cities, with the city of São Paulo being the largest and most important in economic and political terms, with a population of about 10 million people in 1996 (see figure 1).
The paper describes the main social and economic characteristics of the region and the transport demand as well. The paper is divided into 4 parts: Part 1 makes a brief description of the social and economic development of the region and the associated transport development. Part 2 describes current social and economic conditions. Part 3 describes current transport and traffic conditions. Part 4 summarizes main conclusions and recommendations.

Urban development and transport policies in Sao Paulo

In this century, the city of São Paulo and its metropolitan region experienced large physical, economic and social changes, reflecting the major changes at the country and world levels. At the metropolitan scale, the first large transformation initiated in the 30's, when the coffee based economy began to be replaced by the industrial economy. In the two following decades, the region would consolidate as the most important industrial area in the country, with the city of São Paulo itself occupying a prominent position. Nearby cities known as the "ABC region" would in the 50's concentrate large investments in the newly organized Brazilian automotive industry and the related supply industries. The large industrial development would place the region, in this period, as responsible for 35% of the country's industrial production (Emplasa, 1994).
In the 70's, the large growth of Brazilian economy was paralleled by important changes in the region's economy, with a sharp increase in the overall economic output and in the tertiary sector as well: the region consolidated itself as the most important financial and technological center of the country. During the 80's, the region suffered from the economic recession in the country and the industrial sector experienced negative growths in production and employment around –1% to -2% a year, while the overall economy was influenced by the decrease in average income. In the same period, the first industrial decentralization started to occur, with the nearby countryside in the state of São Paulo receiving large investments and increasing its share in the overall industrial output. However, the metropolitan region and the city of São Paulo never lost their prominent position as centers of strategic decisions.
With transport policies, the first major plans and investments initiated in the 30's, when the Prestes Maia arterial system started to be built. In the 40's, the main public transportation system provided by the Canadian Light & Power street cars started to be replaced by diesel buses, in a movement that would last until the 70's, when the

last street car line was terminated. The buses, initially operated by many small private firms, were eventually run also by a special public (city) company created in 1947, the Cia Municipal de Transportes Coletivos, CMTC, which shared the market with private operators. Between 1960 and 1980, mobility increased, road capacity was greatly improved and efficient traffic operation was organized, while keeping public transportation in poor conditions (Vasconcellos, 1997). Space was occupied in conflicting ways - often irregularly - and the urbanized area increased rapidly. Conurbation began to spread, however without changing the dominant role of Sao Paulo. In the 1980's, following the economic depression that reduced activities, mobility decreased and the fiscal crisis of the state deeply reduced investment capacity, making mass transportation infrastructures even less viable. No special policy was adopted and average transportation conditions remained the same. Democratization of the political system in 1982 helped little to alleviate poor transport conditions, in face of diminishing state resources and high inflation rates. Recently, as the new "Real" economic plan succeeded in lowering inflation, the country and the region experienced an intense traffic growth, with parallel high increases in the number of automobiles (see table 5). Mobility conditions started to deteriorate rapidly, with major daily congestion becoming the normal rule, followed by severe environmental pollution and by a steady decrease on overall transport efficiency, for both people and merchandise. The evidence of the crisis brought about a public discussion about alternative transportation policies, which is still under way.

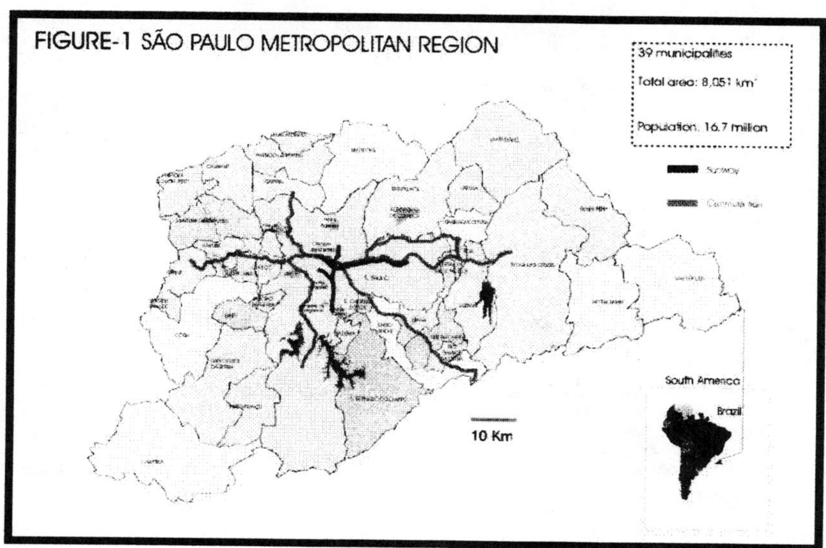

FIGURE-1 SÃO PAULO METROPOLITAN REGION

Population

Table 1: Population increase, São Paulo Metropolitan Region, 1970-96.

Year	Metropolitan Region		City of São Paulo	
	Population	Increase (%)	Population	Increase (%)
1970	8,139,730	-	5,924,615	-
1980	12,588,725	54.7	8,493,226	43.3
1991	15,416,416	22.5	9,626,894	13.3
1996	16,583,234	7.3	9,836,129	2.2

Ref: IBGE (1991) and Seade (1996)

Metropolitan population doubled from 1970 to 1996. The city of São Paulo has grown more slowly and its population has recently stabilized (table1). The annual average growth has also been different among the subregions. Between 1980 and 1991, the North and Southwest areas grew at around 5% a year, while the metropolitan region and the city of São Paulo grew, respectively, at 1.9% and 1.2% a year.

Age distribution

Table 2: Age distribution, São Paulo Metropolitan Region, 1996.

Age (years)	People	%
0 – 9	2,927,972	17.7
10 – 19	3,297,009	20.8
20 – 29	3,153,404	19.1
30 – 39	2,781,154	16.7
40 – 49	2,011,669	12.1
50 – 59	1,172,410	7.1
60 – 69	749,879	4.5
> 69	489,737	3.0
Total	16,583,234	100.0

Ref: SEADE (1996)

Table 2 shows that the population is relatively young, with 58% of people with less than 29 years of age.

Employment

In face of the described deep changes in the economic structure, the tertiary sector (commerce, services) is the dominant employer. Its share has grown from 62% in 1985 to 70% in 1992 (table 3) and it is estimated that it has already reached the 77% level (CMSP, 1998).

Table 3: Employment by sector, São Paulo metropolitan region, 1985/1992.

Sector	Employment (%)	
	1985	1992
Primary	0.9	0.5
Secondary	36.9	29.1
Tertiary	62.2	70.4

Ref: Emplasa (1994)

Income

Table 4: Average individual income, São Paulo metropolitan region, 1995.

Individual monthly income (U$)	People (%)
< 100	5.5
101 – 300	30.6
301 – 500	22.3
501 – 1,000	20.9
1,001 – 2,000	9.6
> 2,000	5.2
Total [1]	100.0

(1) includes people with no income (3.4%) and people with unknown income (2.5%);
(2) Ref: IBGE, 1996.

Average income, although high for Brazilian standards, is relatively low, with 58% of people earning less than U$ 500 a month (table 4). Income is badly distributed, highly skewed towards the upper strata: the first 10% of families in the scale receive just 1.76% of all income, while the highest 10% receive 35.48% (IBGE, 1991).

Vehicle ownership

The number of motorized vehicles in the city of São Paulo was multiplied by six from 1970 to 1996 (table 5). From 1980 to 1995, the number of persons per vehicle has decreased from 5.4 to 2.1 (CET, 1998 a).

Table 5: Growth in the number of motorized vehicles, city of São Paulo, 1970 – 1996.

Year	Motorized Vehicles(1)	Increase (%)
1970	731,728	-
1980	1,585,986	117
1990	3,421,059	116
1996	4,671,362	36 (6 years)

(1): automobiles, buses, trucks, motorcycles,vans;
Ref: CET, 1998a.

Travel Patterns

Table 6: Social and transport characteristics, São Paulo metropolitan region, 1967 – 1997.

Subject	1967	1977	1987	1997
Population (1,000s)	7,097	10,273	14,248	16,792
Annual pop. growth(%)	-	3.77	3.33	1.66
Motorized trips (1,000s)	7,163	15,758	18,749	20,267
Mobility rate [1]	1.01	1.53	1.32	1.21
Automobile fleet (1,000s)	493	1,384	2,014	3,436
Automobile rate [2]	70	135	141	205
Jobs (1,000s)	-----	3,960	5,647	6,920
School enrollments (1,000s)	1,088	2,523	3,676	4,986

(1) motorized trips per person, per day.
(2) Automobiles/1,000 people.
Ref: CMSP, 1998.

Data from the household travel surveys (table 6) show that the metropolitan overall population more than doubled in 30 years, however with decreasing rates. The mobility rate presented sharp changes and is currently at the 1.2 level. Automobile fleet was multiplied by seven.

The use of transport modes

The use of motorized transport modes has changed dramatically in the last three decades, with the major change occurring with the use of private transportation, that increased from 26% in 1967 to 48% in 1997. Accordingly, bus use has decreased from 59% to 39% in the same period. The subway, which was opened in 1974 with its first line (10 km), progressively enlarged its network (although slowly) to the current 45 km. The subway attracted a higher share of the trips. The train system, subjected to decreasing investments and offering low levels of service, remained serving a small part of the demand (table 7). Table 8 compares the share of foot trips to those by motorized travel.

Table 7: Change in the use of motorized transport modes, 1967-1997.

Transport mode	Trips/day (%)			
	1967	1977	1987	1997
Public	63.5	60.7	54.8	50.8
Train	4.4	3.2	4.4	3.2
Subway	--	3.4	7.6	8.3
Bus	59.1	54.1	42.8	39.3 (1)
Private (auto and taxi)	25.9	34.8	41.9	47.3
Other	10.6	4.4	3.3	0.9
Total	100.0	100.0	100.0	100.0

(1) includes 1% of declared trips on illegal minivans.
Ref: CMSP, 1998.

Table 8: Share of foot and motorized trips, São Paulo metropolitan region, 1977-1997.

Mode	Trips (%)		
	1977[1]	1987	1997
Motorized	74.8	64.0	65.6
Foot[2]	25.2	36.0	34.4
Total	100.0	100.0	100.0

(1) the 1967 survey did not include foot trips.
(2) trips longer than 500 meters only.
Ref: CMSP, 1998.

Current division of daily trips according to mode is shown in table 9. It can be seen that public modes, private modes and foot trips account each for approximately one third of trips.

With respect to trip purposes, work and school are the dominant ones. Considering just non-home trips, these two motives account for 74% of all trips (table 10). Within work trips, 61% are related to the service sector, 22% to commerce and 17% to industry jobs.

Table 9: Trips (main mode), São Paulo metropolitan region, 1997.

Main mode[1]	Trips per day	
	Number (1,000s)	%
Public	10,307	33.4
Bus [2]	7,965	25.8
Subway	1,688	5.5
Train	654	2.1
Private [3]	9,578	31.0
Other[4]	382	1.2
Motorized – total	20,267	65.6
Foot	10,615	34.4
Grand total	30,882	100.0

(1) the mode of highest capacity within all modes used (in combined trips).
(2) includes regular transit, hired buses, hired school buses and 200,000 trips per day on illegal minivans.
(3) automobile and taxis.
(4) motorcycle and bicycle.
Ref: CMSP, 1998.

Table 10: Purpose of trips, São Paulo metropolitan region, 1997.

Purpose	Figure		
	Trips/day (1,000s')	Share in all trips (%)	Share without home trips (%)
Work	6,874	22.2	41.0
School	5,525	17.9	32.9
Shopping	746	2.4	4.5
Health	637	2.1	3.8
Leisure	1,145	3.7	6.8
Other	1,852	6.0	11.0
Home	14,103	45.7	--
Total	30,882	100.0	100.0

Ref: CMSP, 1998.

Travel conditions

a) Traffic speeds and congestion
Automobile speeds increased from 25 km/h in the last years of the 70's to 27-28 km/h in the 1980 -1984 period and then dropped to less than 20 km/h in the 90's. Currently

it is 17 km/h in the PM peak and 27 km/h in the AM peak (IPEA/ANTP, 1998). The declining trend of the latest years is directly related to the sharp increase in the automobile fleet that followed the success of the "Real" economic plan. In the afternoon peak, total length of congested roads tripled between 1992 and 1996 - from 39km to 122km - (CET, 1998a) and the percentage of congested roads in the main system is currently 80%. Automobile speed in the main arterial system in the afternoon peak hour is 17 km/h and bus speed is 12 km/h (IPEA/ANTP, 1998). It is estimated that 3,000 among the 11,000 buses used could be taken out of service if severe congestion was eliminated, and that this extra supply causes an extra cost of 16% on bus fares. Congestion under such conditions is causing a waste of 316 million passenger-hours per year in the city, for bus and auto trips (ANTP/IPEA, 1998).

b) Travel times and access to transport

The occupation of the outskirts of the city increased average travel distances. The bus system was not expanded into the outskirts quickly. In addition, tight fare controls led private operators to constantly adapt supply to ensure minimum profitability, often at the expense of service frequency and service to low density areas. Suburban railways offered extremely low levels of service, reproducing the same conditions found in other Brazilian towns. The result was a poor public transportation system, characterized by service irregularity, unreliability and discomfort, and with very limited integration. A sharp contrast with respect to private transportation was clear. Public transportation users also faced traffic problems. Most of the new arterial streets did not have any special physical and operational devices to ease the circulation of buses, and few special traffic priority schemes were organized to improve bus operation. Some of the resulting differences in the quality of public and private transportation may be seen on table 11.

Table 11: Access and travel times to transportation, Sao Paulo, 1997.

Mode	Access time (min) (1)	Travel time (min) (2)
Automobile	1	29
Bus	6	57
Metro	7	77
Train	11	93

(1)walking (one-way);
(2)one-way, from origin to destination (includes walking links)
Ref: CMSP (1997)

c)Traffic safety

The city of São Paulo presents some of the highest traffic accident figures among large cities in the developing world. The yearly number of fatalities has been falling around the 2,500 level since 1980 and the majority is composed by pedestrians (table 12). In 1995, there were about 60,000 victims of these accidents. Among them, it is estimated that 9,000 were seriously injured and that 6,000 remained with permanent injuries (CET, 1996). Currently there are more than 200,000 accidents per year (94% vehicular accidents) (table 13).

Table 12: Traffic fatalities, 1980 – 1997, city of São Paulo.

Year	Vehicle occupants	Pedestrians	Total	Fatalities/10^5 pop
1980	750	1,580	2,330	27.4
1985	1,044	1,515	2,559	27.8
1990	1,094	1,621	2,715	28.3
1995	846	1,432	2,278	23.0
1996	906	1,339	2,245	22.4
1997	933	1,109	2,042	20.4

Ref: Cia de Engenharia de Tráfego – CET (1996 and 1998b)

Table 13: Traffic accidents by type, city of São Paulo,1997.

Type of accident	Number	%
Vehicle-only	189,911	94.1
Pedestrian	11,876	5.9
Total	201,787	100.0

Ref: Cia de Engenharia de Tráfego - CET (1998b)

Tables 12 and 13 show that streets are inherently dangerous to pedestrians (Vasconcellos, 1996). As most of pedestrian trips are made by the lowest income levels, this extremely grave externality is mostly imposed by the few with access to vehicles on the majority.

d) Pollution

The air in the city of Sao Paulo also shows high concentration levels for some important polluters (table 14). In addition to long term effects to the atmosphere, pollutants in São Paulo have already shown a negative effect on people's health: air pollution and mortality of elderly people (over 65 years) were found statistically associated with respirable particles (Saldiva et all, 1995). In addition, a plausible relationship between child poverty-related malnutrition and respiratory diseases was found in the city (Saldiva et all, 1994).

Table 14: Days with inadequate pollutant concentration, 1994.

Pollutant	Days with inadequate concentration (%)
Carbon monoxide (city average)	13.1
Nitrogen oxides (city average)	26.2
All pollutant (metropolitan area average)	10.8

Ref: Cetesb, 1994.

Summary and conclusions

The São Paulo metropolitan region has been experiencing intense growth in the last decades and has consolidated itself as the most important economic and political region in the country. In parallel, the region has been experiencing increasing transportation problems, related to accessibility, speed, safety, comfort and environment conditions. These conditions are unevenly distributed among social groups and classes, in face of the large social and economic differences among social strata. While private transportation users are supported by a series of transport policy actions, public transportation captive users face unfavorable travel conditions, characterized by discomfort, unreliability and inefficiency.
Current conditions are deteriorating quickly, in face of diminishing investments in subway and trains, lack of proper priority treatment for buses and increasing automobile based congestion, leading to high travel times and excess pollution. Current problems are challenging the region's economic efficiency and its related position as a candidate world city and have been promoting an intense debate on alternative transportation solutions.

The experience of the region leads to some basic recommendations. First, at the institutional and organizational sides, decisions on land use, transport and traffic are highly interdependent and agencies in charge of these affairs have therefore permanent crossing paths. Therefore, major efforts should be taken to supersede the historic disconnection between the metropolitan-scale transport actions and local transport policies. Also, inside the cities themselves, urban, transportation and traffic actions should be permanently coordinated. In the same token, it is important to work in order to avoid the occupation of the outskirts or developing areas without adequate urban planning for job and public service location, and provide good public transportation provision. Second, the supply of large, integrated public transportation means should be promoted, offering high quality services. Accordingly, road construction should be reassessed, in order to examine who is going to pay and benefit from it. Third, at the operational side, the management of the bus service has to be greatly improved, by organizing surveillance systems to control service quality in a comprehensive and permanent way and by offering different services to different market sectors. Fourth, at the social side, the irresponsible and uncivilized use of the space by motorized transportation – specially the automobile - has to be reversed, trough better traffic management, pervasive enforcement on user behavior and vehicle pollutant emission, fiscal and economic deterrents to automobile use and large scale traffic educational programs.

References

CET – Cia de Engenharia de Tráfego (1996), Vítimas de acidentes de trânsito – pesquisa, São Paulo.

____________________(1998a), *Operação horário de pico – relatório de avaliação*, São Paulo.

____________________(1998b), *Fatos e estatísticas de acidentes de trânsito em São Paulo*, 1997, São Paulo.

Cetesb - Cia de Tecnologia Ambiental (1994*) Relatório de qualidade do ar no estado de São Paulo - 1994*, São Paulo.

CMSP, Cia do Metropolitano de São Paulo (1988) *Pesquisa OD/1987*, São Paulo.

____________________(1998)*Pesquisa OD/1997 1997*, São Paulo.

Emplasa, Empresa Metropolitana de Planejamento (1994) *Plano metropolitano da Grande São Paulo 1994/2010*, São Paulo.

IBGE – Instituto Brasileiro de Geografia e Estatística (1991), *Censo geral do Brasil*, Brasília.

____________________ (1996), *Pesquisa domiciliar*, Brasília.

IPEA/ANTP (1998), *Redução das deseconomias urbanas com a melhoria do transporte público, relatório síntese*, Brasília.

Saldiva et all (1994) Association between air pollution and mortality due to respiratory diseases in children in São Paulo, Brazil: a preliminary report, *Environmental Research* 65, pp 218- 225.

__________ (1995) Air pollution and mortality in elderly people: a time-series study in São Paulo, Brazil, *Archives of Environmental Health* 50 (2), pp 159-163.

SEADE (1996), Censo demográfico de São Paulo, São Paulo

Vasconcellos, E. A., (1996) Reassessing traffic accidents in developing countries, *Transport Policy* 2 (4) pp 263-269.

________________(1997) The making of the middle class city: transportation policy in São Paulo, *Environment and Planning A* 29:293-310.

Urban Public Transit System in the city of São Paulo, Brazil, its Operations and Use by People

Pedro Luiz de Brito Machado[1]
Francisco Armando Noschang Christovam[2]

Abstract

This paper describes the current situation of the public transport system in the city of São Paulo, Bazil, and its metropolitan area, with emphasis on current developments. A concise evaluation follows, upon which some critical issues, requiring further consideration, are highlighted. In conclusion, an overview is presented of a novel medium-capacity mode currently under development by SPTrans, the local public transit management agency.

Introduction

The city of São Paulo, Brazil, core of the largest metropolitan area in South America, is undergoing a slowdown in its so far dramatic population growth rate. Even as this happens, however, the city´s development process still poses dramatic challenges to traffic and public transit operations, which daily cater to a universe of over 30 million daily person trips.

Existing rail-based transportation which include a regional railway network and an urban heavy-rail subway (*metrô*), carry a much smaller load than buses, even though the very modern and reliable *metrô* presents one of the highest productivity rates in the world. Regional railways play a most important structuring role as they link the metropolitan western, northeastern, and eastern sectors, but the number of passengers carried is relatively low.

[1] Chief Executive Officer, São Paulo Transporte S.A.-SPTrans, R.13 de Maio 1376, S. Paulo, Brazil
[2] Former Chief Executive Officer, São Paulo Transporte S.A.-SPTrans, R.13 de Maio 1376, S. Paulo, Brazil

The growing share of trips borne by those still designated “non conventional” transit services -- jitney vans, shared taxis and minibuses – keeps pace with mounting vehicle ownership and with a perceivable raise in average family income, which grew by ten percent, between 1987 and 1997. It can also be associated to behavioral changes and new travel habits, many related to specific social strata and age groups, for which the conventional modes do not provide adequate service.

The increasing importance of those heretofore lesser public transit services, adds up to the already established “special” transit modes, such as school buses and *hired buses*, private bus services hired by companies or groups of users, in most cases to provide home-to-work-and-return transportation. The most recent trip origin-and-destination survey, concluded in 1997 (see Table 1), reports those nonscheduled services to be far from unimportant, as 200 thousand daily trips are provided by minibus and jitney services, and 800 thousand more by school buses and hired buses.

Subject to milder regulations, as compared to standard bus transit services, many are clandestine, either for running non-authorized routes or for operating vehicles unlicensed for the service they provide. As new routes are illegally established, many run in direct competition with the conventional public modes, mainly the scheduled bus services.

General System Features

About one third of all daily person trips which occur in the São Paulo metropolitan area (SPMA), are made by public transit (see table 1). Within the capital city itself, that proportion rises to about 36%. Most of the public transit trips utilize one or more of the 1300 privately run bus routes. Less than thirty percent of those make use of the rail-based modes and many are linked trips, which include a bus leg. Table 1 indicates current (1997) person trip modal split, according to the main transportation mode used, and highlights the situation observed in the core city, as opposed to what occurs in the whole 38-municipality metropolitan region, core city included.

The heavy rail urban subway is run by *Companhia do Metropolitano de São Paulo*, the São Paulo Metro Company, which, in spite of its name, still does not reach beyond the city of Sâo Paulo municipal border. Metro Company is jointly owned by city, state and federal governments, with the major stake held by the state. Like CPTM (*Companhia Paulista de Trens Metropolitanos*), the State Regional Railways Company, it is subordinated to STM (*Secretaria dos Transportes Metropolitanos*), the State Metropolitan Transport Secretariat.

There are three urban heavy rail lines, catering to about 1.7 million person trips each day, while CPTM runs four regional rail lines, with well below one million daily person trips. Metropolitan inter-city urban bus routes are also controlled by STM through EMTU (*Empresa Metropolitana de Transportes Urbanos*), the Metropolitan Bus Transit Agency. Inter-city bus services between cities in the metropolitan area are totally private. They include a special busway link, with both ends in the city of São Paulo, with mid-route terminals and sheltered stops in Santo André, São Bernardo do Campo and Diadema, cities located Southeast of the Capital. There are also special bus services to the São Paulo International Airport, located in adjacent of Guarulhos, still within the metropolitan region

Table 1
São Paulo Metropolitan Area Person Trip Modal Split – 1997
(thousands of daily trips by main transportation mode)

Grouped Modes	Metropolitan Area	% in group	% in total	City of São Paulo	% in group	% in total
Subway	***1 698***	*16*		*1 533*	22	
Metropolitan Rail	***649***	***6***		*322*	5	
Scheduled Bus	***7 128***	***68***		*5 035*	71	
Non-scheduled Bus	***800***	***8***		*included above*	--	
Minibus, Jitney, etc.	***199***	***2***		*143*	2	
Motorized – Public	**10 474**	*100*	*33*	**7 033**	*100*	*36*
Car+Taxi	*9 741*	***96***		*6225*	*97*	
Others	***405***	***4***		***199***	*3*	
Motorized – Private	**10 146**	*100*	*33*	**6 424**	*100*	*33*
Motorized - All	**20 620**		*66*	**13 457**		*69*
Walk Trips	**10 812**		*34*	**6 158**		*31*
GRAND TOTAL	**31 432**		*100*	**19 615**		*100*

Source: SPMA Trip Origin-and-Destination Household Survey - 1997

In the city of São Paulo, traffic and transit management are the responsibility of SMT (*Secretaria Municipal de Transportes),* the Municipal Transport Secretariat. DTP (*Departamento de Transportes Públicos*), the SMT Transit Department, sets regulations for bus, minibus, taxicab and urban freight services within the city borders, while SPTrans (*São Paulo Transporte S. A.),* a city owned concern, acts on behalf of SMT/DTP in what refers to bus transit planning, operations and contract management, as well as to the registration and inspection of taxicabs, paratransit vans and minibuses. SPTrans issues service contracts to operators, through which lots of buses are hired to run specific routes, under strict operating parameters. Fare revenue is retained by the Municipality, which pays for the service rendered, as per the contract, with discounts to compensate for inadequate fulfillment of contract requirements. 70 service contracts, with 50 private operators, total 11,000 buses plus 530 trolleybuses and employ a staff of 61,000. Public transit services in the other 38 municipalities which make up the São Paulo metropolitan region of São Paulo are predominantly run by private bus companies. However, minibuses and paratransit vans also grab a growing share of that clientele.

General coverage of municipal bus routes within the city of São Paulo is shown in figure 1, below.

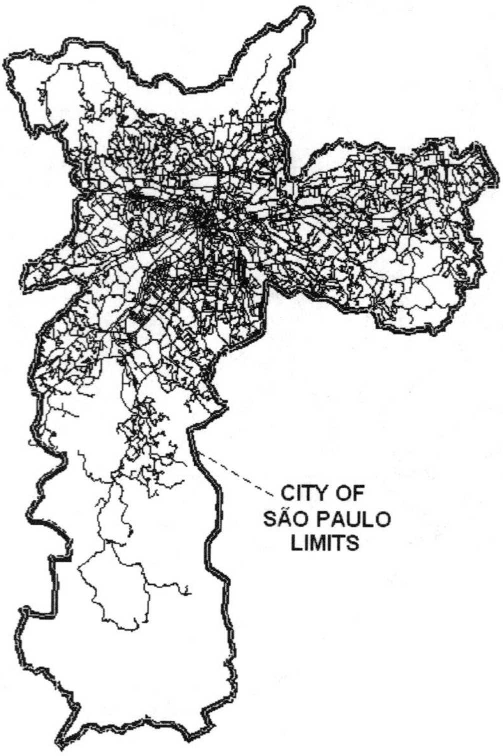

Figure 1 – Scheduled bus route coverage in the city of São Paulo

Essential System Facts and Figures

Some basic facts and figures relative to the scheduled public transit services available in the São Paulo Metropolitan Area are shown in Table 2.

Table 2
São Paulo Metropolitan Area Public Transit Information

SERVICE TYPE	URBAN HEAVY RAIL (SUBWAY)
OPERATOR	COMPANHIA DO METROPOLITANO DE SÃO PAULO – METRÔ
OPERATION TYPE	PUBLIC
EXISTING LINES	BLUE (NORTH-SOUTH) LINE
	GREEN (PAULISTA) LINE
	RED (EAST-WEST) LINE
TOTAL LENGTH	47 Km
Nº OF STATIONS	44 TOTAL (20 WITH BUS OR TRAIN INTERCHANGE)
PASSENGERS/DAY	AVERAGE 1.73 MILLION

SERVICE TYPE	METROPOLITAN RAILWAYS
OPERATOR	COMPANHIA PAULISTA DE TRENS METROPOLITANOS – CPTM
OPERATION TYPE	PUBLIC
EXISTING LINES	NORTHWEST –SOUTHEAST LINE
	EASTERN LINE
	WESTERN LINE AND
	SOUTHERN LINE
TOTAL LENGTH	270 Km
Nº OF STATIONS	91 TOTAL (25 WITH BUS OR METRO INTERCHANGE)
PASSENGERS/DAY	AVERAGE 0.78 MILLION

SERVICE TYPE	SÃO PAULO MUNICIPAL BUS AND TROLLEYBUS SERVICES
MGMT. AGENCY	SÃO PAULO TRANSPORTE S.A - SPTRANS
OPERATION TYPE	PRIVATE
Nº of ROUTES	800 MAIN ROUTES
	500 COMPLEMENTARY ROUTES
TOTAL LENGTH	42,000 Km
Nº OF BUS STOPS	14,500
PASSENGERS/DAY	AVERAGE 5.00 MILLION

SERVICE TYPE	METROPOLITAN INTER-CITY BUS AND TROLLEY SERVICES
MGMT. AGENCY	EMPR. METROPOLITANA DE TRANSPORTES URBANOS–EMTU
OPERATION TYPE	PRIVATE
Nº of ROUTES	420
TOTAL LENGTH	25,000 Km
Nº OF BUS STOPS	8,600
PASSENGERS/DAY	AVERAGE 1.25 MILLION

Source: Brazilian Public Transit Association (ANTP)

Public Transit Usage

- *Demand*

While population growth rates decrease, economic activity levels have been sustained. Still, functional changes have affected metropolitan economy, with marked effects on the overall urban trip patterns. The predominance of service delivery activities, which absorb about 75% of the total workforce, has influenced the trip distribution patterns between job concentration regions and residential areas. This is the main motive for person trips, followed by school-related activities. Even though less pronounced than before, demand still peaks sharply a few hours each day, requiring additional transport supply.

- *Mobility*

Mobility rates in São Paulo have been consistently diminishing, in spite of car ownership evolution along the last three decades. The typical public transit user in São Paulo is a person between 18 and 50 years old, who travels to work or to school. Another important group of travelers are students aged from 7 to 15, but most of the trips they make are walk trips, which do not affect public transit services.

- *Special Passengers*

Some specific population groups are granted special public transport benefits. Students pay only half the regular fare, and people above 65 and the physically handicapped travel free. Moreover, special measures have been adopted to allow easier access to public transport services. Some, like user information, vehicle comfort features and elevated station platforms, while benefiting users at large, are vital for the aged and the handicapped. Others are more specific, like the 200 routes which use lift-equipped buses and the ATENDE service, with 100 lift-equipped vans which provide free-of-charge door-to-door transport to eligible patrons.

Figure 2- Elevated bus stop platform

Figure 3- Lift-equipped ATENDE van

Recent Developments

- *Bus route arrangement*

Bus routes are being reconfigured by SPTrans, as trunk/feeder systems gradually replace conventional radial links. Trunk routes run along specialized bus corridors, and it should be noted that the total length of streets with bus priority grew by 45 percent (110 to 160 km) between 1995 and 1997. Trolleybuses are undergoing a comeback, as new routes are proposed and some diesel routes revert to electric traction. An important addition to the system is the VLP, described in the next chapter.

Transit services throughout the Metropolitan Area have developed along the same technical and operational concepts which apply to the São Paulo local transit system. A network of interurban corridors has been proposed, which is to be integrated with the metropolitan railway, also to be improved, and with the urban heavy-rail subway. It is also to be integrated with the São Paulo municipal corridors, by sharing the use of terminals and exclusive bus lanes.

- *Busways*

Bus-only corridors and passenger interchange terminals, which exist in São Paulo since 1975, have been upgraded and improved to make up for deteriorating traffic and the growing use of the private car. After several attempts at developing a standard corridor cross-section, the current choice, modeled after a 1991 implementation, is to run buses on the leftmost lane, nearest to the median strip, in double carriageway avenues. Stops are located at raised platforms (see figure 2) level with bus floor, allowing easy access for passengers while boarding and alighting, and reducing stop dwell time. Buses operating on such routes are required to have doors also on the left side, a development which required structural redesign.
Local feeder buses connect with the line-haul trunk routes at interchange terminals. Larger vehicles, including trolleybuses, articulated buses or standard Padron commuter buses, serve trunk routes. In addition to these specialized corridors, part of the original radial bus network has been revamped to act as feeders, although some are to survive, to cater for specific flows. Smart-card ticketing, currently under implementation, will introduce a whole new host of intermodal and intramodal interchange possibilities.

Current plans call for 11 bus corridors and 25 interchange terminals, of which 3 and 15, respectively, are already in operation. SPTrans has discussed with the private sector on a possible joint venture to rapidly implement the remainder, but funding problems have forced this idea to be abandoned. The city is, therefore, building them at a slower pace, while requiring current operators to upgrade their fleets for higher capacity running.

Integrated public transit plans for the metropolitan area include the implementation of about 300 Km of preferential lanes and will require 32 passenger interchange terminals, some to be shared with the São Paulo local system.

- *Trolleybus Network*

Although the trolleybus overhead lines and electrification systems and the major part of the fleet are publicly owned, services are provided by three private companies. By contract, each operator was required to buy new vehicles, and to revamp part of the older cars, as part of a comprehensive effort to revive the trolleybus in São Paulo. Three new central loop routes started operations early in 1998, to interconnect the three main downtown interchange terminals. Switch between routes is free at those terminals. Three other routes where trolleybuses had been replaced by diesels are reverting to electric traction. Plans include the building of short overhead sections to inter-connect present routes, thus allowing for new trolleybus options.

- *Subway and Metropolitan Rail*

A new subway line is under construction and another is scheduled to start works before 2000. Furthermore, two of the four existing metropolitan rail lines are to revert to subway operating standards, with additional stations, ATC upgrade and additional train sets (see figure 4)

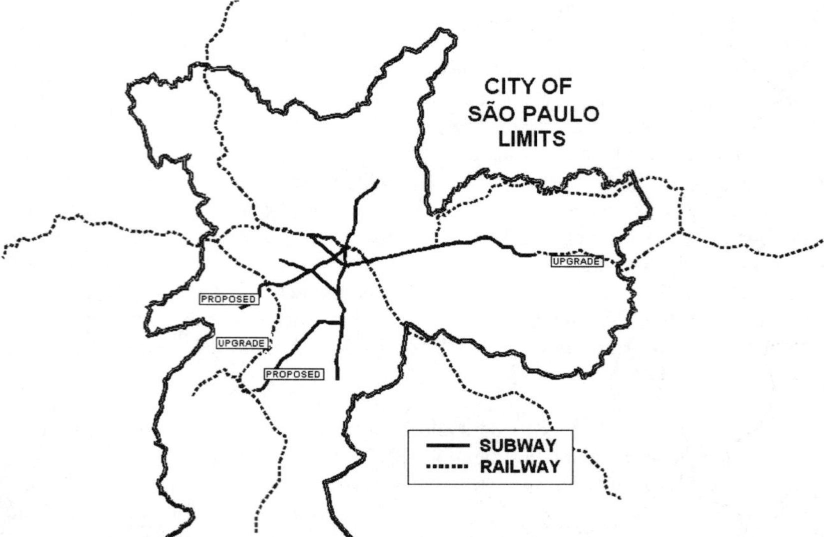

Figure 4 – Subway and metropolitan rail systems showing new developments

Public Transportation Issues

The major public transportation problems in São Paulo, are: hierarchic and functional inadequacies of the public transit system, high proportion of radial routes to serve a no-longer-so-much-radial demand, increasing car ownership and diseconomies due to traffic congestion. These issues are discussed below.

- *Inadequacy of the Public Transit System*

The fact that the bus plays such a major role in urban and metropolitan public, and that rail modes are of less importance, is surely a basic drawback to be faced. Future developments (see above) may contribute to revert such situation, albeit not in the short term.

- *High Proportion of Radial Bus Routes*

During an average weekday, about one third of the Sâo Paulo municipal bus fleet is forced to come downtown at least once, as the routes they serve concentrate on the few accesses to that area. Although many trip ends have moved elsewhere, bus routes have remained radial. As turnaround schemes are normally not provided, most buses must run their routes full length, which results in many low-occupancy vehicles crawling towards their downtown terminal stops along heavily congested streets. Meanwhile, other transport needs are not satisfactorily met.

- *Car Ownership and Jitney Services Boom*

As in other Brazilian cities, public transit has been losing patronage in São Paulo, particularly affecting buses, even though surveys indicate that the subway and trains have also been hurt. The present economic slump is surely one reason, among others. Restriced job opportunities and a momentary drop in average family income (as opposed to the long term prospect) have caused an increase in walking trips.

Changes in the socioeconomic activity pattern also contribute to it, as the regular mobility patterns have definitely changed. The difficulties met by the traditional bus system in its attempts to adapt to the new demand profile, clearly contrast with the nimbleness displayed by jitney paratransit and microbus services, faced with the same challenge, with emphasis on their ability to pinpoint and conquer new and profitable demand segments.

Regardless of other intervening factors, the growth of those alternative transit modes reflects a community reaction to the rigid and impersonal features of regular services, and their refusal to be kept hostage of a transport system which, in spite of all efforts, consistently appears to disregard their requests and needs.

The more and more favorable conditions offered by auto dealers and banking institutions to prospective new or used car buyers, should also not to be disregarded as a potential factor to influence demand to shift away from regular public transit. In fact, they have already led to the explosive ratio of one vehicle for every two inhabitants. Such rate has not fully impacted upon population commuting habits, as many of those cars are still mostly used for non-work trips, but future possibilities are awesome.

- *Diseconomies*

If allowed to continue, the present trend of demand shift from buses to smaller public transport vehicles and to the private car may result in a traffic condition far more deteriorated than the present, during business hours. Moreover, if allowed to replace regular bus routes, microbus operators would certainly greatly reduce transport supply during low demand periods of the day, holidays, weekends, etc, as has been observed elsewhere. It should be borne in mind that traffic congestion is already so serious that authorities have resorted to the extreme measure of adopting a car rotation scheme, through which twenty percent of the regular auto fleet are banned from the streets, from Mondays to Fridays, during peak traffic hours. More than any other vehicle, buses are affected by traffic congestion, specially in view of the limited priority given to them.

Maintaining headway in spite of lower speeds has long been a favored solution, whence the significant fleet and bus mileage increases observed in latter years, which now seem paradoxical, as faced with a dwindling patronage. Furthermore, in an attempt to re-conquer part of their now voluble clientele, new bus routes are created, frequencies of existing routes are increased and new services are provided, which again require further additions to the already disproportional bus fleet.

Closing Remarks

The long term strategy proposed for São Paulo is to significantly improve, both in coverage and performance, the existing rail based services. Current plans include extensive additions to, and renewal of the train and subway networks, as mentioned before. However, funding difficulties have led to a situation where any significant expansion of this system within less than fifteen years does not seem realistic. Thus, in order to meet city goals, work has concentrated in implementing a system of high and medium capacity transit corridors to support line-haul trunk routes, integrated to a feeder system.

This trunk route system will necessarily include segments of the metro-rail network, both those already existing and under construction. Where rail services are not and will not be available, trunk service will be provided by line-haul bus routes, operating on exclusive ways to be constructed or on reserved traffic lanes subtracted from the general traffic. Passengers will be able to transfer between any two of the system components, either at the closed free-interchange integration terminals or anywhere in the system, using smart-card integration.

- *The São Paulo Approach to a Busway Network*

➢ *The System*

VLP is a rubber-tired medium-capacity public transit system. The VLP program for São Paulo city calls for the implementation of 125 km of grade-separated radial and orbital busways. These are to be located across areas where demands between 13,000 and 30,000 passenger per hour per direction are expected, which are above the scope of common buses but do not warrant heavy rail transit. Stops will be at every 800 m, all equipped with platforms level with the vehicle floor, at which passengers will be able to interchange with other transport modes. The full system will require 300 vehicles, to serve a demand of around 1.5 million daily trips. VLP operations are expected to reduce present diesel bus need by at least 2,000 vehicles.

Figure 5- Artist´s perspective of a typical VLP elevated section

➢ *The Vehicle*

A key feature of the VLP vehicle is its steering mechanism, based on the "O-BAHN" system technology, which was developed in Germany, by Daimler Benz, in the nineteen-seventies. Buses are guided by the way of side rails, upon which run rollers connected to the front-wheel steering system. The normal steering mechanism is maintained, which allows vehicles to be driven off the segregated tracks, on regular roads. In order to meet the projected demand, the standard vehicle will be a double articulated unit, to be enabled for convoy operation during the peak demand periods. Since most tracks will be elevated (see figure 5), an extremely high reliability is required, which is to be achieved by the provision of redundancy in all of the main vehicle systems. A two-way radio link with the dispatcher unit and video monitoring will also be provided.

Figure 6- VLP prototype vehicle operating off-track

➢ *Project Feasibility*

Project activities are being conducted by the São Paulo Municipality in close partnership with the private sector. At system inception, city administration will assume such activities as area expropriation, project development and infra-structure implementation. Private sector input shall be 15% of the total of the investment. Multi-lateral financial institution and transnational concern participation shall be acquired through international bidding procedures.

➢ *Research and Development*

A test track has been in operation since early 1998, and a conventional articulated trolleybus was rigged-up for use as a test bed. Items under evaluation include side rollers, traction systems, energy pick-up systems, overhead, etc. The test track was specially designed to include several types of pavement, side guides and flexible electrical overhead, to allow checking vehicle behavior under different operating conditions. A VLP prototype (see figure 6) was also developed – a double-articulated vehicle built around a Volvo do Brasil specially built chassis, with Marcopolo body and Powertronics electrical and electronic equipment -- which has been undergoing test track load essays and component evaluation since May 1998.

DOES CONGESTION MANAGEMENT IMPROVE PUBLIC TRANSIT?

Ladi Biezus[1]
Antonio João Oliveira Rocha[2]

Abstract: Traffic congestion, one of the main urban problems of Sao Paulo, is quantified thru the monitoring of "the congestion queue lengths" (CQL). Due to the growth of peak CQL in recent years, City's authorities enforced traffic restriction policies, e. g. the prohibition of circulation of each auto during one working day per week. The restriction, called "Rodizio", is controlled by the last digit of the car plate. Experience developed since 1995 demonstrates the effectiveness of the "Rodizio", alleviating the congestion and improving the speeds of bus transit. These positive outcomes will be soon offset by the growth of the numbers of cars in circulation. This conclusion shows that the City needs to adopt more stringent policies, including congestion pricing. The revenues derived from these policies should be invested in the most important public transit projects.

1. TRAFFIC COMPLEXITY

The Sao Paulo metropolitan region comprises 39 municipalities. Its most important participant is the City of Sao Paulo, with a population of about 10 million people.

Traffic complexity is one of the main problems of Brazil's largest city. This situation is harmful to the city's public transportation system and results partly from its modal split, which is atypical of an emerging economy. Other structural conditions affect traffic performance, like the radial configuration of the road framework - lacking enough perimetral links - and the fact that the main city's avenues work as an important element for the integration of the state's highway network. Other relevant

[1] Vice President, Logos Engenharia S.A. , R Libero Badaro 377, 6° and , 01009-000 São Paulo (SP) Brazil
[2] Director, Logos Engenharia S.A. , R Libero Badaro 377, 6° and , 01009-000 São Paulo (SP) Brazil

FIGURE 1 - AUTOMOBILE OWNERSHIP IN THE CITY OF SAO PAULO - source CET (*Companhia de Engenharia de Tráfego de Sao Paulo*, the City's traffic management agency)

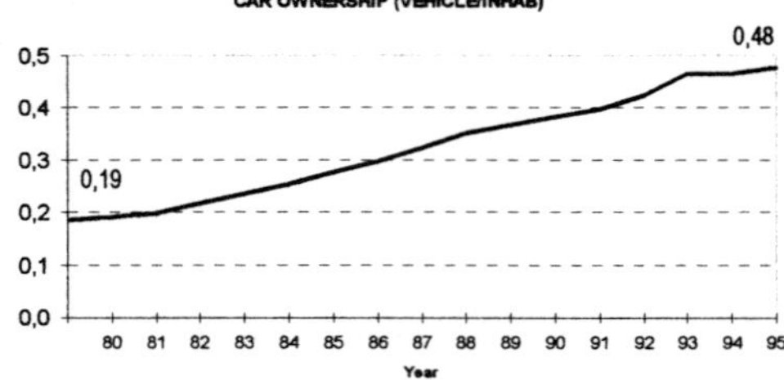

causes of the city traffic complexity are the growth rate of auto ownership and mixed operation of bus transit and privately owned cars on the streets.

The very high rate of automobile ownership and the consequent generation of auto trips probably stems more from inadequate public transportation supply than from high income or any idiosyncratic custom of the city's inhabitants (Figure 1).

One can say that in Sao Paulo, traffic and public transportation are tied together in a vicious circle. The automobile trips grow and saturate the streets because public transportation is not satisfactory; consequently, the performance of transit further deteriorates[i]. This fact causes an additional growth in the auto trips. The perverse cycle continues and results in heavy traffic congestion.

2. TRAFFIC CONGESTION IN THE CITY OF SAO PAULO

FIGURE 2 – AVERAGE EXTENSION OF "CONGESTION QUEUES LENGTH" (CQL)

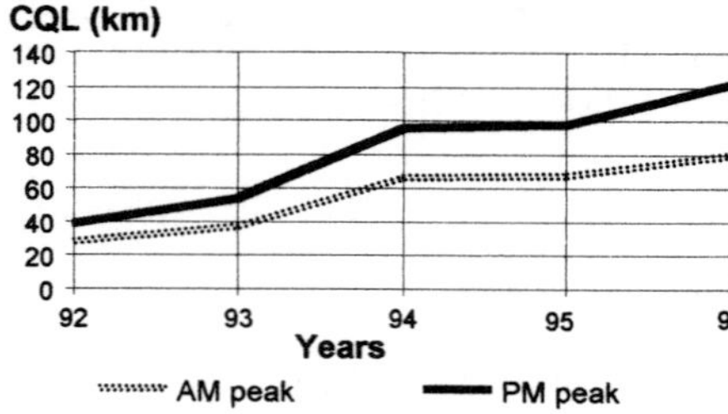

Figure 2 shows the growth of "congestion queues length" in the City of Sao Paulo in recent years.

The concept of "congestion queues length" (CQL) was developed by CET (*Companhia de Engenharia de Tráfego de Sao Paulo*, the City's traffic management agency) and implemented in 1991 to quantify traffic congestion. The measurement is based on the empirical distinction between traffic categories: free-moving, slow[ii], stop-and-go and standing still. The CQL is formed by the sum of the queue lengths of the three latter categories.

CQL measurements are made in 47 advanced field observation posts. These posts located on top of strategically situated buildings monitor the City critical areas comprising 570km of the main road network. (Figure 3 - next page).

Based on visual references selected in the streets, the operator of each observation post equipped with binoculars and radio transmission sets estimates the extension of

FIGURE 3 - INTERNAL AND EXTERNAL BELTWAYS - SAO PAULO METROPOLITAN REGION (the construction of the external beltway will be startina in the nest months)

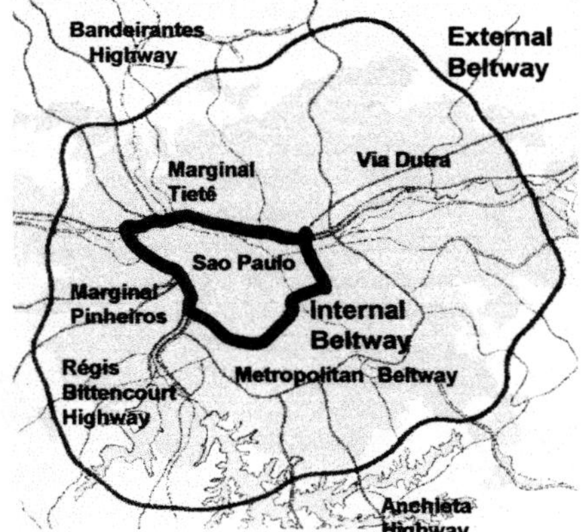

CQL and informs the Operations Center of CET.

The average growth of peak CQL was 200% in the four years before 1997 (Figure 2). In 1996 a record of 242km of CQL took place, during the PM peak, corresponding to 43% of the extension of the monitored road network.

Figure 4 illustrates the hourly evolution of average CQL in a typical month, October 1996, prior to the implementation of the private cars circulation restriction policy known as "Rodizio"

Figure 5 (next page) shows the monthly variation of average peak CQL for 1995 and 1996, also before the implementation of the above-mentioned "Rodizio".

3. THE "RODIZIOS" – NUMBERPLATE CIRCULATION RESTRICTION

FIGURE 4 - HOURLY VARIATION OF AVERAGE CQL IN A TYPICAL MONTH - source CET/SP

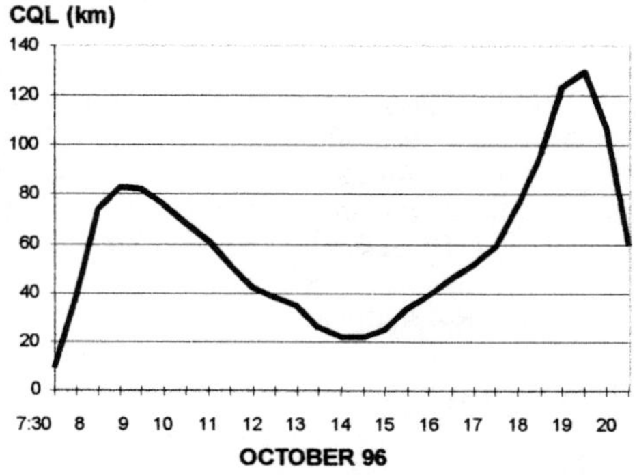

The progressive worsening of traffic congestion contributed to the deterioration of atmospheric conditions. Surveys conducted by CETESB (*Companhia de Tecnologia e Saneamento Ambiental do Estado de Sao Paulo* - Sao Paulo State's environmental agency) in the City's central region, using the amount of carbon monoxide in the atmosphere as indicator, showed that during the five year-period (1992-1997) only in 36.0% of the winter days the air quality was considered good. To tackle the problem CETESB enforced, in 1995, a circulation restriction program called "Rodizio" or Numberplate Restriction Operation, applicable to part of the metropolitan region during the winter.

The City in its turn had implemented policies to curb traffic congestion. After the exhaustion of these policies, the City Administration adopted a similar initiative – the City "Rodizio". Differently from the CETESB "Rodizio", the City "Rodizio" is enforced only during peak hours and is applicable to a restricted downtown area. The CETESB "Rodizio" covers the Metropolitan Area and is enforced from 7am to 8pm.

FIGURE 5 - MONTHLY AVERAGE PEAK CQL - source CET/SP

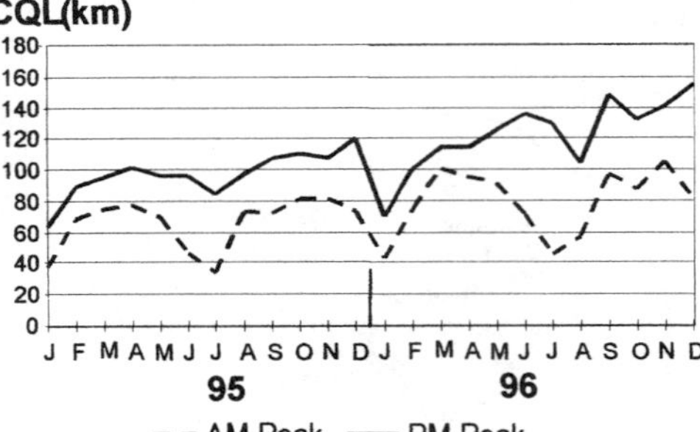

3.1. The CETESB "Rodizio"

Area covered: Part of the Sao Paulo Metropolitan Region (10 municipalities of the Greater Sao Paulo)

Enforcement Period: from July to September,

Restrictions in force: 20% of the licensed vehicles for each workday, using as reference the last digit on the numberplate.

Hours of enforcement: from 7:00 am to 8:00 PM

Implementation:

a. One free-willing experience was carried out during five days, from August 28 to September 1 1995.
b. A second experience, enforced instead of free willing, with fines applied to offenders, was carried out during 25 days in 1996 (from August 5 to 30th).
c. From 1997 on the policy was extended for the entire winter period - from June 23 to September 26.

The "Rodizio" consists in the restriction of circulation for each car during one day every week. The restriction is based on the last digit of the numberplate. Every working day corresponds to the restriction of two digits. Monday for instance prevents the circulation of cars whose numberplates have either digit 1 or digit 2 as its last digit. The restriction day is shifted every month.

3.2. The City "Rodizio"

Area covered: the expanded center of downtown Sao Paulo (area within the Inner Beltway)

Enforcement Period: February to June and October to December.

Restrictions in force: identical to the CETESB program

Hours of enforcement: from 7:00 am to 10:00 am and from 5:00 PM to 8:00 PM

Start of enforcement: October 13, 1997

4. IMPACTS OF THE "RODIZIO"

4.1. Impacts on CQL

Figure 6 compares the average CQL before (Oct96-Mar97) and after (Oct97-Mar98) the implementation of the City's "Rodizio".
Figure 7 compares the average CQL without (in Jun97) and with (in Jun98) the application of the CETESB "Rodizio".

Figure 8 shows the variation of the annual average CQL during AM and PM peaks from 1994 to mid-1998. The period covers the joint action of both types of "Rodizios" and the period immediately prior to them. This diagram shows a reduction of 29,5% in the average CQL.

According to CET, average CQL during the City's "Rodizio" was 37% (AM peak) and 26% (PM peak) lower than CQL before the adoption of the circulation restriction. The average reduction found for the period between 7:00 am and 8:00 PM was 17.7% (Source - *CET's Evaluation Report, April 1998*).

FIGURE 6 – IMPACT OF THE CITY "RODIZIO" ON CQL - source CET/SP

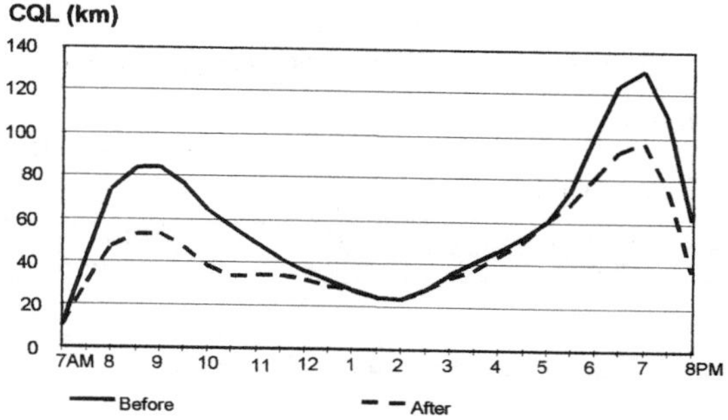

FIGURE 7 - IMPACT OF THE CETESB "RODIZIO" ON CQL

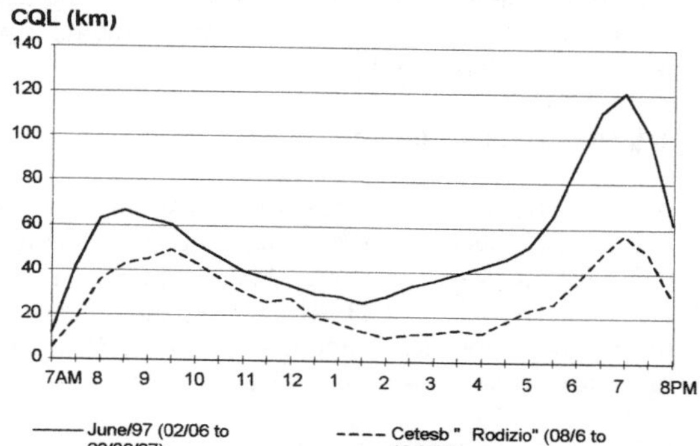

FIGURE 8 - AVERAGE CQL BETWEEN 1994 AND 1998 - source CET/SP

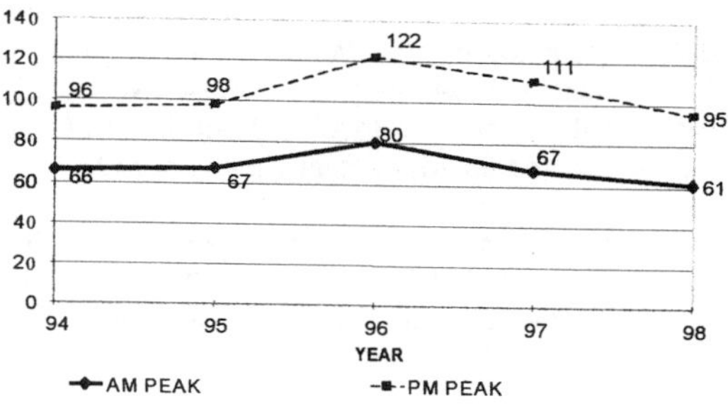

4.2. Impacts on travel time and average speed

CET conducted a field survey to monitor traffic performance on two important city avenues, between October 1997 and March 1998. An automobile equipped with chronometer was driven at peak periods along the predominant direction of the flow.

Table 1 below indicates the improvements in the average speed and travel time.

TABLE 1

		Before the Rodizio	During the Rodizio	Variation
Travel time	Morning	21' 27"	17' 37"	- 18%
	Afternoon	22' 46"	18' 42"	- 18%
Average Speed	Morning	18.6 km/h	22.8 km/h	+ 23%
	Afternoon	17.5 km/h	21.6 km/h	+ 24%

4.3. Impacts on the volume of vehicles

During the same period - October 1997 to March 1998 - CET conducted a volumetric AM and PM peaks survey at seven important avenues of the city. Comparing the survey results with corresponding data obtained before the adoption of the Rodizio, CET found a reduction in the hourly volume of 2% during the AM peak and 5% during PM peak.

4.4. Impact on air quality

Atmospheric Pollution Measuring Stations located in the City's central region indicated a reduction of the carbon monoxide amount released by vehicles during the CETESB "Rodizio" in winter 1997, compared to the average of the five previous years. According to CETESB the "Rodizio" reduced the occurrence of negative and inadequate atmospheric conditions from 14.7% to 3.9% and increased the occurrence of positive conditions from 36.0% to 57.8% of the time during winter.

4.5. Impact on public transit

Figure 9 (next page) shows the variation of the mobility in the metropolitan region between 1967 e 1997 (Source - *surveys conducted by Companhia do Metropolitano de Sao Paulo – City Subway Company)*; the 'mobility' rate meaning the number of daily trips made by the average City inhabitant excluding walking displacements. The reduction of mobility in the last two decades corresponds to the difference between the number of trips made in the 1997, had the mobility rate of 1967 been maintained and the trips actually made in 1997.

Virtual daily trips in 1997: 1.53 x 16,800,000 = 25,704,000
Actual daily trips in 1997: 10,300,000 + 9,960,000 = 20,260,000
25,704,000 – 20,260,000 = 5,444,000
(See Table 2 - next page)

Table 2 shows the changes in the modal split, demonstrating a reduction of the share of public transportation from 45.6% (in 1977) to 33.36% (in 1997). At the same time the share of auto person trips rose from 29.16% to 32.24%, while pedestrian trips rose from 25.23% to 34.36%.

Table 2 also shows a hypothetical distribution of trips for 1997, assuming the 1977 modal split. These figures indicate the following migrations among modes, from 1977 to 1997:

FIGURE 9 - VARIATION OF THE MOBILITY IN THE METROPOLITAN REGION

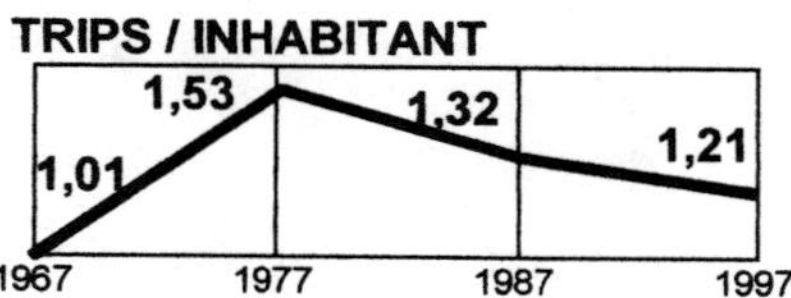

Suppressed trips:	5,444,000
Trips lost by public transportation:	3,780,000
Trips gained by cars:	958,000
Increase in pedestrian trips:	2,822,000

TABLE 2 - MODAL SPLIT IN THE METROPOLITAN REGION - 1977/1997 (Number of Daily Trips in Thousands)

Year/ Modal	1977		1987		1997		1997 w/ distr of 1977	Migration
	Nr. trips/day	%	Nr. trips/day	%	Nr. trips/day	%	Nr. trips/day	Nr. trips/day
Publ transp	9,759	45.61	10,343	35.17	10,300	33.36	14,080	- 3,780
Auto	6,240	29.16	8,473	28.83	9,960	32.24	9,002	+ 958
Pedestrian	5,400	25.23	10,591	36.00	10,610	34.36	7,788	+ 2,822
Total	**21,399**	**100**	**29,407**	**100**	**30,870**	**100**	**30,870**	**0**

Statistical data covering the years of 1996, 1997 and 1998 - the "Rodizios" period - show a recurrent ridership reduction, on the City buses and the Metro (subway), which is consistent with the structural tendency of the last decades. This phenomenon, in fact, is observed not only in the City of Sao Paulo but also in all large Brazilian cities and in other countries. According to the PDU (Plan de Déplacements Urbains – developed for the Ile-de-France region) it is a global trend: in France, for example, between 1976 and 1991, private car trips grew 33%, and estimates for the future tend to indicate an increasing use of private car resulting from what has been called "peri-urbanization", should no measures be adopted to make the collective transportation system meet the population's needs.

This means that the "Rodizio" was not efficient in reverting this trend, that is, the structural reasons of the phenomenon outweigh the action of the "Rodizio". We can imagine that part of these reasons stem from the effects of the vicious circle mentioned in Section 1.

The 958,000 trips/day that migrated to individual transportation correspond to an additional of 639,000 vehicles (using the conventional auto occupancy rate of 1.5 passengers/vehicle). This additional number of cars is equivalent, roughly, to the number of vehicles withdrawn from circulation by the "Rodizio". Studies conducted by CET associated to broader analyses developed by IPEA (Instituto de Pesquisa

Econômica Aplicada, an agency of the Brazilian federal government) provide further evidence of the impact of the Rodizio on the public transportation system. According to the study by CET[iii], of 2.70 million passenger hours spent on trips (data from the metropolitan region, obtained from the O-D 1987 survey) by bus in workdays without "Rodizio", 900,000 are lost in traffic congestion. This figure equals approximately 180 million hours for these losses.

As CET's analyses on the effect of the "Rodizio" showed reductions in travel time of approximately 18%, we may assume that between 21 and 32 million hours a year may have been "recovered" for the users of the public transportation system. The monetary impact of such savings will depend on the value attributed to the user's time. Another positive effect of the "Rodizio" in public transportation was the reduction observed in the total number of trips made by buses amounting to 2,400 a day (corresponding to 5% of the total of 48,000 trips made in average, according to information provided by SPTrans - Institution that manages the bus transit system in the City of Sao Paulo). This reduction contributes to control the excess of bus supply provoked by congestion, which is represented by additional vehicles in the bus fleet, between 7.4% - AM peak - and 30.3% - PM peak (source IPEA study). As mentioned in the IPEA study, the additional buses cause an increase of 16% in the operational cost of the whole bus system. This cost is either supported by the users or covered by government subsidies.

5. THE BALANCE OF THE "RODIZIO"

The "Rodizio" demonstrated clearly to the population that circulation restrictions to the private cars are essential to ensure the mobility that is every citizen's right. The "Rodizio" obtained widespread support because it distributed the burden of the limitation of the right of circulation among all those who own a private car.

As previously explained, the positive effects of the "Rodizio" on the bus transit are also significant. These benefits, however, tend to be extinguished since the "Rodizio" will loose effectiveness soon. Indeed, after 1994 the CQL increase rate is higher than that of the total number of automobiles growth, because the latter has attained a critical size (For a 14% growth of the licensed autos between 1993 and 1997, the CQL rose 140% in average). On the other hand, the number of licensed vehicles continues to grow at a rate of 300,000 vehicles a year. Therefore, for the "Rodizio" to maintain the present traffic flow conditions–since there are no projects to expand significantly the road network – it will be necessary to compensate the future vehicles growth by inhibiting the circulation of additional quantities of autos. At the very least, should no other measures to manage traffic congestion be adopted or if no major expansion of the public transportation system is feasible, the existing vehicles should remain at the current level. This implies inhibiting the circulation of:

- 20% of the newly licensed cars, already covered by the current "Rodizios"

- an additional amount, equivalent to the remaining 80% newly licensed cars

It is estimated that the average number of vehicles in circulation on workdays is approximately 60% of the total number of licensed autos. Thus the intensification of the "Rodizio" should attain, each year, an additional:

80% x 60% x 300,000 = 144,000 vehicles

Since at present 588,000 vehicles (4.9 million vehicles x 60% x 20%) are withdrawn from circulation, the necessary annual intensification of the "Rodizio" would be of 144/588 = 0.24, or 24%. This means that the "Rodizio" should be doubled in approximately four years, extending to two days a week, or 40% of the workdays. Inversely, if there is no intensification, the expansion of the licensed vehicles will return in four years to circulation all the vehicles withdrawn by the "Rodizio", canceling its effect (Figure 10).

FIGURE 10 - VEHICLES WITHDRAWN BY THE "RODÍZIO" X RETURNED BY THE EXPANSION OF THE VEHICLES

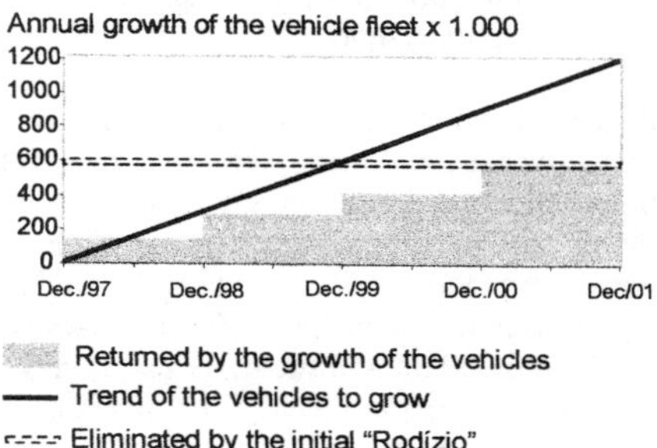

The further intensification of the "Rodizio" will require a complex control and management mechanism; different of the procedures easily assimilated by the population, based on numberplate digits. Besides, this policy would have political limitations. Stronger circulation restrictions reaching 40% of the working days will be seen as unacceptable prohibitions, only imaginable in case of calamities.

The adoption of the "Rodizio" on a permanent basis with progressively harsher measures, will very likely lead to an increasingly number of compensatory measures adopted by auto owners, which tend to make the "Rodizio" not effective, like:

- New individual plans that will revert the non-circulating vehicles (approximating 100% - 60% = 40% of the total) to circulation;
- The mobilization of an additional car (for each household) with a different numberplate, probably of lesser quality and more polluting.

The "Rodizio" must therefore be considered an emergency tool to be used temporarily until other measures can be implemented to provide the required level of mobility.

6. NEW MEASURES TO IMPROVE MOBILITY

A series of measures are still feasible to improve the traffic conditions in Sao Paulo, benefiting the public transportation system. Some are being implemented progressively; others can be adopted based on urban development policies. To mention but a few:

- Changes in the parking price policy, inducing modal transfers in peripheral regions served by stations of the mass transportation system. Inversely, penalties for parking in critical areas;
- alteration of the so called "Blue Zones" (paid parking lanes on both sides of selected streets and avenues) policies, to vascularize the arterial grid;
- changes in land-use legislation to reduce the extent of trips and/or suppress part of them;
- expansion of the road capacity through regulation, minor geometric adjustments or use of ITS resources

These and other measures that can be described will, jointly, be beneficial, and will help sustain the benefits generated by the "Rodizio", but it is easy to understand that they are unlikely to have the necessary efficiency to solve the whole problem. Breaking the previously mentioned vicious cycle implies the introduction of new mechanisms to discipline the circulation of private vehicles, managing the scarcity of road capacity. This policy will support the actions in progress in the municipal and metropolitan jurisdictions of Sao Paulo, funding the existing proposals to expand the mass transport system.

A mechanism that can attain this objective is a congestion pricing policy on the main roads during peak hours. The revenues thus obtained should be reinvested directly into public transportation, giving priority to the modes operated in totally segregated routes both above and below the surface. Congestion pricing will be able to recover the huge economic losses caused by traffic congestion, turning them into resources and reinvesting them for useful purposes without letting them become a new burden on the population. On the other hand, they will free the government budget from its role as main funding source for investments in public transit. This solution may be promising for developing countries' mega-metropolis, given their fiscal base limitation. One way to ensure the allocation of net revenues from congestion pricing to improve public transit consists in conceiving combined road pricing and public transportation concession packages. In other words: to associate each Congestion Pricing Area, supposedly with high revenues and a short implementation time, to one or more specific projects with a long implementation time and a slow return of investment. Congestion pricing lots can be associated to subway lines, to LVT (Light Vehicles on Tires) or LRT lines, to the External Beltway segments, to commuter rail branches, busways, and so on.

7. THE CONGESTION PRICING PROGRAM

A concept for a Congestion pricing[iv] program for the City's central region was devised in 1995. The aim of the proposal is to withdraw from circulation during peak periods 20% of the vehicles in that critical region. The subsequent adoption of the municipal "Rodizio" supposedly had the same objective and came very close to achieving that target. Once 21 critical routes with an extension of 210 km had been selected and considering the hypothesis of a US$ 2,00 - 2,50 tariff to drive along one of these routes during peak traffic hours, the net revenue generated, discounting the capital and operational costs of the Congestion Pricing System, was estimated in the range of US$ 550 million - US$1 billion a year.

According to that proposal, the US$ 550m would correspond to 16% of the losses caused by traffic congestion and to 11% of the losses generated jointly by traffic congestion and by the induced use of cars as an alternative to public transportation. The revenues would be injected in the mass transportation system. Until the achievement of a significant increase in transit supply, a temporary public transportation element would be implemented. It will consist of a fleet of comfortable vans operating on a structured network with a large geographical coverage. The tariffs should be affordable to those low-income car owners affected by the road pricing financial impacts. Once the traffic congestion has diminished and public transportation services with the required extent and quality have been provided, the proposal would have completed its equitable objectives:

- eliminating the burden that the entire population is paying for the losses caused by traffic congestion;
- turning this socialized cost into payment for its use: the users of this scarce commodity - road space - would have the opportunity of choosing either to use or use not their cars through their individual evaluation of the cost/benefit ratio.

8. A PROGRAM TO REVERT THE VICIOUS CYCLE

The Congestion Pricing Program will lead to the productive mobilization of resources currently lost as idle costs. It will permit the allocation of such resources in a more equitable manner, correlating them to the use of public assets, instead of passing them on indistinctly. It will allow integrated transportation planning, linking factors that are interdependent by their very nature and approaching them with a strategic vision.

This proposal presents huge political obstacles. However, in the large towns of the developing countries there are no other effective alternatives to solve the public transportation/traffic deadlock. If adopted, the Program will become a constructive mechanism to improve accessibility, atmospheric conditions, as well as the work and leisure opportunities. In ten years' time a joint concession <Road Pricing - Selected Projects> Program, undertaken under a project finance scheme, will leverage investments for the implementation of:

- the remaining 120 km of the Metropolitan External Beltway;
- 100 km of new subway lines (underground);
- 100 km LVT lines (Light Vehicles on Tires);
- 350 km of pavement upgrading of dual carriage ways;
- leisure areas and public parks worth US$ 1.5 billion;
- an urban rehabilitation program US$ 1 billion;
- an environmental education, transit and use of the public transportation system program worth US$ 500 million.

These investments will allow the recovery of the mobility that the Sao Paulo population had in 1977, that is, 1.53 trips/inhabitant/day in five years. Massive investment in public transport will benefit the low-income groups, which suffered the most from the loss of mobility.

Notes

[i] Public transit is performed basically by buses, as demonstrated in the previous papers on Sao Paulo.

[ii] By definition, slow traffic occurs when the green light time at intersections is not able to absorb the queue of vehicles.

[iii] *Relatório de Avaliação do Rodizio*, April de 1998

[iv] Scaringella, Roberto - *Sistema de Trânsito Tarifado, uma solução abrangente - Revista Perspectiva – Fundação SEADE* – June 1995
A asfixia do trânsito e a decadência urbana em São Paulo – Sistema de Trânsito Tarifado, February. 1995 – Published by Logos Engenharia S.A.

Bibliography

1. ANTP (Associação Nacional de Transporte Públicos) IPEA (Instituto de Pesquisa Econômica Aplicada) - *Redução das Deseconomias Urbanas com a Melhoria do Transporte Público - Relatório Síntese - May 1998*
2. Biezus L. - *Abordagens integradas dos problemas urbanos devem substituir soluções tópicas - Informativo SINAENCO – Ano VI, nr. 17, August 1997*
3. CET (Companhia de Engenharia de Tráfego) Prefeitura do Município de Sao Paulo - *Operação Horária de Pico - Relatório de Avaliação - April 1998*
4. CETESB (Companhia de Tecnologia de Saneamento Ambiental) Governo do Estado de Sao Paulo - *Relatório de Qualidade do Ar no Estado de Sao Paulo - 1996*
5. Logos Engenharia S.A. - *A Asfixia do Trânsito e a Decadência Urbana em Sao Paulo - Uma solução Abrangente - February 1995*
6. Scaringella R., Garcia M. et al. - *A Melhoria do Trânsito pela Tarifação de Vias Congestionadas - Proposta de uma política Integrada de Investimento e Gestão (Anais do Congresso da Associação Internacional de Transporte Público) - Curitiba, 1996*

Socioeconomic Characteristics, Land Use and Travel Patterns in Seoul

Jongho Rhee[1], Keeyeon Hwang[2]

I. Introduction

Seoul is notorious for its severe traffic congestion. To address the problem, the policy focus of Seoul Metropolitan Government (SMG) had been placed on the supply side (new highway and subway construction) until the early 1990's. The road construction, TSM (transportation system management), and subway line expansion were key words of transportation policy. However, after 1993, the city started a different approach to the traffic problems in Seoul. The approach was transportation demand management (TDM) in association with the improvement of service quality of public transit system operated in Seoul. The policy concern shifted from concern about transportation system supply to its users. Beginning in 1995, another important policy item was implanted in the policy list, the green transportation which placed emphasis on encouraging walking and biking.

The purpose of this paper is to understand the nature of the city in terms of the need for public transportation. The public transportation modes to be dealt with are subway, bus, and taxi. This paper consists of three parts. At first, it will discuss the socioeconomic and land use elements, which affect the transportation environment in Seoul. In the second part, historic reviews on the travel patterns will be presented. Finally, it will discuss specific policy alternatives applied for solving the problems in Seoul.

II. Land Use & Socioeconomic Characteristics

The very densely populated city of Seoul is over 600 years old. In 1997, a population of 10.6 million resided within an area of 605 km^2, only 374.5 km^2 was used for human activities. The Han River flows in the middle of the Seoul Metropolitan Area, which is surrounded by mountains. Only a small area, the southwestern part of the Han River, is zoned for industries. Numerous shopping centers and about 30 colleges and universities are spread all over Seoul. The roadway system of Seoul is a radial structure, and therefore there is heavy traffic congestion in the central area of Seoul. Over 25 percent of total trips commute toward the CBD. Around 6 percent of total trips is through-traffic just passing through the central area (see Picture 1).

---

[1]. Kyonggi University, Associate Prof. Department of Traffic Engineering, Kyonggi-Do, S.Korea

[2]. Seoul Development Institute, Research Fellow, Department of Urban Transportation, Seoul, S. Korea

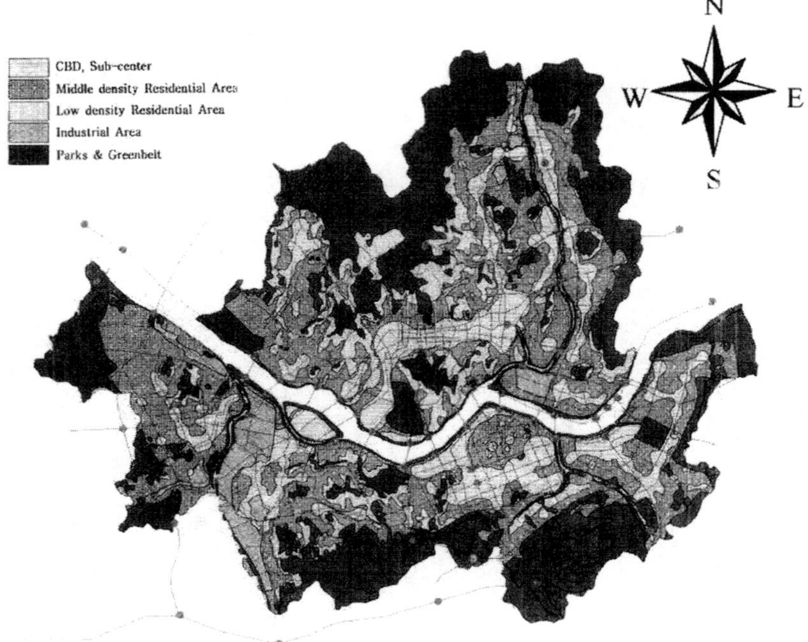

Picture 1: Land Use Map of Seoul

The population of Seoul has increased continuously over the last 25 years until it peaked in 1995 (See table 1). The number has risen from 5.433 million in 1970 to 10.595 million in 1995. As the residential movement to newly built satellite cities began in 1990, the residential population was down to 10.418 million in 1996. Seoul Development Institute (SDI) forecasts the population will be diminishing to 9.839 million in 2006 (SDI, 1998).

On the other hand, the employment has been increasing continuously, which implies the movers to satellite cities still commute to Seoul for work. SDI forecasts the employment number in Seoul will grow up to 6.295 million in 2006 in spite of the declines of residential population. This trend implies that Seoul would have more long-distance commuters in the future.

Table 1: The trends of residents and employment

	1970	1980	1985	1990	1995	1996	1997
City residents (in thousands)	5,433	8,366	9,626	10,726	10,596	10,464	10,389
Employment (in thousand)	-	-	2,926	4,505	5,014	5,073	-

Source: Seoul Metropolitan Government, 1998

The number of daily trips increased over four times from 1970 to 1995 (See table 2). It jumped from 5.750 million in 1970 to 27.099 million in 1995. The number has been increasing steadily due to continuing growth of vehicle ownership in the region and long-distance commuters from satellite cities. While the average daily person trips were 2.29 in 1990, it jumped to 2.62 in 1997. This increase is related with both car-ownership increase and tripled household income during the period between 1985 and 1995.

Table 2: The Trends of Daily Trips

	1970	1980	1990	1995	1996	1997
Daily trips (in thousands)	5,750	12,600	24,638	27,099	27,762	27,203

Source: Seoul Metropolitan Government, 1998

III. Travel Patterns

1. Roadways

The total length of roadways in Seoul was 7,689 km in 1996 (See table 3). The length of two-way streets whose width is wider than 12m is only 1,590km. It is only 20.5 percent of the total length. The size of the street over 12m width is 42.7 km^2, 57% of the total, and the road rate is about 20.42% in 1997 (picture 2).

Table 3: Roadways

Year	Length(km)	Area(km^2)	Paved road ratio(%)
90	7,427	69.3	18.5
95	7,675	74.4	19.85
96	7,689	75.6	20.19
97	-	-	20.42

Source: Seoul Metropolitan Government, 1998

The Olympic Freeway, the Riverside Freeway and the Dongbu Freeway are the backbone of the Seoul Freeway and roadway system. The total length of these three freeways is only 135km. Thus, they are heavily congested in all sections throughout a day. A signal operation system is essential to road efficiency. However, Seoul's signal operation system does not respond automatically to changes in traffic volumes. Instead, it is fixed or actuated to predetermined cycles.

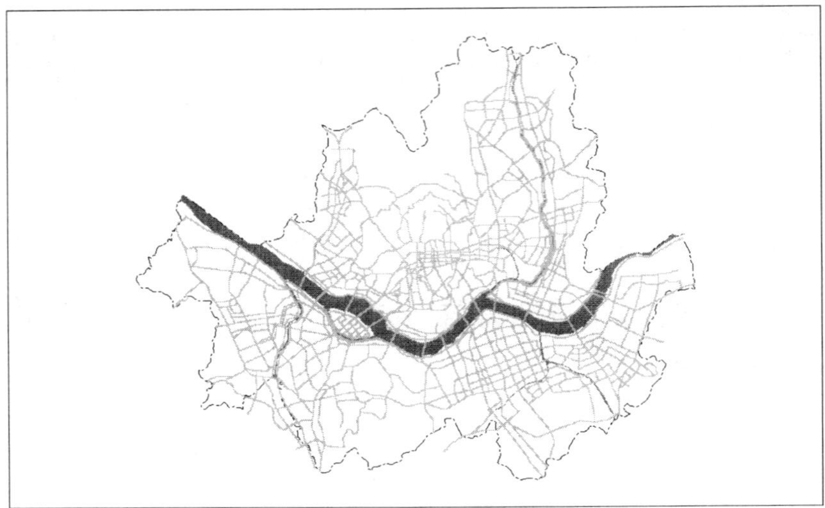

Picture 2: Arterial Roadways

2. Main Travel Modes

The trend of vehicle ownership has been in line with the rapidly increasing trends of daily trips in Seoul. The number of vehicles increased from 60,000 vehicles in 1970 to 207,000 in 1980 (see table 4). There was a 3.5% increase in vehicle ownership during the period. The number increased more rapidly beginning the late 1980s. They reached one million in January 1990. There was a 5.7% increase in vehicle ownership for the last ten years. The yearly growth rate has been decreasing since 1990, however, the total number still keeps rising. In 1997, over 2.2 million vehicles were registered in Seoul. They are quickly turning the city into auto oriented society.

Table 4: The Trend of Vehicle Ownership

	1970	1980	1990	1995	1996	1997
Ttotal number of vehicles (in thousands)	60	207	1,193	2,043	2,168	2,249
Passenger cars (in thousands)	18	99	883	1,595	1,704	1,698
Vehicles per 1000 Population	10.9	24.7	109.4	192.8	-	216.5
Vehicles per 1000 house holds	54.7	112.4	358.4	592.5	-	642.7

Source: Seoul Metropolitan Government, 1998
Note: In 1998, the number has been decreasing by 50 per day due to the IMF economic crisis.

The city of Seoul has two subway systems; Seoul-subways and Metro-rails. The Seoul Metropolitan Government is fully in charge of the system construction. In 1997, the Seoul subways consisted of 4 routes with 133km in length, while the Metro-rails were comprised of 3 routes with 83.5km in length. In addition to the subways, 57.3 km of Korea-Rails are served within the city boundary. They are constructed and operated by the Korea National Rail Company.

Seoul's bus system consists of company buses and community buses (see table 6). The total 89 private bus companies operated 398 routes with 8,655 vehicles in 1997. In 1997, the company buses consist of 4 different types; 6,399 regular buses, 1,960 deluxe buses, 296 circular buses. While the number of company buses s decreased, the number of community buses is increasing. In 1997, the total 232 routes were operating with 1,260 vehicles and average route length was 7 km.

There were 69,635 taxis in 1997, which consisted of 23,187 company taxis and 46,448 private taxis. The private taxis included 4,652 deluxe taxis. The supply of taxi was restricted to 70,000 vehicles. Currently, the customer-in-taxi running ratio is 70% out of total running km. Therefore, it is still not easy to catch a taxi in Seoul. The fare of regular taxi is 1,300 won ($1.09) up to 2 km, and increases 100 won ($0.08) by 210m and increases 100 won ($0.08) per 51 second when running under 15km/h. On average, customers pay 3,500 won ($2.94) for a 6km ride. The fare of deluxe taxi is 3,000 won ($2.52) up to 3km, and increases 200 won ($0.17) by 250m and 60 second.

3. Mode Share

Overall, the ridership of buses and taxis is declining, while subway and auto vehicles have been increasing (see table 5). In 1997, the subway has become, for the first time, the most predominant travel mode in Seoul, accounting for 30.8 percent of daily trips. Until 1996, buses were the most heavily used travel mode in Seoul. The bus patronage has kept declining both because of worsening road traffic congestion and widening subway services. Taxis served 10.1 percent of person trips in 1997, but the share has been decreasing continuously. Other vehicles include two-wheeled vehicles, rental vehicles, and passenger cars. A sharp increase of passenger car share between 1995 and 1996 is partially attributed to a disclosure of new O-D survey results conducted by Seoul Development Institute in 1996. It is noteworthy that inspite of heavy investment on subway construction, it contributed more to the reduction of bus ridership rather than reducing auto vehicle use.

Table 5: Percent of Daily Trips by Travel Modes

	1980	1985	1990	1995	1996	1997
Buses	66.0	58.0	43.3	36.7	30.1	29.5
Subways	7.0	14.0	18.8	29.8	29.4	30.8
Taxis	19.0	16.5	12.8	10.7	10.4	10.1
Others (Pass. Veh)	8.0 (NA)	12.5 (NA)	25.1 (14.0)	22.8 (14.5)	30.1 (21.1)	29.7 (20.6)

Source: Seoul Metropolitan Government, 1998
Note: For the subway case, transfers between different lines were counted as another independent travel.

4. Road Traffic Conditions

Seoul's traffic conditions were normal, without a serious congestion problem up to the early 1980s, except during the period of rush hours. However, traffic patterns changed from rush-hour peaks to all-day peaks by the end of 1980s. The overall traffic speed on major arterials in Seoul kept declined until 1996 and bounced back in 1997 (see table 6). This trend resulted from significant changes of the traffic environment in Seoul. The economy started sliding down in 1997 and passenger vehicle drivers using Namsan 1&3 Tunnels linked to the CBD were charged 2,000 won ($1.68) congestion fee beginning in the late 1996. Furthermore, the most recent speed survey shows a drastic improvement of road speed. The sudden change is associated with the sharp increase in both the unemployment rate caused by IMF economic crisis and an increase of oil price by about 30% increase of oil price as of December 1997 (SMG 1998b).

Table 6: The Trend of Travel Speed Changes (unit : km/hr)

		1989	1990	1991	1992	1993	1994	1995	1996	1997	1998
auto	All	32.60	24.22	24.57	22.62	23.53	23.18	21.69	20.09	21.06	25.41
	CBD	18.70	16.40	17.66	19.28	19.97	20.04	18.25	16.44	16.85	17.72
	other	37.17	25.78	21.89	22.87	23.79	23.40	21.93	21.23	21.33	25.90
bus		18.60	18.80	18.15	16.88	17.02	18.42	18.79	18.35	18.69	20.07

Source: Seoul Metropolitan Government, 1997b, 1998b, 1998c

The trend of road traffic speed changes is closely associated with the rapid increase of vehicle ownership and uses. The traffic volumes on major arterials have increased continuously over the last ten years (see table 7). Traffic crossing the boundary of Seoul has almost tripled during this period. The majority of them are long-distance trips, and therefore must have had a sizable negative impact on the traffic speed. Considering this tendency, the future transportation policy of Seoul Metropolitan Government should be focused on dealing effectively with these long-distant auto trips.

Table 7: The Trend of Traffic Volume on Cordon Lines (unit : 1,000 trips)

	1989	1990	1991	1992	1993	1994	1995	1996	1997
Total	3,792	4,257	4,516	4,578	5,402	5,894	5,601	6,069	6,345
CBD	1,288	1,270	1,286	1,277	1,706	1,819	1,709	1,730	1,657
Han river bridges	1,394	1,562	1,541	1,611	1,721	1,763	1,804	1,906	1,927
Border lines of Seoul	1,170	1,425	1,689	1,740	1,975	2,433	2,088	2,433	2,761

Source: Seoul Metropolitan Government, 1997a

Note : There is a slight difference in the amount of CBD trips in 1993 because trips were surveyed from 06:00 hours to 22:00 hours until 1992 and from 1993, surveys were conducted for 24-hour periods.

5. Prospects for the Future

Traffic conditions in Seoul are projected to worsen. The travel demand is expected to increase continuously until the beginning of the year 2000 if the city successfully overcomes the IMF economic crisis in a near future. The number of vehicles is expected to increase to 3 million by 2001. The number of vehicles per one thousand households is expected to increase from 593 in 1995 to 810 by 2001.

The rates of trip generation and trip population will keep growing because economic, social, and leisure activities are expected to increase in the region. People will be more attracted to private transportation modes because they prefer more convenient and comfortable trips. Further, they will prefer faster public transportation modes because their value of personal time is increasing. The schedule of the second-phase subway construction is delayed until 1999 (planned in 1996). The plan for the third-phase subway construction is still in progress. Thus, the passenger transport capability will be insufficient in the near future.

IV. Conclusions

The ancient city of Seoul is not a planned, but a naturally formed city centering around the downtown. Seoul's radiated structure makes the traffic problems difficult to deal with. Since 1988 when the Seoul Olympics were held, the number of vehicles has been increasing steadily. The new satellite cities generated a large number of long-distance auto drivers. The land price is high and land supply is limited for road expansion. To make it worse, the nationwide economic crisis is restraining the government financing for subway building. In every respect, fighting traffic in Seoul will be a difficult task in a near future.

To respond to transportation problems under the circumstances, SMG sets its first priority of transportation policy as promoting public transit (for instance, bus, subway, and taxi) by improving their service quality. To encourage public transit uses, however, SMG thinks that TDM should be combined with transit improvement simultaneously. In Seoul, the share of private autos accounts for 60 percent of the total volumes on the streets, but they carry only 20 percent of the total daily trips. Because of excessive private vehicle uses, the total amount of social costs is estimated to be 2,400 billion won ($2 billion) per year, consuming 2.9 billion liters of gasoline per year. Auto use is held responsible for 72 percent of the air pollution in Seoul.

In order to increase the subway ridership, the SMG will provide more Park & Ride facilities from the current 21 lots with 4700 spaces to 61 lots with 28,700 spaces, and ten multi-mode transfer centers.

To improve the carrying capacity and convenience of buses, SMG will expand exclusive bus lanes from 37 routes of 154km in length to 64 routes of 270km by 1998, and expand the central exclusive bus lanes. For securing consistent high quality service, the city has under consideration the evaluation of the service level of each bus company based on the results of customer interviews and service monitoring. According to the evaluation results, it plans to differentiate the financial supports among bus companies. To make transfers more convenient among buses and subways, the SMG plans to integrate different fare collection systems into.

To have a more convenient and comfortable taxi service (easy to catch), the city plans to provide more deluxe taxis, increasing from now 4,652 vehicles to 20,000 vehicles, and permit increasing fares of regular taxis. It plans to expand the Call-Service system to the entire taxi fleet. As a result, the share of passenger mileage is expected to drop from 70% in 1997 to 60% in a near future. In addition, to prevent illegal taxi business near subway stations and remote areas, and to strengthen subway station accessibility, the modified taxi systems are under consideration, for instance, route taxi, van taxi, and so on. For securing qualified drivers, SMG will invest fare amount funds to build a welfare center for taxi drivers.

To cut excessive auto use, SMG started charging, starting from November 1996, 2,000 won ($1.68) congestion pricing for 1-2 occupant vehicles in Namsan 1,3 tunnels, and on major arterials linking the southern part of Han River with the CBD. The results of year-long implementation show traffic volumes reduced 13.6%, and that the average vehicle speed improved 38% from 21.6km/h to 29.8km/h. Therefore, the city plans to expand it in other major congested arterials in the near future.

The employer-base trip reduction program has been implemented since 1995, and intends to reduce the Traffic Impact Fee (compulsory tax burdened on large size building owners) over 20% if employers (or building owners) implement trip reduction programs, and reduce travel demand over 20%. It is similar to the Regulation XV program applied in Southern California, USA.

Also, the SMG plans to levy a local gasoline tax on the use of gasoline to control the excessive use (22 thousand km per car per year) of private autos. It is estimated that a 30% gasoline price increase lead to 7% reduction in auto traffic volumes (SDI, 1998). To prevent the opposition of city residents, the city is considering reducing the burden of the Automobile tax.

SMG introduced a Parking Ceiling system in 1997. It will increase the existing parking lot requirement for buildings from 20 percent to 40 percent in heavily congested areas and overpopulated areas, including the downtown.

Reference

1. Seoul Regional Police Bureau, '97 Cordon Line Traffic Volume Study, 1997a
2. Seoul Metropolitan Government, '97 Road Traffic Speed Study, 1997b
3. Seoul Metropolitan Government, General Plan for Bus Operation Reform, 1997c
4. Seoul Metropolitan Government, Seoul Yearly Statistics Book, 1998a
5. Seoul Metropolitan Government, Traffic Survey Results under IMF Economic Crisis, 1998b
6. Seoul Metropolitan Government, Summary of Transportation Bureau Minutes, 1998c
7. Seoul Metropolitan Government, '97 Transportation Indicators released for Press,1998d
8. Seoul Metropolitan Government, 100 Day Policy Plan of SMG, 1998e
9. Seoul Development Institute, A Preliminary Report: the Revision of the 3rd Phase Subway Construction, 1998

Urban Public Transportation in Seoul: System, Operation, Use by People

Keeyeon Hwang[1], Jongho Rhee[2]

I. Introduction

Subways and buses are two of the most important public transit systems served in Seoul. Bus service for Seoul started over 50 years ago. There are two different types of bus services in Seoul, which are 8,700 company buses and 1,300 community buses. Subway service in Seoul started in 1974. In 1997, subways and buses together accounted for over 60% of the mode share. Ultimately, the Seoul Metropolitan Government (SMG) hopes to increase this to 75 percent of total trips by 2001.

This paper describes details of these subway and bus systems. In this paper, however, community buses are not dealt with in detail because the firms operating them do not produce statistical data on their operations. Also, since company buses are owned by private firms, it is difficult to get the information requested. The data used is from 1996 and 1997. The paper consists of three parts: system characteristics, operations, and usage. The system part discusses system and route information. The operations part discusses organization, operation characteristics, and facilities. Finally, the part on usage includes mode shares, ridership trends, and user characteristics.

II. System & Facility Characteristics

Subway System

The Seoul Metropolitan Government is fully in charge of the subway system construction. The city of Seoul has two subway systems: Seoul-subways (line 1-line 4) and Metro-rails (line 5 – line 8). These systems are very similar in vehicles and power systems besides the dates of operations started. Seoul Metropolitan Subway Company (SMSC) is in charge of Seoul-Subway operations and Seoul Metropolitan Rapid Transit Company (SMRTC) is in charge of Metro-Rail operations. In 1997, the Seoul subways consisted of 4 routes with 133km in length, while the Metro-rails were comprised of 3 routes with 83.5km in length (see table 1). In addition to the subways, 57.3 km of Korea-Rails operates within the boundary of Seoul. The Korea-Rails were constructed and operated by the Korea National Rail Company, mainly providing intercity rail travel in the nation.

---

[1] Seoul Development Institute, Research Fellow, Department of Urban Transportation, Seoul, S.Korea

[2]. Kyonggi University, Associate Prof. Department of Traffic Engineering, Kyonggi-Do, S. Korea

Table 1: Subways and Urban Rails within Seoul (in 1997)

	Seoul Subway		Korea-Rails
	Seoul-Subways	Metro-Rails	
Length(km)	133	83.5	57.3
Construction Period	1971- 96	1990-96	Korea-Rails were constructed by Korea National Rail Company
No. of Stations	114	83	
No. of Trains	1,944	83	

Source: Seoul Metropolitan Government, 1998

Reviewing system characteristics by line, line 2 is the longest (see table 2 & picture 1). The construction of line 1 started in 1997, and line 2 took the longest time to build. There are 197 stations in total, and line 5 includes the highest number of stations (51). There are 2,778 subway cars amounting to an average of 13 cars/km.

Table 2: System Characteristics of Subway System by Line

	#1 line	#2 line	#3 line	#4 line	#5 line	#7 line	#8 line	total
Length (km)	7.8	54.5	35.7	32.3	52	16	15.5	216.5
Construction Period	71-74	78-96	80-93	80-94	90-96	90-96	90-96	-
No. of Stations	9	49	31	25	51	19	13	197
No. of Cars	160	834	480	470	608	136	90	2,778

Source: Seoul Metropolitan Subway Co., 1997

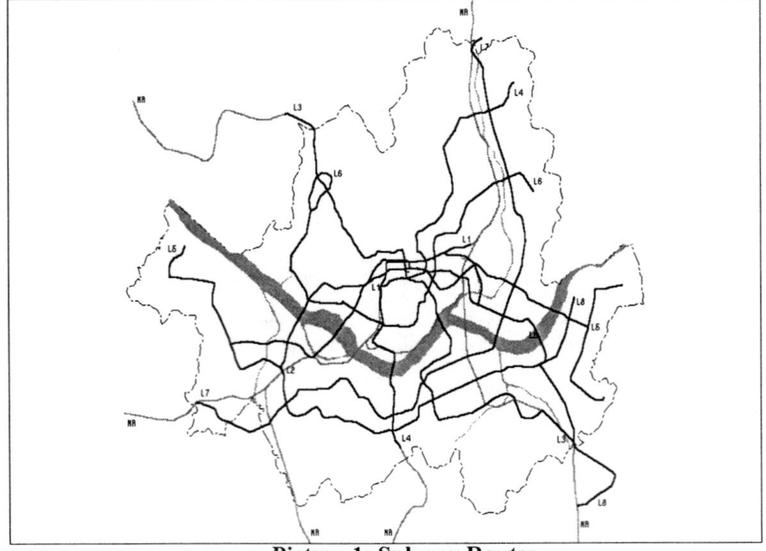

Picture 1: Subway Routes

Bus System

Seoul's bus system consists of company buses and community buses (see table 3). The 87 total private bus companies operated company buses with 398 routes and 8,655 vehicles in 1997 (see picture 2). The bus industry has declined in size from 1996 to 1997 by 70 buses, reversing previous growth rate from 1990.

In 1997, the company buses consisted of 3 different types; 6,695 regular buses and 1,960 deluxe buses. The regular buses (length: 9m, seats: 20) run an average of 7.3 times a day on 317 routes. The route length is 31.4km and average round trip time takes 102.7 minutes. There are 3,878 stops for regular buses.

On the other hand, the deluxe buses (length: 9m, seats: 45) run on average only 6.4 times per day on 131 routes. Since the route length is 48.8km on average and 17km longer than the regular buses, they are mainly in charge of serving long-distance users. Also, since running longer distance, the average running time per round trip takes 22 minutes more than the regular buses. There are a total of 892 stops for deluxe buses.

Table 3: Private Bus Operation System

		1985	1990	1995	1996	1997
Company Bus	**No. of Routes**	348	379	460	448	398
	No. of Buses	8,301	8,283	8,725	8,725	8,655
	No. of Companies	90	90	89	89	87
Community Bus	**No. of Routes**	-	112	Na	na	232
	No. of Buses	-	446	Na	na	1,260
Bus-only lane (km)		-	-	161.4	228.5	242.8

Source: Seoul Metropolitan Government, 1997, 1998

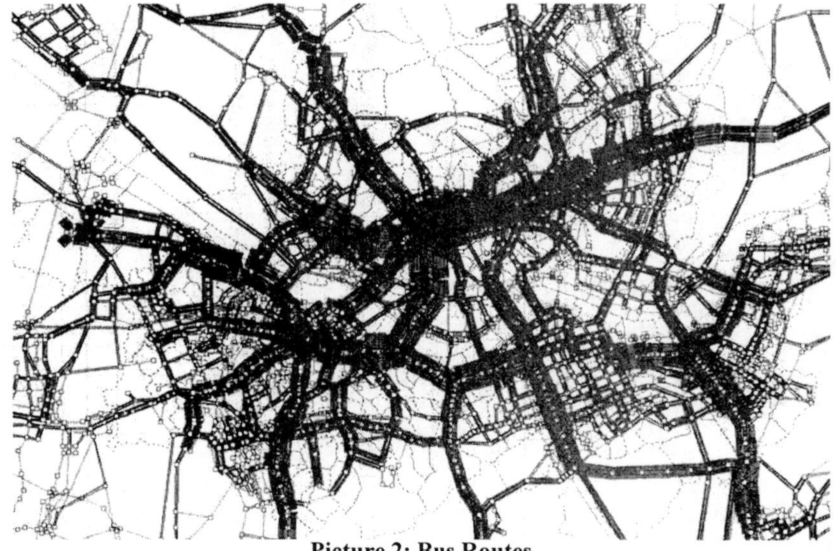

Picture 2: Bus Routes

While the number of company buses is remaining steady, that of community buses is increasing (see table 3) rapidly. In 1997, a total of 232 routes were operating with 1,260 buses and an average route length of 7 km.

Subway Facilities

As the number of passengers using the subway increases and some facilities become old, the subway companies (SMSC, SMRTC, and KORAIL) constantly promote the improvement of services and facilities for more comfortable, efficient and safe operations.

First, for improving system safety on the platform, the SMSC has installed color TV monitor and automatic alarm systems on every platform. The system rings an alarm, lights information signs, and plays information broadcast automatically in sequence whenever trains arrive.

Second, unlike the recently opened 2nd phase subways, the 1st phase subways did not have enough air-conditioning equipment for peak cooling needs in the summer. The SMSC has been moving forward with air-conditioning work since 1985 in stations that were not air-conditioned because they were built in the early stage. Currently, 17 stations are air-conditioned.

Third, the subway company is improving system accessibility and convenience by installing facilities for bike users and for persons with disabilities. Fixed and mobile wheelchair lifts were set up in every station. Also, toilet facilities in twenty stations have been equipped for the handicapped. Floor bumps and dot blocks for blind people were installed in every station. Bike racks were provided in 24 subway stations, and 3,340 bikes can be stored.

Fourth, to promote usage and park and ride service, Seoul operates twenty-three parking lots, with a capacity for 6,044 cars. Besides these, 24 new parking lots with 11,800 spaces were built in 1998.

Bus Facilities

A representative roadway facility for bus is the exclusive bus lane (EBL). To enhance bus operations, improve travel times and service, 218.5km bus-only lanes on 59 road sections were implemented. This accounted for 25.4% of the total length of major arterials in Seoul. The majority of EBLs are designated on the outside lane except one 4.5km EBL designated on the center lane. The 56 sections with 209.6km operate from 06:00 to 21:00, and 3 sections with 8.9km operate during two time periods of the day, 07:00 to10:00, 17:00 to 21:00 during the peak hours. The monitoring results show that EBLs helped in improving bus speeds by 10%.

Almost all the buses are equipped with air-conditioning and heating systems. However, there are no buses equipped with lifts for the handicapped, and no CNG buses have been introduced.

III. Operational & Managerial Characteristics

Subway Operating Characteristics

The average speed of the subway is 35km/h. The headways are different among subway lines. The subway congestion during rush-hours is 207% of its capacity and the peak hour headway is 3 to 5 minutes and non-peak headway 4 to 6 minutes (see table 4). The service frequency (number of trains per day) ranges from 1,009 trains on line 2 to 350 trains on line 7. The subway service starts at 05:30 and operates until 01:00, for 19.5 hours every day.

Table 4: Subway Operating Characteristics

		#1 line	#2 line	#3 line	#4 line	#5 line	#7 line	#8 line	total
Head-way (min)	Peak	3	2.5	3	2.5	2.5	5	4.5	-
	Non-peak	4	5.5	6	5	4	6	6	-
Service frequency		577	1009	423	530	596	350	359	3,844

Source: Seoul Metropolitan Subway Co., 1998

Tickets used in the subway are classified into regular tickets, fixed amount tickets, complimentary tickets, transfer tickets and group tickets. Regular tickets are good for one ride in a designated zone regardless of the purchase date and transfer (see table 5). Fixed-amount tickets are valid for any ride regardless of the purchase date, and they are available in 5,000 won ($4.20), 10,000 won ($8.40), and 20,000 won ($16.80) tickets. There is 20% student discount and 10% general discount on 10,000 ($8.40) and 20,000 won ($16.80) tickets, and one can travel even the longest distance even if only 10 won ($0.01) is left. There are complimentary tickets for people over 65 and people of national merit such as war heroes. Transfer tickets are used for linking subway and bus, and 10 won ($0.01) is discounted from the fare. Group tickets are for groups of more than 20 people going together to the same destination. There is a 20% discount for general group and a 30% discount for student groups of middle school and over. The subway fares are bi-level by sections, 450 won ($0.38) and 550 won ($0.46) per ride. There are no weekly, monthly, or daily free-use tickets.

Table 5: 1st Section Fare Changes of Regular Ticket

Year	Fare level	Fare level reflecting consumer price index	Rate of increase
1981	100	100	-
'85	170	147	-
'87	200	163	17.6%
'90	250	132	25%
'93	300	163	20%
'94	350	179	16.7%
'95	400	196	14.3%
1997	450	na	12.5%

Source: Seoul Metropolitan Subway Co., 1997

Bus Operational Characteristics

The average bus operating speed was 20.07km/h in 1998, which is the fastest average speed for the last ten years (see table 6). This improvement is associated with a continuous extension of exclusive bus lanes and a sharp increase in the unemployment rate. Still, the current bus speed is much lower than the auto vehicle speed, 25.4km/h in 1998. The speed difference between autos and buses had been 2 3 km/h, and therefore the widening difference is likely to shift bus passengers to autos.

Table 6: Trend of Bus Operating Speed Changes (unit : km/hr)

	1989	1990	1991	1992	1993	1994	1995	1996	1997	1998
Bus	18.60	18.80	18.15	16.88	17.02	18.42	18.79	18.35	18.69	20.07

Source: Seoul Metropolitan Government, 1997b, 1998b

The average bus headway is 7.9 minutes for regular bus (including circular bus), and 8.8 minutes for deluxe buses. There are 25 routes licensed with over 15-minute headways. The highest frequency (6 to 9 minutes) of operation is on 168 out of 448 routes. The bus operation starts at 4:00 in the morning, and ends at 2:00 in the morning, or 22 hours per day.

The fares per ride vary by type; 450 ($0.38), 1,000 won ($0.84) for regular bus and deluxe bus, respectively. There are no periodic tickets. When purchasing a pre-paid Bus Card (IC card), there is a discount of 5%.

Subway Organization

The Seoul Metropolitan Subway Corporation (SMSC) management structure was composed of a president, four directors, an auditor, 20 department, 50 sections, 29 field offices and 115 stations in 1996. There were 11,492 staff members in the SMSC. Six of them were officers, 11,275 were in regular government service and 211 were security police. Regular government staff was composed of 4,638 desk workers and 6,637 technicians. More than half of the staff has worked with the SMSC for over 10 years.

The Seoul Metropolitan Rapid Transit Corporation (SMRTC) is composed of one institute, 14 department, 43 divisions, and 13 site offices. As of December 1995, 2,795 staff members worked for SMRTC.

Bus Organization

There are 89 private bus companies operating 8,725 company buses in 1997. The biggest company operates 272 buses and the smallest does 22 buses. According to Korean Auto Transport Industry Law, a company should have 70 buses to be licensed, but the 21 companies have less than 70 buses. About 24 thousand employees work for the industry. Among them 83% are drivers and mechanics s, and the rest of them are office workers and officers. On average, a company hires 268 employees in total, 197 drivers, and 24.6 repair workers. There are two drivers and 0.44 office workers per bus.

Subway Financing

The budget of SMSC in 1996 was 1.02 trillion won ($857 million). The debt budget was 47.7 billion won ($40 million). The revenue budget by item: 512.6 billion won ($431 million) came from the fare revenue, 25.4 billion won ($21 million) came from non-operating revenue, 1.3 billion won ($1.09 million) came from disposal of fixed assets, and 41.3 billion won ($34.7 million) was balance brought forward. In the case of the auxiliary fundraising: 363.6 billion won ($306 million) came from the bonds and 100 billion won ($84 million) came from the investment by the Seoul Metropolitan Government. In the expenditure budget, 418.8 billion won ($352 million) was spent on operating expenses, two billion won ($1.68 million) on non-operating expenses, 483.3 billion won ($406million) on redemption of principal, 85.8 billion won ($72 million) on investment and 54.3 billion won ($46 million) as a balance to carry forward.

Bus Financing

The total asset of 89 bus companies is 354 billion won ($297 million) and on average 4 billion won ($3.36 million) per firm. The total debt is 400 billion won ($336 million) and on average 4.6 billion won ($3.86 million) per firm. The debts exceed the assets by 47 billion won ($39.5 million) in aggregate. The average capital per firm is only 540 million won ($0.45 million). These numbers imply how serious the financial situation of bus industry is. The 53% of firms recorded financial deficit in 1997, on average 760 million won ($0.64 million) per firm. The deficit-ridden industry tends to be less attractive to reinvest resulting in continuous degradation of the services.

IV. System Usage

Mode Share

Bus ridership has been declining, while subway ridership has been increasing. The 1997 bus patronage was half of 1985's level. The sharp decline is attributed to both worsening road traffic congestion and expanded subway system. On the other hand, in 1997 the subways were, for the first time, the most predominant travel mode in Seoul, accounting for 30.8 percent of daily trips in Seoul. It is noteworthy that in spite of heavy investment, the expanded system has only contributed to the reduction of bus ridership in Seoul. SMG expects and worries that the bus share may reduce to 20% in the early 2000's.

Table 7: The Trend of Mode Share Changes

	1980	1985	1990	1995	1996	1997
Buses	66.0	58.0	43.3	36.7	30.1	29.5
Subways	7.0	14.0	18.8	29.8	29.4	30.8
Taxis	19.0	16.5	12.8	10.7	10.4	10.1
Others	8.0	12.5	25.1	22.8	30.1	29.7
(Pass. Veh)	(NA)	(NA)	(14.0)	(14.5)	(21.1)	(20.6)

Source: Seoul Metropolitan Government, 1998c

Subway Ridership

The subway extensions and ridership have been improving over the last 7 years as the new subway lines keep opening (see table 8). The ridership of the 1 through 4 lines operated by SMSC peaked in 1995, but declined slightly in 1996 and again in 1997. The declines are caused by several factors. Most importantly, they are attributed to the opening of 2nd phase subway lines operated by SMRTC at the end of 1996. It is noteworthy from the effectiveness point of view that the opening of new subway lines does not have synergy impacts in terms of ridership among existing services. This phenomenon implies that, from now on, subway policy in Seoul should place emphasis more on encouraging ridership of existing services rather than on building new lines. The most congested time period in a day is 08:00 to 09:00, which accounts for 14% of the daily ridership. Around 50% of the ridership is concentrated during the time periods of 07:00 to 10:00 and 17:00 to 20:00, and therefore ridership during non-peak hours needs to be increased.

A customer survey disclosed the following major complaints that tend to discourage subway usage. These are: inconvenient transfer facilities such as insufficient elevators and escalators, and poor access to subway stations.

Table 8: Subway Ridership by Year

		1991	92	93	94	95	96	1997
SMSC	Car km (1,000)	133,188	160,098	174,025	193,056	212,242	216,623	206,489
	Ridership (1,000)	1,241,157	1,354,150	1,388,037	1,404,233	1,476,788	1,422,570	1,354,818
	Person-km	9.23	8.46	7.98	7.27	6.96	6.57	6.56
SMRTC	Car km (1,000)	N/A	N/A	N/A	N/A	N/A	N/A	101,712
	Ridership (1,000)	N/A	N/A	N/A	N/A	N/A	N/A	282,041
	Person-km	N/A	N/A	N/A	N/A	N/A	N/A	2.77

Source: Seoul Metropolitan Subway Co. 1998
Note: SMRTC started revenue service in 1997

Bus Ridership

There is no detailed official bus ridership data, because the privately owned bus companies are not required to report this information and do not want to disclose the information. The only way to estimate bus ridership is through O-D survey conducted by SMG every 5 to 10 years. As shown in the above table 7, it is clear that the ridership is decreasing fast.

The decrease can only be explained indirectly by user surveys. According to a survey conducted by SMG in 1996, the most important reason not to ride the bus is the excessive waiting times. Wait times range from less than 5 minutes to over 40 minutes mainly because of road congestion. Over 51% of bus users experienced waiting times over 15 minutes. Reckless driving and rush hour in-vehicle crowds are second and third reasons for not using bus. About 60% of bus users responded during the survey stated that they are not satisfied with the current bus services.

V. Conclusions

This paper reviewed the two most predominant public transit modes in Seoul, the subway and bus systems. The contents include important characteristics of these modes in terms of system characteristics, operations, and usage.

The SMG plans the second-stage subways with the length of 145km to be completed by year 2,000. The third-stage subways with the length of 120km will start construction in 1999. In total, the subways will include a total of 12 routes with a route length of 400km by the middle of year 2000. For service improvement, SMG plans to provide more accommodation facilities, user information program, and the elevator guide system for the disabled.

The SMG plans to keep the privately owned bus systems, but intends to reorganize bus route systems, and to support private bus companies for securing good quality public services. In addition, SMG plans to reduce long distance routes and winding lines, to increase the number of luxury-seat-buses and circular buses, and to replace old buses with air-conditioned buses. For securing consistent high quality services, the city has under consideration a plan to evaluate the service level of each bus company based on the results of customer interview and service monitoring. According to evaluation results, it plans to graduate the financial supports among bus companies. To make transfers between buses and subways more convenient, the SMG plans to integrate different fare collection systems into one by using the above mentioned Bus card system.

References

1. Seoul Metropolitan Government, General Plan for Bus Operation Reform, 1997a
2. Seoul Metropolitan Government, '97 Road Traffic Speed Study, 1997b
3. Seoul Metropolitan Government, Seoul Yearly Statistics Book, 1998a
4. Seoul Metropolitan Government, Traffic Survey Results under IMF Economic Crisis, 1998b
5. Seoul Metropolitan Government, '97 Transportation Indicators released for Press, 1998
6. Seoul Metropolitan Subway Co., Seoul Subway, 1997
7. Seoul Metropolitan Subway Co., 1997 Major Business Data, 1998.

Development of the New Transit System Concepts in Korea

Seongsoon Yun[1]

Abstract

New transit systems technologies including the Automated Guideway Transit (AGT), stand out in the world's major cities in respect to its promise of relatively lower construction costs compared to existing subway systems. In Korea, the development of new transit systems technologies has become necessary to cope with growing urban transportation problems. The Korea Transport Institute has completed a master plan for constructing the new transit system in the metropolitan areas of Seoul and Pusan. Currently, the basic design is on the board and the registration for RFP (request for proposal) has been made. In this paper, we will discuss the strategies for developing the new transit system technologies, especially automated guideway systems (AGT) in transportation planning, construction, operation, and financing aspects in Korea.

Introduction

In Korea, rapid economic growth and urbanization have been accompanied by over population and traffic congestion problems. Presently, transportation problems are the top issues to be solved at the government level. In order to solve the transportation problems, a balanced transportation plan, in demand and supply side, is required. To do so, comprehensive transportation plans are required to cope with

[1]. Director, Department of Urban Transportation, The Korea Transport Institute, Ildong Bldg., 968-5, Taechi-Dong, Kangnam-Gu, Seoul 135-280, Korea

various travel demand patterns.

The necessity of shifting urban transportation systems from automobile to urban rail transit has been recognized and addressed by constructing subways in large metropolitan areas such as Seoul and Pusan. However, in the long-term view, it is questionable whether the transit policy of single mindedly constructing subway systems is the best alternative due to its high construction costs. Presently, the development of the new transit systems in many countries around the world offer a better alternative in its lower construction costs and effective investment than the current subway.

The purpose of this paper is to derive a policy for developing the new transit system technologies in Korea through the review of construction and operation examples of the new transit systems technologies in other countries. In section two, we will review the characteristics of the new transit system. In section three, the construction examples of new transit systems in other countries along with Korea are evaluated. In section four, the background for introducing the new transit system and progress of development in Korea are presented. The provisions of future prospects for the new transit system in Korea are also discussed. In section five, a brief introduction of background and construction plan of the Kimhae corridor in Pusan metropolitan area and expected difficulties of the construction project will be discussed.

Definition of New Transit System

Urban rail transits can be classified into three types, based on their capacity: Heavy Rail Transit (HRT), Light Rail Transit (LRT), and Personal Rail Transit (PRT). Among these rail systems, the new transit system is defined as an advanced technology which has adopted the modern transit system on existing rail transit systems, using smaller vehicles than heavy rail transit systems and is run by the automated guideway system. This study has focused on the capacity and operation type of transit systems rather than vehicle type or physical appearance. The focus of the study is on the fully automated guideway transit (AGT) system.

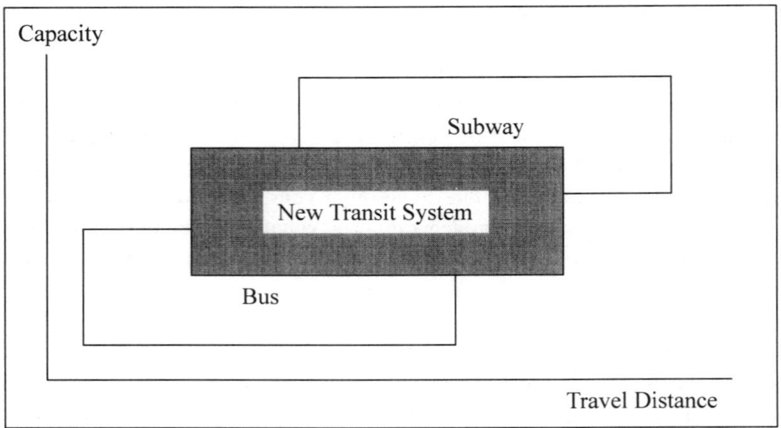

Figure 1: Application Range of New Transit System

Characteristics of New Transit System

Constructing and operating the new transit systems (AGT) in terms of transportation planning, construction costs, and operation aspects are evaluated as follows:

Transportation Planning Aspects

The major characteristic of the new transit system is its "medium" passenger capacity, moving 5,000-40,000 passengers per hour per direction (pphpd) compared to bus capacity (1,000-5,000 pphpd) and heavy rail capacity (40,000-70,000 pphpd). Also, this system appears to be more cost-effective due to its lower construction costs and operational costs as discussed below. Due to the flexible running conditions of the vehicle such as handling tight turns and a steeper gradient, planning alignments and constructing feeder systems with other modes are easier. It can provide a better level of service for passengers in overall impacts of increasing transit riderships. However, its limited capacity makes it difficult to replace the function of trunk line railroads in the large cities.

Construction Cost Aspects

The components of construction costs of the new transit system are rails, guideways, stations, train depots, station facilities, rolling stocks and various system facilities for civil works. Due to the smaller size of its vehicle, civil construction costs are significantly less for infrastructures such as tunnels, elevated guideways, and track. Civil construction cost saving could vary such as saving of 40% for tunnel cross section of excavation and of 45% for elevated structure. Lightweight of vehicle can contribute to construction cost savings. Significant reasons are that it saves land compensation costs and civil construction costs through constructing elevated structures over existing roads.

Since Korea has no prior construction experience with the new transport system, there may be an increase in construction costs. Most new transit systems are patented technology, so vehicle and system technology costs are relatively high and the transfer and localization of the new technology take time. Table 1 shows the cost experience in Japan.

Table 1: Construction Cost (in Japan)
(in Millions f US$)

Travel Mode	Subway	New transit system	Guideway Bus
Cost per Km	200-300	60-100	30

Source: New Transport System of Japan, 3rd International Conference on APM, 1991.

Operational Aspects

The new transit system is fully automated with ATO (Automatic Train Operation), ATC (Automatic Train Control), and ATP (Automatic Train Protection) systems. During peak hours or other situations, adding vehicles is relatively simple and compared to existing subway systems, the number of workers can be reduced to over 50%.

Labor costs of the Seoul subway systems take up 66% of total operation costs but the new transit system is expected to take up only 40%. The new transit system has a significant advantage in operational cost saving. The percentage distribution of operating costs among various operating areas is shown in Table 2.

Table 2: Percentage Distribution of Operational Costs by Area

Items	Personnel Cost	Power Cost	Maintenance Cost	Management Cost
New Transit system 1	39	14	39	8
New Transit system 2	42	17	31	10
Existing subways (Seoul)	66	8	22	4

Note: New Transit 1, France, Lille city's VAL
New Transit 2, Japan, Osaka City's New Tram

A comparison of annual transit passengers per worker of existing subways and new transit systems shows that new transit systems have a higher operational efficiency. Figure 2 shows the comparison of number of annual passengers served per worker. In Lille, France, 260 workers served 44.2 million passengers in 1990. This shows that 170,000 passengers per worker were carried, which is almost twice the operational efficiency of the existing subways.

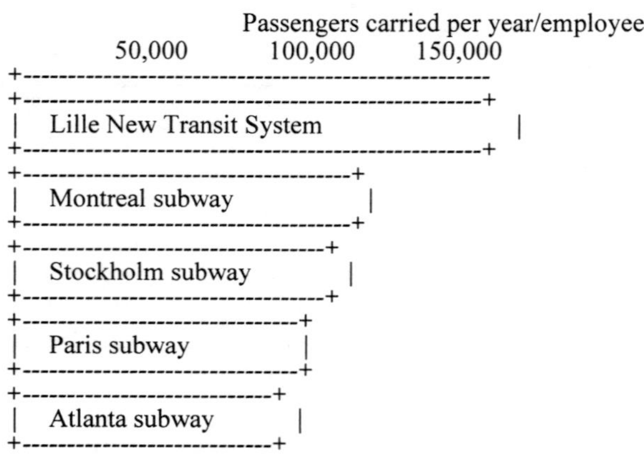

Figure 2: Comparison of Operational Efficiency

Although the new transit system has relatively less capacity than existing subway systems, the fully automated system shortens headway and can compensate for its smaller carrying capacity by running more frequently. Seoul and Pusan's subways operate at headway of 150 seconds but Lille's VAL and Berlin's M-Bahn operate at headway of 60 seconds, and Vancouver's Skytrain operate at headway of 75 seconds.

Passenger Aspects

The new transit system's significant benefits for passengers are shorter waiting times at stations and an overall reduction in travel time. Due to its tighter turns and steeper gradients, it can serve passengers even to the shortest destinations. In the case of constructing large railroad networks, it can easily connect with existing subways. New transit system can make up for the weak points of existing subway functions by improving accessibility to facilitate shifting of automobile users to mass transit. Finally, guideway types of new transit system with rubber tires and linear induction motors can reduce environmental pollution such as noise and vibration.

Developing the New Transit System in Korea

Shortage of Transportation Facilities and Transportation Problems

In 1995, the number of automobiles in Korea had already passed 8 million and the projected number of automobiles by 2001 is 14 million, which means one car per household. Due to Korea's limit on usable land and resources, the road expansion rate is only 1.5% in developed urban areas. In order to overcome the urban transportation problems, subway construction projects are being processed, but remarkable financing requirements make it difficult to expand subway networks. In Seoul, construction costs for the subway system has increased from US$31.3 million/km in the '80s to US$48.1 million/km in the '90s. The average travel speed of vehicles in Seoul and Pusan is estimated to be 10 km/hour in 2001. Thus, development of a new transit system as an alternative of mass transit is required.

Table 3: Rates of Road Expansion and Increase in Cars in Major Cities (1986- 2001)

City	Seoul	Pusan	Taegu	Kwangju	Taejeon
Rate of Road Increase (%)	1.29	1.17	3.37	2.11	1.67
Rate of Car Increase (%)	13.8	13.3	17.8	15.5	17.6

Increasing the Efficiency of Transportation Investment

In order to construct a comprehensive transportation system, multilateral investments and management policies for roads (passenger car and bus), subways, and new transit systems are required. The new transit system will help to overcome the limitations of existing transportation modes. Elevated structural construction can reduce land compensation and civil costs.

New Financing Scheme for Transportation Investment

Recently, the Korean government has enacted the "Regulation of Promoting Investment of Private Sector to Social Overhead Capital" to raise investment funds and technological improvements in transportation facilities by actively encouraging participation by private sectors. It has been decided that constructing the new transit system through private sector's participation is necessary to construct large fixed guideway networks to provide connections between central cities and satellite cities in major metropolitan areas of Korea.

Development Plan for the New Transit System in Korea

Followed by a feasibility study in 1993, the Korea Transport Institute has completed a master plan for constructing the new transit system in Seoul and Pusan. With a total length of 191 km, the construction of the expected new transit system will take five years for completion and the total construction cost is estimated at US$4.6 billion. However, most of projects shown in Table 4 were postponed due to the financial problems.

Table 4: Prospects of Developing New Transit System in Korea

Projects	Construction Period	Route length (Km)	Stations	Project Cost (Million US$)
Seoul--Hanam	1995 to 2000	18.65	20	409
Pusan--Kimhae	1995 to 2000	26.00	16	615
Seoul--Uijongbu	1995 to 2000	14.85	16	300
Suwan-Bundang-Yongin	1step : 1995 to 1999 2step : 2000 to 2004	35.40	15	885
Masan-Changwon-Jinhae	1998 to 2011	48.20	61	1,205
Mikum-Kuri-Sangbang-Hoeki	-	15.96	14	399
Sihug-Puchon	-	6.75	6	169
Yangsan	1996 to 2000	10.30	5	258
Pusan station-TaeJongDae	1996 to 1999	11.25	16	281
Solak Mountain-Sokcho city	1995 to 1998	6.30	6	158
Total	-	190.81	-	4,679

Development Policy for the New Transit System

The development policy for constructing and establishing a comprehensive transportation system requires feeder systems with each travel mode in order to provide convenient transferring systems. Also, maximum travel effectiveness should be obtained under limited fund, and for these purposes, mass transit priority policy should be implemented. Rearrangement of road systems and rail transit systems is necessary to respond to the increasing travel distances, population mobility, and urbanization started by large-scale housing development and new town development. It is necessary to clearly define the function of the new transit system and adjust the function of each travel mode in the mass transit system. Table 5 shows a concept plan of role sharing of modes in the mass transit system, including new transit system.

Table 5: A Concept Plan of Role Sharing of Mass Transit System

Travel Mode	Function	Service Type	Role
Subway	major transit	trunk line service	- major corridor in service in large cities
Bus	supporting transit	connection to subway and circulation	- connection between residential area - connection between residential area and district centroid
Taxi	short distance assistant transit	connection to other mode	residential areas and mass transit network connection
New Transit System	major transit and feeder transit	feeder line and connection with subway	connection between central city and satellite city
		circulation of trunk line and assistant trunk line	major corridor in medium and small cities
	paratransit	demand responsive high level of service	connections between major activity centers

Tasks for New Transit System Development

Planning and Construction Aspect

The new transit system should maximize connectivity with existing railroad, subway, and road systems by considering future transportation networks and construction plan. In order to maintain the profits in operation, the route should be chosen to maintain enough travel demand during both peak hours and all day. Also, major trip generating facilities should be placed near routes and comprehensive land use policies should be considered very carefully.

The performance specification of the transit system needs to be provided by considering transportation and geographic conditions of target area. The RFP (request for proposals) will be made based on performance specification. In order to minimize land compensations, maximum use of public owned lands for the construction of the new transit system is needed.

Financing Aspect

It is difficult to construct and operate the new transit system exclusively by the private sector without intervention by the public sector for ensuring public interests. Therefore, it is desirable to proceed under a joint venture system with private and public agencies working together on the construction of a new transit system. With the assumption of no financial support from central government and considering the current fare levels of public transit, operating profits would be relatively low. Therefore, funds must be sought through innovative strategies. Attracting private funds for public projects makes it possible to supply transportation services efficiently and reduce the burden of government in large-scale investment.

Technology Aspect

Technology transfer of a new transit system from foreign companies to domestic companies by individuals is not a good approach. Introduction of new technology by government is desirable for maintaining efficient management of technology transfer in the future.

Case study: Developing New Transit System in Pusan

Background

Kimhae City, a satellite city of Pusan, has a relatively high potential for development in the northwest area of Pusan. There are only two transportation corridors, local road #14 and Namhae Expressway, in this area. These two major roads are insufficient to handle the rapidly increasing travel demands and this generates serious traffic congestion. As part of constructing rail transit network plan for the Pusan metropolitan area, the construction of new transit system in Pusan-Kimhae corridor is proceeding in order to tackle the regional traffic congestion problem.

Construction Plan

The route of the new transit system connecting Kimhae and Pusan consists of 16 stations with a total length 26km.

Time schedule for implementing the project is as follows:
Basic design and detail design: 1995 – 1996 (18 months)
Land acquisition and compensation: 1996 (6 months).
Construction and test track operation: 1997- 2000 (4 years).
Opening: 2001

The estimated cost of the project is shown in the following table.

Table 6: Costs of Project
(Millions of US$)

Design Cost	Land Compensation Cost	Civil Construction Cost	Rolling Stock Cost	Total
19.5	25.0	382.4	187.8	614.7

Difficulties in Processing Project

The major difficulties in processing this project are financing and choosing an appropriate implementation body for the project.

Government Subsidy

In most countries, government subsidies for urban railroad construction amount to over 50% but in Korea, subsidies provided by the governments come around to 40-50% according to "Urban Railroad Act" only for subway (metro), not for new transit system.

Subsidiary Development Revenues

The "Regulation of Promoting Investment of Private Sector Act" prevents the scale of housing development in subsidiary developments from exceeding 100% of total construction costs, which causes difficulties in project financing.

Sink Investment

From the total project cost of US$614.7 million, supposing that 40% is invested from private and public sectors, initially, the private sector cost burden from private will be $196.8 million and pubic investment will be $49.3 million. This may restrict the willingness of private sector to participate in the project.

Selecting Implementation Body

Selecting an implementation body for the project could be delayed due to different interests among different participants. At the earliest moment, vehicle system and implementation body for the project should be selected for basic and detail design.

References

1. Armstrong, Wright. A. (1986), Urban Transit System: Guidelines for Examine Options, World Bank Technical Paper No. 52, Washington DC: The World Bank.
2. Boyce, David E. (1979), "Impact of Federal Rail Transit Investment Program on Urban Spatial Structure", in the Urban Impacts of Federal Programs, edited by NJ Glickman, Baltimore, The John Hopkins Press.
3. Gary, Dennis A. (1993), "A Vision for 21st Century Guided Transit", Automated People Mover IV- Proceedings of 4th International Conference, pp. 88-98.
4. Kieffer Jarold A. (1993), "The Fundamental Gap in Urban Transportation", Automated People Mover IV-Proceedings of 4th International Conference, pp. 109-118.

5. Kikuchi, Koso (1991), "New Transit System in Tokyo", Automated People Movers III - Proceedings of 3rd International Conference, pp. 43-61.
6. Shen, David L. (1993), "The Role of APM in a Multi-Modal Transit System," Automated People Movers IV- Proceedings of 4th International Conference, pp. 230-248.
7. Transportation Research Board (1989), Light Rail Transit: New System Successes at Affordable Prices, Special Report 221, Washington DC, TRB.
8. Yun, Seongsoon (1994), Feasibility Study and Basic Plan of New Transit System for Pusan Metropolitan Area, The Korea Transport Institute: Seoul.

Socioeconomic, Land Use and Travel Patterns of Shanghai

Eric Pengkuan Ho[1], Member
Ximing Lu[2] and Junhao Li[3]

Introduction

Shanghai is a very important city in eastern China and also a major commercial and industrial center of the nation. It is located at the mid-point of the East China Sea's coastline, and near the outlet of the Yangtze River, the longest river and the most important inland waterway in China. Shanghai was developed as a trading center as early as the 1850s due to its prime location. In the early 1920s, Shanghai had already become a modernized international city. Today, Shanghai is a municipal district with an area of 6,340 square kilometers and with more than 13 million residents.

Shanghai is situated on the Yangtze Delta, which provides a large flat area for the city to expand. There are two rivers cutting through the downtown area: the Wusong River to the north and the Huangpu River to the east. Before World War II, the city had expanded to the north of the Wusong River. In the last few decades, the city has expanded to the west of the city along several external corridors. The area east of Huangpu River, now called Pudong New District, was identified in early 1990s as the most important urban development project in eastern China. Many infrastructure projects, commercial, industrial and residential areas have been or are being developed there.

Due to economic reform, Shanghai has experienced tremendous economic growth in the past decade. Numerous urban development projects have been implemented. Following the economic growth and the extensive urban

[1] Principal, Gallop Corporation, 451 Hungerford Dr., Suite 612, Rockville, Maryland 20850

[2] Director, Shanghai City Comprehensive Transportation Planning Institute, 331 Tong Ren Road, Shanghai, 200040, People's Republic of China

[3] Chief Engineer, Shanghai City Comprehensive Transportation Planning Institute, 331 Tong Ren Road, Shanghai, 200040, People's Republic of China

development, travel characteristics have changed substantially. People in Shanghai now demand fast, reliable and comfortable transit service. The traditional bus system, which has been the only transit system in Shanghai for the last few decades, can no longer satisfy the people's desire for high quality transit service. This resulted in a recent significant decrease of the transit ridership. In 1995, only 15% of daily travelers were served by the transit system.

This paper provides an overview of the economic and urban development in Shanghai, and further investigates the impact on the urban travel characteristics. It further summarizes the issues affecting the service and ridership of the transit system. Finally, it suggests a number of strategies to strengthen the development of the transit system in Shanghai.

Economic And Demographic Development

Economic Development

Shanghai has achieved tremendous economic development in the past several years due to economic reform. As shown in Table 1, the Gross Domestic Product (GDP) increased by more than 400 percent between 1985 and 1995.

Table 1 Major Socioeconomic Measures in Years 1985, 1990 and 1995

	1985	1990		1995	
	Value	Value	% Growth From 1985	Value	% Growth From 1985
Population (million)	12.17	12.83	5.42	13.01	6.90
Employees (million)	7.76	7.88	1.55	7.94	2.32
GDP (billion yuans)	46.68	75.65	62.06	246.26	427.55
Agriculture Sector (billion yuans)	1.95	3.26	67.18	6.17	216.41
Industrial Sector (billion yuans)	32.56	48.27	48.25	140.99	333.02
Service Sector (billion yuans)	12.16	24.12	98.36	99.1	714.97

* The currency exchange rate is approximately 8 yuans per US dollar.
Source: (1)

The economic structure also changed significantly. In 1985, the industrial GDP accounted for about 64 percent of the total GDP value. In 1995, the weight of the industrial GDP dropped to 57 percent due to the rapid development in the service sector. Because of the economic growth, most of the people in Shanghai are much wealthier now than they were a decade ago. The average annual salary of an employee increased from 1,400 yuans (US$180) in 1985 to 9,300 (US$1,160) yuans in 1995.

Demographic Growth

Because of strict population growth policy imposed by the Chinese government, the native resident population grew marginally between 1985 and 1995. Due to the economic growth, a large number of people moved to Shanghai to seek employment. In 1995, the resident population of Shanghai was about 13 million. The majority of them were employed in industries like construction and textile manufacturing, and lived in temporary housing or dormitories provided by the employers.

In Shanghai, most of the people are employed in government agencies or state enterprises. The number of state employees increased slightly from 7.6 million to 8.0 million between 1985 to 1995. However, there was a major growth in non-state employees. In 1995, there were about 2 million non-state employees in Shanghai. Most of them were non-resident, non-skilled laborers employed temporarily by state or non-state enterprises. In addition, the economic reform created a new class of high-income people who are executives or managers of non-state enterprises.

Land Use Development

Before 1985, major development in Shanghai was confined to the region west of the Huangpu River. The center city was a very high density, mixed land use area with narrow streets. In the past ten years, urban development has focused on the following four areas:

1. Pudong New District: An area east of the Huangpu River, with a size of about 250 km^2, was planned in early 1990s as a new urbanized area that would house 2 million people and provide 1.2 million employment opportunities in 2020. The core development project is the Lujiazui Financial District, which is an area of 30 square kilometers on the east bank of the Huangpu River, across from the existing Central Business District (CBD) on the other side of the river.
2. Expansion of the urban area: Many residential and industrial development projects have taken place in the urban fringe area, in particular along the external highway corridors in the western part of the city.
3. Renewal of the center city area: A number of districts with specific functions were developed in various parts of the center city. Also, several retail and commercial areas were re-developed. More than a thousand high-rise commercial and residential buildings were constructed. Many high-density residential communities and factories were demolished for the development.
4. Satellite towns: A number of satellite towns with a population ranging from 100,000 to 300,000 have been developed to serve specific industries, such as automobiles, textiles, pharmaceutical products, etc. These towns are located 30-70 kilometers from the center city.

The land use development impacted distribution of population and employment in different areas in the city. As shown in Table 2, the population and the number of employees dropped in the Center City District, while they increased significantly in the Urban Fringe and Pudong New Districts.

Table 2 Distributions of Population and Employees

	Area (km^2)		Population (million)*				Employees (million)**			
			1985		1996		1985		1996	
	Value	%	Value	%	Value	%	Value	%	Value	%
Center City District	93	1	4.9	40	4.3	32	2.6	34	2.4	30
Urban Fringe District	251	4	1.5	12	2.1	16	0.9	12	1.1	14
Pudong New District	252	4	0.6	5	1.2	9	0.3	4	0.6	8
New Develop. District	1,445	23	1.9	16	2.1	16	1.6	21	1.4	18
Rural Area	4,283	68	3.4	27	3.4	27	2.2	29	2.4	30
Total	6,324	100	12.3	100	13.1	100	7.6	100	7.9	100

* Resident population only
** State-employees only
Source: (4)

Transportation System Development

Vehicle Growth

The people in Shanghai use a variety of transportation means for their daily travel, including walking, bicycling, mopeds, motorcycles, regular bus, premium bus, subway, taxis, passenger cars, agency vans, etc. At present the majority of travel in Shanghai is made with non-motorized modes (i.e., walking and bicycling), that together accounts for 75% of daily travel. However, motorized vehicle use has grown significantly.

Table 3 presents the growth of motorized vehicles from 1985 to 1995. Shanghai has adopted a relatively restrictive policy on owning and operating motorized vehicles, as compared with other cities in China(5). Still, the numbers of passenger cars and motorcycles have grown very rapidly, almost fourfold in ten years. It should be noted that the numbers of motorcycles shown in Table 3 actually include only those which can be operated in both urban and rural areas. There are another 160,000 motorcycles which, in 1995, could only be operated in rural areas.

Table 3 Growth of Motorized Vehicles From 1986 to 1995

	1986	1992		1995	
	Value	Value	% Growth From 1986	Value	% Growth From 1986
Passenger Car	35,000	78,700	124.9	169,400	384.0
Truck	61,200	90,800	48.4	119,300	94.9
Motorcycle	19,500	56,800	191.3	89,100	356.9
Total	115,700	226,300	95.6	377,800	226.5

Source: (4)

The number of taxis also grew dramatically, from 7,000 in 1985 to 37,000 in 1995, accounting for about 20% of all passenger cars. This has resulted in too many taxis with very low utilization rates in Shanghai. This has worsened the congestion situation and created a parking problem in the streets.

"Agency vehicle" is another popular means of transportation in Shanghai. Many public and private agencies provide passenger vehicles for their employees' daily commuting. A 1996 home interview survey showed that 5 percent of daily travel was made with agency vehicles. In 1996, there were more than 19,000 passenger vehicles (i.e., vans and privately used buses), most of which were agency vehicles. These “paratransit” services are more convenient and reliable than the regular bus service.

Highway System

The city of Shanghai was developed almost a century ago. The streets in the center city, most of which were built before modern automotive technology developed, are very narrow. During its early development period, Shanghai was divided into several territories governed by foreign powers. Each territory independently developed its infrastructure, without effective coordination with other neighboring territories. As a result, the roadway network in Shanghai does not follow a systematic pattern. Many streets are circuitous. In the last decade, the Shanghai City Government devoted a significant amount of effort to improving the highway system. The development can be characterized into five areas as discussed below:

1. Building the urban expressway system: A 47 kilometer-long ring road (with a 30-mile elevated segment west of the Huangpu River), an elevated north-south expressway, and an elevated east-west expressway were implemented. An outer ring road will be built in the near future.
2. Improving the surface streets in the center city area: This includes widening and upgrading existing streets and constructing new streets with direct routes along major transportation corridors.

3. Building major arterials to new development areas: Several urban arterials were constructed to accommodate the traffic generated from the newly developed areas.
4. Building bridges and tunnels crossing the Huangpu River: To enhance the connection between the Puxi (West Huangpu) and Pudong (East Huangpu) regions, a number of bridges and tunnels have been constructed. In 1985, there was only one bridge and one tunnel in the urban area, with a total of 4 traffic lanes in both directions crossing the Huangpu River. By 1995, there were three bridges and two tunnels, with a total of 18 traffic lanes in both directions. There are plans for construction of 3 to 5 more tunnels to be built.
5. Building external highways: Several freeways and arterials were constructed from the western part of the city to satellite towns and other cities.

From 1985 to 1995, highway mileage increased by more than 60 percent (from 3,200 km to 5,400 km), while the pavement area increased by about 200 percent (from 2,700 sq. km to 7,400 sq. km). This is an indicator of the Governments significant effort devoted to the widening of existing roadways.

Transit System

The development of the transit system lagged behind the development of the highway system. Until a few years ago, the only transit service in Shanghai was the "regular bus system" operated by the municipal bus company. Table 4 shows the changes in the operating statistics of the regular bus system. There was moderate growth in the scale of the regular bus service between 1990 and 1995. However, the system ridership decreased significantly during the ten-year period, from 13 million in 1986 to 9 million in 1995(3). The bus service deteriorated due to a number of reasons. These include: 1) an aging bus fleet and insufficient system capacity; 2) extremely slow and unreliable schedules due to the congested roadway traffic; and 3) failure to adjust bus routes to accommodate the emerging transit markets in the new development areas.

Table 4 Operating Statistics of the Regular Bus System

	1985	1990		1995	
	Value	Value	% Growth From 1986	Value	% Growth From 1986
Number of Buses	4,700	6,100	29.8	7,500	59.6
Number of Routes	300	390	30.0	520	73.3
Total Operating Mileage (kilometers)	10,100	18,600	84.2	24,500	142.6

Source: (2)

In the last few years, the bus service in Shanghai has improved substantially. Some premium bus services were introduced and operated by separate enterprises. These included mini-bus service with flexible routes and stop locations, air-conditioned bus routes, and express bus routes. The ridership of these services increased significantly as they satisfied the increasing demands for high-quality transit service. In 1992, there were about 110 routes of these premium bus services available, carrying less than one percent of all bus passengers. In 1995, the premium bus routes increased to 420, and they carried almost 15 percent of all bus passengers.

The first subway line in Shanghai, with a length of 16 kilometers and 13 stations, opened in 1995. That same year, the subway line carried about 240,000 passengers daily. The second subway line, with a length of 16.3 kilometers and 10 stations, is under construction and is expected to open in 1999. The third line is in the design stage and is scheduled for construction soon.

It is worth noting that the premium bus and subway services are three to five times more expensive than the regular bus service. The increase of ridership for these transit services implies that people in Shanghai are willing to pay higher prices for higher quality of transit services.

Changes of Travel Characteristics

Because of the economic growth and land development, there were significant changes in passenger travel characteristics between 1985 and 1995. The average per person daily trip rate increased from 1.79 in 1985 to 1.95 in 1995. In 1995, there were about 26 million person trips generated on a typical weekday. Table 5 shows the changes of travel characteristics between 1985 and 1996.

As shown in Table 5, the proportions of home based work and home-based school trips increased significantly from 1985 to 1995. In 1995, these two types of trips together accounted from about 70% of total travel. It may be due to the fact that the traffic conditions and bus services in 1995 were so poor that people avoided making non-commute trips. It should be noted that in China most of the adults, both male and female, participate in the labor force. This limits the time available to non-routine travel. However, non-commute trips are expected to increase significantly with extensive subway service. For example, a lot of non-home based trips may be generated during the lunch break if there is convenient subway service.

Table 5 also shows that mode shares changed dramatically between 1986 and 1995. The shares of walk trips and bus trips dropped almost 10 percentage points. The share of bicycle trips increased from 31 percent to 45 percent. This reflects that people were wealthier in 1995 than they were in 1985. In 1995, more people could afford to purchase bicycles and use them for their travel. Also, it indicates that in general, people prefered riding a bicycle to taking the bus since riding a bicycle was usually faster and more reliable than riding the bus. Table 5 also indicates that except for passenger vehicles, the average travel times by all modes were much longer in 1995 than they were in 1985. This indicates that people had to travel further or at slower speeds in 1995 than in 1985.

Table 5 Changes of Selected Travel Characteristics Between 1986 and 1995

	1986	1995
Trip Purpose Proportions (Percent)		
Home Based Work	48.9	51.3
Home Based School	12.6	17.7
Home Based Other	29.9	24.4
Non-Home Based	8.6	6.5
All Purposes	100.0	100.0
Mode Shares (Percent)		
Walk	41.3	32.8
Bicycle	31.3	45.1
Transit	24.1	15.1
Automobile/Motorcycle	3.3	7.0
All Modes	100.0	100.0
Average Trip Lengths by Mode (min)		
Walk	13	19
Bicycle	21	35
Bus/Subway	48	62
Passenger Vehicle	55	55
All	33	36

Source: (2)

Transit System Development Issues

As described above, while Shanghai has experienced rapid economic growth and urban development in the last several years, the ridership of its transit system has dropped significantly. In this section, we examine from different perspectives various issues which affect the development of the transit system.

Socioeconomic Development

1. Demand for high quality transit service: Because of economic growth, people now demand fast, reliable and comfortable transit services, such as premium bus, subway, etc.
2. Heterogeneous demand characteristics: The economic development magnifies the heterogeneity of socioeconomic conditions of the people, thus demanding different types of transit services.

Land Use Development

1. More dependent on motorized modes for daily travel: With urban expansion, people have to travel farther to work and thus are more dependent on motorized travel modes like bus, motorcycle, and automobile.
2. Increase of trips crossing the Huangpo River: The development of Pudong New District will increase the river crossing trips dramatically and thus would require a high capacity transit system to serve the new travel demand.
3. Service to new development Area: The new development areas along radial corridors in the western part of the city will constitute major transit markets. It needs to adjust the bus routes on a timely basis to respond to the land use development.
4. Impacts of ring roads: The development of the ring roads would encourage a low-density area-wide land use development pattern and hence may restrain the future development of efficient transit systems.

Competition with Other Travel Modes

1. Private vehicle growth: Without restrictive policies on the number of motorcycles and automobiles, people using non-motorized modes now may shift to using private motorized mode directly, hence hampering the development of the transit system.
2. Emerging transit services: The traditional bus service faces strong competition from the emerging transit and paratransit services, such as the minibus and the agency van. These services are much more convenient and reliable under the congested traffic conditions.

Integration of Transit Services

1. Coordination with the subway system: Coordination between the subway system and other transportation modes is necessary for providing convenient transit services to people in various parts of the city. The designs of the feeder bus network and transfer facilities at subway stations are of particular importance.
2. Bus transfer centers: The expansion of the urban area has increased the travel distances and thus caused more transfers to be made by bus passengers. It thus needs to establish a number of bus transfer centers at major activity and distribution centers to allow people to transfer between bus routes conveniently.
3. Inter-modal terminals: The economic growth has caused a major increase in traffic volumes at major transportation terminals such as the airport, maritime terminals, and train stations. It thus requires more connective transit services to these terminals.

Roadway Network

1. Deteriorated roadway traffic conditions: Congested roadway traffic has greatly reduced the speed and reliability of the bus service.

2. Physical roadway conditions: The mixed traffic on the narrow streets in the center city reduces the maneuverability of bus vehicles, in particular the articulated trolley buses.

Transit System Development Strategies

To improve the service and increase the ridership of the transit system, there needs to be substantial changes in the management and operations of the transit system. It also requires some changes in the transportation policies to facilitate the transit system development. This section outlines various strategies that would improve the transit service and attract more riders using the transit system.

Operational Strategies

1. Integrated fixed guideway transit system: In order to carry a large amount of passengers under congested roadway environment, it is necessary to develop a rail-based transit system. Such a system consists of a number of routes along major transportation corridors, and connects efficiently with other transit modes to provide an integrated transit service to various parts of the cities.
2. Variety of transit services: A variety of transit services should be provided to satisfy the needs of various types of people with different socioeconomic characteristics.
3. Bus fleet improvement: The bus fleet needs to be upgraded and should consist of various types of vehicles to serve various transit markets.
4. Traffic management measures: Some bus preferential measures, such as bus lanes, bus priority signals, etc., can be considered to improve the operating speeds of the buses on congested roadways.
5. Transit service monitoring: The transit operators can consider the use of management information systems (MIS) to effectively monitor the performance of various transit routes, and thus allow for making timing adjustments to the transit service to respond to any changes in transit markets.
6. Transit planning: The transit operators should apply systematic approaches for designing the transit network as well as determining the alignments, frequencies and vehicle types of specific transit routes.

Policy and Organizational Strategies

1. Urban transportation policy: An urban transportation policy needs to be developed to identify the long-term vision of the urban transportation in Shanghai. The policy also should clearly defines the roles and functions of various transportation modes, and provide guidelines for implementing measures to encourage or restrict the uses of individual transportation modes.
2. Urban land use policy: The land use development should consider the impact of land use development on the effectiveness of the transit system. For example, a high-density corridor-oriented development strategy would be preferable to a

low density area-wide development strategy. Also, major land use development projects should consider plans to encourage the use of transit systems.

3. Urban transportation organizations: Various transportation organizations should be operated in an environment to facilitate the coordination and integration of various transportation modes. It would be appropriate to create an agency like the "Department of Transportation" in other countries. Such an agency should be responsible for the coordination and integration of various transportation modes, as well as for monitoring, regulation and provision of a balance among transportation modes in planning and operation.
4. Regulation to transit service providers: The government should devise effective regulations to ensure transit operators to provide acceptable levels of the transit services. It should also consider other types of enterprises (e.g., private companies) for providing transit services.
5. Transit fare policy: Due to the economic reform, it may be necessary to review and revise the current fare structure to effectively reflect the balance between market demand and service level. Also, it should examine fare structure and subsidy policy to ensure fair competition among transit operators.

Conclusions

Due to the economic and urban development in Shanghai, there have been substantial changes in the urban travel characteristics. Since the transit system failed to keep pace with the rapid economic and urban development, it cannot provide the appropriate transit services to satisfy people's demands. The transit ridership has dropped dramatically in the last several years. With economic growth and expansion of the urban area, more people will rely on motorized modes for their daily commuting. To attract more people to use public transit service, the city needs to develop an integrated rail-based transit system that can carry large amount of passengers with fast, convenient, and comfortable service. To facilitate the development of the transit system, the government needs to review and revise certain transportation and land use policies, and adjust the functions of government organizations for coordinating, integrating and regulating the operations of various transportation modes in the city.

References

1. Shanghai People Publishing Company, Shanghai Almanac, 1997.
2. Shanghai City Comprehensive Transportation Planning Institute, The Second Citywide Comprehensive Transportation Survey in Shanghai: Transit Equipment Survey Report, 1996.
3. Shanghai City Comprehensive Transportation Planning Institute, The Second Citywide Comprehensive Transportation Survey in Shanghai: Transit Ridership Survey Report, 1996.
4. Shanghai City Comprehensive Transportation Planning Institute, The Second Citywide Comprehensive Transportation Survey in Shanghai: Final Report, 1997.
5. Stars, Stephen and Liu Zhi, "Motorization in Chinese Cities; Issues and Actions," Proceedings, China's Urban Transport Development Strategy, 1995.

URBAN PUBLIC TRANSPORTATION SYSTEM OF SHANGHAI AND ITS OPERATIONS AND USE BY PEOPLE

Weici Xu[1], Jian John Lu[2], and William Zhang[3]

ABSTRACT

Shanghai, one of the largest cities in the world and with a population of 13.5 million, has been campaigning for reliable public transportation since late 1979's. Today, it becomes one of the greatest place to live and work, not only because of its strong economic opportunities, but also its commitment to enhance the city's public realm, providing a strong social support structure and finding new ways to meet residents' need. Public transportation is the major transportation mode used by most population in Shanghai. In past years, major development of public transportation has occurred and new development for future has been planned. This paper presents some background about the public transit systems in Shanghai. Statistical data regarding the current situation and future prediction are summarized and presented. Current challenges and policy issues are also discussed in the paper.

INTRODUCTION

In 1958, Shanghai built the first trolley bus system in China. Since that time, public transportation system has grown rapidly in both size and service. At the end of 1986, the public transit system attracted over 14 millions passengers everyday and became one of the largest public transit in the world. The transit network consists of 5,500 buses along 331 routes, which amounts to total service mileage of 14,000 km. The aggressive urban economic development and planing policy that started in late 1970's was the key to this boom of public transit in Shanghai. However, since the late 1980's, the demand for public transit has decreased. Eventually, bicycle

---

[1] Professor, Department of Highway and Traffic Engineering, Tongji University, Shanghai, China

[2] Associate Professor, Department of Civil and Environmental Engineering, University of South Florida, Tampa, Florida 33620

[3] Senior Structural Engineer, Vanasse Hangen Brustlin, Inc., Watertown, MA 02471

transportation mode exceeded public transit and became the major transportation mode in 1990's. For instance, the ratio of public transit to bicycle mode was 69 to 31 in 1981, 59 to 41 in 1986 and 40 to 60 in 1995. Table 1 and 2 are the statistical data of the supply and demand for last two decades.

TABLE 1. The Supply and Demand of Public Transit System in Shanghai in the 1970's and 1980's

Year	1977	1980	1986	1989
Number of Vehicles	2768	2901	5500	6264
Total Passengers (Billion)	2.258	3.398	5.182	5.500

TABLE 2. The Supply and Demand of Public Transit System in Shanghai in the 1990's

Year	1991	1992	1993	1994	1995	1996
Number of Vehicles	6562	6837	N/A	N/A	N/A	1702
Total Passengers (Billion)	5.694	5.868	5.596	5.225	4.871	1.909

The following are explanations for this decrease of demand for public transit:

(1) Since the 1980's, the organization and management of public transit were still operated by the old state-controlled system while transportation services became the marketing-oriented business.

(2) The delayed bus schedule and traffic jam due to limited highway capacity and bad roadway condition have dramatically affected the quality of service and forced people to seek other mode of transportation.

Recent efforts to reduce the congestion with massive expressway construction by the city and state government have resulted in a significant improvement to the flow of traffic into and out of the city. Unfortunately, the effect on inner city traffic has been marginal. Shanghai desperately needs an alternate form of public transportation. In 1998, the city government started a design and construction project of a 62 km elevated mass transit system around city. This project will be completed by 2000 and should dramatically improve the public transportation system.

CHALLENGES

For many years, the following have been observed as main obstacles for building an efficient integrated public transit system in Shanghai:

(1) The more money the city spent in improving transit system and increasing service capacity, the more money they lost. The service did not generate any profit.

(2) The investment for higher productivity has been decreased while management cost has been dramatically increased.

(3) The citizen's transit demand has been continuously shrinking while urban economic and land development keeps growing.

(4) No action has been taken to implement the plan of developing the efficient public transit system.

Recently, the public transit system has faced even more challenges:

(1) **The urban expansion and outflow of the population** - Steady economic and business development around the suburbs has generated many job opportunities and attracted many residents from the inner city. The outbound extending of residential area around the suburbs has increased demand for transportation outside the city and reduced transportation needs in the inner city. Thus, the existing public transit system that was built primarily for in-city services does not meet the need of many residents who lived outside of city.

(2) **The competition from private transportation** - Bicycle, motorcycles, cars, and taxis have become main competitor of public transit. Not only because people become richer and can afford private transportation, but also because public transit provides only limited service areas with limited transfer connections between the routes and unreliable schedules.

(3) **Market-oriented business environment** - There is a long learning curve for the state enterprise to operate public transit as a business system - providing good services and making a profit.

POLICY, REFORM, AND EFFICIENCY

There are different opinions about government subsidy. Some believe government should use tax income to improve the public transit system. Others believe tax income can only provide the limited budget and resource, and the potential for further development cannot be explored when governmental subsidy is applied. In China, different approaches have been attempted to improve the service of public transit and make some profit to reduce dependence on governmental subsidy. In general, the reasons for loss in the public transit system fall into two categories: policy related and bad operations related. These two reasons have coexisted for a long time and should be re-assessed.

In conclusion, governmental subsidy, private investment, and, self-development should work together. The government should aggressively increase infrastructure investment and establish fair competition environment. Top priority and favorable policy should be given to an effective public transit system. In the meantime, the public transit organization should improve their management and operations by reducing overhead and work towards self-sufficiency.

The top priority and favorable policy in financial support, construction project, use of right of way, and traffic management will be secured if the city does the following: (1) through legislation, secure funding resources for the public transit system; (2) improving quality of government management and control; (3) accelerate railway construction; (4) develop and build intelligent highway and transit system to improve network, bus stops, and passenger transfer connections. Qualitative and quantitative criteria could be used to evaluate progress in transit enterprise. The qualitative criteria could be both passenger's and enterprises' satisfaction level, while the quantitative criteria may be determined by the number of passengers and the ratio of benefit over cost.

PREDICTION OF PASSENGER FLOW AND MODE DISTRIBUTION

In recent years, Shanghai has experienced significant social, economic and urban development that has dramatically affected the passenger demand and thus changed the distribution of mode selection. Public and private transportation systems are currently major transportation modes in Shanghai. The public system consists of train, subway, traditional and special transit lines, and taxi. Traditional transit line means the passengers can get in and off the bus at regular stations, while the special transit lines are normally the express line with very few stops. The past demand and projected demand of passengers are shown in Table 3.

TABLE 3. Past Demand and Projection of Total Passengers (Billion) Using Different Public Transportation Modes in Shanghai

Year	1991	1992	1993	1994	1995	1996	1997	1998	1999	2000	2010
Traditional Public Transit	5.694	5.868	5.596	5.225	4.871	1.909	1.878	1.920	2.000	2.410	4.990
Special Transit Lines	0.029	0.050	0.109	0.209	0.301	0.427	0.500	0.550	0.560	0.590	1.020
Subway	N/A	N/A	N/A	0.005	0.065	0.089	0.112	0.160	0.220	0.360	1.510
Taxi	0.137	0.168	0.269	0.448	0.511	0.598	0.471	0.610	0.650	0.720	1.110
Total	5.860	6.080	5.974	5.887	5.748	3.023	2.961	3.230	3.430	4.080	8.630

Private bicycles, motorcycles, motorbicycles, and cars constitute the private transportation. Because of the rapid changes in social and economic activity, the conventional prediction model may not be appropriate for predicting passenger volumes, distribution, and mode selection. A new model, a satisfaction level model, has been developed to replace the conventional prediction model. This model is simple, but very useful for qualitative and quantitative prediction of mode selection distribution.

The purpose for developing such a model is to analyze passengers' behavior in selecting public transportation modes and monitor changes in passenger distribution. This model will help to build the functional relationship between passengers' satisfaction level and the distribution of passengers' public transportation modes. Passengers' satisfaction level is the kind of combined characteristics that reflect the reliability, convenience, cost-effectiveness, and comfort level of public transit. The satisfaction level is represented by the values from zero to one, where zero reflects the lowest usage of public transportation and one indicates the highest usage of public transportation. Survey data are calibrated to obtain the information on weights of its model function. The model curve is presented in Figure 1. Previous analysis indicated that the model value could reach 0.627 by 2000. The values for passengers' distribution in selecting public transportation (transit modal split) will be 0.58 to 0.63. The esults are justified by different models used or developed in other projects.

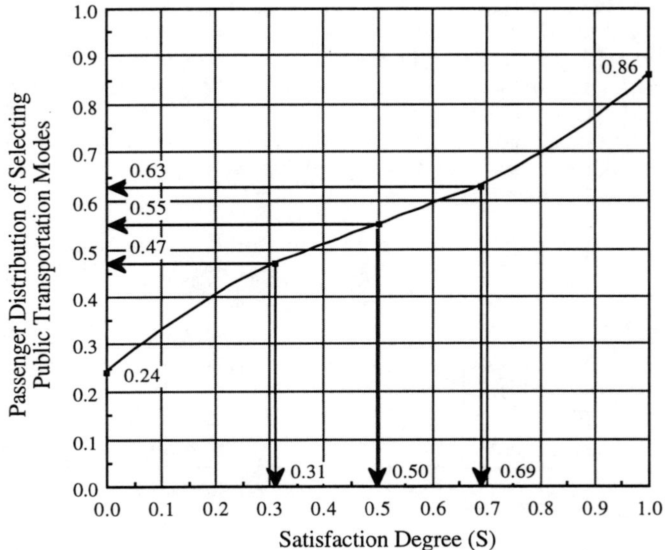

Figure 1. The Relationship of Passenger Satisfaction Degree and Distribution of Selecting Public Transportation Modes

Table 4 presents the prediction results for mode selection by work-related travelers from 1996 to 2000. To get the transit modal split for each different year, passengers' satisfaction level values need to be determined first. The model in Figure 1 can be used to predict the corresponding transit modal split.

TABLE 4. Projection of Transportation Mode Distribution for Work-Related Trips (Transit and Bicycle only)

Year	1996	1997	1998	1999	2000
Public Transit (%)	43 - 45	44 - 48	46 - 51	51 - 56	58 - 63
Bicycle (%)	53 - 57	52 - 56	49 - 54	44 - 49	37 - 42
Total (%)	100	100	100	100	100

CONSIDERATIONS IN PUBLIC TRANSPORTATION NETWORK

Recently, the public transportation system in Shanghai has significantly improved in both its network distribution and density. The coverage of travel demand from new residential areas has dramatically increased; many new remote public transit lines and express lines have been built; the capacity of the existing route have been increased and many elevated viaducts have been constructed to release traffic jams on surface streets; and more transfer stations have been built at the connections of different transportation systems. Table 5 presents the operational status of public transit system in 1996 and 1997.

TABLE 5. Comparison of Public Transit System Operational Status in 1996 & 1997

	Number of Buses		Number of Lines		Total Length of Routes (10^6Km)		Passenger Demand (10^9persons/yr)		Operational Income (Million USD)	
	Normal Transit Lines	Special Transit Lines	Normal Transit Lines	Special Transit Lines	Normal Transit Lines	Special Transit Lines	Normal Transit Lines	Special Transit Lines	Normal Transit Lines	Special Transit Lines
1996	8,042	5,281	510	548	332.03	243.35	1.90947	0.39796	$150.47	$86.60
	Total: 13,323		Total: 1,058		Total: 575.38		Total: 2.30743		Total: $237.07	
1997	8,733	5,474	562	516	363.03	278.61	1.87762	0.50005	$167.52	$110.00
	Total: 14,207		Total: 1,078		Total: 641.64		Total: 2.37767		Total: $277.52	

Note: A normal transit line is a transit line with stops, and a special transit line is a non-stop transit line.

PLANNING AND CONSTRUCTION OF RAIL TRANSPORTATION

Shanghai is planing to build a 75-km long high-speed railway system by 2000 and a 190-km long system by 2020. At that time, the average travel speed in Shanghai shall approach 24 km/h; travelers could arrive any place inside the city within 40 minutes; and 70% travelers entering and leaving the central business district would be carried by train. The final goal is to build a railway network of 400 km in length. The planned railway network is shown in Figure 2.

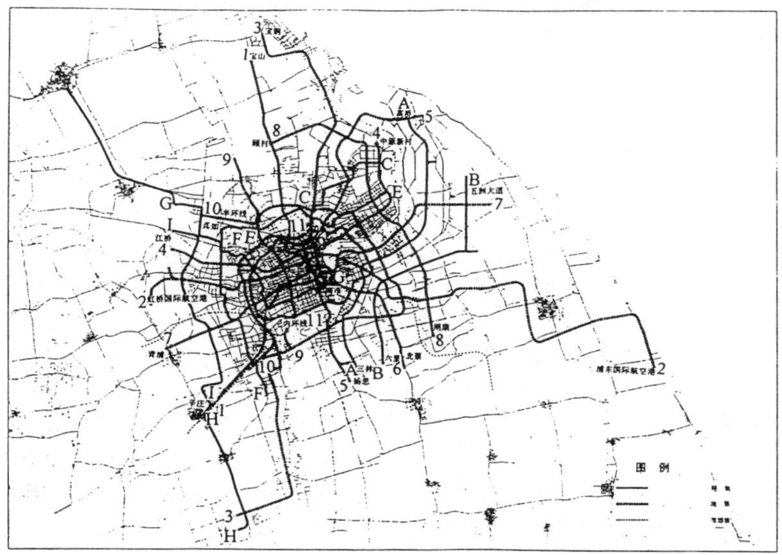

Figure 2. The Railway Network (including subway and light rail transportation)

On April 10, 1995, the completed 20-km long subway No. 1 started its operation. The 17-km long subway No.2 is scheduled to be completed early 1999. The 38-km long subway No.3 is scheduled to be completed by the end of 1999. Survey data between January 1993 and June 1998 indicated that 300 million passengers had taken the subway No.1. In 1997 alone, there were 112 million passengers. These subways served for about 3.67% and 4.5% of the total public transportation riderships (taxis not included) in 1996 and 1997, respectively. It is estimated that this percentage reached to 6.5% in 1998. The average passenger demand for subway No.1 during holidays in the first six months of 1998 reached 510,000 persons/day. The peak demand reached 620,000 persons/day on October 1, 1998 (China National Holiday Day).

CONCLUSIONS

People living in big cities prefer to make their own choice; whether to drive, walk, ride a bicycle or take public transportation. The government should concentrate on doing the following:

(1) Invest in convenient alternatives for people who don't drive, who can't afford to drive, or who would rather not have to drive.

(2) Promote alternatives to help people who are constantly stuck in traffic jams because although building more roads relieves traffic congestion temporarily, experience shows congestion soon returns.

(3) Invest in convenient transportation alternatives to save money in the long run.

(4) Make transportation investments that protect the environment and reduce air pollution from traffic jams, trucks and buses which millions people are forced to breathe.

(5) Plan for a better future for the city by building a connected network of neighborhoods with sidewalks, roads, public transit and open spaces so that shops, schools, parks and day care are within short distances to homes and work places.

(6) Transportation plays a vital role in the economic prosperity, cultural options and quality of life in all the communities. In a time of severe budget constraints, transportation programs should be affordable, oriented to providing those choices, and should allow more local control in setting priorities.

(7) Transportation options that offer real choices and local control will save money and ensure reliable, affordable transportation for everyone.

Finally, the city should strive for a reliable, cost effective, environmentally friendly public transportation system.

Public Transport Reform and Development in China: World Bank Perspective and Experience

D. Tilly Chang and Zong Yan[1]

Abstract

It is observed that traditional bus shares have declined precipitously in many cities across China during the past 15 years. This paper reviews the factors underlying this trend, the goals and strategies of public transport reform, and an overview of World Bank assistance strategy for the sector.

Public Transport Then and Now

In discussing the path of public transport reform in China, it is useful to begin with snapshots of the sector comparing the period just preceding reform and the situation today.

Historically, China's public transport sector was characterized by a single state-owned operator, providing regular bus services with simple fare structures. Typically, these state-owned enterprises (SOEs) exhibited labor intensive operations and low-tech management and operating methods, common inefficiencies characteristic of monopoly operators. To some extent, this reflected the use of planning targets rather than efficiency benchmarks for staff planning, capital planning and operations. For example, staff to bus ratios in Louyang, Shanghai and Tianjin were as high as 9.4, 10 and 10.5:1 respectively in 1995.

Aside from fare revenue, municipal budgets were the only source of finance for public transit, and large subsidies were needed for both operating and capital expenses. In terms of mode share, public bus operators held a steady 25% - 35% share of trips until around the mid- 1980s when they began losing shares to non-traditional (minibus) operators and other modes.

Today, several years into reform, the public operator is still the dominant provider of public transport services, although a more diversified supply of operators is competing in, and in some cases, for the market. Moreover, we are seeing more diversified production and ownership arrangements for municipal bus companies which are now corporatized and increasingly commercialized. As a result, operating efficiencies are much improved as staff to bus ratios have reduced from 10:1 to near or below 6:1 in some cities (although progress on this is uneven across cities). Public transportation choices are more differentiated with the development of premium services and fares to match.

---

[1] Tilly Chang and Zong Yan are Operations Officers specializing in urban transport within the World Bank, East Asia Region, Transport Sector Unit. World Bank, 1818 H Street, N.W., Washington, D.C. 20433.

Public operating subsidies are reducing, and there is more private participation in the sector, in particular in the finance of vehicles (short -term bank loans, private investors). Over time, this will allow cities to redirect public investment to other investments to support public transport, such as terminals and other infrastructure. While it is still early to tell, there are encouraging signs that the benefits of reform, including the provision of more and better choices for users, will re-attract passengers to public transport over time.

How did this transformation come about? Policies and actions implemented in the late 1980s and early 1990s under are just now yielding results. The following section provides background on the conditions which first brought about need for reform.

Trends Underlying Need for Public Transport Reforms

For a time before the economic reforms which transformed China in the 1980s, traditional bus transit enjoyed significant mode share (up to 30%) within the "big three" modes of walk, bike, bus which typically made up 90%+ of all trips. Ridership was steady, or growing slowly. Input costs were regulated under the planned economy. Cost-recovery was fair and operating subsidies were limited and stable.

In the mid-1980s, in an attempt to protect public transport mode shares, the Chinese government (GOC) established a policy that public transport should be the dominant mode of urban passenger transport (*1*). The policy emphasized development of public transport systems and containment of growth of privately-owned vehicles. Despite increased public investment between 1985 and 1994 however, traditional bus public transport loss mode share. Table 1 shows the mode shifts over time for Shanghai (urban population 13 million) and Shijiazhaung, the capital city of Hebei Province (urban population 1.5 million) over this period.

Table 1: Public Transport Mode Share Loss (*2*)

	Shanghai	Shanghai	Shijiazhuang	Shijiazhuang
Year	1986	1995	1985	1998
Mode Share				
Non-motorized	73%	78%	53.3%	88%
Walk	41%	33%	34%	34%
Bicycle	31%	45%	58%	54%
Motorized	27%	22%	8%	13%
Auto/Motorcycles	3%	7%	2%	5%
Taxi				2%
Company car			1%	2%
Transit	24%	15%	5%	3%
Other				2%
Total	100%	100%	100%	100%

Several factors contributed to public transport mode share loss:

Urbanization. From 1981 to 1994, the number of cities in China increased from 225 to 622 (2.7 x) with major growth (1.5 x) of large cities (over 1 million population). As cities grew in size, and as the decline of the "danwei" or state-owned enterprise system caused jobs/housing dispersal, urban densities fell and public transport services were not able to meet demand. This problem presents a major challenge for cities throughout China even today. For example, an official of the Beijing Public Transport Company (BPTC) recently reported that, of the 300 new residential developments in Beijing in 1998, only 90 are well served by public transport (*3*). Standards for minimum provision of parking for residential developments in outlying areas exacerbate this problem.

Motorization. More importantly perhaps, the growing affluence of households and demand for private modes of transport was being met by government industrial policies supporting motorization. In stark contrast to the 1985 policy promoting public transport, the Government declared the automotive industry a "pillar" industry in 1994, introducing at that time the notion of the "household car".

Figure 1: Motorization In Asia (1960-1990) (*4*)

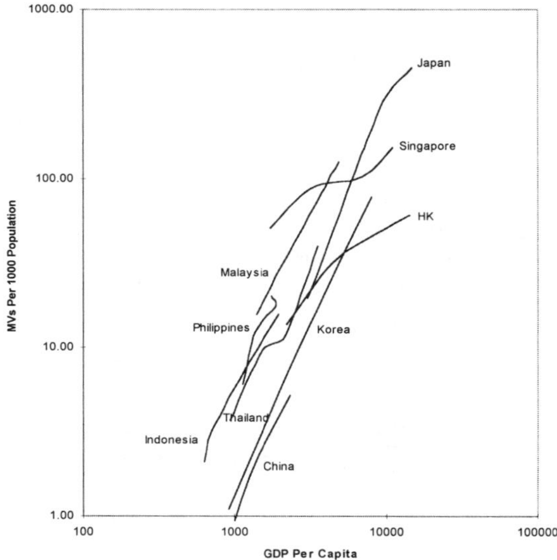

As shown in Figure 1, China is in the early stages of motorization. With passenger vehicle growth rates of 18% per annum between 1985-1995, however, growth of motor vehicle ownership and use has the potential to create massive problems in urban areas in the coming period. At this time, most car ownership is under enterprises and not private individuals; but visions of the "household car" remain on the minds of many.

Congestion. Cities were not well equipped for the problems associated with motorization, however, and fragmented government agencies struggled with various challenges in licensing, providing new road space and managing existing capacity. Growth of taxis, motorcycles and other non-public urban transportation services was rapid during this period. Examples of the latter include unlicensed or poorly regulated transport enterprises, mini-bus and long-distance bus companies operating in the urban area. As a result, traffic congestion grew rapidly, and in the absence of priority measures, caused a reduction in average bus operating speeds, for example to an observed 8-10 km./hr. in Guangzhou and Shanghai in the early 1990s.

Inadequate Public Transport. Poor route efficiency, deteriorating vehicle conditions, and underdevelopment of passenger facilities worsened public transit levels of service and added to travel times.

Thus, despite stated government policy support for public transport, backed by investment in vehicles and services, there was a general mode shift from transit to private or "premium" modes amongst those able to afford better services, and a choice between reduced level of service or a shift back to bicycles (which began to provide faster door to door service than transit) for those who could not. The effectiveness of increased capacity in public transport was therefore eroded by a decrease of transport efficiency and quality of services, relative to other modes.

Figure 2: China Public Transport Trends (*3*)

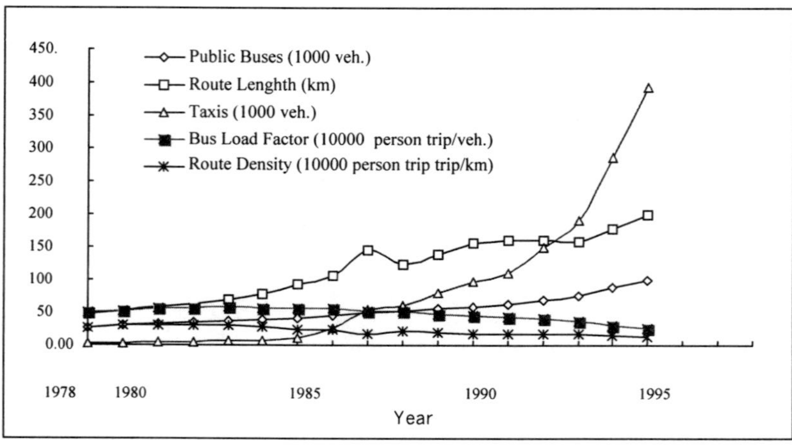

As shown in Figure 2, while route length and vehicles increased and vehicle load factors decreased slightly (implying service improvements), passenger density fell from 1985-

1995 in China. Meanwhile, the growth of premium modes such as taxis and mini-buses exploded.

The Effects of Mode Share Loss

With loss of ridership, many municipalities and public transport companies faced a severe revenue crisis. Some localities attempted to institute real fare increases with limited success, as emerging competition from minibus and other premium services presented an upper-bound on sustainable fare increases. Fare evasion became a growing concern. Alternatively, some bus companies tried to institute fare structure changes such as canceling monthly passes. This was tried in Shanghai in 1995, not anticipating that diversions could engender the opposite of the desired revenue effect due to elastic demands.

The ability among public transport companies to control costs was severely constrained by employment regulations limiting their flexibility to shed surplus labor. As a result, cost reductions were found in variable cost items, such as lowered fuel quality and deferred maintenance(5). In turn, the perceived and actual quality of services deteriorated, further exacerbating ridership loss and resulting in increased need for greater operating subsidies. The result was the classic and vicious downward cycle for transit.

As losses grew and local budgets began to hemorrhage from subsidies, local officials soon determined the need for serious reform of public transport. A national survey of 124 cities indicates the worsening of the fiscal crises in the early 1990s (see Table 2):

Table 2: China Public Transport Losses and Subsidies (*6*)

	Increase of Losses over Previous Year	Increase of Subsidies Over Previous Year
1992	60.5%	27.7%
1993	74.9%	61.8%

By 1994, losses in urban public transport totaled Yuan (Y) 1.0 billion and Government assistance to public transport had reached Y2.9 billion nationally (*6*).

Public Transport Reform Agenda

In an effort to retain and expand public transport mode share, while decreasing public subsidies to transit, many cities sought a two-fold approach to reform:

1) In the near term: raise operating efficiencies and lower costs, e.g. via operating subsidy regulation; and

2) In the longer-term, the goal was to improve and expand services with minimum public finance, thereby reversing mode share losses in a sustainable way.

While the former goal targets productive efficiency, the latter refers to increased allocative efficiency, a notion related to the scale economies of public transport. These goals are symbiotic: As operators adopt a variety of measures to reduce costs and increase efficiencies, the savings in operating subsidy and other resources can be redirected toward improvement of public transport facilities and expansion of new services. Early public transport reform efforts tended to focus on operators themselves and the problem of labor redundancy or inefficiency. Increasingly, cities are also recognizing the value of commercialization, competition and liberalization of supply to raise efficiencies and expand the choices for users.

Early reforms

The major problem facing most public transport companies is labor inefficiency and redundancy. One early strategy to reduce costs, commonly employed by many Chinese government enterprises, was to spin off excess staff to "tertiary enterprises", in the hopes that these would generate supplemental income for unprofitable bus operations. However, these tended to divert management attention from core operational issues. Moreover, the heavy social obligations associated with employment were often too difficult to overcome, an unsurprising outcome of transferring inefficiency from one industry to another.

Another strategy aimed at both reducing operating inefficiency and expanding services was to retrain conductors and other staff as bus drivers. While the move to single-operator buses reduces costs, given the lack of modern fare collection equipment, a potential disadvantage is the trade-off in lower average speeds. Operators are learning that bus priority measures, proper scheduling and modernized fare payment systems can help mitigate this problem.

Today, relationships between companies and their employees are also changing. In Urumqi, new hires are now ostensibly hired on a short-term (1-3 year) basis, although it is not likely they would be "let go" at the end of their hire periods since new hires are generally amongst the more productive. Also, many cities have converted staff to contractors who provide (mostly mini-bus) services under various arrangements (see below). This strategy reflects bus companies' attempts to regain some lost market share and cross-subsidize their regular operations.

Economic Reforms

Consistent with general SOE reform, corporatization emerged as a major strategy to commercialize bus operations and attract private investment. Typically, this involves organizational restructuring into a main "head" company which manages several operating companies distinguished by function or lines of business. This model has more aggressive and less aggressive variations.

Those companies in earlier stages of reform are choosing a tighter internal structure and commercializing relationships between units as a means to prepare them for competition. For example, in Urumqi, the three-tiered contract responsibility system is used to strengthen performance of each layer of the organization (company, sub-company, and vehicle team). Various incentive and penalty systems are used to improve performance rather than subjecting units to external competition, although internal competition for route-concessions may not be far off. One reason for this is that regulation of "private" public transport services is extremely weak as yet.

In Shanghai, there has been more deregulation with the result that operating companies are much more independent and actually compete for service contracts. Shanghai and Guangzhou operators also exhibit more diversified ownership arrangements, e.g. Pudong Bus Company which is a joint stock company and the development of five joint-venture companies in Guangzhou.

More recently, municipalities have begun to experiment with various forms of concessioning in an effort to manage subsidies and expand services without additional public finance. Examples include "joint-operations" arrangements, as found in Shijiazhuang which instituted net and gross-cost contracting with former staff and in Shenyang, Liaoning, which contracts with owner-operators and transport enterprises for services. In the Liaoning cities of Anshan and Fushun, we see these strategies in combination with concession of scheduled, route-based services to private Hong Kong-based operators. As mentioned above, Shanghai has been the most aggressive in terms of deregulation of public transport services and competitive tender of service contracts.

Shanghai Case Study

The Shanghai experience in public transport reform can be regarded as quite advanced and successful. In early 1996, the Shanghai Public Transport Company (SPTC) was reorganized into Shanghai Public Transport General Company, an umbrella company, and 13 operating companies, all of which were to be financially and operationally independent. A Shanghai Public Transport Limited Joint-Stock Company was established to hold the state-owned assets for all 13 companies and took on middle level management staff. The asset-company continues to be supported financially by government (it does not earn revenues), though leasing arrangements for these assets will be established.

The city's municipal finance bureau then contracted with each of the 13 companies on an area basis, including provisions for subsidies. These allowed subsidies to be given on a declining basis for losses incurred over the three year period (1996 to 1998). The savings from operating subsidies would be re-directed to increased capital investment. A passenger transport management division was established under Shanghai Public Utility Bureau in order to regulate the passenger transport sector and manage the tendering of all new bus routes to the 13 companies and 100 or so other eligible public transport companies. In other words, the 13 operating companies competed amongst themselves and with other operators for the operating rights to new routes.

Within a three-year period, Shanghai's public transport sector has transformed into one characterized by a competitive market with many players. From 1995-1998, passengers benefited from route restructuring and expansion and improved vehicles. Length of operations has increased about 11.5% per annum. As a result, following the fare structure debacle of canceling monthly passes, which caused a 50% drop, ridership has increased annually about 3.9% per year. Across all 13 operating companies, revenue has increased about 16.5%, public subsidies for operating losses have reduced 10-fold and capital investment has increased 3-fold (see Table 3).

Table 3: Shanghai Public Transport General Company Statistics (Millions Yuan) (*7*)

Year	Operating Subsidy	Investment in Buses	Investment in Depots
1995	800	45	100
1996	168	200	150
1997	84	300	160
1998	84	300	156

Table 4 provides some data from one of the 13 operating companies. Note that route length has increased, staff to bus ratio has reduced, and the company was able to record an operating "profit" given its subsidy contract.

Table 4: Shanghai Number One Public Transport Company Statistics (*6*)

Item	1995	1996	1997	1998
Routes	38	49	55	57
Length (km)	370	468	540	555
Daily Ridership	1,500,000	750,000	900,000	1,000,000
Staff/Bus ratio	7.8:1	7.2:1	6.4:1	5:1
Subsidy (Millions Yuan)	100	18.8	9.4	9.4
Net profit (Millions Yuan)		-9.8	0.1	1.0

A Comment on Fare Regulation

Fare regulation is a difficult and sensitive issue as it often relates to the question of how to finance the costs of reform. Given that a major political, management and financial challenge of reform is labor redundancy the question arises to what extent fares (as user charges) should cover the costs of operations? At the heart of this question is the question of whether users paying their fair share for services received.

Increasingly, the answer in China is "yes", if we consider the cross-subsidy between premium and regular bus services being run by most operators. Bus companies in China are making profits on their premium services, e.g. minibuses, to the degree that they are willing to borrow to finance the cost of new vehicles. Meanwhile, they are making losses, though not huge ones, on their regular bus operations. As a result, on a system-level, many public bus companies are covering a significant share (70% to 90%) of their operating expenses (inclusive of depreciation on vehicles) from the farebox.

Implications of this include the importance of the cross-subsidy mechanism for balanced development of the sector and the view that Government should continue to bear the burden of financing operating losses to the extent that these stem from the vestiges of past, government policies. As companies continue to be constrained by structural and operational inefficiencies from past employment and other policies, it can be argued that it is the remnants of these employment and other policies which are being subsidized, and not public transport passengers, who are generally making a fair contribution for services received (*5*).

Assessment of Reform

Cities across China are experimenting with various strategies to reform and develop public transport. Results are mixed, depending upon the particular economic conditions in the locality. The following are some general observations on the current situation:

- Commercialization: Good. Although the instruments and pace of commercialization differs across cities, companies are becoming stronger. There is need for greater assistance to bus companies for corporate planning.

- Economic Regulation: Fair to Poor. Regulation of supply and competition is generally weak, especially in terms of coordination of motor vehicle licensing between government agencies, and establishment of a "fair playing field" between public and "private" operators. Complaints are common of chaotic on-street competition and traffic congestion at stops resulting from cherry picking and hanging back behaviors.

- Finance: Good to Fair. Improved farebox-recovery and private participation is providing needed new investment in vehicles. Successful subsidy regulation has been demonstrated in the larger cities, but this is weaker elsewhere.

- Patronage: Not Clear. Due to current ridership estimate practices (on the basis of revenues) most companies do not have an accurate picture of ridership. Neither is there good knowledge of ridership carried by non-public formal and informal operators. It is likely that overall ridership is in a maintenance stage; though experimentation with market reforms and bus priority is promising.

Remaining Challenges

As trip lengths continue to increase with city growth and development patterns, and as motorization threatens the accessibility of cities, municipalities are faced with many challenges in developing public transport.

- Policy/Vision: First and foremost, inter-modal policies need to be sharpened, especially with respect to motorization in all of its facets (licensing, user charges, cross-subsidies, road development and traffic management, land use policy)

- Planning and Economic Regulation: City functions in planning and regulation need to be strengthened and integrated significantly. Policies to liberalize supply and capacity to regulate competition policy should improve as lessons of experience are disseminated.

- Finance and Investment: Fare and subsidy policies should be clarified and adequate provision made for public transport infrastructure and facilities, including priority measures. Investment in new technologies should be considered carefully in terms of cost-effectiveness and local capacity.

Public Transport operators also must continue to prepare themselves to operate in a market environment. This requires better understanding of financial management, cost-effective investment in technology to improve management and operations, and improved knowledge of customer preferences and the markets in which they operate.

Residents and public transport users could also play more active roles as consumers and planners of services. Citizens already do provide views and feedback via "hotlines" and in some cities, participate in "price" committees reviewing fare increase proposals.

Public Transport Assistance in World Bank Projects

The World Bank's experience in China's public transport sector dates back to the early 1990s, when public transport assistance was first provided in the context of urban and environment projects. These were followed by 5 urban transport projects, in which assistance to public transport was focused within a more comprehensive approach emphasizing the need for balanced development of the urban transport system.

In terms of strategy, World Bank support to public transport development in China has evolved alongside the public transport reform process. Early urban transport operations targeted modernization of public transport companies and therefore included finance of vehicles, depots and other direct assistance to public transport companies.

More recent projects deepen assistance to companies (e.g. corporate planning) and expand support to investments designed to benefit users of public transport services in general, whether publicly or privately operated, in anticipation of increased liberalization within the sector. These include planning studies, policy and institutional development, and passenger and bus priority facilities.

As with all World Bank lending operations, technical analyses underpinning project design are supported by social, environmental and economic evaluation of project activities. Recent projects have included consultative participation with stakeholders (non-public operators, residents/users) and environmental emissions control strategies building upon local initiatives including conversion of vehicles to natural gas power.

All views, findings, interpretations and conclusions expressed in this paper are entirely those of the authors and should not be attributed in any way to the World Bank, its affiliated organizations, members of the Board of Executive Directors or the countries they represent.

References:

(1) State Council Document No. 59 of 1985; 1985 Blue Book of Technology Policy; Sector Policy, 1989, in Wang Jingxia, et. al., "Theme Paper 7: Reform and Development of China's Urban Public Transport Enterprises", in Stares and Zhi, ed., China's Urban Transport Development Strategy, World Bank Discussion Paper No. 352, Washington, D.C., 1996.

(2) Eric Ho, et. al., "Socioeconomic, Land Use and Travel Patterns of Shanghai", paper delivered at ASCE 1st International Conference on Public Transport, Miami, March, 1999 and World Bank.

(3) Zhao Boping et. al. "Prospect and Characteristics of Urban Public Transport in China", paper delivered at ASCE 1st International Conference on Public Transport, Miami, March, 1999.

(4) Stephen Stares and Liu Zhi, "Theme Paper 1: Motorization in Chinese Cities: Issues and Actions", in Stares and Zhi, ed., China's Urban Transport Development Strategy, World Bank Discussion Paper No. 352, Washington, D.C., 1996.

(5) Lupton, David, "Case Study: China Medium Cities Project", 8th World Conference on Transport Research, Antwerp.: Jointly with Overington DJ, July 1998.

(6) Wang Jingxia, et. al., "Theme Paper 7: Reform and Development of China's Urban Public Transport Enterprises", in Stares and Zhi, ed., China's Urban Transport Development Strategy, World Bank Discussion Paper No. 352, Washington, D.C., 1996.

(7) World Bank Shanghai Metropolitan Transportation Project Office, 1999.

Paper 1: Social Characteristics, Land Use and Travel Patterns of Singapore

Peter Bow
Manager (Planning)
Land Transport Authority

1.0 INTRODUCTION

In 1995, the Singapore Government formed the Land Transport Authority (LTA) by merging all the public sector entities in charge of transportation-i.e. the Roads Division of the Public Works Department (PWD), the Mass Rapid Transit Corporation (MRTC), the Registry of Vehicles (ROV) and the Land Transport Division in the Ministry of Communications. While the LTA is basically an amalgamation of the four organisations, it has an enhanced role that of a central cohesive body which removes the boundaries between road, rail and various forms of land transport, thereby providing for an integrated and efficient land transport system which meets the needs and expectations of Singaporeans. The formation of the LTA will allow better integration of the functions of planning, development, implementation and management of all transport infrastructure and policies. LTA's mission is to provide Singaporeans with a world class transport system in the next 10 to 15 years. The White Paper sets out how the Authority intends to achieve its mission, its transport vision for Singapore, its operating philosophy, and the initiatives it will undertake in the short and long term.

Peter Bow, Manager (Planning), Land Transport Authority, 460 Alexandra Road #19-00, PSA Building, Singapore 119963

Singapore, with a population of 3.8 million and land area of 648 square km (about 80% the size of New York City), is one of the most densely populated and urbanised countries in the world. Over the last 3 decades since independence, Singapore has grown from a third world country with a per capita GNP of US$300(S$507) to US$21,400(S$36,200) as at 1997. And one of the highest standard of living in the world, according to the Swiss based World Economic Forum (WEF). Financial and business remained the leading contributors to the overall economic expansion. Despite its small size, high population density, rapid economic growth and intensive urbanisation, Singapore stands out as a city that is relatively free from urban traffic congestion. This has been realised through the careful forward planning of land use and the corresponding transportation framework and the sustained implementation of demand-curbing policies over the years consistent with the transportation strategy for a small urban country.

Figure 1 – The city centre's towering office and residential blocks overlooking the Singapore River

The economic progress was reflected in the transformation of the physical appearance of Singapore through land reclamation, urban redevelopment, the construction of residential and commercial complexes, major new highways and MRT/LRT systems. With a small population base and no natural resources, Singapore gave priority to developing its human resources and talent. A sense of Singaporean identity among a people of diverse values and culture had emerged.

Over the last 10 to 15 years, the increase in demand for transportation service has been explosive. The number of vehicular trips grew annually by 7% from 2.6 million trips per day in 1981 to about 7.0 million trips per day presently. Out of this, over 60% of the trips are by public transport; 3.0 millions by bus, 1.0 million by MRT and another 1.0 million by taxi and are expecting 10 million trips per day by 2010. The increase is not surprising, given rising levels of income, educational attainment, changes in lifestyle and a host of other demographic and social developments. In 1995, 1 out of 10 people (10%) owned a car as opposed to 1 out of 15 (6.7%) in 1981. Car ownership is expected to continually rise despite demand management policies to restrain car ownership and usage.

The transportation system for Singapore in the next century will have to meet not only the increased demand to travel, but also the rising expectations for a quality transport system. The challenge is to provide a high quality public transport system good enough to serve as an alternative mode, replacing private car usage.

For transportation policies to be effective, it is necessary to maintain a long term perspective, anticipate change, understand the forces driving the demand and supply and adopt a holistic approach when responding to the challenges. In addition, policies have to take into account changing aspirations, new lifestyles and the psyche of the society.

2.0 CONCEPT PLAN

Land use planning plays a key role in determining the land transport network as it can influence the need for travel, even the mode of travel. The need for a comprehensive approach incorporating integrated landuse and transport planning to guide Singapore's development was recognised with the country's independence in 1965 as major changes in development policies were then put into effect. It was then recognised that the Master Plan adopted in 1958 was clearly inadequate and inappropriate to cope with the changed conditions of a rapidly expanding and diversified economy. The long-range landuse and transport plan was prepared under the State/City Planning Project that started in 1967. The Concept Plan guides how we can develop our land and transportation concurrently. Our New Towns, the airports, expressway, main roads and MRT are shaped by the Concept Plan. In the 1971 Concept Plan, the strategy was to decentralise the population by building residential towns away from the city centre, and connecting them by an efficient system of roads, expressways and public transport. To avoid the serious problems of inner city congestion and urban sprawl that plague many cities, provisions are made for the increasing demand for public transport.

The economy and population have been steadily growing and so are the changing wants and needs of Singaporean. In 1991 transport considerations resulted in a revised Concept Plan which includes:

- Reducing the percentage share of commercial quantum in the Central Area by adopting a strategy of decentralisation to Regional Centres and other commercial nodes served by Mass Rapid Transit (MRT);
- Providing a more balanced distribution between jobs and homes in each region to minimise travel distance and reduce total travel time. This will help ease congestion in the Central Area, particularly during peak periods;
- Extending urbanised corridors and new towns along MRT lines to enhance the role of the rail network as the main form of moving masses, better utilising public transportation infrastructure; and
- Retaining the current percentage share of live-in population within the Central Area to provide a greater balance between job opportunities and residential population in the city.

A key component of the revised Concept Plan was the Strategic Transportation Plan (STP) which provided a framework for a viable transport network that will meet the ever increasing travel demand of a population and economic growth by the year 2030.

A White Paper published by the Land Transport Authority in 1995, outline the guiding principles to achieve a world class public transport system that provides commuters with highly efficient, comfortable and convenient rides in free-flowing traffic. It aims to improve public transport means by not only improving bus or train travel, but also improving all the intermediate and end-point facilities that make for a complete door-to-door journey. This includes linkways, service information and improved customer service.

The overall key is to integrate urban development with transport planning. Having a proper mix of development and the highest building densities concentrated at and around MRT stations ensures maximum accessibility by public transport for commuters to key nodes of employment, housing and other social activities. Commuters facilities and building developments will also be fully integrated. For example, the Sengkang MRT station is designed with extensive property developments and transport nodes. It will be built in the Sengkang Town Centre, a new residential town in the North East sector of Singapore, adjacent to a bus interchange with an LRT stations. Passengers transfering from node to node will essentially travel vertically, in air-conditioned comfort, with minimal horizontal distances to traverse. Above and adjacent to this multi-modal interchange will be commercial and retail facilities that will make the interchange the focal point for all activities in the Town Centre.

Figure 2 – MRT/LRT system integration at Sengkang

Figure 3 – Cross section of MRT/LRT system integration at Sengkang

3.0 TRANSPORTATION SYSTEM

Over the last 10 years, the road network increased by 27%. However the car population grew even faster. It is clearly not tenable to continue building more and more roads to accommodate new cars, given our limited land area. Roads now constitute 11% of the island's area, about the same percentage as housing. Currently, more than 80% of population living in public housing. (about 0.64 million dwelling units).

Singapore relies on ownership measures to moderate the demand for cars, and on usage measures to restrain utilisation. What is perhaps still unique is the Vehicle Quota System (VQS). Introduced in 1990, VQS requires the buyer of any vehicle to bid for a Certificate of Entitlement (COE), the bid is for the right to own a vehicle for 10 years. VQS, has achieved the transport objective of controlling total vehicle population by effectively reducing its annual growth from 6% to 3%, in tandem with road capacity, and the dividend has kept our roads relatively congestion free.

Other strategies include managing the traffic in the city. Singapore introduced the manual Area Licensing Scheme (ALS) in 1975 effectively controlling congestion in the CBD, even though vehicle population has more than doubled since 1981. Road pricing allows motorists to be more aware of the cost of congestion they impose on other people every time they use their vehicles. Road pricing will encourage motorists to consciously plan their trips and consider public transportation alternatives. The Electronic Road Pricing (ERP), the world's first, was introduced in September 1998, replacing the existing manual road pricing. The ERP system, which utilises a sophisticated combination of radio frequency, optical detection, imaging and smart card technologies, is capable of handling multiple vehicle travelling. The system is able to communicate with the In-vehicle Unit, identify the type of vehicle, deduct charges and if a violation occurs, freeze in rear image at the vehicle. ERP charges will be deducted from the motorist's smart card as and when he crosses a gantry during operational hours. The ERP rates for different classes of vehicles are based on the Passenger Car Unit (PCU) rating of their vehicles. ERP allows us to shift from ownership towards usage-based charges. The ERP is more flexible in responding to changing travel patterns and more efficient in charging according to road usage.

As a result of the long-term aspiration, coupled with the persistence to implement stringent management measures, the land transport system in Singapore is now in a relatively healthy state. There is minimal inner city congestion, a comprenhensive road network and efficient and affordable MRT and comprehension bus systems. The fare structure of bus and MRT are distance related. The fares ranges from US$0.35(S$0.60) for distance travelled up to 3.2Km to US$0.94(S$1.60) for distance travelled for more than 26Km.

The strategic thrust is to shift more trips towards efficient modes like rail and bus. While not every Singaporean owns a car, their need for greater mobility can be satisfied through an efficient, reliable and integrated pubic transport system. Singapore needs a public transport system that provides a high quality service, offering an attractive alternative to the private car. The priority is therefore given to improving public transport and creating a wider range of public transport options.

Rail transport can meet the transport needs of heavy demand corridors while maintaining high travel speeds and predictability of arrival and departure times. This explains why cities of high density are usually dependent on rapid transit systems with dedicated rights-of-way. Singapore has opted for rail even though it is more expensive option than bus. The target is to have a high percentage of trips on a quality transport system where 75% or more of trips into the city centre are by public transport.

MRT systems are planned to serve heavy corridors of passenger traffic such as high density housing. For lighter corridors, LRT system, which require less capital investment, are a more practical alternative.

Singapore plans to expand the rapid transit network to become the backbone of the public transport system. Currently, there are 83 km of MRT rail line, 63 km above ground and 20 km underground, with 48 stations. The system carries about one million passengers trips per day. The North-South and East-West Lines are joined at Raffles Place, City Hall and Jurong East Interchange stations.

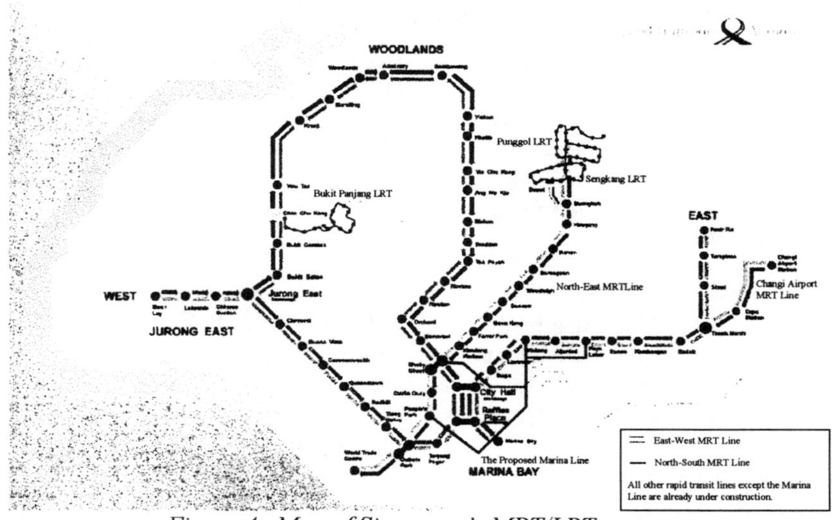

Figure 4 –Map of Singapore's MRT/LRT system

Singapore has many upcoming projects in developing the MRT and LRT system. By the end of 2004, Singapore will have around 154 kilometres of rail lines.

The North-East Line (NEL), currently under construction, is the biggest and most expensive transport infrastructure project in Singapore since the first MRT system was finished in 1990. The 20-km line with 16 underground stations at an estimated cost of US$3 billion (S$5 billion), is expected to be completed in 2002. This system runs from the World Trade Centre in the south to the new residential town Punggol in the north-eastern part of the island.

The 6 km Changi Airport Line (CAL) is an extension of the current MRT system and will serve as a vital link between the airport and other parts of Singapore. The Changi Airport Station is strategically located underground, between Terminal 2 and 3 will serve as a vital link between the two terminals. Extra attention has been put into making the stations user-friendly through the necessary facilities and conveniences that international travellers need for a smooth transit to and from airport. The CAL is currently under construction.

The Marina Line, currently at a design stage, serves Singapore's Central Business District. The Marina Line will improve accessibility to the MRT system, thus encouraging more people to switch to public transport.

By 1999, Singapore will have its first Light Rapid Transit (LRT) system to MRT network. This driverless system will serve the Bukit Panjang Town and link this town to the MRT system. The US$169 million (S$285 million) line, currently under construction, is being built on 7.8 km of elevated guideway. As the LRT stations will be located less than 400 metres to most of the apartment blocks of Bukit Panjang Town, this will make public transportation more accessible to residents.

By 2002 and 2004 respectively, both Sengkang and Punggol new towns will also have a LRT lines that feed the NEL. The US$385 million (S$650 million) system, currently at a design stage, will have a total length of 24 km tracks that includes 33 stations. The Sengkang and Punggol LRT lines will have its stations integrated with the facilities at the neighbourhood centres.

- **Financing Framework**

The financing policy for public transit projects in Singapore is based on the concept of partnership. The Government will provide the infrastructure that includes capital costs of the MRT/LRT lines and stations and the first set of operating assets. There are no operating subsidies given to the operators.
A private company, Singapore MRT Pte Ltd, currently operates and maintains the MRT system.

The main bus services in Singapore are provided by privately owned companies, The Singapore Bus Service and Trans-Island Bus Service, which operate under licence from the Government. There are also other privately operated peak-hour commuter bus services. These bus services compete in terms of efficiency, cost-effectiveness and service levels. The bus companies are obliged to ply all routes, even less profitable ones, at prescribed frequencies and regulated fares. Consumers' interest are safeguarded in the duopolistic market by the Public Transport Council (PTC), which regulates bus routes, service standards and public transport fares while balancing commuters' interests with the need for the operators to remain financially viable.

Bus services will continue to be a major mode of public transport in Singapore. They are very efficient carriers of passengers and are 39 times more efficient than cars in terms of road usage. Buses priority over cars in road usage, so that bus rides are as smooth as possible. Priority is given to buses at traffic light junctions, enabling them to move off before the rest of the vehicles during the green, and along the roads by demarcating bus lanes.

While an efficient and comprenhensive bus network is adequate to serve the lighter corridors, buses cannot be a solution for a compact city like Singapore. This is because bus service levels deteriorates sharply once demand exceeds a certain threshold. For compact cities, the solution is to provide a good rail network, offering services with high frequency, reliability, speed and comfort.

4.0 FUTURE DIRECTIONS

Singapore vision is to aim for a system that meets the needs and demands of a dynamic and growing city with a population that will increasingly expect high standards in service and infrastructure. Given our circumstances, rapid transit will form the backbone of the public transport system. The Rapid Transit System will eventually be within easy reach of the majority of the population. Singapore will plan for MRTs to serve the heavy corridors of traffic. For lighter corridors, LRTs or people movers are more practical alternatives. Preliminary studies indicate that a comprehensive rail network of 160 km over the long term is possible.

To provide commuters with greater convenience, accessibility and comfort, we intend to fully integrate urban developments with transport facilities. One such example is at the Woodlands MRT Station where other transport facilities like bus interchange and taxi and drop-off points are well integrated with the station. Commuters can interchange easily, in comfort, even in inclement weather.

Figure 5 -The Woodlands MRT station integrates other transport features like bus interchange at basement; taxi stand and car drop off point at grade and MRT at platform

Singapore wants to provide a comprenhensive range of public transport services, each being developed to the highest quality commensurate with the fares charged, well integrated for the commuter, offering a seamless journey.

The road network must also be comprehensive to sustain economic activities and to provide better connectivity for all Singaporeans. Good connectivity will benefit not only private transport, but also public transport such as buses and taxis. Over the next five years, Government will spend more than US$0.65 billion (S$1.1 billion) to expand our road network by another 330 lane-km. It will include new expressways, expansion of major roads and upgrading of key junctions.

In addition to developing a comprehensive road network, LTA will also maximise network capacity by adopting advance traffic management systems. The Expressway Monitoring and Advisory System (EMAS) consists of a network of computers and video cameras placed on site. They are connected to an intelligent computer system. Television, radio stations and emergency crews providing ambulance services are linked to the system. Both detection and surveillance cameras are switched on 24 hours and early detection and quick reaction to accidents and breakdowns will minimise congestion and danger to road users. LTA plans to electronically link to all intelligent transport systems at national levels so that the various transport agencies can respond accordingly through the Integrated Transport Management System (ITMS). It has been conceived as a means to provide real-time traffic and public transport information to motorists, commercial vehicle operators, public transport operators and government agencies. This way, motorists and commuters can be advised on the best routes and modes of transport available based on real-time information.

5.0 CONCLUSION

In the long run, Singapore intends to increase the percentage of public transport trips, by making public transport more accessible, more convenient, more comfortable and faster.

Therefore Singapore starts solving tomorrow's transport problems today. It is managed all the time, and the improvements made continually to achieve LTA's mission of providing Singaporeans with a world class transport system in the next 10 to 15 years.

URBAN PUBLIC TRANSPORTATION OF SINGAPORE

Fong Seck Kong

I Background

Singapore is a city state with a land area of 648 sq km and a population of 3.8 million. As at 1997, the per capita GNP was US$21,400 (S$36,400).

The total daily trips generated is about 7 million. Of these, around 5 million trips are on public transport which includes Mass Rapid Transit (MRT), bus and taxis. The remainder of the trips is mainly by private vehicles.

The public buses account for the largest portion of public transport ridership, followed by the MRT and taxis. Seee Table I below

Table I : Usage of Public Transport In Singapore

Mode	Daily Passenger Trips	Ave Trip Length Km/Passenger-Trip	Passenger-km	% Passenger-km
Public Bus	3.0 million	6.0 km	18 million	45.9%
MRT (Aug 98)	1 million	12.6 km	12.6 million	32.1%
Taxi	1 million	8.6 km	8.6 million	22%

The public bus system is run by two separate bus operators, both of which are private companies. They operate 2 basic types of bus services - feeder services and trunk services. Feeder services usually operate in housing estates to convey passengers from their homes to a bus interchange, where they can transfer to a trunk bus or the MRT.

Deputy Managing Director, Singapore MRT Ltd, 251 North Bridge Road, Singapore 179102

Trunk bus services provide links between major housing townships, and between housing townships to major destinations such as the Central Business District, the tourist belt and industrial estates.

Between them, the two bus operators own a fleet of 3,300 buses, and run 240 bus service routes with a daily average of 63 million place kilometres.

II Financial Arrangements In The Construction and Operation Of The MRT System

The current MRT system was funded and built by the Singapore government through a statutory board called the Mass Rapid Transit Corporation (MRTC). In 1995, the MRTC was absorbed into a new statutory board called the Land Transport Authority (LTA). The LTA has jurisdiction on all land transport matters in Singapore, and is the authority which will continue to build new fixed guide way systems in Singapore.

The current MRT system is operated by Singapore MRT Ltd (SMRT), which is a private company. The shares of the company are currently entirely owned by the Singapore Government through its holding company.

The current MRT system was funded and built by the Government through a statutory board called Mass Rapid Transit Corporation, now absorbed into a new statutory board called the Land Transport Authority (LTA).

SMRT now operates the MRT system under a new 30-year lease from April 1998. Under the terms of the new lease:

- LTA continues to own the infrastructure. SMRT leases the infrastructure and will maintain and repair the infrastructure.

- SMRT owns the operating assets. It will pay the historical cost of the operating assets when they are replaced. LTA will pay for the inflation in cost, if any.

The LTA also has a regulatory role over SMRT. It sets operating performance standards through the lease agreement which SMRT has to meet, failing which SMRT has to pay a penalty.

The organisation structure of SMRT is shown in Chart 1 below:

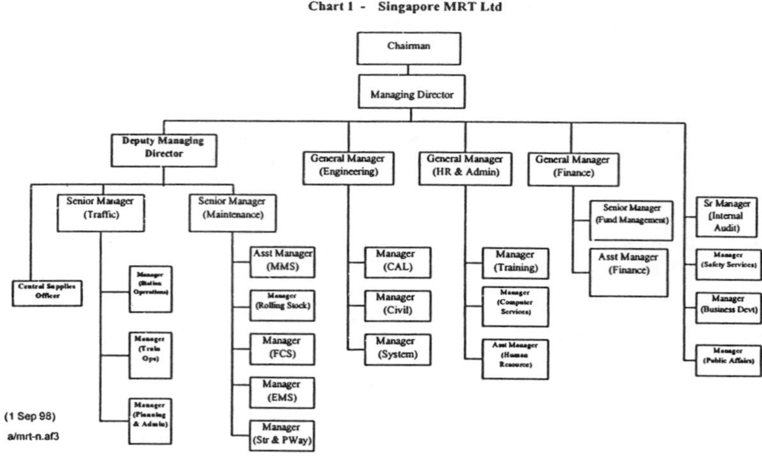

The total staff strength is about 2,800 with the majority in the operation of the railway. There are about 1,400 employees in Maintenance, and about 1,000 in Traffic Operations.

III The MRT System

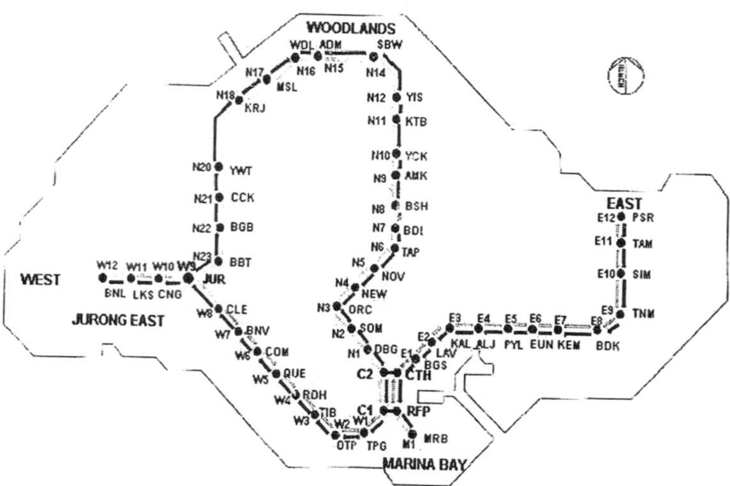

The current MRT system in Singapore has a route length of 83 kilometres, with 48 stations. Twenty kilometres of the route is underground, mainly in the Central Business District. It is a steel on steel system, with a rail gauge of 1435mm. It has 3 depots with the main depot at Bishan where major maintenance, repairs and train overhauls can be carried out. Bishan Depot also has a 1.8 km test track.

Route

The MRT system links major public housing estates to major economic centres such as the tourist belt, the Central Business District, and industrial estates. Three major regional centres, which are planned to decentralise some of the economic activities from the Central Business District, will be developed adjacent to 3 of the larger MRT stations.

MRT Stations

There are 48 MRT stations. Fifteen of them are underground, one is at-grade and the others are above ground. All the underground stations are air-conditioned (cooled) to maintain a comfortable ambience.

Platform screen doors (PSDs) are installed at the platform edge of all underground stations, to conserve energy by containing the cooled air within the station. The spin off is that the platform becomes safer for passengers who cannot fall onto the tracks. The stations are also cleaner as the moving trains do not blow dust into the stations.

PSDs have reduced the electricity consumption for airconditioning the stations by more than 50%. However, PSDs do require a high train stopping accuracy. In practice, trains have been able to stop within the target of ±0.5 metres in excess of 99.9% of train stops.

In all except 4 stations, the platforms are island platforms. In 3 stations, there is a centre track while one station has side-platforms because of geographical limitations.

Automatic Fare Collection (AFC) System

A fully closed AFC system was in use from the start of operation in 1987. Passengers have to use a magnetic ticket to enter and exit stations, through automatic gates.

There are two main types of tickets - single trip tickets and stored value tickets. Single trip tickets are useable only on the MRT. They are sold through passenger operated ticket vending machines.

Stored value tickets, called farecards, are useable on MRT and buses. A fare is deducted for each trip until the stored value is exhausted or too low to start a new journey. The passenger then has to revalue the farecard.

Farecards are marketed and revalued by a company called Transit Link Pte Ltd formed by SMRT and the 2 bus operators.

Signalling System

The signalling system is computer controlled, and has the 3 main components of:

- Automatic Train Protection
- Automatic Train Supervision
- Automatic Train Operation

which enable trains to operate in a fail safe manner, at close headways according to a set timetable, and with minimal manual operation by the train operator.

The maximum train speed is 80 kph, with an average line speed of 45 kph.

Rolling Stock

There are 85 trains of 6 cars each. The trains have through gangways that will allow passengers to walk throughout the length of the train.

Each car is 23 metres long and 3.2 metres wide, and able to accommodate 250 passengers, with 62 passengers seated on longitudinally mounted seats.

Traction current is supplied by a third rail, at 750V DC. The first batch of 66 trains have DC traction motors. The second batch of 19 trains have AC motors.

All trains have regenerative braking, generating up to 35% of the traction current used by trains, and about 10% (of the traction current) are inverted to AC to power station facilities.

Operating Characteristics

The MRT system operates as 2 separate lines, the East-West Line and the North-South Line. Passengers can transfer between the 2 lines at 3 stations, namely City Hall, Raffles Place and Jurong East. Transfers are convenient cross platform transfers. Train can cross lines, but this is done only when there are disruptions along certain sections of the lines.

The main operating characteristics are summarised below:

Average Weekday Ridership	980,000 daily
Peak Hour Headways	2 mins to 4 mins
Morning peak	(8.00 am to 9.00 am at Raffles Place Stn)
Evening peak	(5.15 pm to 6.30 pm at Raffles Place Stn)
Off Peak headways	6 mins.
Maximum Train speeds	80 kph
Average speed	45 kph
Average daily train-km operated	34,600
Operating Hours	19 hours

Maintenance

SMRT carries out most of the maintenance with its resources. The exceptions are the cleaning of stations and trains and horticultural maintenance, repairs of specialised components, and statutory inspections which must be carried out by independent engineers.

The Maintenance Division has four main departments. These are:

- *Rolling Stock Department.* It carries out preventive, corrective maintenance and overhaul of trains and track-borne maintenance vehicles.

- *Fares, Communications and Signals Department.* It maintains the Automatic Fare Collection, Communications and Signalling Systems.

- *Electrical and Mechanical Systems Department.* It maintains all the power supply equipment, fire safety equipment, lifts and escalators, stations and tunnel ventilation equipment, airconditioning systems in stations, and lighting.

- *Structures and Permanent Way Department.* It maintains the civil structure of stations, tunnels and viaducts, and the permanent way. It also administers contracts for cleaning stations and station toilets.

  Maintenance of the Permanent Way is carried out after traffic hours. There is no redundancy which will allow the MRT to operate while maintenance is done on the tracks.

Financial Performance
Revenue

SMRT owns the operating assets from 31 March 1999. There are no financial accounts based on the new financial regime as yet. Broad indicative figures are used to show SMRT's possible current financial performance for 1998.

The main sources of revenue for SMRT are passenger fares, advertising space and shop rental. Additional revenue comes from the investment of funds, but this is not considered as part of the revenue from railway operation. There is no operating subsidy from the Government.

Passenger revenue, advertising income and rental income amounts to about $200 million (S$357,544,163).

Operating Cost

Operating cost is mainly made up of manpower cost, maintenance materials, maintenance contracts and electricity. This amounts to about $175 million (S$295.5 million) and depreciation. This yields a profit of $25 million (S$42.5 million) before tax.

Use by People

The MRT ridership has reached almost 1 million per day.

The trend in MRT ridership between 1987 and 1998 is shown graphically in Chart 2 below.

Chart 2 - MRT Ridership 1987 - 1997

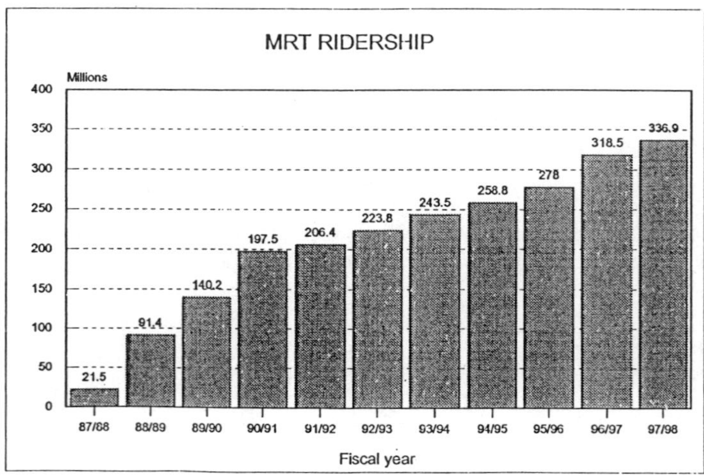

The rapid growth between 1987 and 1990 was due to the opening of new sections of the MRT. Between 1990 and 1996, the growth was attributed to the effects of the bus/MRT integration described later in this paper, and the development of new housing and economic centres along the MRT corridor.

Ridership had a quantum jump in 1996 again with the opening of a new 16 km extension with 6 stations called the Woodlands Line.

Ridership has now slowed to near the average annual growth in public transport ridership (3%).

IV Importance of Good Public Transport in Singapore

Singapore has a population of 3.8 million living in a land of 648 sq km. Sustained economic growth has raised per capita GNP to US$21,400 (S$36,400) in 1997. The limited land has to be allocated for housing, industries, water catchment recreation, transport, among others. Congestion-free roads is considered an important factor in the economic growth for Singapore. Already 11% of land area is used for roads. To slow down the demand for more roads, and maintaining the roads congestion-free, comprehensive measures to restrain car ownership and usage are in place.

Restraint measures have kept car ownership at 1 car per 10 persons as at 1995.

The operators of the public transport systems have an important role to play in providing an acceptable alternative to private cars, particularly for SMRT, as the MRT is perceived as the superior mode of public transport.

Soon after the completion of the first two phases of the MRT system, SMRT and the 2 bus operators worked with the government to implement a series of projects to provide an integrated public transport system, which will be described below.

Network Integration

The MRT system can move large numbers of passengers through densely developed corridors. However, its route is fixed, and cannot readily tap new passenger growth areas that are away from the MRT route. Buses on the other hand can be readily redeployed to serve new passenger demands.

In the late 1980s, the bus operators and SMRT agreed to rationalise the bus network with the following guiding principles:

(a) the MRT will serve as the backbone of the public transport network in its catchment.

(b) Bus routes which parallel the MRT route will be rerouted to new catchments.

(c) buses will provide feeders to bring passengers to and from the MRT system.

Fare Integration

The fare structures of bus and MRT are distance related, with a relatively high boarding charge (see Chart 3). With the network integration, more passengers will take multi-mode journeys which could become quite expensive if each transfer results in the payment of a new boarding charge.

Chart 3 - MRT and Bus Fare Structure

Distance	MRT Fare		Bus Fare			
	Farecard	Farecard Single Trip	Air-Con F/C		Non Air-Con ST	
Up to 3.2 km	35¢ (60¢)	41¢ (70¢)	35¢ (60¢)	41¢ (70¢)	32¢ (55¢)	35¢ (60¢)
3.2 to 4.4 km	41¢ (70¢)	53¢ (90¢)	47¢ (80¢)	53¢ (90¢)	38¢ (65¢)	41¢ (70¢)
4.4 to 5.6 km	47¢ (80¢)	53¢ (90¢)				
5.6 to 7.2 km	53¢ (90¢)	65¢ ($1.10)	59¢ ($1.00)	65¢ ($1.10)	44¢ (75¢)	47¢ (80¢)
7.2 to 8.0km	59¢ ($1.00)	65¢ ($1.10)				
8.0 to 9.4 km	62¢ ($1.05)	71¢ ($1.20)	65¢($1.20)	71¢ ($1.20)	50¢ (85¢)	53¢ (90¢)
9.4 to 10.4 km	65¢ ($1.10)	71¢ ($1.20)				
10.4 to 12.4 km	68¢ ($1.15)	76¢ ($1.30)	71¢ ($1.20)	76¢ ($1.30)	53¢ (95¢)	59¢ ($1.00)
12.4 to 14.4 km	71¢ ($1.70)	76¢ ($1.30)				
14.4 to 16.5 km	74¢ ($1.25)	82¢ ($1.40)	76¢ ($1.30)	82¢ ($1.40)	62¢ ($1.05)	65¢ ($1.10)
16.5 to 18.6 km	76¢ ($1.30)	82¢ ($1.40)				
18.6 to 21.1 km	80¢ ($1.35)	88¢ ($1.50)				
21.1 to 23.6 km	82¢ ($1.40)	88¢ ($1.50)				
23.6 to 26 km	85¢ ($1.45)	94¢ ($1.60)				
26.0 to 28.5 km	88¢ ($1.50)	94¢ ($1.60)				
Over 28.5 km	88¢ ($1.50)	94¢ ($1.60)	82¢ ($1.40)	88¢ ($1.50)	68¢ ($1.15)	71¢ ($1.20)

An integrated fare scheme was implemented. The MRT AFC was extended to include buses. A stored value ticket could then be used on the MRT and bus for intermodal journeys. At each intermodal transfer, a rebate of 15¢ (25¢) is given so that an intermodal journey is only marginally more than a single mode fare for the same distance.

Integrated Information and Marketing

The operators through Transit Link publishes information that will enable passengers to select a most convenient intermodal journey. It is also responsible for marketing the farecard through a chain of ticket sales offices and automatic vending machines in MRT stations and bus interchanges.

Closer Physical Integration

The three operators and the Government have an ongoing programme to bring bus and MRT services closer together at intermodal exchange points. When the MRT system was built, every effort was already made to site MRT stations close to bus interchanges. There was even closer integration when the Woodlands Extension Line was built. Bus and taxi shelters were built as part of the station structure. A bus interchange was built beneath one of the stations, at Woodlands Station to provide maximum convenience for intermodal transfers.

Covered linkways are progressively being built to provide all weather walkways between housing flats, town centres, major buildings and bus stops, bus interchanges and MRT stations.

V The Challenge for SMRT - Improving Customer Service

All the public transport operators are committed to continually improve customer service. The programmes of SMRT to improve customer service will be described here.

(a) Service Reliability

SMRT is required to meet Operating Performance Standards set by LTA which cover service quality, reliability and safety. The standards, and SMRT's performance for 1996 and 1997 is shown below:

OPERATING PERFORMANCE STANDARDS	ACTUAL PERFORMANCE 1996		ACTUAL PERFORMANCE 1997	
(1) Percentage of arrivals and departures at terminal stations within 2 minutes of time table exceeds 94% for arrival and 96% for departure.	Arrival Departure	95.55% 98.67%	Arrival Departure	95.54% 98.37%
(2) Train service availability is at least 98%.	99.88%		99.89%	
(3) Corruption of tickets by automatic fare gates does not exceed 0.03% of transactions	0.0131%		0.0153%	
(4) Performance standards for the following equipment are met:				
a) Ticket vending machine (>15,000 MTBF)	35,825 MTBF		44,843 MTBF	
b) Automatic fare gate (>20,000 MTBF)	289,421 MTBF		263,390 MTBF	
c) Signalling equipment (>1,500 MSBF)	1,630 MSBF		2,257 MSBF	
d) Escalator (<200 hrs downtime per 100,000 hrs operation)	134 Hrs Downtime per 100,000 hrs operation		103 Hrs Downtime per 100,000 hrs operation	
(5) Total passenger injury rate (excluding first aid only cases) does not exceed 0.4 per million passenger trips	0.18 per million passenger trips		0.17 per million passenger trips	

These operating performance standards formed the basis of a Customer Charter which SMRT published in November 1997, pledging its commitment to a reliable, safe and quality train service.

In order to maintain a reliable and high quality service, SMRT closely monitors faults and delays in train service. Long delays are thoroughly investigated in order to improve maintenance or operation.

(b) Customer Service

Customer surveys are conducted regularly to gather customer perceptions of the service quality provided. Feedback forms are also available at all stations for customers to provide feedback all the time. The feedback is carefully studied.

Where improvements can be made, SMRT will invest or reinvest to upgrade service. For example, based on public feedback, SMRT have begun a programme to upgrade air-conditioner chillers in underground stations in order to lower the temperature in such stations for a more comfortable ambience. Toilets of all older stations are also being upgraded to provide better ventilation and lighting, and higher quality fittings.

SMRT also realises that ultimately its employees have a key role to play in delivering customer service, and for SMRT to live up to its pledges in the Customer Charter. A series of programmes were implemented in 1997 to improve the mindset of the frontline employees towards customer service.

Training

All frontline staff will be progressively sent for a course on customer service. The course will teach them how to provide a good customer service, but more importantly, how to deal with difficult customers and situations.

Interstation Competition

Staff among the 48 stations compete for a Best Station Award that is given every 2 months with a grand award each year. The competition encourages staff to take ownership of the station they operate, to ensure they are clean and equipment functions well.

Points are deducted for complaints received, and conversely points are awarded for compliments earned.

VI The Future

The constraints of a small land area, high population density, affluent population with high expectations for good quality of life including car ownership will not change. It is essential to continually improve the public transport system.

The MRT system is being extended to the international airport, with an intermediate station to serve a new exposition centre. A third station will be built in the existing system, mainly to serve a polytechnic. These new stations will come into operation at the turn of the century.

SMRT has also been appointed to operate a new LRT system which links a housing estate to an MRT station. The LRT system will be operational towards the end of 1999.

SMRT will continue to work with the LTA to provide a world class public transport system for Singapore so that usage of private cars is minimised.

APPLICATION OF NEW TECHNOLOGY IN PUBLIC TRANSPORTATION SYSTEMS IN SINGAPORE

Leong Kwok Weng
Land Transport Authority
Singapore

INTRODUCTION

Public transport in Singapore is essentially served by a mass rapid transit (MRT) system, the buses and the taxis. The MRT system consists of two lines viz, North-South and East-West Lines on an 83 km route length of twin tracks. With 48 MRT stations and a fleet size of 85 6-car trains the MRT system provides a service close to 2 minutes headway during the peak hours. Buses which are operated by two major bus companies provide an alternative mode of transport for commuters in Singapore. These two companies together own a fleet of 3,300 buses which ply the roads of Singapore. For a more personalised service, passengers can turn to taxis and there are some 18,000 taxis on the Singapore roads.

Every day, about 7 millions trips were made in Singapore. This comprises about 1 million MRT trips, 1 million taxi trips, 3 million public bus trips and the rest on cars, motorcycles and other vehicles.

With economic expansion, population growth, the high rate of household formation and population dispersion, the number of daily trips made is expected to increase. To meet this rising travel demand, the Land Transport Authority (LTA) strategy is to rely on public transport to allow masses of people to travel efficiently and minimise road congestion. It is the aim of LTA to build a world class land transport system which provides quality of service at affordable rate to the public. With the advances in technology in telecommunication and computers, it will be possible to realise this aspiration with the use of technology in the various public transportation systems.

This paper will provide an insight on the use of technology in the various mode of public transport system in Singapore in order to bring about a better quality of service to the commuters.

TRANSIT SYSTEMS

Fully automated operation has been actively pursued in Singapore and it refers to the feature in which vehicles in the transit systems are able to operate without the need for drivers or attendants on board. Present day technology has made possible the ability to operate transit systems in this manner. There are several reasons why fully automated operation is desirable in transit systems. Manpower cost constitutes to quite a significant proportion of the operating costs of the transit system as shown in the present MRT system operated by Singapore MRT Ltd. In the years to come, labour cost will continue to increase and hence any measure to reduce manpower in the operation of a transit system will invariably lead to cost savings.

There will also be greater flexibility of operation of the transit system as service can easily be ramped up or down with comparative ease to meet passenger load demands with greater expediency and convenience.

The quality of service is also expected to increase because automatic operation would be able to provide a smoother trip with better regulation. It is also a well known fact that most errors made in conventional railways are due to human errors. Thus a fully automated transit system will able to provide a more reliable system for the commuters.

Several transit projects in Singapore such as the North East Line, Bukit Panjang LRT and Sengkang/Punggol LRT will be fully automated systems. From the operator's point of view, fully automated transit system will be able bring with it cost savings, better regulation, increased flexibility of operation and a more comfortable level of service to passengers.

ROLLING STOCKS

Rolling stock traction drives have for many years been characterised by the need for robustness and longevity that has resulted in a general conservatism in the application of technology advances in other fields. As a result, developments have moved forward but almost always lagging the general developments in the other electronics industry and never at the cutting edge of the developing technology.

Power Electronics

The developments in power electronic semiconductors have had a major impact on transit vehicles notably in the area of propulsion systems. The first major traction application of power semiconductors was the deployment of thyristor devices applied to chopper systems that were used to control the DC motor traction drive systems. A major breakthrough was achieved with semiconductor technology with the advent of the Gate Turn Off (GTO) Thyristor. This device was able to be controlled both on and off by low power signals and immediately offered the prospect of more compact chopper control of DC motors dispensing the need for high power commutation circuits. This technology was used very successfully on the first fleet of vehicles in the present MRT system. However GTO offered a much more significant benefit since it permitted the development of a compact three phase power invertor that would permit trains to be driven by ac motors. For a long time the traction industry had looked for the benefits of brushless machines and the lighter weight afforded by AC motors. The control circuits for such equipment inevitably required the use of microprocessor technology and the marrying together of light and heavy electronics proved on the second fleet of the MRT trains to be a very efficient and reliable AC drive system. The AC drive system also enable better control of the train particularly in braking as it enables electric braking to lower speeds. This maximises the electric braking and thus the energy recovered during braking.

Recently there have been developments in the technology that further reduce the size and complexity of the power semiconductor circuits. The Integrated Gate Bipolar Transistor (IGBT) device requires much simpler drive and protection circuits thereby reducing the size and complexity of the traction package with consequent increase in the reliability of the systems. This technology will be used on the next generation of transit vehicles in Singapore

Control Electronics

Along with the power semiconductor developments there have been similar development in the control electronic circuits. The advent of microprocessors has enabled the more complex functions and algorithms to be used. Microprocessors are used extensively on the second fleet of the MRT trains where they are also able to provide a high degree of self diagnostics, so important to ensure high train availability.

Surface mounted technology which gives much smaller sizes for the control equipment has now matured enough for application to the rail vehicle and this is increasingly being used on the rail vehicles. This technology will be used extensively on the North East Line vehicles.

These developments in electronics have had a major impact in the way the train is controlled and monitored and more sophisticated techniques can now be used. Traditionally, discrete hardwired systems used on vehicles required a great deal of train wires but gave limited functionality. Vehicles nowadays are supplied with their own built-in diagnosis systems which lead to the development of a train central system called the Train Integrated Management System (TIMS) where the information can be managed and controlled to be readily accessed by the operator. The TIMS is presently conceived primarily as a monitoring function but as the trend develops the TIMS may be increasingly used for control functions in addition to monitoring.

Information Systems

Traditionally information systems on trains have consisted of voice announcements over a Public Address (PA) system and card advertisement panel. LED technology has enabled next station and courtesy messages to be displayed. For the future the deployment of full graphics LCD displays similar to those used in computer laptop screens to display full colour video images and still pictures is now practicable. These enable a limitless amount of information to be displayed, from next station video messages to live updates and news headlines. The quality of the messages can be improved with the ability to use prestored standard video and or audio messages.

SIGNALLING SYSTEM

As a safety critical system with a direct impact on the safe operation of the vehicles in the transit system, signalling systems have lagged on the development as compared to the others areas of the transit system such as the communication. However, in recent years, there is a surge in the development effort in signalling system leading to new and innovative technology being employed.

At the heart of the signalling system is the interlocking which govern the safe routing of vehicles in a transit system. The present MRT system utilises interlockings which are relay based. This type of interlockings is easy to test and, once commissioned, is very reliable. But subsequent modifications and extensions in the transit involving relay interlockings will be time consuming, tedious and costly to carry out. For these reasons, there is an emphasis on the use of computer based interlockings in future transit systems. Computer based interlocking is not a recent technology as the first of such interlocking was installed in Leamington SPA in the UK in mid 1980s'. But over the years with more powerful processors available and reliability of processor chips, computer based interlocking has found widespread acceptance. The merits for computer based interlockings in transit systems are that they do not take up as much space as the relay interlocking which come at a premium, are easy to install and provide the flexibility and ease for future expansion and modification.

For an automated transit system, automatic vehicle operation (AVO) and automatic vehicle protection (AVP) systems are required. The first MRT system has such automatic systems to provide a fair degree of automation in the trains. Trains in the present MRT system are propelled automatically under the control of the automatic train operation system. Trains are safely separated using fixed block signalling system and protected by the automatic train protection system using speed code signals which are transmitted to the tracks. The use of high performance microprocessor technology and communication technology has made possible the real time control of vehicles on the track. It is technically feasible for trackside computers to broadcast information on the locations of all vehicles within its control sectors and the distributed intelligence in the vehicle onboard computer systems to determine where the vehicle is and when the vehicle is required to apply brake. This is the concept of transmission based signalling and the new North East Line under construction will use this mode of signalling and train control systems.

Transmission based signalling system will require the transmission of vehicle positioning data to be transmitted over an air interface to be received by the vehicles. In underground transit systems a leaky feeder will be essential to overcome the constraint in the propagation of radio signals inside the tunnels. In the North East Line, a new medium will be used. This comes in the form of a transmission waveguide which is a rectangular waveguide with slots and is laid along the tracks to permit the transit vehicles to receive and transmit data to the trackside computers. Presently transmission based signalling system is used together with track circuits as a fall back. It is anticipated that as the technology is proven in reliability in the years to come, track circuits which account for a major portion of the trackside signalling faults can be dispensed with thereby further improving the reliability of the signalling system and the transit system as a whole.

Computer technology has also advanced to a stage where it is possible for diagnostics application and health monitoring to be incorporated within the critical equipment in the transit system. With this application, it will be possible to provide transit systems which a higher level of reliability and availability as it is possible to pre-empt failures through preventive/predictive maintenance.

COMMUNICATION SYSTEM

Communication has always played an important role in transit systems. The growth in communications over the past decades has been phenomenal. With the rapid advance of technology, higher bandwidth transmission has become possible.

The cost relative to transmission capacity has become lower and more sophisticated methods of communication have been developed.

Data Transmission System

Data transmission system forms the backbone of the transit communication system where the signalling system, the fare and control system, the supervisory, control and data acquisition (SCADA) system, the passenger information display system, etc all ride on it for their audio, data and video signal transmission. Its reliability and availability are therefore crucial to the operation of a transit system. In addition, adopting standards-based systems to allow open system connectivity is also a key consideration. The system should be upgradable to accommodate new services such as the broadband multimedia services which are 'bandwidth-hungry' applications.

There are several technologies available for a data transmission system such as fibre distributed data interface (FDDI), 10BaseT, 100BaseT, gigabits Ethernet, plesiochronous digital hierarchy (PDH) and synchronous digital hierarchy (SDH).

FDDI and Ethernet which offer bandwidth capacity of up to 100 Mbps are technologies widely used in the local area networks. The newly evolved gigabit Ethernet is not a product commercially available at this point in time. PDH which evolved in response to the demand for plain old telephone service is not ideally suited for the efficient delivery and management of high bandwidth connections required for the sophisticated telecommunication services today.

The SDH was introduced in the late 1980s and there was a concerted effort in its standards development. The synchronous transmission has overcome several limitations of plesiochronous transmission. The network is simplified with the simple channel 'drop and insert' design, the resilience of the network is improved through the automatic protection switching transparent to users and efficient network management and control is now realisable through the network management channels built within the SDH frame structure. But importantly, multi-vendor SDH inter-working is now possible with the definition of the fibre-to-fibre interfaces at the physical level.

New transit systems in Singapore such as the North East Line, the Bukit Panjang LRT, the Sengkang and Punggol LRT which are currently being built will have SDH systems installed as their data transmission systems. At the same time, the present MRT system operator is also in the process of upgrading their present pulse code modulation (PCM) data transmission system to a SDH system. Essentially, the high resiliency, open-system design and the capacity which the SDH offers are the main reasons for their implementation.

Radio Communication System

Radio system plays an important role in the operation of a transit system in extending the communication coverage to areas that cannot be served by wireline transmission system. Radio system is used in transit systems to provide the wireless communication path for a diverse range of functions from transmission of voice to data and eventually graphics. Comparing this with the earlier forms of private mobile radio communication which were mainly voice-based, there is marked difference.

The **TE**rrestrial **T**runked **RA**dio (TETRA) standard is an advanced TDMA based digital trunked radio system. Besides the benefits of a digital system such as signal re-generation for better quality transmission and enhanced security, the TETRA standard makes efficient use of the scarce frequency spectrum by interleaving 4 channels onto one single 25 kHz carrier.

TETRA offers a wide range of services and functionality such as Inter System Interface (ISI) which allows users to roam across networks, integration with telecommunication networks, standardised advanced services and facilities such as seamless roaming, direct mode operation, etc. These are some features which are particularly useful when it comes to integration of operation of different transit systems. The bandwidth on demand feature of the TETRA of up to 28.8 kbps unprotected data transmission and two-way wireless broadband communication is another interesting area being looked into. Real time transmission of surveillance closed circuit television video images from trains back to the central control to allow operators to monitor passengers in the trains as well as for interactive audio and video communication between train passengers and control centre are some envisaged applications.

The radio system installed in the present MRT system is a conventional VHF analogue system. With the demand for greater functionality and the ageing of the VHF radio system, it would seem timely for an upgrade of the radio system. Indeed it is the operator's plan to retire its more than 10 years old conventional analogue radio system to make way for a TETRA system. The North East Line will also have its TETRA system operational when the transit system opens for commercial service in 2002.

SUPERVISORY CONTROL SYSTEM

Central to a modern transit system is a supervisory control system to provide overall supervision and control. The approach to the control system design of most of the older generation of supervisory control systems is generally not integrated. They are designed with separate systems for the supervision of the various disciplines such as signalling, power, environment control, etc. Each has its own unique supervisory control with different types of hardware and usually proprietary software but co-located within the same control centre. Operations are often complicated by the fact that staff have to be accustomed to different human machine interface and maintenance has to contend with the diversity of hardware and software used.

With the advent of high performance processors and LAN technology, it is now possible to integrate the supervision of the various disciplines of the transit system together using one supervisory control system. The control centre design of transit systems thus will be integrated not only in physical form but both in the hardware and software using open system architecture and commercially-off-the-shelf computer products. A typical schematic of the supervisory control system which will be used in the new line such as the North East Line is shown in Figure 1 to illustrate this concept.

This design will greatly facilitate operations of the transit system in that it will be possible to use the same hardware and software for the supervision and control of all the disciplines in the transit system. Operations staff will only need to assess the various E&M systems of the transit system using a single terminal. Such a design also greatly facilitates maintenance as there is commonality in the use of equipment and this would lead to lower maintenance cost for the system. As the design uses commercially-off-the-shelf equipment, the system can be better supported both in hardware and software and the scalability of the equipment will permit flexibility for any future upgrade or extension of the transit system.

PUBLIC VEHICLES

With several thousands of buses and taxis on the roads each day, there is a need for the bus and taxi operators to better manage the fleet of public vehicles to improve operational efficiency. The bus operators are reviewing an automatic vehicle location system capable of tracking the location of the buses in real time via a digital radio network. The automatic vehicle location system will be based on satellite GPS linked to dead reckoning calculated through the vehicle gyroscope and odometer data. The bus location information will also be relayed from the buses to the Bus Travel Information System (see Section 5) using the same radio networks at regular intervals. The Bus Travel Information System will compute and estimate the arrival times of the buses at the bus stops. The estimated arrival times will be displayed on variable message signs at bus stops, interchanges and rapid transit stations.

Taxi operators are also improving their service by employing GPS technology to better manage their fleet so that the taxis are able to be more quickly despatched and to respond to commuters needs based on the location of the taxis.

FARE COLLECTION SYSTEM

While the present Integrated Ticketing System (ITS) has served the public transport commuters well over the past 8 years, it faces many technical constraints. For example, the magnetic farecard can only accommodate 16 fare stages and minimum fare increment of 5 cents. This restricts flexibility and innovation for fare revisions to effectively influence travel pattern.

More importantly, the ITS cannot cater to the demands of the future transit network. The rapid transit systems have been designed such that MRT/LRT transfers will be seamless and passengers need not exit the paid area for the transfers. This requires trips on the rapid transit systems to be recorded as one trip regardless of number of transfers and the entire fare to be deducted only once upon exit. To do this, the numbers of fare stages have to be increased. However, the ITS has fully utilised the 16 fare stages and cannot be expanded to accommodate future rail systems. Hence an Enhanced Integrated Fare System (EIFS) using Contactless Smart Card (CSC) technology has to be developed to ensure that it can meet the longer-term needs of the public transport network.

CSC can store more information and accommodate a far greater number of fare stages required for the future transit network. It also brings about greater flexibility in setting fare structure e.g. it can accommodate fare increment in smaller unit of 1 cent. Besides, the CSC will bring with it many more benefits which include lower maintenance cost, reduced fare leakage, faster boarding time, greater convenience for bus commuters and loyalty schemes.

The system architecture is such that it is easily expandable and can accommodate expansion with minimal changes up to the year 2010. The system has been designed to allow for easy scalability to serve a multiple of the specified capacity.

The system consists of the following sub-systems:

a. *The Integrated Fare Central Computer*: It is a central clearing system and handles all the transactions from the bus and transit networks. It also manages the smart card readers, sales of cards, fraud detection and the add-value machines and terminals. It is also linked to banks (currently POSB) for GIRO transaction.

b. *Rail Ticketing System*: It contains the rail network central computer (RNCC) which acts as the clearing house for all transactions in the transit systems. It enables seamless transfers within the transit network.

c. *Bus Ticketing System*: It functions in the same manner to the corresponding rail ticketing system. Inspectors may board buses to check against fare evasion. Hand held ticket analysers will be used to analyse the smart cards to determine the validity of the travel.

d. *Smart Card System*: It consists of CSC, their readers and antennae.

The technologies used in the system are:

a. Advanced smart card technology such as CSC or Combi Smart Card will be used as the ticket media. This facilitates multi-applications in one card with high speed and secured transaction.

b. Bidirectional digital RF communication technology will be used for automatic downloading and uploading of bus parameters and transaction records.

c. GPS will be used for automatic updating of bus stages data which resides in the bus equipment. It also provides a means for Travel Information System.

d. Structured software analysis/design and object-oriented methodology will be used. Software development cycle and CASE tool will be used for all software development activities.

e. All fare collection equipment will be linked via high speed LANs and WANs.

The EIFS will be designed based on open system architecture using LAN and servers. A schematic of the proposed EIFS system architecture is shown in Figure 2. Tender for the EIFS is now in progress with completion targetted for 2001.

TRAVEL INFORMATION SYSTEM

To provide standardised real time travel information to assist road and public transport users to make informed travel decisions an Integrated Traffic Management System (ITMS) is currently being implemented. Information from the public transport system will be transmitted to the ITMS network for dissemination to the users. A central Bus Travel Information System (TIS) will provide bus information to the ITMS and information on transit systems will be provided by the various Rail TIS (RATIS). The RATIS system comprises the following main components:

Station Management System (SMS) - The SMS comprises a station hub, station operating consoles, display controllers and passenger information displays (PID). The station hub interfaces with the Central Management System (CMS) and the signalling system where the transit service information are collected and distributed to the designated PIDs.

Central Management System (CMS) - Each transit system has a Central Management System (CMS) located at the Operations Control Centre (OCC). It consists of a RATIS server and a content workstation. The RATIS Server provides a means to interface with the RATIS Exchange Hub and interfaces with the associated Station Management System (SMS) via a LAN.

RATIS Exchange Hub - Central to the RATIS is the RATIS Exchange Hub which links up all the OCC RATIS servers. It is responsible for managing the information between the various transit systems.

PIDs comprising LED and plasmas types will be located at strategic locations in the stations and bus interchanges. The types of information to be displayed at these locations will determine the types of PIDs to be used. Schematics of the RATIS setup and system architecture are shown in Figure 3. Tender for the RATIS is in progress and the completion of the whole of the works is targeted for 2002.

CONCLUSION

Developments will continue at an increasing pace in the future and a review of the emerging technologies reveals what could be applied to public transport systems in the future when the technology has matured enough. A few possibilities of what these could be are digital TV reception on vehicles increasing use of composite materials which could include bodyshells, novel and smaller shape motors, use of GPS technology to track and control train movements, remote diagnosis, telemetry, etc.

The application of new technology in public transport system will only be considered and implemented if this bring about increased features which lead to a better quality of service to the commuters at affordable prices. Several projects involving the use of latest technology in the areas of power electronics, computer technology, communication, etc are now in progress. Many of these projects will be completed by the turn of the century. The harnessing of technology for use in public transport system will continue to be one of the several key areas which the Land Transport Authority will focus on in its quest to build a world class land transportation system in Singapore.

Figure 1 INTEGRATED SUPERVISORY CONTROL SYSTEM

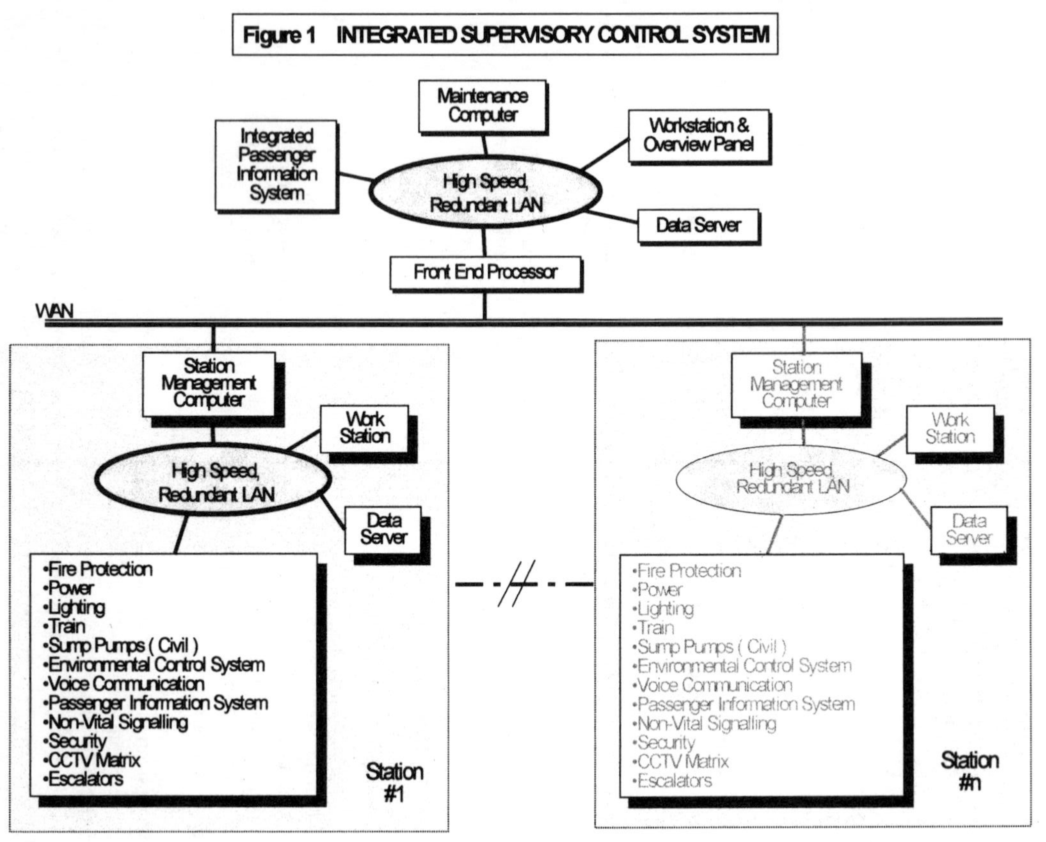

Figure 2 EIFS SYSTEM ARCHITECTURE

Integrated Fares CC
Bus Exit Processors
Bus Network CC
Rail Network CC
PSM
AVM
LRT CC
NEL CC
SMRT CC
Bus Central Server
Interchange Computer System
Bus Depot Computer System
GTM
GTM
GTM
NETS CC
Gate
•PSM •Gate
•PSM •Gate
Banks
Bus On-board Equipment
CC - Central Computer
PSM - Passenger Service Machine
AVM - Automatic Vending Machine
GTM - General Ticketing Machine

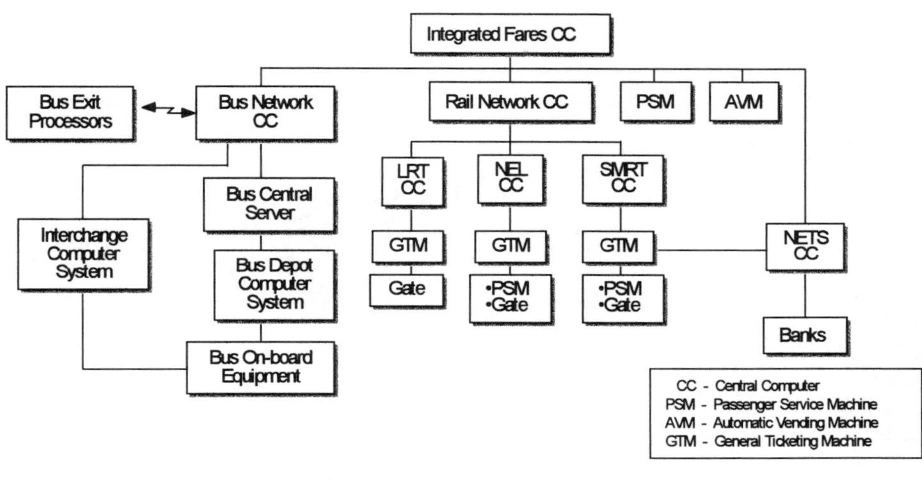

Figure 3 RATIS SYSTEM ARCHITECTURE

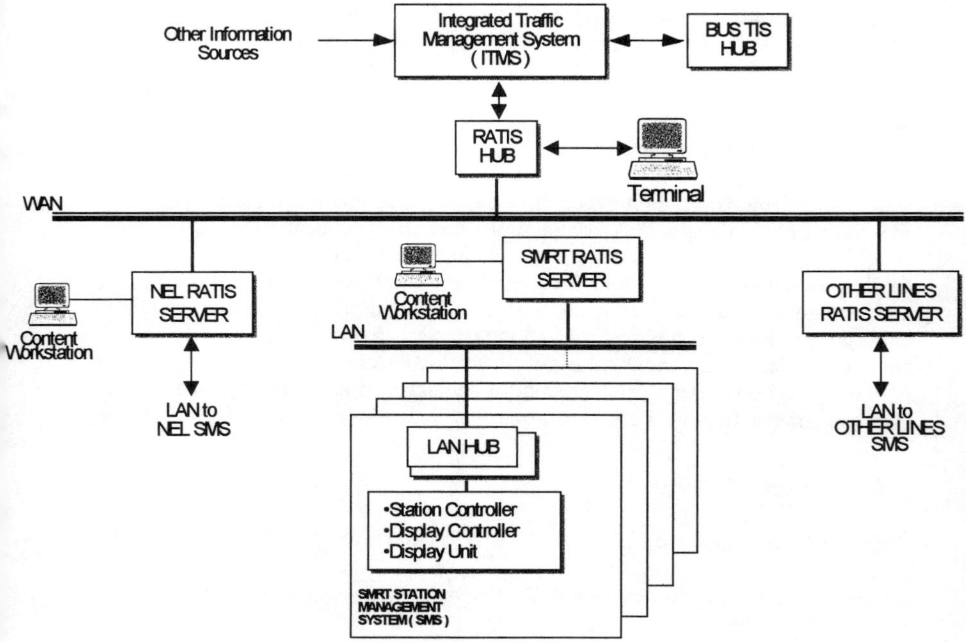

Other Information Sources
Integrated Traffic Management System (ITMS)
BUS TIS HUB
RATIS HUB
Terminal
WAN
SMRT RATIS SERVER
Content Workstation
NEL RATIS SERVER
OTHER LINES RATIS SERVER
Content Workstation
LAN
LAN to NEL SMS
LAN to OTHER LINES SMS
LAN HUB
•Station Controller
•Display Controller
•Display Unit
SMRT STATION MANAGEMENT SYSTEM (SMS)

Socioeconomic Characteristics, Land Use, and Travel Patterns in Tokyo Metropolitan Area

Shigeru MORICHI[1]

1. Introduction

Tokyo Metropolitan Area (TMA), the largest urban agglomeration in the world, is composed of four prefectures (administrative districts), namely Tokyo, Kanagawa, Chiba and Saitama and some part of other prefectures. The Metropolitan Area extends up to 50 km from the central Tokyo. Public and private offices' concentration in central Tokyo and spreading of residential areas to the suburb is the characteristic pattern of land use in TMA. This is mainly because of people's preference for individual housing units, excessively high land prices in central Tokyo and Japanese companies' preference to locate their headquarters in the central Tokyo. This pattern of land use has continuously increased concentration of economic activities in central Tokyo and, as a result, generated a high volume of commuter traffic with long commuting time (average 69 minutes). As a policy response to this problem, government has formulated a master plan with the objectives of creating a multi-polar urban structure and decentralizing the activities to sub-centers. Also, government has taken various measures to develop adequate transport infrastructure. Following sections give a brief account of socioeconomic condition, land use and travel patterns in Tokyo Metropolitan Area.

2. Socioeconomic Elements

With a total population of 33 million, TMA accounts for 25.7 percent of Japan's

[1] Professor, Department of Civil Engineering, The University of Tokyo, 7-3-1 Hongo, Bunkyo-ku, 113-8656 Tokyo, Japan.

population and 32 percent of the country's GDP while it covers only 3.5 percent of Japan's land area. As a majority of people reside in the suburbs and work in central Tokyo, the daytime population of Tokyo prefecture (central Tokyo) is increasing despite the decreasing trend in resident population. Government has taken various measures to decentralize the population from TMA to local cities and business activities from central Tokyo to other sub-centers. There has been strict control on new establishment or expansion of manufacturing plants in Tokyo. For offices and universities, government has provided incentives to relocate them in the suburban areas. Such relocation is expected to facilitate the emergence of a multi-polar urban structure. Rapidly changing demographic characteristic is another key socioeconomic element. It is estimated that Japan will witness a negative population growth in around 2010. The effect of declining birth rate is now more visible in terms of rapidly falling student population in TMA. This trend coupled with the trend of increasing life expectancy is expected to push the ratio of population over 65 year up to 25 percent.

The automobile population in Tokyo Prefecture is about 4.5 million, the highest of all prefectures, while Tokyo's figures for number of automobile per capita is the lowest among the prefectures. This is because of high degree of road traffic congestion, the highest density of the railway network, and high cost of parking. The data on population, employment type, age distribution, income and car ownership are shown in Table 1 and Figure 1 to 4.

Table 1. Population, income per capita and structure of output in TMA(1994)

	Tokyo Pref.	Kanagawa Pref.	Chiba Pref.	Saitama Pref.
Population (thousands)	11,573	8,104	5,718	6,642
Income per capita (US $)	36,625	28,550	25,975	25,492
GDP (US $ million)				
First sector (public)	450 (0.1%)	717 (0.3%)	2,925 (1.9%)	1,400 (0.8 %)
Second sector (private)	177,925 (23.9%)	94,167 (37.9%)	49,392 (32.1%)	62,100 (37.1%)
Third sector (public &private)	564,425 (76%)	153,458 (61.8%)	101,500 (67%)	104,100 (62.1%)
Total	742,800 (100%)	248,342 (100%)	153,817 (100%)	167,600 (100%)

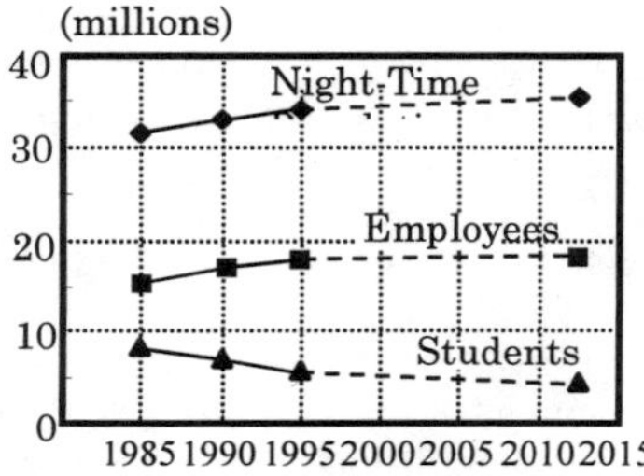

Figure 1. Trends of population by category in TMA

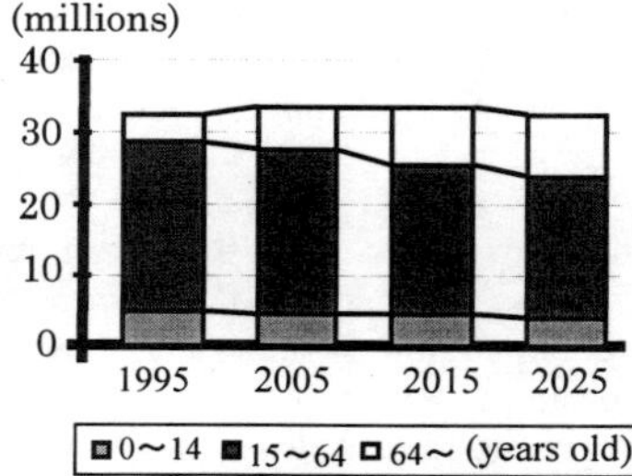

Figure 2. Trends of population by age groups in TMA

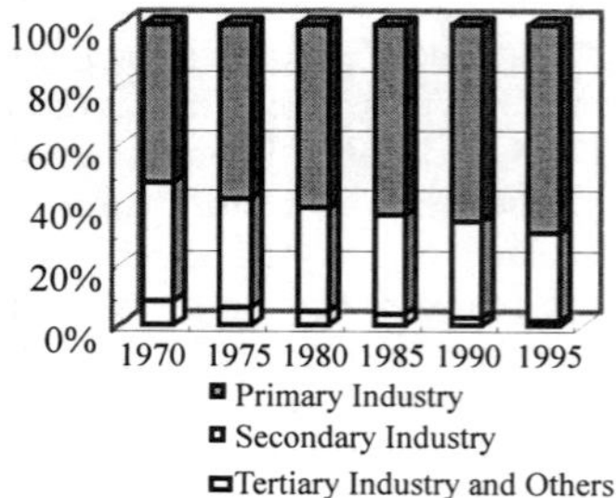

Figure 3. Trends of employment by sector in TMA

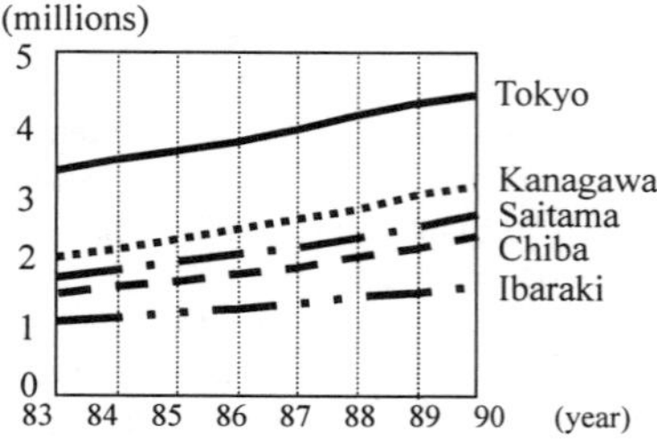

Figure 4. Trends of car ownership

3. Land Use Elements:

The basic land use pattern in TMA includes;

- Manufacturing plants are located around the coastal area of Tokyo Bay
- Government office buildings and major company's headquarters are located in central Tokyo
- Prefectural capital of suburban prefectures, namely Kanagawa, Chiba and Saitama, has office buildings for local companies and branch offices of major companies.
- Residential area has been spread over the surrounding suburban area

As mentioned above, to reduce the over-concentration in central Tokyo, government has resorted to various policy measures. Relocation of government research institutes to Tsukuba City, 60 km far from the center of Tokyo, is one of such measures. Likewise, government offices with regional jurisdiction over the Kanto Region (comprising eight prefectures including Tokyo Prefecture), which were located in central Tokyo, are now being relocated to other sub-centers such as Yokohama, Chiba and Omiya. Recently, government has constituted a task force to assess the possibility of relocating the national capital function of Tokyo to somewhere in the suburbs. By the end of 1999, the task force is expected to submit its recommendations to the government and the government is to take a decision on the matter.

The center of Tokyo is losing nighttime population and gaining the daytime population. As a result, in the surrounding prefectures, the ratio of daytime to nighttime population is less than one (Figure 5). This has created peak hour congestion in railway traffic and excessively long commuting time. To mitigate this problem, private railway lines in sub-urban areas are connected directly to the sub-way line in the center of the Tokyo, and the hierarchy of railway network was established. Long trip makers can use the faster railway line or express train. For

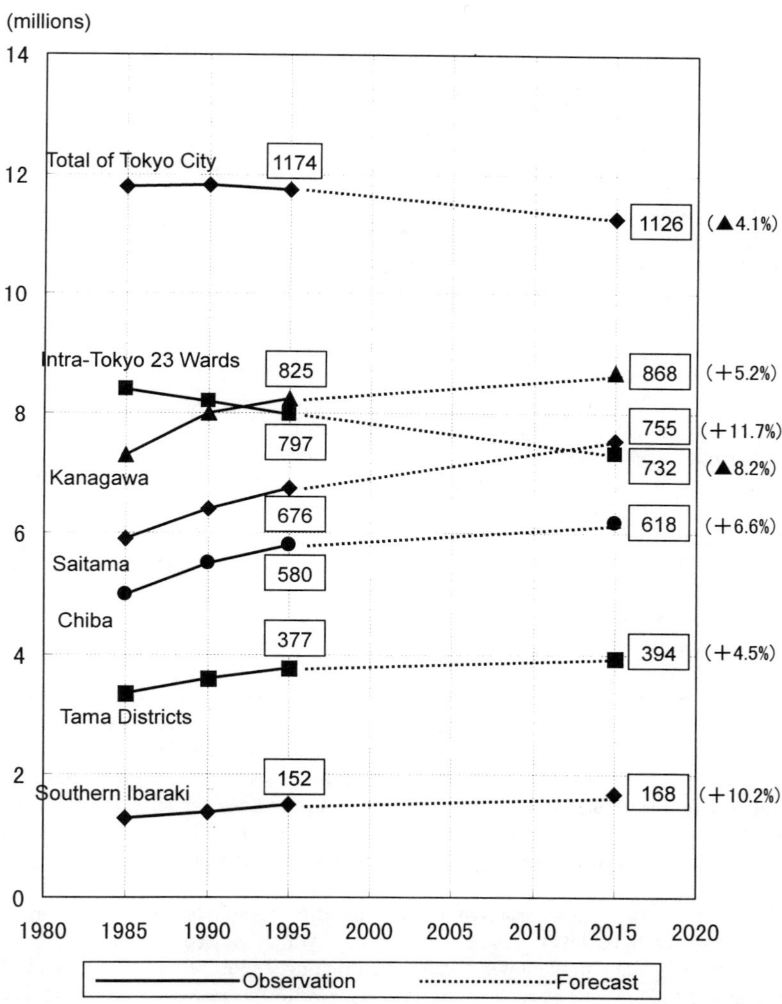

Figure 5. Trends of nighttime population by Prefectures

example, 50-100 km trip length uses Shinkansen (Bullet train), 30-50 km trip length uses the second category and shorter trips use ordinary trains or sub-way (Table 2)

Table 2. Hierarchy of Urban Railway in Tokyo Metropolitan Area
-Station Spacing and Operation Speed of Different Transit Modes-

Railway Type	Station Distance	Operating Speed*
Shinkansen Railway (Bullet Train)	30-50 km	120-130 km/hr
Inter-city Train (Japan Railways) Express Train (Private Railways)	5-6 km	50-60 km/hr
Ordinary Train (Private Railways)	1-2 km	40-45 km/hr
Subway	0.5-1 km	30-35 km/hr
Monorail/AGT	0.5-1 km	20-30 km/hr

* Includes stoppage time

4. Travel Patterns

More than 5.5 million commuters and students are making their journey to central Tokyo every morning (Figure 6, 7 and 8). To serve such a high travel demand, TMA has depended on the railway-dominated public transport system. The metropolitan area has been criss-crossed by railway lines of total length 2973 km, which is almost double the railway length in Paris (1603 km), New York (1473) and London (1192 km). In terms of density of railway network, TMA accounts for 22 km per 100 square km of land area while corresponding figures for New York and London are only 14.2 km and 13.3 km respectively. However, for railway length per capita, Tokyo's rank is lowest as 0.9 km whereas New York and Paris have railway length of 1.1 km and 1.5 km per capita respectively.

Despite an extensive railway network, peak hour commuters experience severe congestion in Tokyo urban trains. Figure 9 shows the degree of congestion in various railway lines. Here, the 100 % congestion rate means four persons standing in one square meter of the train vehicle floor area. Many sections of railway lines are facing a congestion ratio of over 200 percent. To expand the capacity, the frequency has been increased by establishing a with 2 or 3 minutes headway. Such a high frequency has, in turn, caused a lower operating speed during peak hours.

An additional congestion problem exists in major terminals. Many sections of surface railway have level crossing with roads, which caused severe road traffic congestion as a high frequency of trains means longer closure time for a road. To solve this problem, reconstruction projects to elevate the railway track have been initiated. A substantial part of construction cost (90 percent in case of private railway and 70 percent in case of JR) is covered by government subsidies (using fuel tax revenue).

Figure 10 shows the modal split for all-purpose trips in Tokyo metropolitan Area. In central Tokyo, the railway has a share of 43 percent. However, if the modes of bicycle and walking are excluded, the railway share is 69 percent. The railway share

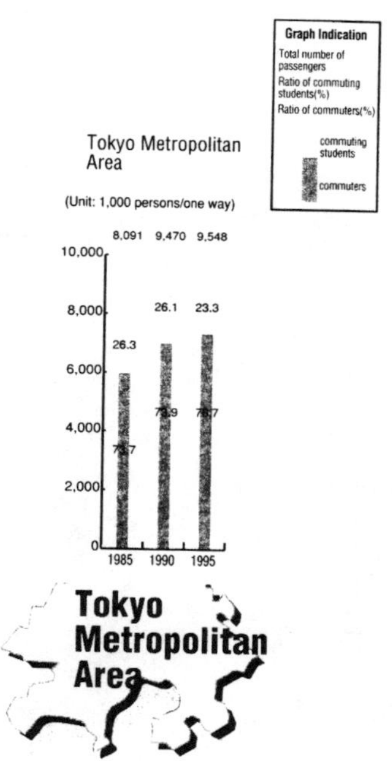

Rush-hour situation due to commuting students and commuters in the center of Tokyo (Photo: JR Shinjuku Station on the Yamanote Line)

Rush-hour situation due to commuting students and commuters in the center of Osaka City (Photo: JR Osaka Station on the Tokaido Line)

Figure 6. Trends of traffic volume commuters and commuting students arriving in the central Tokyo

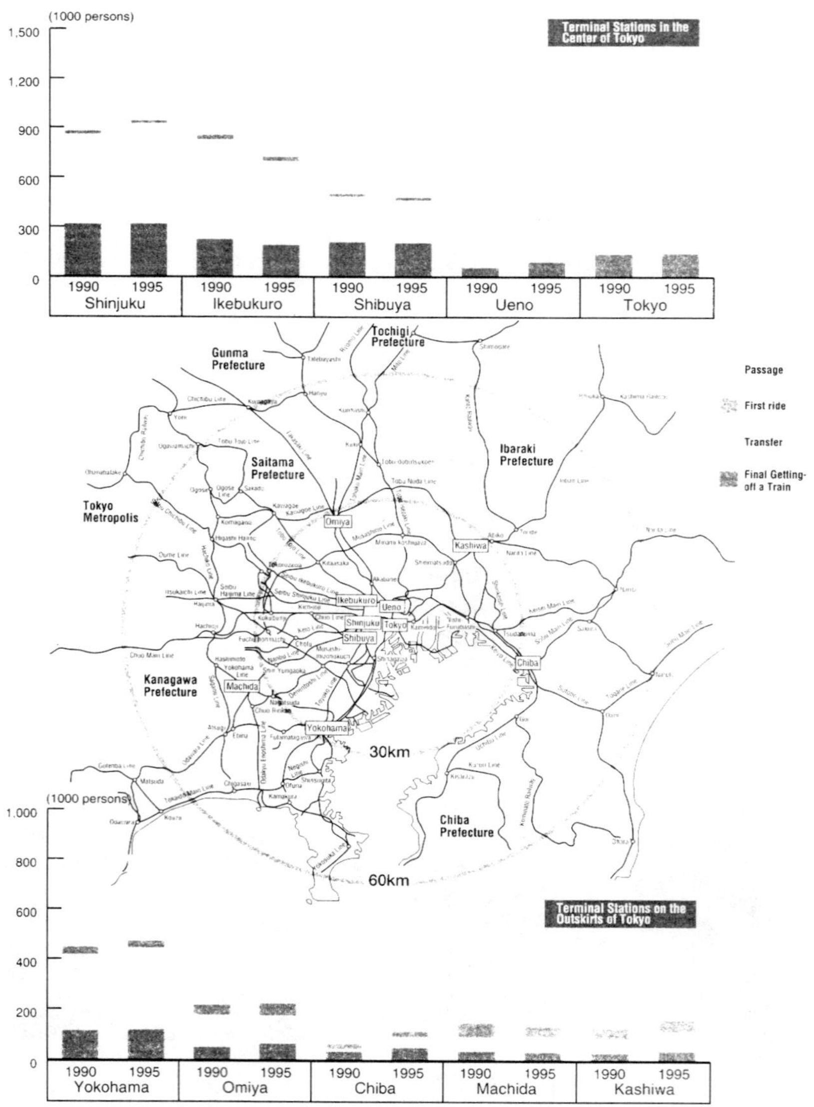

Figure 7. Daily ridership at major terminals

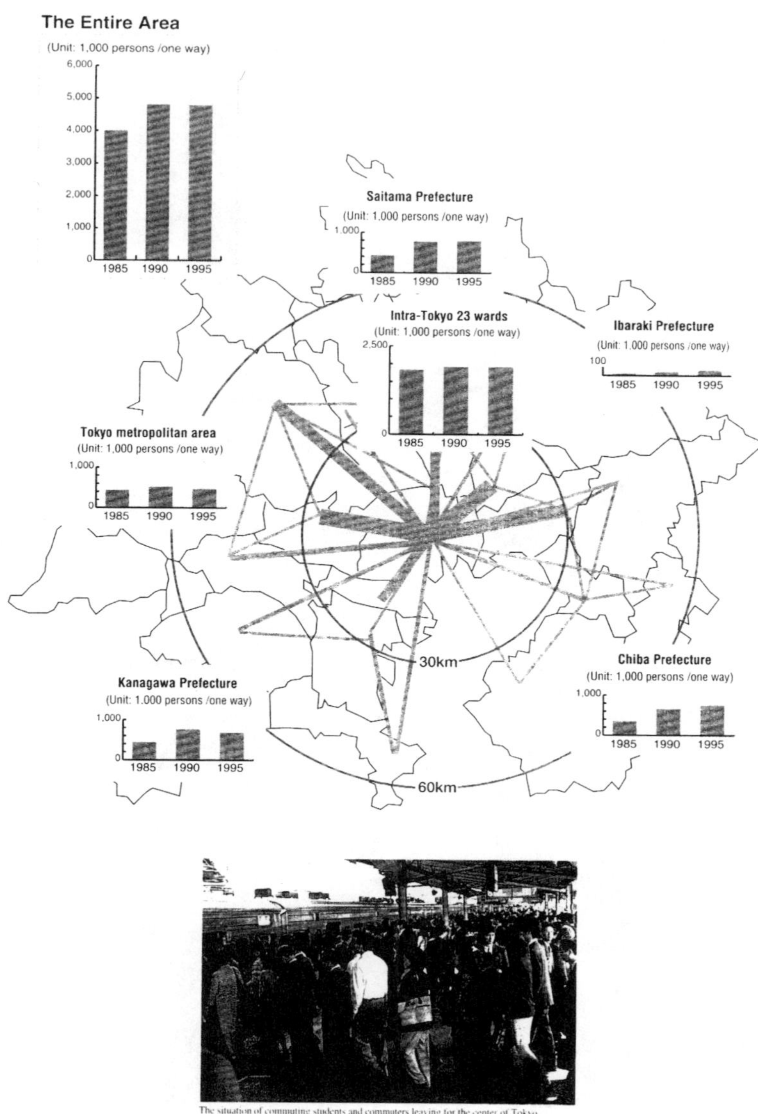

The situation of commuting students and commuters leaving for the center of Tokyo (Photo: JR Yokohama Station)

Figure 8. Trends of rail transit demand in TMA

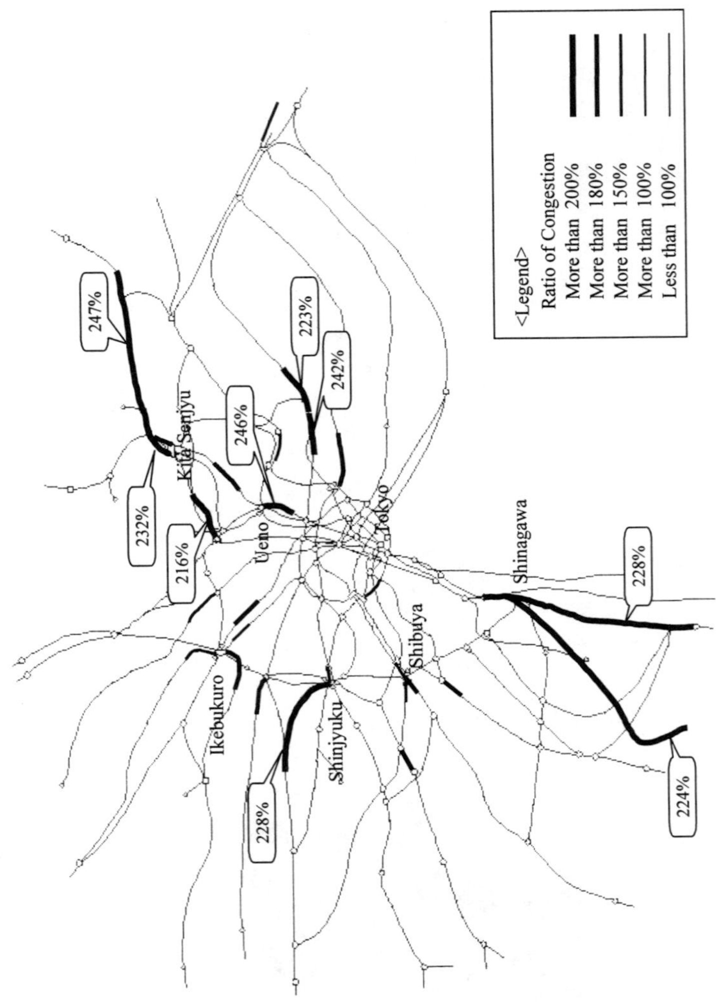

Figure 9. Ratio of congestion (volume/capacity) in major 32 sections in TMA (1995)

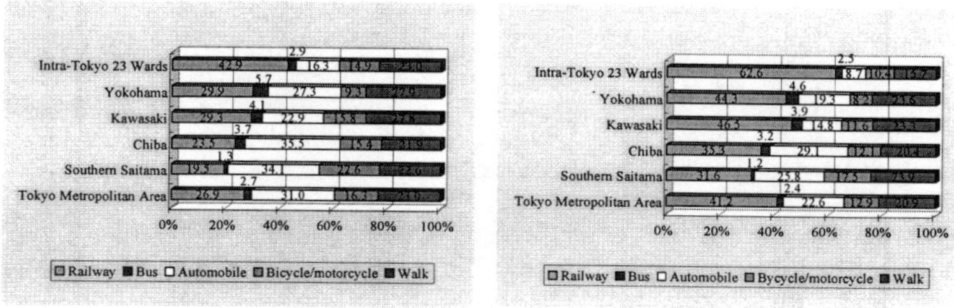

Figure 10. Modal Share - All Trip Purposes

Figure 11. Modal Share - Commuters

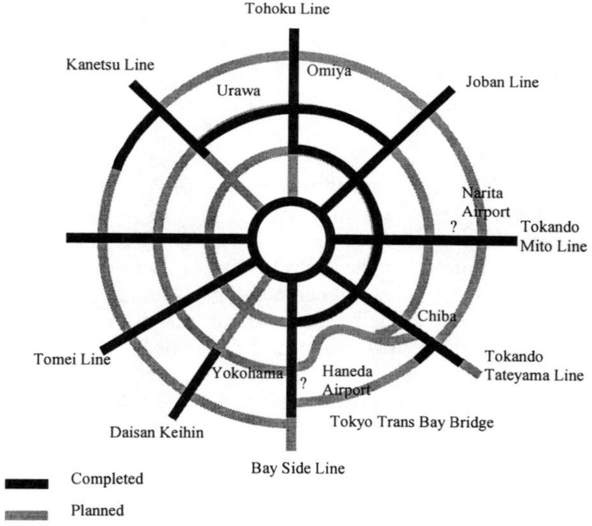

Figure 12. Metropolitan expressway network in TMA

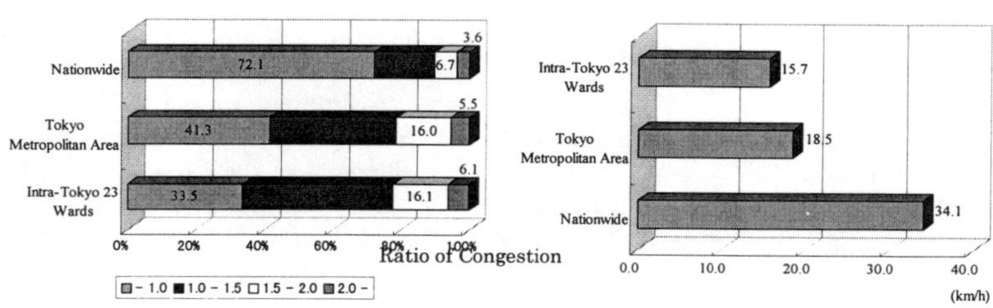

Figure 13. Ratio of Congestion by Region

Figure 14. Average Speeds in Peak Hour

in commuter's trip in central Tokyo is 62.6 percent which jumps to 84 percent if bicycle and walking are excluded (Figure 11).

Figures 12 shows the expressway network in Tokyo Metropolitan Area. Seven radial expressways are under operation, while almost all planned ring roads are not yet completed. Almost all routes experience daytime congestion (Figure 13 and 14).

Bus networks cover the whole metropolitan area but are mainly used to provide feeder services to the railroad. Because of lower speeds and higher fare level, the demand for bus trips is decreasing. The role of a bus as feeder service is still there as the "Park & Ride" practice in TMA is limited due to lack of parking space. The increasing trend of "Kiss-&-Ride in recent years has, however, diverted part of the demand for private cars. Likewise, the use of bicycle has also substituted for bus use. The municipal government provides parking spaces for bicycles, but it is far inadequate to meet the demand. As a result, illegal bicycle parking on sidewalks is an emerging social problems.

5. Conclusion

The urban structure of TMA, for practical proposes, can be characterized as mono-centric. Therefore, the population and economic growth in the metropolitan area generate a corresponding increase in trip volumes from suburb to central Tokyo. Over the time, the concentration in the central Tokyo has further reinforced and other sub-urban centers are left far behind in competition. The huge traffic generated as a result has put a severe stress on the public transport system. As the road network in the Metropolitan Area is not adequate to meet the peak hour demand for road traffic, it is only the railway service, which the commuter can rely upon. But even with a very dense railway network and maximum possible train frequency, the congestion in the trains during peak hours is very severe. On the other hand, the forecasting of negative population growth in near future has made the railway companies hesitant to attempt capacity expansion. In addition such a demographic trend has posed a new challenge for public transport operators to cater an increasing number of aged passengers.

Given the sheer size of Tokyo's population, an efficient mass transit system is indispensable to ensure smooth movement of people. And despite the overcrowding problem of peak hour trains and terminals, railway will continue to be the major mode of travel in TMA. It is expected that the government measures aimed at reducing the over-concentration in central Tokyo, will also work towards improving the traffic situation. In addition, government has emphasized the expansion of the capacity of road and railway infrastructure. For instance the master plan for urban railway (covering 1985-2000) in TMA proposes construction of 389 km new railway line and capacity expansion of existing lines in the Metropolitan Area. 157 km of new line is already opened for service and 267 km is under construction. Such addition of capacity is expected to give a relief to congestion. The next railway master plan for 2000-2015 is under discussion. Likewise, construction of various sections of the urban expressway network is underway. However, in recent years, these government initiatives to expand transport infrastructure facilities have confronted serious problems such as higher construction cost, difficulty in acquisition of right-of way, and citizen's resistance due to environmental impacts. Such constraints have, in turn, caused significant delay in project completion.

The government has also offered various kinds of incentives to attract the private sector to make investments in transport infrastructure. Private railway companies have been major service providers for a long time. The government also made a special provision to allow railway companies to be involved in other business (such as land development and housing, department store, hotel, etc) that made it possible to capitalize on the external benefits of railway development. But now, private railways are not interested in expanding capacity to ease congestion. As a result, the government has made additional provisions of external subsidy and relaxation in fare regulation (railway company is allowed to have a fare raise of 10 % under the condition that the additional revenue so generated is used for capacity expansion). Similarly, government has recently allocated a fund (5 billion yen for 1999) to subside the installation cost of escalators and elevators in the railway terminals to serve the elderly population.

In spite of poor service condition due to over crowding in urban trains, the railway system in TMA has been able to meet the high demand of travel. In a way the present state of the system is the outcome of a wide range of policy interventions in the past, among which, the following seem to be successful.

- Establishment of hierarchy of railway networks, making it possible to match the train operation with trip characteristics.
- Coordination among different railway companies to have direct train operation between subways and suburban lines.
- Provision of cross-subsidy scheme making it possible to subsidize the construction of less profitable (but economically viable) routes from revenue of more profitable routes.
- Various incentives to private sector to invest in railway infrastructure.

These policy measures could be suggested for other large metropolitan areas in the world.

References

Ministry of Transport: Public Transformation Census for Large Urban Areas, 1995 (in Japanese).

Tokyo Metropolitan Region Transport Planning Commission: Present and Future of Transport in Tokyo Metropolitan Area, 1993 (in Japanese).

Urban Public Transportation System and Its Operation and Use by People in Tokyo

Takashi YAJIMA[1], F.JSCE

§ 1. Urban Public Transportation System

1-1 Major Cities in Tokyo Metropolitan Region

The Tokyo Metropolitan Region (TMR) consists of four prefectures; Tokyo Metropolis, Kanagawa Pref., Chiba Pref. and Saitama Pref.. The TMR covers land area of 13,556 km^2, 3.6 percent of the total national territory. The population residing in the TMR is 32.5 million, 25.8 percent of the national total population in 1995. The GDP (Gross Domestic Product) in the TMR accounts for US$1,238 billion, 31.5 percent of the national total, in 1994. The 23 wards of Tokyo Metropolis are not only the center of the TMR in terms of employment, but also the largest national center in terms of politics, economy, culture and education.

Throughout the decades of 1960 and 70, the TMR has attracted large number of population from rural area. However, the population increase in TMR has been substantially slowed down in the 1980's and 90's. The population in TMR is estimated to reach 33.7mil. in and around 2010, and to

----

[1] Managing Director, Teito Rapid Transit Authority, 3-19-6 Higashi-Ueno, Taito-ku Tokyo, JAPAN

Director, City Planning Institute of Japan

decrease slowly thereafter. The percentage of aged population (more than 65 years old) has been increasing constantly so far, and has reached the level of 11.6 percent in 1995. The figure would exceed 20 percent in 2010.

1-2 Public Transportation Modes

In the TMR, varieties of public transportation modes exist; i.e. heavy rail, suburban commuter rail, subway, monorail & AGT (Automated Guideway Transit), bus, tram and taxi. Trolley-buses once existed at a limited scale, but not at present. So-called para-transit once existed, but not at present. In the following pages, public transportation system is described.

1-3 Public Transportation System

The TMR is covered by a well-developed heavy and suburban commuter rail network, operated by the Japan Rail East (JR-East) company and the other 9 private rail companies. The Tokyo's 23 wards are covered by two separate subway networks.

The tram system once served as a dominant intra-urban public transportation system in Tokyo's 23 wards. From 1967 to 1972, the tram was abolished and replaced by buses and subways. Now, only two tram lines are operating. First monorail system appeared in 1964 as an airport access mode from downtown Tokyo to the Haneda Airport (the former Tokyo International Airport, presently serving for domestic flights). Subsequently, several monorail and AGT systems have been introduced in TMR. The role of buses has gradually faded down since 1970's, especially in Tokyo's 23 wards. Buses provide feeder service complementing rail and subways.

1-4 Public Transit Network

Railway network in TMR basically forms a radial pattern extending from the Tokyo's 23 wards. Two ring routes operated by JR-East interconnect numerous radial routes operated by JR and the other private companies. An inner ring, called the Yamanote line, serves within the Tokyo's 23 wards. The route length of the oval-shaped line is about 35 KM (north-south diameter is about 12 KM and east-west diameter is about 6.5 KM). Most of the radial lines operated by the other private companies originate from the major stations on the Yamanote line. An outer ring, consisting of the Musashino line and the Nanbu line, serves suburban part of TMR. The route length of the semi-circle line is about 100 KM, and its radius is about 20KM.

The subways in Tokyo basically cover the Tokyo's 23 wards,

connecting JR's major stations and terminals of the other private railways on the Yamanote line. Monorail and AGT serve the areas that are not covered directly by rail/subway. Their routes typically originate from rail stations, run though hinterlands, and, in many cases, terminate at other rail stations. The subway and monorail/AGT network in Tokyo is shown in Figure 1.

1-5 Public Transit Routes

The total route length of heavy and suburban commuter rail network in TMR is about 3,500 KM. Two third of the total network is owned by JR-East, and the remaining one third is owned by the nine private passenger rail companies which provide suburban commuter services. The average station spacing is about 7 KM for JR-East and about 2 KM for the other private rail companies. The total route length of subway in Tokyo's 23 wards is about 250 KM. Out of the total system, about two third is operated by the TRTA (Teito Rapid Transit Authority), and the remaining one third is operated by the TBTMG (Transportation Bureau of the Tokyo Metropolitan Government). The average station spacing is about 1 KM for the both parts.

The total length of monorail and AGT in Tokyo's 23 wards is 29 KM with 23 stations. Those systems provide more accessible service. Average station spacing is 1.0 ～ 1.5 KM, shorter than that of railways. The total bus route length accounts for about 21,300KM. Out of the total, less than 5% is operated by TBTMG, and the rest is operated by eleven private companies. Average distance between bus stops is about 1 KM for the eleven private companies operating mainly in suburbs, and much shorter for the TBTMG operating mainly in Tokyo's 23 wards.

§ 2. Operations

2-1 JR-East Rail

The largest rail operator in TMR is the JR-East passenger railway company, one of the successors of the former Japan National Railways (JNR). In 1987, the JNR was divided into six passenger rail companies, one cargo rail company and one public corporation for liquidation. Privatization process of the six passenger rail companies has been progressed steadily since then. The level of privatization at present, however, differs significantly among the companies. The JR-east is one of the most advanced. More than 60 % of its stock held by the government was sold out at the open market in 1993 FY. The

Fig-1 Subway Network in and around Tokyo's 23 words (as of 1998)

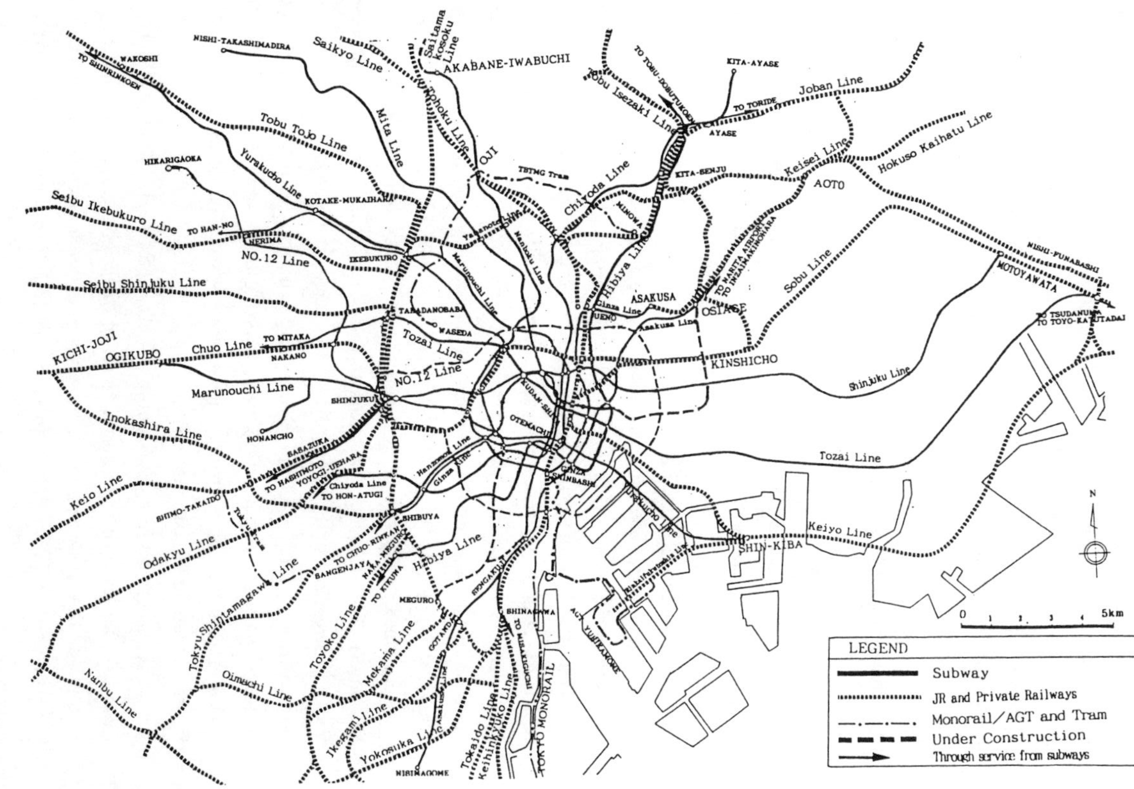

JR-East's Tokyo Metropolitan area network comprises 32 lines extending to 876 route-km, located within a radius of approximately 50 KM from downtown Tokyo. All lines are 1,067mm gauge. For 16 trunk lines, high-frequency inner-suburban service is served by a large standardized fleet of EMUs (Electric Multiple Units), colour-coded according to the line of operation. Headway on the Chuo line (rapid service) is 2 min and on the Yamanote and Keihin-Tohoku lines is 2.5min. ATS (Automatic Train Stop) of an improved type is introduced throughout the Tokyo Metropolitan area. Automatic fare collection gates have been installed at most metropolitan area stations. The JR-East is the pioneer introducing card system in rail sector. A pre-paid card was introduced in March 1986. The card is to be used for buying ticket. A stored-fare card system is introduced, subsequently in 1991. The passengers are allowed to feed such cards directly into automatic fare collection gates.

2-2 Other Private Rail Companies

The operational characteristics of the other nine private rail companies are summarized in Table 1. The size of those rail companies differs significantly. The Tobu Railway Company with its route length of some 460 KM is the largest one. Each company operates within its exclusive franchise which has been actuated in the course of history. Each franchise typically fans out from its terminals on the JR Yamanote line to its suburban hinterland.

Those rail companies adopt either 1,067, 1,372, or 1,435 mm gauge. In terms of route-km, the majority is 1,067mm gauge. All are provided with 1.5 kV· DC power. High-frequency services are provided by different sized fleets of EMUs, designed and standardized by each company. The size and performance of EMU cars operated by each company is more or less same. Headway in peak periods differs from 1min 22sec to 4min. Headway in off-peak differs significantly from 3min to 9min 30sec. The latter represents the figure for outer-suburban line.

Although all lines are equipped with ATS or ATC (Automatic Train Control), the level of their sophistication differs by companies and lines. Automated fare collection gates have been installed at most stations. Either pre-paid card or stored-fare card system is introduced by most companies.

2-3 Subways

Two operational agencies exist for Tokyo subway network. The one is

Table-1 Operational Characteristics of Rail/Subway in Tokyo (1996 FY)

		Route length (km)	Through -service (km)	Car km (million-km)	Fleet size	Min. Head way (min・sec)	
						peak	off-peak
JR-east rail		876.4	(45.2)	2,144	6,689	2'00"	3'30"
9 private rail	Tobu	463.3	(74.0)	252	1,887	1'22"	3'00"
	Seibu	176.6	(2.6)	164	1,215	2'04"	4'00"
	Keisei	91.5	(62.5)	79	494	3'10"	5'00"
	Keio	84.8	(38.1)	107	848	2'09"	3'30"
	Odakyu	120.5	(41.9)	147	1,042	2'04"	3'00"
	Tokyu	95.5	(48.1)	106	1,036	2'10"	3'30"
	Keikyu	83.8	(70.2)	94	758	2'13"	4'00"
	Sagami	32.8	(−)	44	430	2'00"	4'00"
	Sin-Keisei	26.5	(−)	17	202	4'00"	9'30"
Sub way	TRTA	169.3	(114.2)	236. 2	2,401	1'55"	3'00"
	TBTMG	68.1	(41.8)	66. 9	636	2'30"	5'00"

Note(1) : Figure in parenthesis represents reciprocal through service route length with subways.

Note(2) : Figures for JR-east represent its Tokyo Metropolitan area network lines in 1995.

Source : Ministry of Transport

the TRTA, whose initial line was partially opened in 1927. The other is the TBTMG, whose initial line opened in 1960. The TRTA is a public corporation, established solely for subway construction, maintenance and operation, the capital of which is paid in exclusively by the National and the Tokyo Metropolitan Governments. By mid 1960's, the TBTMG operated extensive tram network within the Yamanote line, as well as bus network in the Tokyo's 23 wards. As a substitute of discontinued tram operation in 1960's, the TBTMG initiated subway construction, maintenance and operation.

Regarding the size of operation as shown in Table-1, the TRTA is substantially larger than the other. The TRTA adopts 1,067 (on 6 lines) and 1,435 (on older 2 lines) mm gauge. The TBTMG, on the other hand, adopts 1,067 (on one line), 1,372 (on one line), and 1,435 (on two lines) mm gauge. In terms of electrification, the TRTA uses 1.5KV・DC by overhead catenary or trolley bar (for 6 lines), and 600V・DC by third rail (for older 2 lines). Average train speed is 34～44KM/h, comparable for all lines of the TRTA and TBTMG. In case of rapid service provided in one line each for TRTA and

TBTMG, average speed reaches 49~50KM/h. Time of operation is from 5:00 am to 00:30am for both agencies. Headways in peak period on main lines vary from 1min 50sec (on one TRTA line) to 5min (on one line each for TRTA and TBTMG). Headways in off-peak period on main lines vary from 3min (on one TRTA line) to 7min 30sec (on one line each for the both). The TRTA lines are equipped with ATC or Cab-Signal Automatic Train Control (CS-ATC). The TBTMG lines are equipped with an advanced type of ATS or CS-ATC. Automated fare collection gates have been installed at all stations of the both agencies. An integrated stored-fare card system is being used by both TRTA and TBTMG since March 1996.

Fare structure is graduated by kilometric-section, like other railway companies. Tariff of TRTA is relatively cheaper. Multi-tickets, passes (for 1, 3 and 6 months), day-tickets and other forms of tickets are sold by the both agencies. TRTA's discount rates for one month pass for commuters and students are 37.4% and 66.1%, respectively.

An outstanding characteristic of the subway operation in Tokyo is its reciprocal through operations with the other railway lines. By the reciprocal through operation, suburban commuter trains sun directly into subway lines, and *vice versa*. And thus commuters need not change trains at terminal stations. The through operation was initiated in August 1964. At present, 7 subway lines (5 TRTA lines, and 2 TBTMG lines) provide through operations. The total route length of their partner lines (on the JR-east and the 7 private rails) extends to some 380KM, as indicated in Table 1. On the TRTA's older 2 lines, through operation is impossible basically because of its current collection system by third rail.

2-4 Financial Position of Rail/Subway Operation

The financial position of the JR-east and the other nine private rail companies are as shown is Table-2. Every company managed to retain gross profit in 1996 FY. The outstanding features of the financial position of these companies are twofold. The first feature is the revenue from non-operating business (i.e. realestate, food and drinks, advertisement etc.) amounts to considerably sizable figures. For four out of the nine private rail companies, non-operating revenue even exceeds fare revenue. For the predecessor of the JR-East, the former JNR, non-operating business was restricted by law. Since the division and privatization of the JNR, the JR-East quickly built up its non-

Table-2 Financial Position of Rail/Subway in Tokyo (1996 FY)

		Capital (US$ mill)	Emplo-yees (1,000)	Revenue (US$mill)			Gross profit (US$ mill)
				Fare	Subsides	Non-operating	
JR-east rail		1667	65.1	15,663	0	598	879
9 private rail	Tobu	552	7.6	1,315	0	766	78
	Seibu	181	4.0	778	0	871	45
	Keisei	193	2.3	454	0	269	36
	Keio	492	2.6	670	0	370	70
	Odakyu	498	3.6	918	0	459	64
	Tokyu	896	3.5	1,024	0	1,413	89
	Keikyu	266	2.7	564	0	643	54
	Sagami	258	1.2	272	0	718	32
	Sin-Keisei	49	0.5	99	0	73	13
Sub way	TRTA	484	10.4	2,455	59	43	99
	TBTMG	—	3.4	706	89	442	△162

Note : the figure for JR-east is for all company

Source : extracted from statistics by Ministry of Transport

operating business and succeeded in strengthening its financial base. The second feature is that no subsidy is expended for those companies. Under the traditional policy, the government subsidies have not been directed for operation of metropolitan rails, but only for construction of new lines.

Financial structure for two subway agencies (TRTA and TBTMG) is summarized also in Table 2. Two points can be raised. The first point is the percentage of revenue from non-operating business is about 10%, considerably smaller than that of the nine private rail companies. The second point is the government subsidy expended for construction of new lines and stations. Average construction cost of TRTA's new subway lines (1973～) is substantially large, US$240 mil. per kilometer including rolling stock. Due to the limited amount of subsidy, both agencies bare a heavy burden of interest payment required for the investment for on-going new lines. The TRTA managed to make profit, partly due to tariff increase actuated in September 1995.

2-5 Bus Operations

In Tokyo Metropolis, public bus service is provided by the TBTMG and ten private companies. The total size of fleet is about 8,600 vehicles as of 1995 FY. The total number of passengers carried was 1.3 billion, out of which the share of TBTMG was 23%.

Bus ridership decreased sharply since its peak in 1970, and has gradually decreased for these ten years, despite of various counter measures tested and implemented so far. The measures tested include, among others, (i) provision of exclusive / priority bus lanes, (ii) better service for passengers (bus location information system, introduction of air-conditioned coaches, etc.), and (iii) restructuring bus operations (route change, extension of operational hours, etc.). Thus, financial status of bus operations is quite tight. Very few bus companies manage to make profit.

Bus operation, however, is recently reevaluated as desirable mode of urban transportation, from the point of serving transportation disadvantaged and in reducing CO and CO_2 emission from transport sector.

2-6 Facilities for the Transportation Disadvantaged

As the percentage of aged population is increasing, provision of more friendly public transport system is a growing concern. Facilities for the physically handicapped people, escalators in rail/subway stations, have been promoted in a limited scale under the initiative of MOT (the Ministry of Transport). Recent policy change is to accelerate the program with broader scope covering not only the handicapped, but also the aged people. The guideline recently set forth by MOT calls for the provision of elevator and escalator (including special type allowing wheel chair) at such rail/subway stations as; ① passengers' vertical movement more than 5M and ② daily passengers more than 5,000. Among such target stations, only 11% is equipped with elevators, and 45% is equipped with escalators. Rail/subway operators initiated their effort to follow the guideline.

A mini-bus operation, recently launched, demonstrates the possibility of buses serving transportation disadvantaged. The mini-bus is called "Moo-bus", as a word play of "Move-us", operated in the city of Musashino (population: 130 thousands) located in the western suburbs of Tokyo Metropolis. Its town center, built-up around a JR-east rail station, is known as one of the busiest commercial business and cultural sub-centers in suburban Tokyo. Residential area around town center is characterized by ① higher percentage of aged/upper middle class population, ② some elementary schools, ③ narrow streets and ④ no public transport service. A four-year study by city hall recommended a mini-bus operation in such residential areas. A private bus company is operating the "Moo-Bus" on commission by the city hall, since November 1995. The one-way bus loop extends about 4.3km, starting from the JR station, running through eastern residential area and returning to origin. The distance between bus stops is about 200M. Scheduled operation time required for a round trip is 25 minutes. The Moo-buses start from the JR station every 15 minutes. Flat fare is US$0.83 (just a coin in Japanese currency) for every passenger. Hours of operation is 8:00am～6:20pm. The vehicles, which are owned by the city hall, are 2.0M width and 7.0M long, with 29 passengers capacity. To help in boarding, auxiliary step automatically extends out at bus stops. The Moo-buses are quite popular among aged people and school children. The average number of daily passengers amounts to 900. The deficit arisen from the Moo-bus operation is being covered by subsidy from the city hall. The amount of subsidy is smaller than expected, due to its popularity.

§3. Use by People

3-1 Modal Split

Modal split for total daily movement in TMR shows that railway usage with slightly increasing trend accounts for 27 percent (on linked-trip base) in 1993. Bus ridership has been diminishing over years and accounts for less than 3 percent. Modal split for home-to-work trips in TMR is as shown in Fig-2. Railway usage accounts for 46 percent in 1993, and bus usage accounts for less than 3 percent. The number of daily commuting trips particularly oriented to the Tokyo's 23 wards is about 7.5 million, and the rail share accounts for more than 70 percent.

3-2 Ridership

Ridership by major system component is as follows. The JR-East carried some 5.4 billion passengers per annum in 1996 FY. The other nine private rail companies carried some 4.4 billion passengers, and the two subway operators carried some 2.6 billion passengers. The bus operators (TBTMG and 11 private companies) carried some 1.3 billion passengers. The share of Monorail/AGT and tram is seemed to be marginal.

Fig-2 Modal split by representative modes (home-to-work trips) in TMR

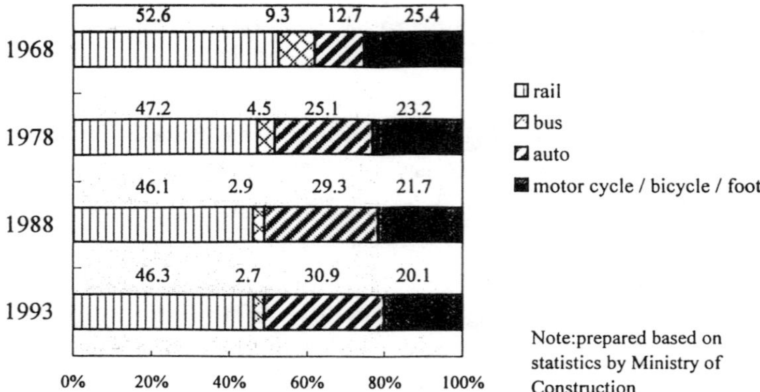

Note:prepared based on statistics by Ministry of Construction

3-3 Rail/Subway Congestion

Because of the above mentioned huge volume of in-flowing commuting trips to the Tokyo's 23 wards, the congestion on rail and subways, particularly in morning peak period, has long been and still is a major issue on urban pubic transport in Tokyo. During these two decades (1975-95), rail/subway transport capacity has increased by 58 percent due to extensive investment on infrastructure and rolling stock, while the volume of passengers has increased by 37 percent at the most critical 31 rail/subway sections in and around Tokyo's 23 wards. As a result, average rate of congestion has decreased from 221 to 192 percent (Fig-3). More particularly, level of congestion is 220 percent for JR-east line and some 180 percent for the other private rail and subway lines. The rate of congestion at 220 percent represents totally jam-packed train. Under the congestion rate at 180 percent, commuters are barely allowed to stand side by side in the cars, reading newspapers.

§4. Conclusive Remarks

The public transportation system in TMR, primarily based on a well-developed rail/subway network, manages to support passenger movement in this huge urban complex. Following are major issues to be tackled with;

(i) to alleviate congestion by increasing rail/subway transport capacity,

Fig-3 Trend of congestion rate in peak period in TMR rail/Subways

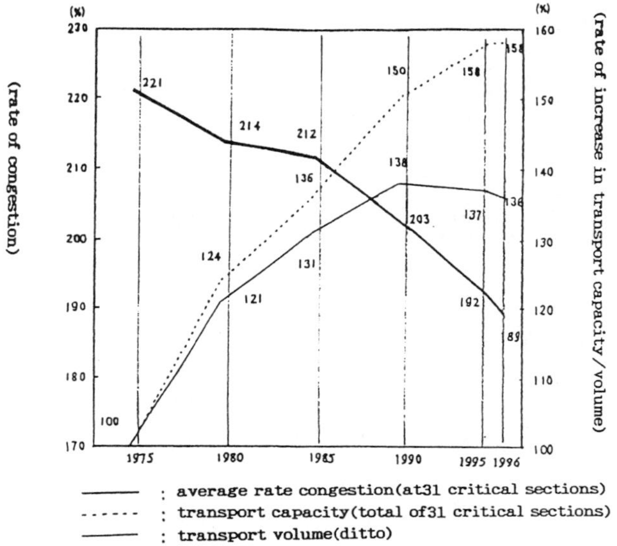

—— : average rate congestion(at31 critical sections)
- - - - : transport capacity(total of31 critical sections)
—— : transport volume(ditto)

Source : Kanto Transport Bureau,Ministry of Transport、1998

(ii) to realize more friendly public transport system by improving facilities and rolling stock,

(iii) to provide more accessible public transport system by introducing medium capacity transit , i.e., monorail or AGT, and

(iv) to revitalize bus transport providing innovative services

For that purpose, adequate support should be strengthened by the government. And extensive efforts should be exerted on the operator side to provide challenging transport services and to pursue better financial basis.

Reference

1. Kanto Transport Bureau, Ministry of Transport, "Trends in Transport in Kanto Region", 1998

2. TBTMG, "1997 Outline of Tokyo Metropolitan Transportation Service", 1998

3. TRTA, "98 TRTA Hand Book", 1998

Practices (Past & Prospects) for Tokyo's Public Transport

Hitoshi IEDA[1]

Abstract

This paper firstly describes the strong and weak points of Tokyo's public transport systems from various perspectives together with the technical and institutional practices adopted. Secondly, the author discusses distinctive features of Tokyo's experience, what problems it faces, and what is required for the future. Finally, recommendations for other large cities are outlined.

Strong and weak points

The author shows the strong and weak points of Tokyo's public transport from various viewpoints. In the following, "S" and "W" denote a strong point and a weak point respectively.

Management

S1: Rail and bus operators are basically private firms. Even in the case of public corporations, they are financially independent from the national/ local government. Most of them are more or less profitably managed (or at least at the break-even status). Few are receiving financial support for their operating cost from the government.

W1-1: The public transport system is independently managed by many private and

[1] Professor, Dept.of Civil Engineering, University of Tokyo, 7-3-1 Hongo, Bunkyo-ku, 113-8656 Tokyo, Japan

semi-public operators; two subway operators, and many suburban rail operators (including East Japan Railway Company, JR in short) and bus operators.
Although "through operation" of trains coordinated by different operators was realized after WWII and became the present de facto standard (presently at 12 connection points), the quality of inter-operator services has been poor. Some examples of this are expensive tickets, complicated ticketing systems, inconvenient transfer facilities, and inconsistent and incompatible sign-systems for guiding passengers.

W1-2: Operators only invest when they can expect a corresponding increase in revenue. Now that natural and social growth in population has been slowing, they have little incentive for investment. Subsidizing unprofitable lines with the revenues from profitable lines, "intra-firm cross subsidy", has been the financial basis both for infrastructure investment and for covering operation cost. From this viewpoint, small operators cannot invest as much as large ones do. Even when they do, the situation could be disadvantageous because the level of fare becomes more expensive.

		Operating Length (km)	Passenger-km (Millon)	Fare rate (USD/person/km)	Fare Income / Operating
Private Operatiors	Tobu	463	14374	0.081	1.07
	Seibu	177	9411	0.073	1.06
	Odakyu	122	10874	0.075	1.08
	Tokyu	101	8739	0.104	1.08
	Tokyo Monorail	17	725	0.175	1.02
Public Operators	TRTA	169	15881	0.137	1.01
	TMS	68	3870	0.162	0.81
	YMS	21	904	0.174	0.56
Third Sectors	Hokuso Kaihatsu	29	341	0.208	0.73
	Chiba Monorail	14	69	0.359	0.80

TRTA: Teito Rapid Traffic Authority
TMS: Tokyo Municipal Subway
YMS: Yokohama Municipal Subway
(1USD=135Yen,1998)

Table 1. Basic Factors in Rail Operators in Tokyo (examples)

W1-3: Transit projects often bring so-called network effects such as the "substitution effect" (the reduction in congestion and in revenue) and the "complement effect" (the increase in congestion and revenue) on operators and users of other lines. These positive and negative effects cannot easily be reflected in the ordinary financial aspect of the projects under the present scheme of non-harmonized independent management of each operators, and often becomes one of the obstacles to the implementation of the projects.

Network

S2-1: The rail network is hierarchically composed of various types of lines; intra-city subway operation, middle distance radial suburban lines of private operators, JR's rapid and massive-scale medium-to-long distance suburban lines with double-double tracks, and high-speed Shinkansen trains for long-distance commuters. Performance of the suburban lines are especially outstanding in quantity and speed.

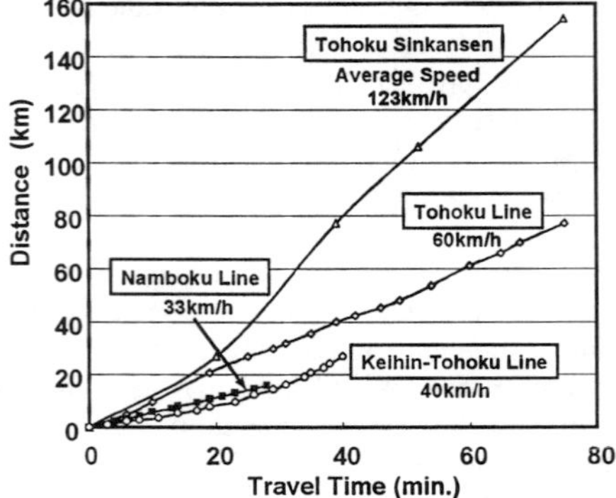

Figure 1. Station-Interval and Travel Speed by Train Type (Tohoku Corridor)

S2-2: The rail oriented transport system in Tokyo enabled both the provision of large amounts of labor from a large suburban area to the city center and the concentrated use of the core as business center which brought economies of scale and agglomeration.

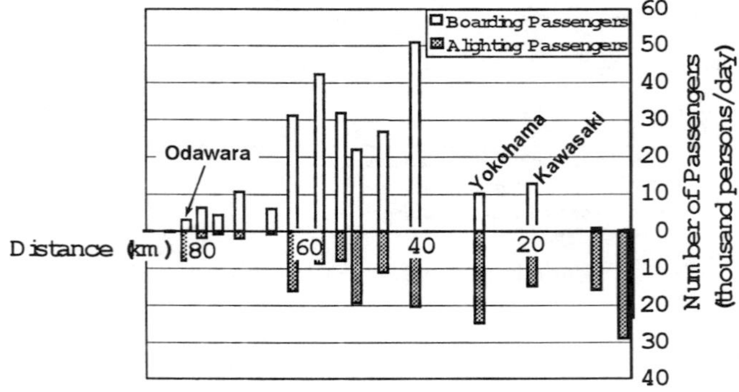

Figure 3. Spacial Distribution of Boading and Alighting Passengers (JR Tokaido Line)

W2-1: Commuting time has been increasing due to the residential suburbs sprawling outward. Circumferencial mobility is comparatively poor, while radial lines are well established.

W2-2: The bus network is not sufficiently harmonized with the rail network. Inter-modal connection facilities and the required institutional system for inter-modal mobility in such as rail-bus, car-rail, and bicycle-rail, are also still unsatisfactory in quantity and in quality, although station plazas (as connection facilities) have been constructed due to the cooperation between rail operators and the government.

W2-3: Transit systems with medium capacity which would supplement the gap between rails and buses are not well provided, although monorails and AGTs were constructed in a few places. The lack of tramways, trolley-buses, and LRTs is a negative feature of the city's transport.

Operation

S3: Huge transport volume is served quite effectively in the main lines by the well-organized rail operators. Their safe, punctual, and reliable operation is doubtlessly the best in the world. Rail operators provide commuter train services in variety, even in some cases on the same track, which differ in speed, stop-pattern, and in seat-availability to meet the various travel needs of passengers.

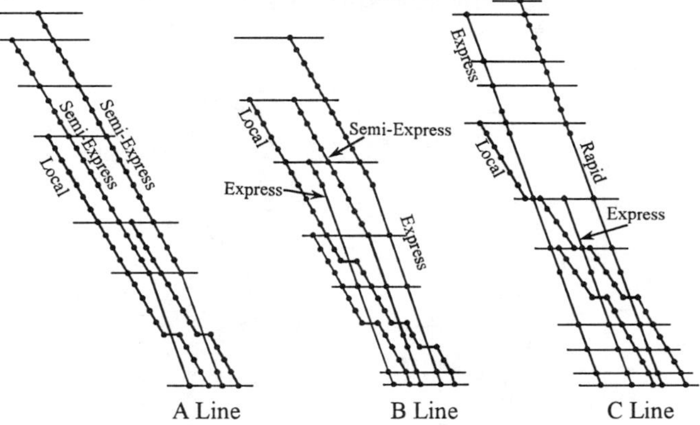

Figure 2. Multifarious Train Operation Patterns

W3-1: The congestion problem is still very serious, especially during peak-hours on commuter lines. It brings the inevitable speed reduction of the train caused by the too short headway of trains as well as the well-known terrible congestion in train vehicles.

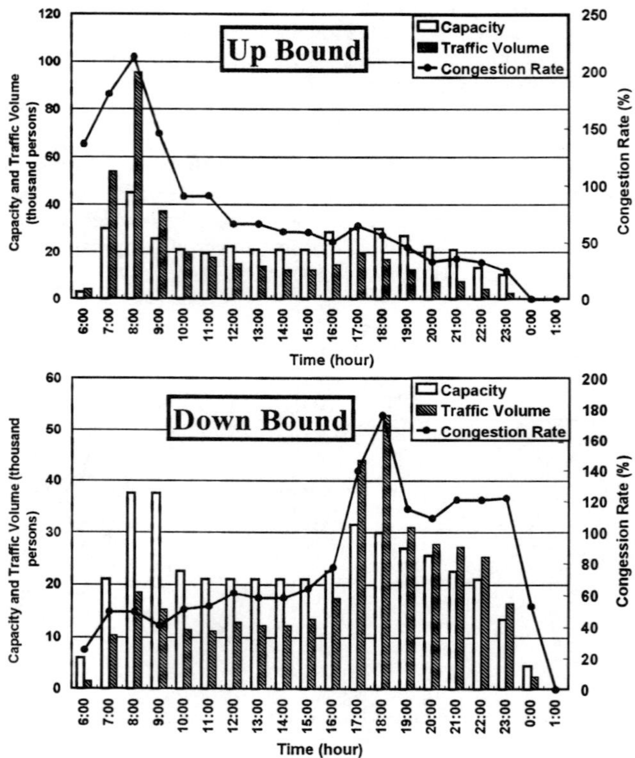

Figure 4. Chronological Change in Capacity, Traffic Volume and Congestion Rate(Chuo Line)

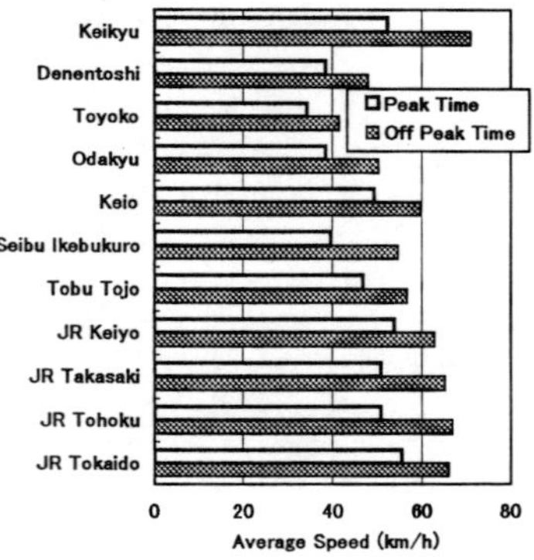

Figure 5. Train Speed in Peak/Off Peak hours by line

W3-2: Bus operation is far less reliable than rail. Moreover, its routes are far less recognized by the people, except in the suburban feeder services between houses and stations. This is due to its weak marketing and comparatively expensive level of fare. Short distance trips within the city center are not appropriately served by bus operation as a result.

Contribution of Public Sector

S4-1: A commuter rail master-plan of Metropolitan Tokyo both for construction and for improvement has been discussed, revised (eight times in the last 70 years), and authorized by the Council for Transport Policies for the Ministry of Transport. It functions fairly well as a guideline for operators as well as for the government for their investment and assistance.

S4-2: Several methods of the public financial assistance were prepared by the national and local government for various investment by operators; public loans for investment, public subsidy (50%) for subway construction by public and semi-public operators, financial assistance for the payment of interest on large-scale-investment by private rail operators, financial assistance for the construction of physical facilities and vehicles for handicapped passengers, the special beforehand allocation from fare-income for the improvement projects of commuter rail, favorable tax treatment, and so on.

W4-1: The measures for financial support by the government depend on the types of operators (public, semi-public, or private) and pay more favor on public operators, although private operators also implement the same sort of projects.

W4-2: Financial resources for the public subsidies to be invested into public transport are quite limited and is one of the serious bottlenecks for the promotion of the improvement of the level of service. Although the urban rail system is doubtlessly reducing traffic on roads and the environmental load on the city, the comparatively plentiful ear-marked budget for road construction collected from fuel tax revenue has been less flexibly applied for transit projects except in the following cases; the grade separation projects of rail tracks which will eliminate level-crossings and will make the road traffic smooth, and the construction of middle-capacity transit systems such as monorails using road space to substitute a part of road traffic.

Urban Development and Transport System Development

S5: It is quite easily observed in the city that houses are spreading along radial suburban railways. Cores of business and commercial activities have grown up around railway stations. Huge-scale public apartment houses are developed in the suburbs where there are fairly good connection to the city center by rail systems. Citizens often refer to the structure of the city by the major rail network, two rings and five major spokes. Tokyo looks as if it is one of the most successful cities in the Transit Oriented Development (TOD) scheme.

W5: It is fact that the transit systems have been constructed in conjunction with residential and commercial development mainly by the hand of the private rail operators. However, the city has limitlessly developed far beyond the capacity of the transit systems due to the insufficient control on land-use and growth. This defect brought various sorts of over-concentration and congestion problems to Tokyo during the rapid growth period in 1960s to 70s.

Distinctive features

The above mentioned strong and weak points of Tokyo's public transport may be summarized in the following four distinctive points.

The most important point of Tokyo's public transport is that it has been based on the private-oriented management of operators in terms of the construction and the operation. This resulted in various un-measurable positive points such as the efficient, profitable, and cost-minded management, and rail development that was commercially combined with various regional developments. On the other hand, this also produced the tendencies that financially non-profitable projects were apt to be left behind, and the level of inter-operator services has become terribly poor. As a result the profitability of public transport systems tended to become positive, since the level of fares were revised according to the change in the cost of operation, and since their networks were developed basically within the limit that their profitability would not become negative! This (superficially) positive profitability distracted people's eyes from the various social effects brought by public transport systems and from the importance of the financial and political contribution of the government to the public transport.

Therefore, the author must point out that the second distinctive feature of Tokyo's public transport is that the contribution of the government is comparatively weak and limited compared to the cities in western Europe in the aspect of the public initiative on the following points: the public transport development plan, the coordination in the activities of multiple operators and multiple modes, the financial assistance on construction and operating cost, and the consideration of public transport in various policies in the city planning.

The technical status in Tokyo's commuter railway system would be the third outstanding features. We can be especially proud of its speed, quantity, reliability and safety.

The comparatively high transport share of rails in Tokyo shows that the rail systems are to some extent in a so-called "regional monopolistic market". This would be the fourth feature. Then, the users tend to become insensitive to the poor level of service though they are actually paying various cost for travel including qualitative inconvenience and discomfort. At the same time the operators have less incentive to improve the level of service for passengers.

Required practices for the future

After considering the above mentioned features of Tokyo's public transport, the author points out the required practices including on-going measures for better future public transport as follows.

<u>Coordination of services</u>

-Promotion of a strong institutional system for the coordination of the level of services, especially in the fare system, for the connected use of the service provided by multiple operators or multiple modes

(Discounted ticket for connected use and stored fare card commonly applied for different operators are partly introduced.)

-Periodical and official implementation of scientific evaluation of the level of service by section, by line, by operator, and by mode, and the publication of the results in order to promote the effective functioning of the market mechanism

-Re-arrangement of routes of bus operators in such a way that they can maximize the performance of the network along with the rail systems

<u>Financial and organizational measures to improve the level of service</u>

-Full-scale peak-load fare system of commuter rails to finance improvement projects

(Discounted *carnet* for off-peak hours was already introduced to promote the chronologically dispersed use)

-Reasonable pooled fund of fare income of multiple rail operators for financing improvement projects

-Introduction of the vertical separation of organizational system of rail operation: public-oriented infrastructure and commercial-based operation in order to remove the management risk involved in large-scale investments from private rail operators, and to enable more flexible and open access of operators to infrastructure (partly realized already)

-Flexible application of the revenue from fuel tax on automobiles to projects for the improvement of public transport systems

-Creation of additional special taxes on firms located in the city center, and/or environmental taxes on various transport modes

<u>Technical means for better pubic transport</u>

-Technical development for the reduction of the construction cost of underground structures, and of the environmentally negative effect (noise, vibration and emission) provided especially by over-ground rail systems and buses

-Application of information technologies to improve the performance of public transport systems including ITS in the information-oriented society equipped with the personal and mobile information devices

-Technical development of an advanced rail traffic control and signal system to enable faster and more frequent train operation

Harmonization with city planning

-Better coordination of public transport policies and urban development policies in terms of both institutional measures and proper design of facilities

(The act of residential development and rail development was enacted and applied, which combines the rail construction project and the readjustment of land-lots in a harmonized manner, in order to enable smooth land-acquisition, to realize orderly urban development around rail stations and along rail lines, and to collect the development profit which would be brought about by rail projects and would partly be reflected by land prices)

-Stricter land-use control from the view-point of TOD

Recommendations

The following are the measures recommended for other large cities based on Tokyo's experiences. However, the author must note that Tokyo still has problems and many things to realize as shown in the previous chapter and that the following recommendations are the ones which should be properly combined with practices Tokyo is now trying to realize.

-Promotion of private-minded management in the public transport sector especially in its operation

-Combination or harmonization of transit system development and residential/ commercial regional development

-Hierarchical design of the function of the public transport systems

-Highly advanced technologies in rail systems

-Master-plans of urban rail systems made by the governmental council and the public initiatives to realize projects which are authorized there

-Others: -Multifarious train operation enabled in double-double tracks

-Coordinated train operation through different rail operators

-Station plaza construction and grade-separation of rail tracks

Acknowledgement

The author appreciates the remarkable contribution of his colleagues, Mr. PHAN Le Binh and Mr. OKAMURA Toshiyuki for their assistance on the numerical analysis and the drawing of figures of this paper.

The Washington Region:
Land Use and Travel Patterns

Ronald F. Kirby[1]

Abstract

The Washington Region is the eighth largest metropolitan area in the United States, with a population of four million people spread over 4,000 square miles. By 2020, the region's population is projected to reach 5.6 million, 43 percent more than in 1990. Likewise, the number of jobs is expected to increase by 43 percent between 1990 and 2020, to more than 3.5 million. Most of this growth will occur outside the regional core, in areas with limited road and transit services. The number of trips made daily by Washington residents is expected to grow by more than 65 percent between 1990 and 2020, and the number of miles driven will increase by more than 75 percent. Meanwhile, the funded improvements in the current long-range transportation plan provide for only a 23 percent increase in the capacity of the region's highway system, and very little expansion of the transit system. Under current plans and policies, traffic congestion on area roads—already among the nation's heaviest—is projected to be more widespread in the future, affecting all major travel corridors and more side roads.

Introduction

Flanked by the Blue Ridge Mountains on the west and the Chesapeake Bay on the east, the Washington metropolitan area has grown from a small collection of communities along the Potomac River to a prominent international region of more than four million persons and two million jobs. In the late 1700s the nation's capital was moved from Philadelphia to this area, thus shaping the destiny of the District of Columbia as a major world capital and the Washington region as a global economic center. Metropolitan Washington is part of the mid-Atlantic region on the eastern seaboard of the nation. It has often been considered the southern terminus of the northeastern "megalopolis," which spans from Washington to Boston and contains other prominent cities including Baltimore, Philadelphia, and New York.

---

[1] Director, Department of Transportation Planning, Metropolitan Washington Council of Governments, 777 N. Capitol Street, N.E., Suite 300, Washington, D.C. 20002; 202-962-3200

The Washington region consists of the District of Columbia and the jurisdictions of Suburban Maryland and Northern Virginia, extending from the urbanized central core through the well-established suburbs and ending in the rural fringe. The District of Columbia along with the City of Alexandria and Arlington County in Virginia are considered the **central jurisdictions** of the Washington area. The **inner suburbs** consist of Montgomery and Prince George's counties in Maryland and Fairfax County and the cities of Fairfax and Falls Church in Virginia. This group of jurisdictions is characterized by heavy growth that has taken place in the past few decades, and it is in this group that the majority of the region's residents live and work. Finally, the **outer suburbs** include Loudoun, Prince William, and Stafford Counties plus the cities of Manassas and Manassas Park in Virginia, and Frederick, Charles and Calvert Counties in Maryland. While officially part of the Baltimore region, Howard and Anne Arundel counties in Maryland act very much like suburbs of the Washington region. A considerable number of the residents of these two counties commute to jobs in the Washington region, and consequently account for a significant level of travel on the area's transportation network. (See Figure 1)

Metropolitan Growth And Development

Recent Trends

The economy that has evolved in the region is inextricably linked to the role Washington plays as the nation's capital. The federal government is the region's largest employer and, along with the services sector, is the engine that drives the economy of metropolitan Washington. Throughout much of the post-World War II period, the federal government was the single largest employment sector among the major industries. During the 1980s, however, the services sector surpassed the federal government in the number of jobs held in the region, reflecting the tremendous growth in the services sector nationwide.

Growth during the 1980s fueled a surge in commercial construction, and with it came the emergence of suburban employment centers throughout the region. Many of the new jobs that were added in the region were located in these suburban areas, and this resulted in shifting commuting patterns region-wide. In addition to many workers traveling to their jobs in the central core, a significant number of workers now commute to jobs located in the suburbs.

The composition of jobs in the region (primarily government and services) has resulted in a highly educated labor force with one of the highest participation rates in the nation. Furthermore, the Washington region has one of the highest labor force participation rates among women nationwide, and households with more than one member holding a full-time position are very common. The unemployment rate in the Washington region has remained approximately two percentage points below the national average, even during the recessionary period of the early 1990s.

Figure 1
TPB Planning Area, Metropolitan Statistical Area (MSA)/ Air Quality Non-Attainment Area, and Surrounding Areas

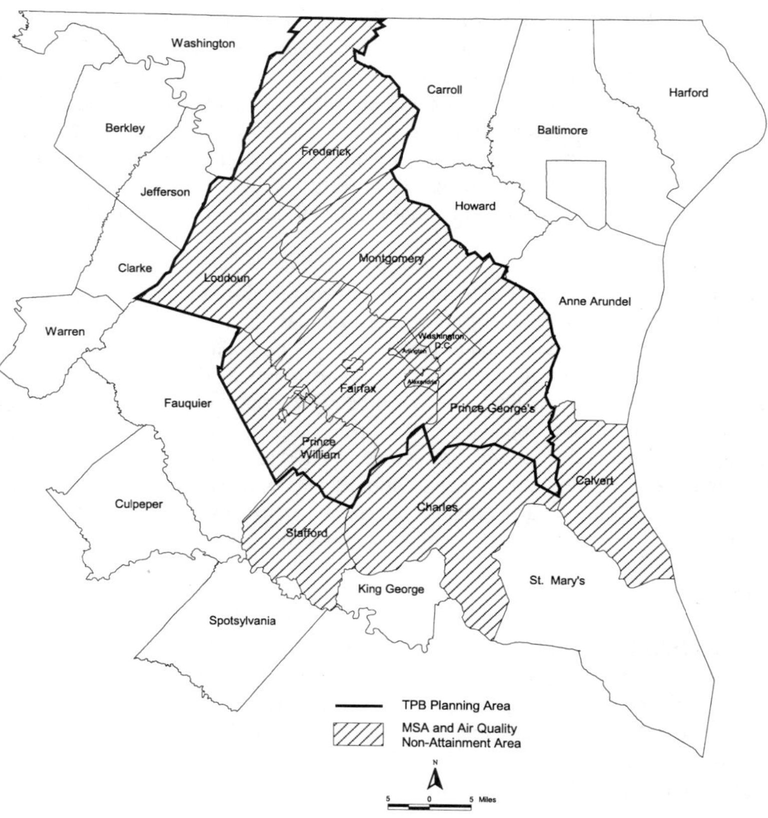

Population Growth

The comparatively healthy economy of the Washington region during the past few decades fueled strong population growth in the region. In 1960, the population of the Washington region was 2.2 million, but by 1990, the population was 3.9 million, a 77 percent increase. Metropolitan Washington is expected to have a population of 5.6 million people by the year 2020, representing a gain of 1.7 million people (43 percent) from the 1990 level. In other words, the population in the Washington region grew at an average annual rate of 1.9 percent between 1960 and 1990, but is expected to increase annually by only 1.2 percent between 1990 and 2020. The largest percentage increase in population will take place in the region's outer suburbs. Population in this group of jurisdictions is forecast to more than double from 700,000 in 1990 to over 1.4 million people in 2020. Anne Arundel and Howard counties in Maryland will also experience considerable growth.

Although the outer suburbs will bear the largest percentage increase, the inner suburbs will remain the population stronghold and will undergo the largest increase in actual numbers of people during this 30-year period. Counties such as Montgomery, Prince George's and Fairfax collectively are projected to grow from over 2.3 million residents in 1990 to 3.2 million residents in 2020, over a 37 percent increase. The inner suburbs of Maryland will increase 32 percent from almost 1.5 million people in 1990 to just under 2 million people in 2020. The inner suburbs of Virginia will expand 46 percent from 847,800 people in 1990 to 1,240,000 in 2020. The central jurisdictions of the region will not experience as dramatic an increase as the inner and outer suburbs. The District of Columbia's population in 2020 will grow by 3 percent, while Arlington County will grow by 21 percent and the City of Alexandria by 19 percent.

Household Growth

Between 1990 and 2020 households in the region are forecast to increase by 47 percent. This growth mirrors population growth in the various jurisdictions. Household growth in the outer suburbs will more than double from 234,000 in 1990 to almost 517,000 in 2020. Overall, households are forecast to increase at a slightly higher rate than the population, reflecting a continued national and regional trend toward smaller households. Figure 2 illustrates population growth throughout the region.

Employment Growth

Employment in the region is forecast to grow by 43 percent between 1990 and 2020. Noteworthy is the fact that while the District of Columbia will maintain a large portion of the employment base, jurisdictions outside the traditional central business core will witness the largest percentage growth and maintain the lion's share of jobs. Employment in the inner suburbs will increase from more than 1.2 million in 1990 to over 1.8 million in 2020, an increase of 50 percent. Although employment in the outer suburbs will remain below that of the central jurisdictions and inner suburbs, it will more than double in the 30-year period. Employment in the outer suburbs will

Figure 2

Distribution of Household Growth

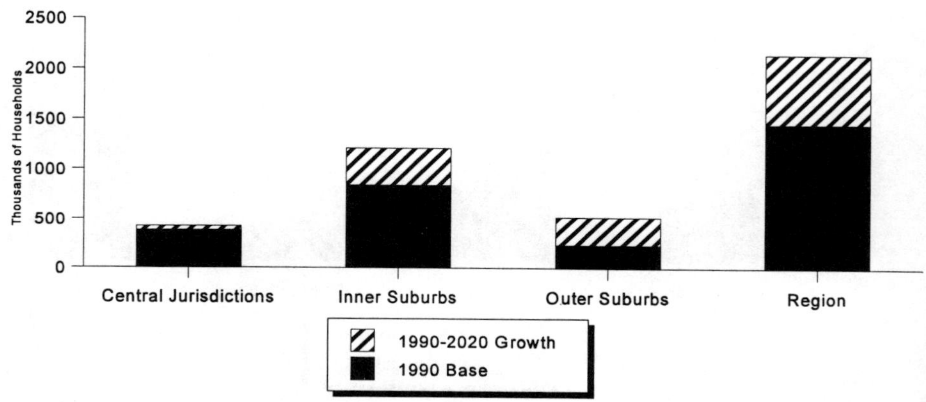

increase from 236,900 jobs in 1990 to 518,300 in the year 2020. Figure 3 illustrates employment growth throughout the region.

Travel Options

Highways

The road network is the foundation of the transportation system in the Washington region. This network consists of freeways, principal arterials, minor arterials, collectors, and local streets, each designed to provide a specific type of service. A large portion of the monies available for the transportation system are used to maintain and utilize this infrastructure as efficiently as possible. The region's highway system includes a number of facilities that are reserved for high-occupancy vehicles (HOVs).

Metrorail, Metrobus, and Other Transit Services

The Washington Metropolitan Area Transit Authority (WMATA) operates the Metrorail and Metrobus service in the region. The Metrorail system radiates out from the downtown core, and Metrobuses feed into the Metrorail stations, creating a comprehensive mass transit network covering 1,500 square miles. About 750,000 trips were made on Metrorail and Metrobus, collectively, on an average weekday in 1997 (this figure is the sum of total bus and rail trips minus combined trips). When completed in 2001, **Metrorail** will expand from its current 92.8-mile system to a 103-mile system with a total of 83 stations, shown in Figure 4. Metrorail's heavy-rail trains operate with three

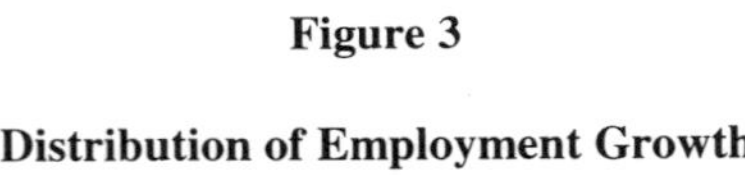

Figure 3

Distribution of Employment Growth

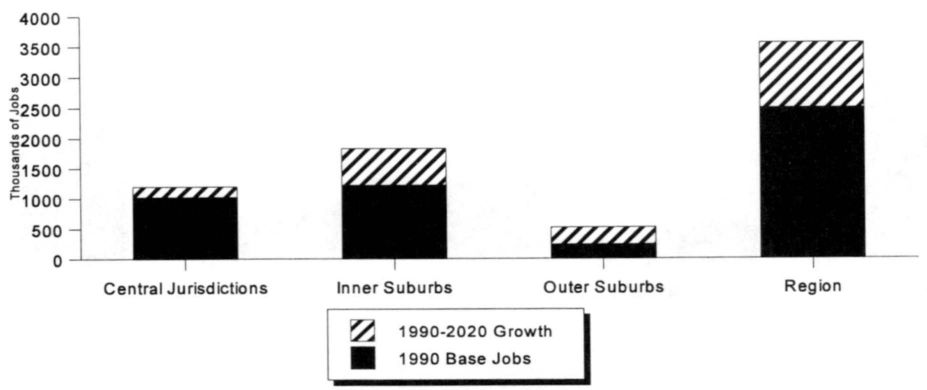

to six minute intervals between trains during peak periods and with six to sixteen minute intervals during off-peak periods. WMATA operates approximately 1,300 **Metrobuses** with routes in the District of Columbia, Alexandria, and Arlington, Fairfax, Montgomery, and Prince George's counties. Since 1975, the Metrobus system has been transformed from a predominantly radial system serving the District of Columbia to a feeder network serving the Metrorail system. Each time a new segment of the rail system has been opened, bus routes in the affected corridor or corridors have been modified either to serve or to turn back at the new stations.

All Metrorail trains are wheelchair accessible, and all stations have elevators for riders who are unable to use the escalators. When an elevator is not working, Metrorail has a van service to transport patrons to the next-closest station. Metrorail runs a telephone information line that details the stations without operating elevators so passengers with disabilities may plan their route in advance. WMATA also provides an on-call bus service as part of its Metrobus system. WMATA expects to have a 100 percent wheelchair-accessible fleet by 2006. In addition, WMATA operates a paratransit system exclusively for persons with disabilities; over 11,000 persons are registered and use the system. Those who qualify are issued an identification card and are able to schedule curb-to-curb travel service.

In addition to Metrobus service, several jurisdictions have their own **local bus service**. These include Montgomery County's Ride-On, Alexandria's DASH, Prince George's County's The Bus, Fairfax County's Connector, and the City of Fairfax's CUE systems. In addition, Prince William County offers commuter bus service through its CommuteRide system. Several private commuter bus companies exist as well.

Two **commuter rail services** operate in the region, Virginia Railway Express (VRE) and Maryland Rail Commuter (MARC). The Virginia Railway Express provides commuter rail service to Union Station in Washington, D.C. on two routes, VRE runs thirteen trains each way every day, and provides about 6,400 trips per day. MARC also provides commuter rail service to Union Station on three routes, with a total of 80 trains on these three lines, and carries about 18,000 persons on an average weekday.

Ridesharing

The Washington region is the carpool capital of the nation. According to the 1990 Census, almost 16 percent of Washington commuters used car or van pools to get to work. The high rate of ridesharing is encouraged by a number of factors, including the area's successful HOV lanes and an abundance of park-and-ride lots, which enable commuters to access a car or van pool or bus or rail service for their commute to work. Another resource that has helped the region attain such a high rate of carpooling is the Commuter Connections Program, which matches commuters for ridesharing, provides a regional Guaranteed Ride Home (GRH) program, and operates a regional system of Traveler Information kiosks, and coordinates regional programs for teleworking.

Bicycle Facilities

Both for the benefit of the environment and for the people they serve, bicycle facilities are important components of the region's transportation system. The Washington region currently enjoys more than 900 miles of on-street and off-street bikeways. Most jurisdictions in the area have developed bicycle transportation plans and have planners on staff to coordinate the bicycle/trail programs of the particular locality.

Airports

Residents of the region have an abundance of airport capacity to meet their travel demands. Three major commercial airports are located in the Washington region. Washington National Airport, located in the central core of the region, serves domestic travel needs, while Washington Dulles International Airport, located in Loudoun County, Virginia serves both domestic and international routes. Baltimore-Washington International Airport, located in northern Anne Arundel County, Maryland near the city of Baltimore, is also accessible to many area residents and provides access to domestic and international destinations. In 1996, these three airports served 41.3 million arriving and departing air passengers. In addition to the major commercial airports, the region features a number of general aviation airports to serve non-commercial air activity such as corporate travel. The three major commercial airports also include general aviation facilities.

Intercity Rail

Amtrak offers intercity passenger service for the Washington region with approximately 60 trains per day. Washington, D.C. is the southern anchor of Amtrak's Northeast Corridor, which extends north

Figure 4

103-Mile Metrorail System

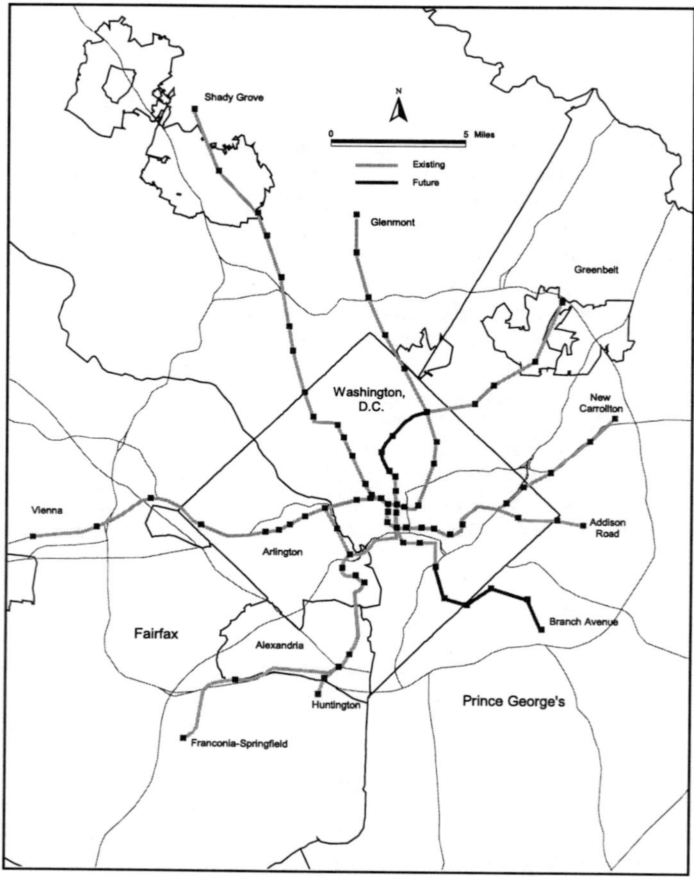

to Boston. High-speed trains run between Washington's Union Station and New York's Pennsylvania Station in this corridor. Amtrak is the largest passenger carrier between New York and Washington and currently commanded a 43 percent of the total air and rail passenger travel between the two cities in 1993. After the turn of the century, once high-speed trains are purchased and line improvements are completed, Amtrak trains will operate at 150 mph between Washington, D.C. and Boston, as well as New York.

Movement of Goods

Most of the Washington region's economy consists of government agencies and service and tourism industries. Freight movement in the region is, therefore, oriented toward delivery of office supplies and equipment and retail goods rather than heavy manufacturing materials. The freight sector plays an important part in the area's economy and is dominated by four modes: trucking, shipping, air cargo, and freight rail. Package express and postal services are also important to the region's economy. On a tonnage basis, trucks carry about 71 percent of the inbound freight and 96 percent of the outbound freight in the region, and represent between 3 percent and 8 percent of the traffic on the major routes in the Washington area. Water cargo accounts for 24 percent of the inbound freight and less than 1 percent of the outbound freight in the region. An additional 1 percent of the inbound freight and 4 percent of the outbound freight are transported by air. Trains carry about 4 percent of the inbound freight and less than 1 percent of the outbound freight.

Travel Patterns

The tremendous growth experienced by the Washington region since 1960, coupled with the increasing suburbanization of both people and jobs as discussed earlier, has had profound implications for travel. Not only has there been an explosion in the overall number of trips made on the region's highways and transit facilities, but travel has shifted away from a predominantly suburbs-to-downtown orientation as trip-making between the suburbs has surged.

Travel to Work

The latest travel estimates for 1990 indicate that more than 13 million person trips were made throughout the region on a typical weekday. About a quarter, approximately 3.6 million person trips, involved travel to and from work. Figure 5 shows the distribution of the different transportation options that residents used to get to work in 1990.

Projections available at the small area level highlight the increasingly suburban focus of work travel in the area, and in particular, the emergence of a significant suburb-to-suburb commuting market. They show, for example, that in 1990 more than half of all work trips in the region were to jobs in the near and far suburbs; commuting to the downtown core accounted for less than a quarter of all trips to and from work. Moreover, most of the trips destined to suburban jobs began from suburban residences. In all, more than half of all commuting trips in 1990 were estimated to both start and end in the suburbs. Forecasts indicate that suburb-to-suburb commuting will account for an even larger

Figure 5

Commuting in the Washington Region in 1990 by Transportation Mode

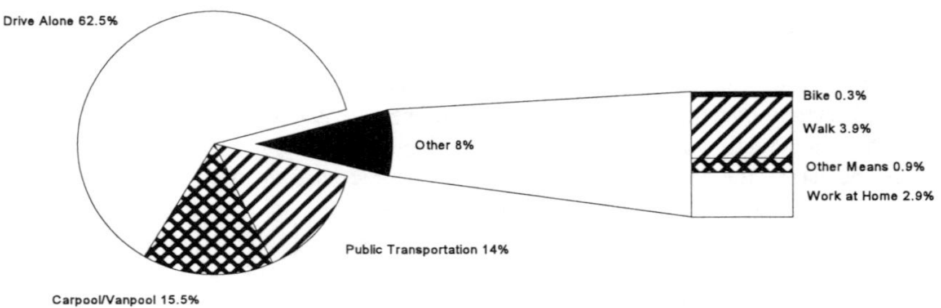

Source: 1990 Census

share of work travel in the future. Another characteristic of work travel that has emerged is "reverse commuting," where residents of the core commute to jobs located in the suburbs. Over a recent eight-year period, reverse commuting from the core increased 45 percent.

Transit Share

In 1990, about 504,000 trips, roughly 14 percent of all commuting trips regionwide, were on transit. Nearly half of all the commuting trips originating in the District of Columbia in 1990 were made on transit. Earlier estimates at the small area level show that although more than one-third of the workers traveling to jobs in the downtown core in 1990 were on transit, less than three percent of those commuting to jobs in the near and far suburbs used this travel mode. Transit captured less than 2 percent of the commuting trips made entirely within the near and far suburbs in 1990 -- a phenomenon that reflects the dispersed nature of the population and jobs in these locations, as well as the lack of cross-suburban transit options. While the absolute number of transit trips within the region will grow significantly in the future, transit's share of all commuting trips is predicted to remain flat.

Car Occupancy

The car occupancy rate for commuting trips averaged 1.16 persons per car for the region as a whole in 1990. Car occupancy rates also vary significantly among different travel markets. For instance, higher-than-average car occupancies were estimated for trips from the near and far suburbs to the downtown core, reflecting the availability of HOV facilities, as well as the higher employment densities and parking charges in the core. Projections for 2020 based on the facilities in this plan suggest a slight increase in car occupancy rates both for the region as a whole and for most commuting markets, including the suburbs.

Key Issues Facing the Region

Maintaining, Operating and Managing the Transportation System

At or near the top of the transportation agenda for every jurisdiction in the region is the challenge of maintaining the extensive transportation system in place today. During the next two decades, the operation and maintenance of the current highway and transit systems will consume about three-quarters of the available transportation revenues for Suburban Maryland and Northern Virginia, and almost all of the District's transportation revenues. Once relatively minor issues in discussions of long-range planning, maintenance and operations costs are now central. They limit the region's ability to finance facility expansions. Indeed, unless major new funding sources are developed, it must be assumed that most of our future transportation system is in place today. The challenge then becomes how to manage that system--and modify it where necessary--for the greatest future benefit.

Actions to better manage existing highway facilities can take many forms, ranging from relatively simple capital investments such as traffic signal improvements to regulatory approaches such as carpool lane restrictions or congestion pricing, which involves the use of fees to discourage unnecessary travel on congested facilities. Other strategies include new HOV facilities to encourage ridesharing, park-and-ride lots at selected commuter rail stations, transit information and marketing initiatives, bicycle connections at Metrorail stations, and pedestrian improvements in areas served by bus or rail.

Limiting Traffic Growth and Reducing Automobile Emissions

Faced with large population and job growth forecasts, the challenge of limiting traffic growth, or mitigating its effects, is an enormous one. Not only will the region house more families and individuals than at present, but on average, their activities will be spread over a larger radius as both housing and employment centers become more decentralized. As travel demand grows, it is no longer possible to increase the supply of roadway capacity to commensurate levels. First, most state and local governments cannot afford to build major new roads. Second, environmental and community concerns about new road construction and regulatory restrictions have reduced the political viability of large-scale road building as a way of "solving" traffic congestion problems.

Because of the limits on new road and transit capacity, it will be necessary to consider a larger range of demand management options that reduce the need for vehicular use. These include travel reduction methods such as telecommuting (working in or near the home), transit and ridesharing incentives, innovative land development and site planning techniques, and more controversial regulatory methods such as increased parking charges, employer-based controls on solo commuting, or direct pricing of road use. The availability of new technologies may make some forms of demand management more acceptable to the public over time. These include electronic pricing systems that could accumulate driving charges, for example, on a "debit card" that could be prepaid by users.

Serving Dispersed Population and Employment Centers

The decentralization or "suburbanization" of travel mentioned earlier and the emergence of "edge cities" are phenomena seen throughout the United States, and pose long-term challenges from every angle. By the year 2020 nearly two-thirds of all work trips in the region are expected to both start and end in the suburbs. The average weekday trip length was nearly 8 miles in 1990 and is projected to grow even longer. The average commute time, which in 1990 was longer in the Washington area than in any other U.S. city except New York, is also projected to lengthen.

In the short term, there is relatively little potential to reduce automobile use by altering the locations currently planned for development because most of the development that will be in place by the year 2020 already exists or will be built in the next few years. Many existing residential settlements are too dispersed to permit much pedestrian travel. During a longer period (beyond 30 to 40 years), however, the placement of new development could begin to affect travel patterns more significantly. In addition, certain design features can be incorporated in new developments to promote walking and transit use, and older areas can be "retrofitted" to improve pedestrian conditions.

Financing New Facilities

One of the key issues that will need to be addressed in future regional transportation plans is how to finance proposed facilities that go beyond those included in this plan. There are proposed major projects in the region that have been identified or desired in the past, but exceed the financial constraints on the plan required by federal regulations. Many of these projects are in the plan under a “study” category (construction unfunded). To implement many of these projects would involve billions of dollars, requiring the region to identify major new sources of funding. This could mean substantial increases in transportation user fees, such as tolls, gas taxes and parking charges. An effort to develop major new revenues would require substantial cooperation among the states and local jurisdictions in the region, and much greater public commitment to transportation improvements.

References

This paper has been excerpted from the following document.

National Capital Region Transportation Planning Board, *1997 Update to the Financially Constrained Long-Range Transportation Plan for the National Capital Region*, July 15, 1998

PUBLIC TRANSPORTATION IN WASHINGTON, D. C.

Charles W. Thomas[1]
R. Wayne Thompson[2]

ABSTRACT

The Washington, D. C. metropolitan area is served by a world class transit bus and heavy rail subway system operated by the Washington Metropolitan Area Transit Authority (WMATA). The system is the third largest in the United States, and works in close coordination with a host of other transit bus and commuter rail services operated by local governments and private contractors.

THE SYSTEM AND THE TRANSIT ZONE

The primary transit service area (the "transit zone") of the U. S. Nation's Capital is composed of eight local government jurisdictions, as depicted in Figure 1: The District of Columbia, two adjacent counties in Maryland, two adjacent counties in Virginia, and three incorporated cities in Virginia. In addition, a number of transit services - primarily express bus and commuter rail - operate into the Washington metropolitan area from numerous locations outside the transit zone. The transit zone is about 3,885 km^2 (1,500 mi^2) in size and contains a population of about 3.2 million residents.

The principal transit operator in this zone is the Washington Metropolitan Area Transit Authority, which operates a heavy rail subway system (Metrorail), an extensive transit bus system (Metrobus), and a demand response system (MetroAccess) that uses 51 small special purpose vehicles serving the physically disabled. Metrorail is a radial route system connecting the suburbs with the downtown employment district. It has been under construction since 1969 and currently has 153 route-km (95 route-miles) and 76 passenger

[1]Deputy General Manager for Operations, Washington Metropolitan Area Transit Authority, 600 Fifth Street, N. W. Washington, D. C. 20001

[2]Program Analyst, Office of Planning, Washington Metropolitan Area Transit Authority, 600 Fifth Street, N. W. Washington, D. C. 20001

stations in service, out of a planned total of 166 km (103 mi) and 83 stations once the system is complete. Figure 2 depicts the Metrorail system.

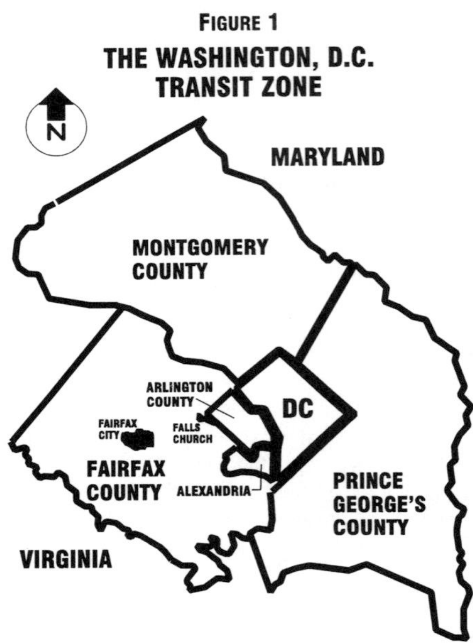

Metrobus operates both line haul and local neighborhood service, much of which is oriented to feed the Metrorail stations. Routes operate in a radial pattern to downtown in areas not served by the Metrorail subway.

Montgomery County, Maryland operates an intra-county local bus service network using a fleet of small and medium size buses. Prince George's County, Maryland has a local bus system, and in Virginia, Fairfax County, the City of Alexandria and the City of Fairfax have been operating local bus services for some years.

The states of Maryland and Virginia operate commuter rail services from localities outside the transit zone. The Maryland MARC system serves Washington from Western Maryland and from Baltimore, about 48 km (30 mi) away. The Virginia Railway Express (VRE) system operates from Fredericksburg and Manassas, Virginia. Washington's Union Station is the southern terminus of AMTRAK's northeast corridor service (Washington to Boston). The MARC and VRE commuter rail systems interface with Metrorail at Union Station, New Carrollton, Rockville, Silver Spring, Greenbelt, College Park, L'Enfant Plaza, Crystal City, King Street, and Franconia-Springfield Metrorail stations. At these locations the two services run parallel to one another and provide transfer opportunities.

Intermodal connections are a major consideration in the design of Metrorail passenger stations. Forty-three of the current 76 operating Metrorail passenger stations have off-street bus terminal/transfer facilities, and 30 stations have high volume parking lots or garages.

Table 1 shows the largest of the publicly-owned transit operators in the transit zone. The passenger miles and fleet size figures provide an indication of their relative size.

In addition to the services named in Table 1, there exists a vast network of small bus and van services that operate throughout the metropolitan area, and especially feeding passengers to the Metrorail system. A recent survey of privately operated transit vehicles accessing Metrorail stations found in excess of 250 different service providers. Among the most frequently observed were vehicles operated by hotels, general transportation services under contract to corporate clients, residential communities such as senior citizen apartment complexes, services directly operated by corporations, universities and schools, federal and

FIGURE 2

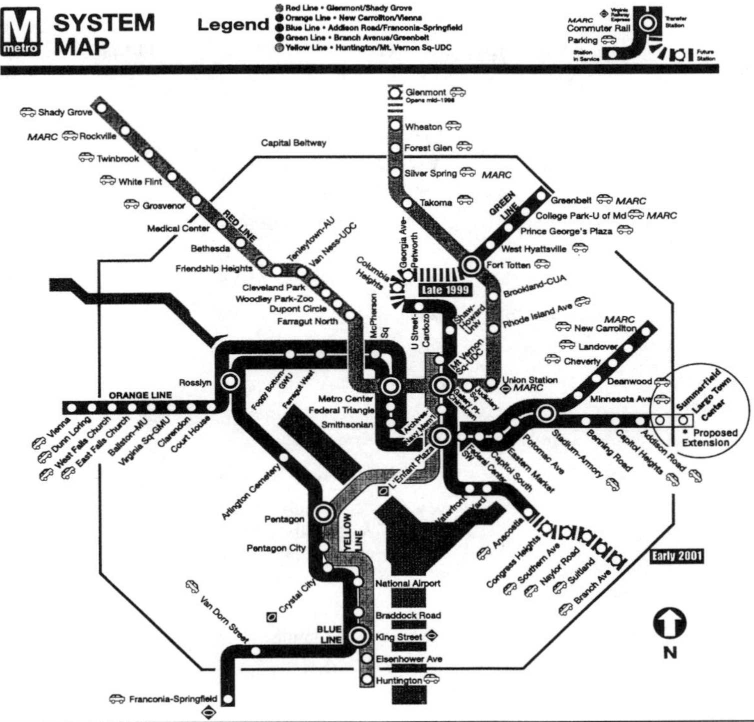

other government agencies, hospitals, shopping centers, automobile dealerships, and airport limousine services. Most of these are operated as courtesy shuttles for employees or paying customers of the sponsoring businesses. Together they transport a minimum of two percent of Metrorail's total riders.[3]

[3]The survey's field observations were made only in peak hours. Passengers arriving at other times were not counted.

TABLE 1: WASHINGTON, D. C. TRANSIT ZONE OPERATORS 1998

SERVICE NAME	MODE TYPE	OPERATOR NAME	FLEET SIZE	ANNUAL PASSENGER MILES (millions)
Metrorail	Heavy Rail	Washington Metropolitan Area Transit Authority	764*	1,077.1*
Metrobus	Transit Bus	Washington Metropolitan Area Transit Authority	1,299*	402.7*
MetroAccess	Demand Response	Washington Metropolitan Area Transit Authority	51*	2.0*
Ride-On	Transit Bus	Montgomery County, MD	348*	59.1*
TheBus	Transit Bus	Prince George's County, MD	46	✪
Connect-A-Ride	Transit Bus	Town of Laurel, MD	20	✪
Fairfax Connector	Transit Bus	Fairfax County, Virginia	127✿	58.9✿
CUE	Transit Bus	City of Fairfax, Virginia	12	✪
DASH	Transit Bus	City of Alexandria, Virginia	52	✪
PRTC✣ OmniRide OmniLink	Transit Bus	WMATA under contract to Potomac & Rappahannock Transportation Commission Prince William County, VA	75	17.0❂
MMTA✣	Transit Bus	State of Maryland Mass Transit Administration	NA	1.1❂
MARC✣	Commuter Rail	State of Maryland	NA	2.9❂
Virginia Railway Express (VRE)✣	Commuter Rail	Commonwealth of Virginia	NA	1.4❂

*Source: 1998 FTA National Transit Database ❂Source: 1997 FTA National Transit Database
✿Source: 1996 FTA National Transit Database ✪Comparable data not available. This is a small local system that does not submit a separate National Transit Database report.
✣Only a small portion of these services are within the transit zone. The miles shown are those allocated to the Washington, D. C. area (Urbanized Area 7).

A BRIEF HISTORY OF PUBLIC TRANSPORTATION IN WASHINGTON, D.C.

Early History: The first regular-route transit service in the Nation's Capital began in 1848, when horse-drawn stagecoach-type vehicles began operating between the port of Georgetown and the Washington Navy Yard via Pennsylvania Avenue and Capitol Hill. In 1862, Congress authorized the construction of the city's first horse-drawn street railway, the Washington & Georgetown, which would operate over the same route as well as on 7th and 14th Streets. Other companies quickly followed, and by the end of the 1880s, horsecar lines extended from Boundary Street (present day Florida Avenue) to Anacostia and from Georgetown to 15th and H Streets N. E.

The world's first successful electrically powered street railway opened in Richmond, Virginia in 1888. Later that year when Congress granted a charter to the Eckington and Soldiers' Home Railway, Washington's first electric service line was born.

The Washington & Georgetown and the Columbia Railway soon converted their horsecar lines to cable drive, and beginning in 1890, Washington had cable cars similar to those in San Francisco. Before long both companies had converted their cable systems to electric propulsion power.

The turn of the century saw a boom in electric railway construction. Electric cars were cheaper to build and operate, and provided faster, more dependable service than other systems. Then, as now, commercial development followed the car lines out of the city from downtown. In fact, the street railways were largely responsible for the growth of Washington from a town to a city.

By the start of World War I, with the city's network of trunk routes complete, Washingtonians were riding to work and play on bouncing four-wheel dinkies. These soon gave way to double-truck cars longer than a modern bus. From special mail cars to the popular open-sided "breezers" that operated in the summer, trolleys of many sizes, shapes, and purposes operated throughout the city.

From Streetcar to Bus: In 1921, the Washington Rapid Transit Co. became the first local streetcar operator to convert to buses when it began operating two bus routes from what is now the Metrobus Northern Division. The fleet soon grew to 46 buses, including 15 double-deckers, and the era of the motorbus was born. A number of other local transit companies soon followed suit and began replacing unprofitable carlines. In 1931, facing considerable expense for track repair, Capital Traction Co. converted to buses. Two years later, Capital Traction and The Washington Railway and Electric Company (WRECO) merged to form Capital Transit Co. Over the next three years, major changes took place in the route network, as Capital Transit and other local operators sought to consolidate their operations and eliminate duplicative service.

Even as the streetcar network was shrinking, Capital Transit Co. demonstrated its commitment to continued trolley operation through support of the Electric Railway Presidents' Conference Committee. Modern PCC cars were purchased and more were on order when World War II broke out. Ridership more than doubled between 1940 and 1944, as every car and bus that could run was pressed into service.

After the war, transit ridership fell as young families began buying automobiles. Buses replaced streetcars because buses could handle operational constraints and serve routes with greater flexibility. Bus service began expanding into neighborhoods beyond the ends of the carlines and provided the crosstown service that tied radial streetcar lines together.

D. C. Transit System: A 45-day strike in the summer of 1955 caused Congress to revoke Capital Transit Co.'s franchise. A clause in the new franchise required the conversion to an all-bus system within seven years. On August 26, 1956, eleven days after takeover, the new company, D. C. Transit System, Inc., implemented a plan to phase out streetcar service. The forced takeover began in earnest in the fall of 1958 when two lines were converted to bus operations. The last day of streetcar service in Washington was Saturday, January 27, 1962.

The first 40-foot buses were placed into service in 1952. These buses marked the beginning of the fleet-wide changeover from gasoline to diesel power. The first air-conditioned buses arrived in 1958, and the first "new look" bus went into service in Washington in 1959. Renovated new look buses are still in service.

D. C. Transit system continued the expansion of the bus route network begun by its predecessors, pushing new services far out into the suburbs and taking over small local operations. As operating costs mounted, D. C. Transit implemented innovative services such as using the Capital Beltway, connecting special routes with AMTRAK train service, and designing counterflow service to carry city residents to outlying employment centers.

Suburban Bus System Consolidation

Washington Virginia and Maryland Coach Co: The Washington Virginia and Maryland Coach Co. (WV&M) was founded in 1926, operating service from suburban Virginia. During World War II, after the Pentagon was built in Virginia just across the Potomac River from Washington, additional service was provided between there and. downtown. Suburban growth in the 1950s and 1960s brought new extensions to Fairfax County. In 1964, D. C. Transit acquired control of WV&M.

Alexandria Barcroft and Washington Bus Co: The Alexandria Barcroft and Washington (AB&W) bus company began service between the Virginia Suburbs and downtown in 1921. In 1969 and 1970 AB&W, under contract to the Northern Virginia Transportation Commission (NVTC), participated in a federally sponsored demonstration project to operate express service via exclusive reversible bus lanes in the median of the Henry G. Shirley Memorial Highway (U. S. Interstate Route 95) in Virginia. This was a pioneer effort to give transit a speed advantage over other traffic on crowded urban highways. Ninety new buses were purchased by NVTC to be used by AB&W for this service. The Authority now operates 29 routes over some portion of the Shirley Highway, carrying 8,158 passengers on an average weekday and logging 10.9 million passenger miles per year in the restricted lanes (Source: FY1997 NTD Report).

Washington Marlboro and Annapolis Motor Lines: In 1926 the Washington Marlboro and Annapolis Motor Lines (WM&A) was established to provide service between the Maryland suburbs and downtown. The operation remained profitable until the early 1970s.

The Washington Metropolitan Area Transit Authority

In the mid-1960s the U. S. Congress recognized the need to improve public transportation in the Nation's Capital. Planners recommended the development of a world class heavy rail transit system operating throughout the region. In response to this imperative, the Washington Metropolitan Area Transit Authority (WMATA) was created in 1967 by an act of the U. S. Congress, signed into law by the President of the United States. The act established a compact among the District of Columbia, the State of Maryland, and the Commonwealth of Virginia creating an interstate authority empowered to acquire property by eminent domain, to issue revenue bonds for the construction of the Metrorail system, and to operate the system once it was built.

The Authority's jurisdictional makeup is unique in the United States. The WMATA Board of Directors is composed of appointees from the three participating compact jurisdictions: two from the District of Columbia, two from Maryland, and two from Virginia. It is the only case in the U. S. in which a local transit operation traverses three states, and as a result, it has been especially important for the functioning of the system that the WMATA Board adopt a regional policy perspective.[4]

Bus System Takeover: Beginning in the late 1960s and early 1970s local private bus companies began struggling to stay in business. They were faced with rising inflation, escalating fuel prices, and declining ridership. Even though WMATA had been established for the purpose of building and operating a subway, by 1973 it had become clear that the Authority would have to operate the bus system as well. Faced with the financial failure of the four largest bus companies in the city, Congress gave WMATA the authority to acquire their assets in order to preserve their service for the citizens of Washington.

The Authority faced a formidable task of integrating four separate route networks into a unified regional system. New WMATA system colors were painted on the buses, duplicate route numbers were eliminated, boarding and alighting restrictions were dropped, fares were stabilized, 620 new AM General buses were purchased, old equipment was retired, service was expanded into areas not served previously, and 700 new bus shelters were installed. In 1975 and 1976 fringe parking lots went into operation and a new system-wide zone fare structure was implemented.

With the opening of each new segment of the Metrorail system, Metrobus service has been restructured to complement rail service. In 1975, Montgomery County, Maryland became the first local government jurisdiction to begin operating its own bus service, with the inauguration of Ride-On. Others followed, and now there are seven local government transit systems in operation in the Washington area. It is a testament to the area's unique commitment to regionalism that these local governments fund their own transit operations while at the same time subsidizing Metrobus and Metrorail.

Metrorail Construction and Operation: On December 9, 1969, the Washington Metropolitan Area Transit Authority broke ground and began construction of the Metrorail system. The first Metrorail trains began operating on the Red Line on March 27, 1976. Since that time Metrorail has grown and matured into the third largest heavy rail rapid transit system in the United States.[5]

---

[4]Although the District of Columbia is a territory of the U. S. Federal Government, in many respects it functions like a state.

[5]The New York City Transit Authority system and the Chicago Transit Authority system are first and second largest in the U. S. in terms of most measures: passenger miles, vehicle miles, track/route miles, ridership, number of stations, number of vehicles, etc. The CTA system and the WMATA system are nearly identical in size, while NYCTA is several times larger than either of the other two.

System Success: The Metrorail and Metrobus system has met or exceeded every expectation of its early planners. It carries about 940,000 passengers per day at the heaviest time of year[6], and at 37 percent, the transit mode split for AM peak period trips to the CBD is one of the highest in the country.[7]

TABLE 2: METRORAIL AND METROBUS OPERATING CHARACTERISTICS 1998

	METRORAIL	METROBUS
Frequency of service	3 to 6 mins peak hours 6 to 12 mins off-peak hours	5 to 60 mins peak hours 10 to 60 mins off-peak hours
Hours of service*	5:30 a.m. to 12:00 midnight	24 hours per day
Fleet size*	Rohr cars .... 298 Breda cars 466	1,299 buses (Flxible, Ikarus & MAN arctic, Orion, RTS)
Route miles (one way)*	Present .... 95 FY 2001 103	Fixed guideway .... 35 Mixed traffic 1,274
Vehicle revenue miles*	44,788,104	33,240,586
Passenger miles*	1,077,145,702	402,675,692
Unlinked passenger trips*	213,044,900	125,968,257
Revenue from fares*	$249,832,205	$83,782,405
Total operating cost*	$367,188,986	$227,845,222
Total income*	Fares, charter & contract service, investments, other $364,468,637	
Subsidy*	Federal (FTA Urbanized Area Formula) .... $3,294,000 State .... $138,004,914 Local $162,082,621	
Operating employees* (full time)	Vehicle operations .. 1,406 Vehicle maintenance .. 745 Non-veh. maintenance 1,494 Gen. Admin. 416	Vehicle operations .. 2,230 Vehicle maintenance .. 790 Non-veh. maintenance . 152 Gen. Admin. 301
Total WMATA employment (FY 1998)**	 8,499	

*Source: WMATA submittal to the 1998 FTA National Transit Database (all annual figures)
**Source: FY 1998 WMATA Management Budget.

Table 2 shows the basic operating characteristics of the Metrorail and Metrobus systems. Metrorail uses a state-of-the-art automatic train control (ATC) system, consisting of wayside and cab signaling components tied to a computerized central control facility. Construction of the system is still underway. The 103 mile original system is scheduled to be complete in FY 2002, and new extensions are already under consideration.

---

[6]*Monthly Report On Metrorail and Metrobus Ridership*, Washington Metropolitan Area Transit Authority 600 Fifth Street, N. W. Washington, D. C. 20001, through FY 1998.

[7]Metropolitan Washington Council of Governments, 1996 cordon count.

WMATA Fare Structure and Fare Collection System: Metrorail fares are collected by an automatic fare collection (AFC) system developed by Cubic Western Data Corp. The AFC system measures the distance traveled by a patron at the exiting station, and charges a fare based on a mileage formula.

Each Metrobus is equipped with a GFI electronic registering farebox, and a radio that permits voice and distress signal communication with the Metrobus Operations Control Center.

The fare structure on Metrorail and Metrobus is complex, as is evident from Table 3. A variety of discounted multiple trip passes are also offered, some of which apply to both modes. Discounted senior citizen and disabled patron fares are charged on both services.

TABLE 3: METRORAIL AND METROBUS FARE STRUCTURE

	METRORAIL	**METROBUS**
Peak Period Weekdays only 05:30 a.m. - 09:30 a.m. 3:00 p.m. - 8:00 p.m.	Mileage-based formula: $1.10 first 3 composite miles $.19 each add'l mile 3 to 6 $.165 each add'l mile >6 $3.25 maximum	Zone fare: $1.10 base fare Various zone charges dependent on local government jurisdiction
Off-Peak Period All other hours/days	Mileage-based formula: $1.10 first 7 composite miles $1.60 >7 miles to 10 miles $2.10 maximum	Zone fare: $1.10 base fare $1.45 DC to VA $1.75 DC to MD
Discounted transfer from Metrorail to Metrobus dependent on local government jurisdiction. No discounted transfer from Metrobus to Metrorail. $.10 transfer from bus to bus.		

Note: A composite mile is the average of the straight line (airline) distance and the rail distance between entering and exiting stations. *Source: WMATA Tariff № 18, revised July 1998.*

The WMATA Board of Directors is studying a simplification of the fare structure scheduled to take effect in the summer of 1999. The most prominent change will be the elimination of all bus fare zones. A $1.10 flat fare will be charged for all regular bus trips. Some express bus routes will charge $2.00 per trip. The Metrorail mileage-based fare structure will remain unchanged, but the evening peak period will be shortened to end at 7:00 p.m.

Parking: WMATA is one of the largest operators of automobile parking facilities in the region. There are over 38,000 parking spaces at lots and garages located adjacent to Metrorail stations, and major additions are planned for the next five years. Parking fees average about $2.00 per day. Monthly rates and reserved spaces are available. Each lot includes a kiss-and-ride section and most have short-term metered parking.

Safety and Security: The Authority has its own 300 member force of Transit Police who are sworn law enforcement officers in every jurisdiction of the transit zone. They not only perform routine patrol duties but also provide security during revenue collection and transport operations. Security on other systems operated by the local governments in the transit zone is provided by their respective jurisdictional police departments.

The Authority has a 16 member Office of Safety that provides inspection and oversight in the areas of occupational safety and health, environmental issues, fire protection, and operating system safety.

MOVING PEOPLE

As in most U. S. cities, the great majority of the transit travel in Washington, D. C. involves the journey to work. A recent system-wide survey of Metrorail passengers determined that the home-based work trip accounted for about 80 percent of all travel on that mode.[8] On Metrobus and the other smaller locally-operated systems in the Washington area, trip purpose is somewhat more balanced between work and non-work.

TABLE 4: RIDERSHIP ON SELECTED SYSTEMS IN THE TRANSIT ZONE (x 1,000)

Service	AM Pk	Mid-day	PM Pk	Other	Avg Wkdy	Avg Sat	Avg Sun	Avg Annual
Metrorail*	238	155	265	68	726	333	207	213,044
Metrobus*	115	112	140	53	422	219	133	125,967
Ride-On*	16	23	15	5	59	38	20	18,149
Fairfax Connector✿					17	5	2	4,512

*Source: 1998 FTA National Transit Database - Unlinked Passenger Trips
✿Source: 1996 FTA National Transit Database - Unlinked Passenger Trips

According to another recent WMATA survey, 69 percent of frequent Metrorail users and 85 percent of occasional users have access to an automobile.[9] The same study found that 49 percent of frequent Metrobus users and 63 percent of occasional users have access to an auto. The Metropolitan Washington Council of Governments estimates that 37 percent of commuter trips to the core area of Washington, D. C. are made using public transportation - the third highest mode split in the nation.[10] Table 4 shows ridership figures for the largest transit service operators in the transit zone.

---

[8] *The 1994 Metrorail Passenger Survey* by George Hoyt and Associates for the Washington Metropolitan Area Transit Authority.

[9] *WMATA Service Area Usage and Attitude Assessment Research,* by QS&A Research and Strategy for the Washington Metropolitan Area Transit Authority, May 1997.

[10] Metropolitan Washington Council of Governments, 1996 cordon count.

The graph in Figure 3 shows the 25 year ridership trend on the Metrorail and Metrobus systems. By mid-1989, Metrorail had surpassed Metrobus in terms of passenger volumes. To some extent this was the result of the opening of new segments of the Metrorail system and the associated bus route modifications to feed the new rail stations. Passengers switching from bus to rail mode produce no increase in total transit ridership.

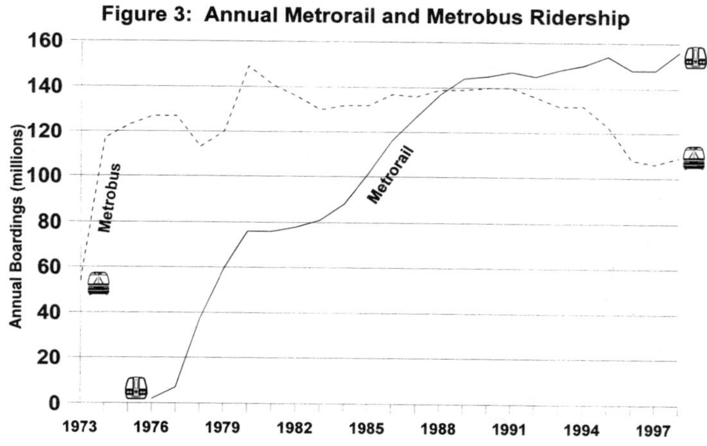

Table 5 shows Metrorail and Metrobus ridership by demographic group.

RECENT INNOVATIONS

In 1999 WMATA will introduce the use of *Smart Card* electronic technology to combine Metrorail fare collection with banking and debit card services. The Smart Card is a permanent plastic device, resembling a credit card. It contains a computer chip memory to which cash value can be repeatedly added by the holder of the card. By touching the card to a "target" object on a Metrorail fare gate, the value of the fare is deducted. Eventually the technology will be applied to Metrobus fareboxes, Metrorail parking facilities, and other transportation services. It is a giant step toward the goal of seamless transportation in the region.

In the summer of 1998 WMATA competitively bid and won a contract to operate the 75 bus system owned by the Potomac and Rappahannock Transportation Commission (PRTC) in Prince William County, Virginia, south of the transit zone. This is the first contract operation ever undertaken by WMATA, and it represents a significant effort to expand service into areas not previously served by the Authority. It recognizes that the National Capital region is growing and that WMATA's services must keep pace. In seeking this contract WMATA demonstrated that, with labor's cooperation, public authorities can be cost competitive in the transit market.

In 1997 the Authority opened a state-of-the-art connection between the National Airport Metrorail Station and the new airline terminal at Reagan National Airport. The connection makes the Metro station part of an integrated airline passenger terminal, automobile parking, car rental, shuttle bus, and subway facility. The Metrorail station is closer to the terminal than the parking garages. Moving walkways on covered pedestrian bridges connect the Metro station with the airline terminal for one of the most convenient airport/rail transit connections in the country.

TABLE 5: DEMOGRAPHIC CHARACTERISTICS OF METRORAIL AND METROBUS RIDERS

Measure	Category	Metrorail	Metrobus
Sex	Male	55%	47%
	Female	45%	53%
Education	High school	24%	41%
	Some college	28%	31%
	College grad	48%	28%
Age	18-29	38%	36%
	30-39	29%	32%
	40-64	32%	32%
	65+	nil	nil
Marital Status	Married	32%	23%
	Single	55%	66%
Race	White	44%	26%
	Black	37%	56%
	Other	19%	10%
Income	Average	$46,800	$37,700
Employment	In D. C.	66%	64%
	In suburbs	34%	36%
Residence	In D. C.	30%	44%
	In suburbs	70%	56%

Source: *Ridership Analysis Study*, by Peter Harris Research Group for the Washington Metropolitan Area Transit Authority, July 1995.

In the next ten years, WMATA expects to increase its rail car fleet to 948 cars to accommodate the anticipated ridership growth that will result from expansion of parking facilities and extension of the system. The Authority has initiated a an accelerated *(fast track)* construction program to ensure that construction of the 166 km (103 mi) Metrorail system is completed by the year 2002. At the same time, the Metrobus fleet will be increased to enable WMATA to extend Metrobus service into untapped transit markets, particularly in the suburbs. In keeping with the Authority's *100% Customer Service Commitment* each employee is pledged to safe, clean, reliable, courteous service, that is of excellent value.

CONCLUSION

Public transportation in the Nation's Capital has matured in the past 25 years to become a world class operation in keeping with the stature of the city. The 21st Century will be bright for the transit industry as the Washington area's local governments continue to expand and improve transit services in close coordination with land use and development. The Washington Metropolitan Area Transit Authority, in coordination with other transit operators in the region, is intent on providing innovative and customer-friendly transit services that are a viable and attractive alternative to the private automobile.

Socio-economic Characteristics, Land Use and Travel Patterns of Mexico City

Alejandro Villegas Lopez[1]

Introduction

Over the last sixty years, Mexico City has grown from an agreeable city with a population of a million people to an expanding megalopolis of around seventeen million.

The metropolitan area extends across two distinct, self-governing, administrative areas (the Federal District and the State of Mexico). The Mexico City Metropolitan Area (MCMA) comprises the sixteen boroughs in the Federal District (DF) plus twenty-eight municipalities in the State of Mexico (SM). This region is known collectively as the Valle Cuautitlán Texcoco (VCT), where about 17 % of the country's total population live in an area of 4,945 km^2.

While these two areas are part of the same urban region, they have different policies and resources, which result in different standards, costs and varieties of transport. Dealing with the economic, environmental and transport problems arising from continuing population growth and severely limited resources is the major challenge facing management of the metropolitan area of Mexico City.

Prior to 1940, Mexico City grew slowly and almost all its population lived in the central area of the DF. With a population of 1.7 million in an area of 100 km^2, the 300-km streetcar network promoted expansion towards the south which, with its associated industrial and tertiary activities, established Mexico City as the center of the nation's economic development. By 1950, the city began extending to the north and east across the DF boundaries into the SM.

[1] Transportation Planning Director, Mexico Capital City Government, Alvaro Obregon 269, 9o Piso, Col. Roma C.P. 06700, Delegacion Cuauhtemoc, Mexico, D.F.

This has continued to such an extent that most of the DF is now urbanized with most growth occurring in the VCT. Average annual growth in the conurbation municipalities over the 1995-2000 period would be approximately 3%, in contrast to a figure of 0.3 per cent in the DF for that same period.

1. Socio-economic and land use elements

The MCMA can be divided into four areas according to its population annual average growth rate (AAGR) forecasted for the period 1995-2000. **(See Chart 1)** Eight of the sixteen boroughs of the DF plus two SM municipalities although containing 43% of the MCMA population, have negative growth and are urbanized at an average density of 137 inhabitants per hectare. These ten jurisdictions are located at the core center of the conurbation area. Seven DF boroughs plus three SM municipalities, located in the first outer ring around the core center, mainly to the south, east and west, are growing at a positive AAGR from 0.3 to 3%. These areas have an average density of 52 inhabitants per hectare and account for 39% of the MCMA population. Although there are some relatively wealthy suburbs with little capacity to absorb further population growth, there are also important poor neighborhoods. Eleven SM municipalities contain 14% of MCMA population and are located on a second outer ring, in the west, north and east. These municipalities are growing at a rate between 3.4 to 6.9% and their low average density of 26 inhabitants per hectare is allowing for rapid urbanization by planned growth of medium-income dwellers and unplanned growth of low-income people. The remaining eleven SM municipalities are largely rural, as well as one DF borough. These jurisdictions located in the third outer ring, account for only 4% of the MCMA population ranging between 21 to 82 thousand people each. Average population density is nine inhabitants per hectare and AAGR is moderate.

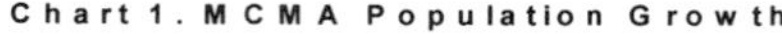

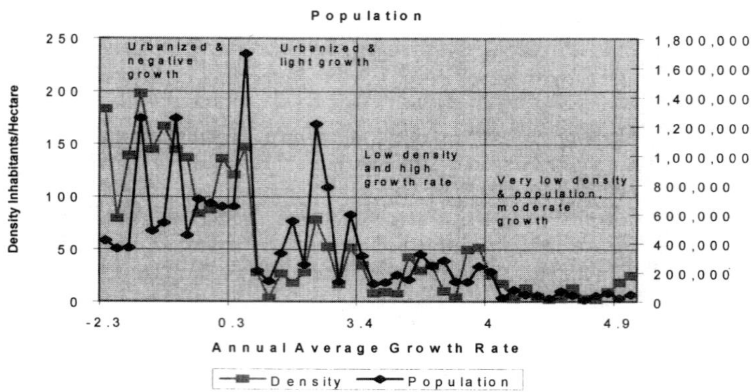

Chart 2 shows the 16 boroughs and the limits of the Federal District. The central area and the three rings mentioned in **Table 1** are illustrated. It is important to note that, within the six administrative areas in gray 37% persons-trips-day are concentrated.

Table 1. Population density by región in the Federal District. (Inhabitants/km2)

Region	1970	1980	1990	1995	2000	2010	2020
Central Área	21780	17840	13900	12674	12530	12520	12500
First ring	14770	14880	13680	14249	14280	14400	14570
Second ring	4410	6650	9500	11101	11650	12340	13110
Third ring	6050	5870	5040	6411	6230	6990	7670

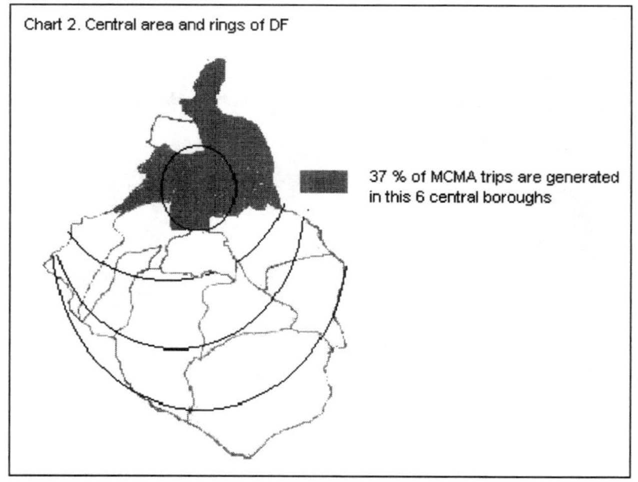
Chart 2. Central area and rings of DF
37 % of MCMA trips are generated in this 6 central boroughs

Population in the DF is aging. From 1970 to 1995 the percentage of people between 1 to 14 years old diminished by 33%, decreasing from 2.8 million to 2.3 million. In the same period, those between 15 to 64 increased by 20 % (**Table 2**)

Table 2. Age Distribution in Mexico D.F. area years 1970, 1980, 1990 and 1995

	1970		1980		1990		1995	
Age Group	**Population (Millions)**	**%**	**Population (Millions)**	**%**	**Population (Millions)**	**%**	**Population (Millions)**	**%**
0-14	2.83	41.3	3.27	37.0	2.51	30.6	2.35	27.7
15-64 years old	3.80	55.4	5.21	59.0	5.30	64.5	5.68	66.9
65 and more	0.23	3.3	0.35	4.0	0.40	4.9	0.46	5.4
No specified	0.00	0.00	0.00	0.00	0.03	0.00	0.00	0.00
Total	6.86	100	8.83	100	8.24	100	8.49	100

Source. INEGI National Census IX, X, XI

Regarding employment, based on the 1995 economic census, 55.8 % of population within labor force (those older than 12 years old) are active employees, whereas 44.1% are not active. Among the latter group, 37% are students, 47.2% work at home and 15.8% do something else (not specified). **Charts 3 and 4** show the share of population older than 12 years old which is considered within potential DF labor force in 1995.

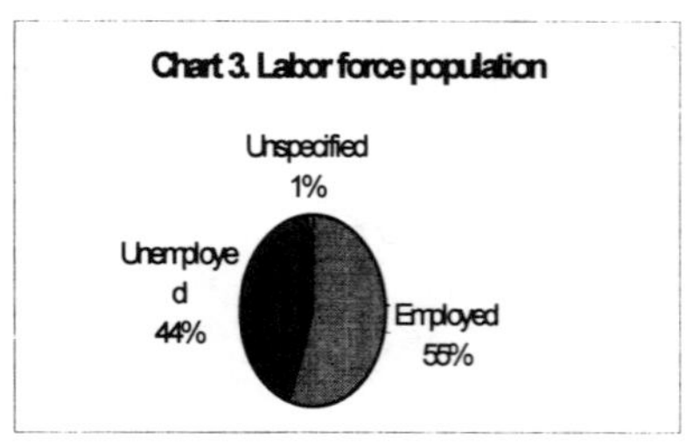

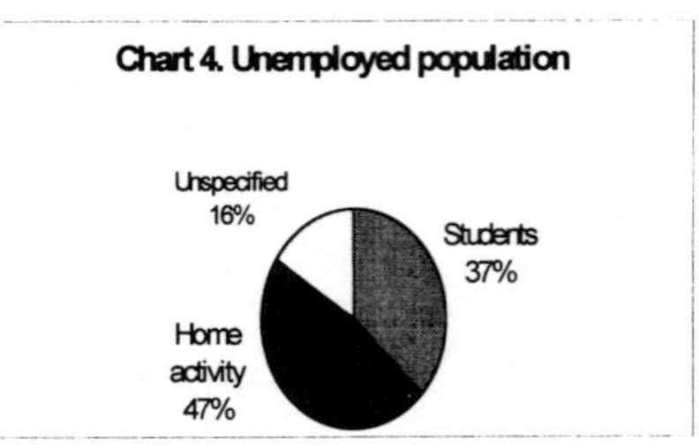

Source. Socioeconomic Census by State. INEGI 1995

In terms of income, 31.84 % of the population earn from 2 to 5 times the minimum wage, while 30.67% earn from 1 to 2 times the minimum wage. According to these figures, 2.47% receive no income and 10.41% receive only 1 or less of a minimum salary, as shown in **Chart 5.**

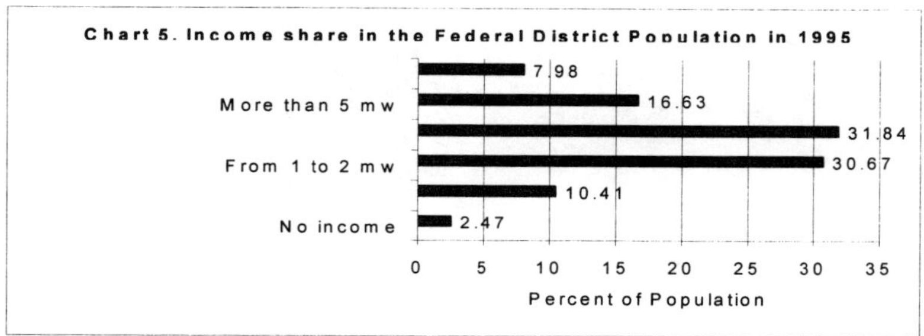

Note: MW means minimum wage that is the indicator to measure income by the Mexican National Census Bureau. In 1995, minimum wage was 18.43 pesos, equal to approximately 3 USD per day of 8 working hours. By 1999 mw is 34.50 pesos a day, equal to 3 USD.

Source. Socio-economic Census in the Federal District, INEGI, 1995

The Mexico City Metropolitan Area generates a third of the Gross National Product (GNP) and 38 % of manufacturing output. Most of this activity is in the DF, where commercial, industrial and retailing activities are intermixed with residential areas **(Map 1).** Commercial areas are dispersed throughout the city. While local markets are located within walking distances in boroughs and neighborhoods, shopping centers and super markets, that produce high mobility throughout the city, are concentrated in the jurisdictions within the MCMA central core area. This is so at the expense of further, and sometimes poorer, *delegaciones* such as Alvaro Obregón, Xochimilco and Iztapalapa in southern MCMA. However, economic activities are becoming decentralized out into the VCT, where a number of distinct centers of secondary and tertiary economic activities are developing. Overall, 32% of the population in the VCT urban areas work in the DF.

Map 1. Location of commercial areas by type in Mexico D.F. and surrounding areas of MCMA

Source. Facilities and infrastructure. Internal document SETRAVI 1997

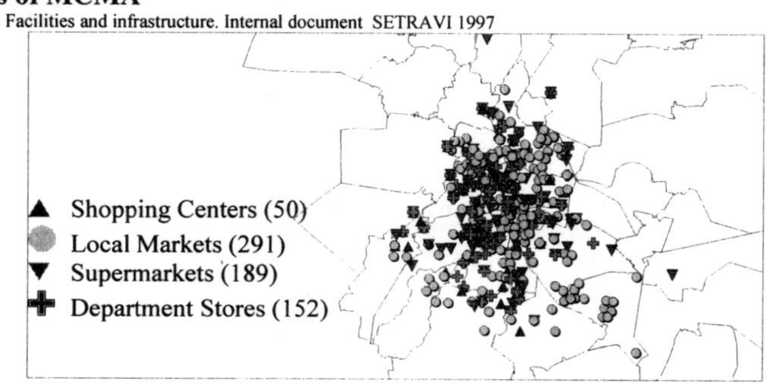

Map 2. Location of major terminals in Mexico D.F. and surrounding areas of MCMA

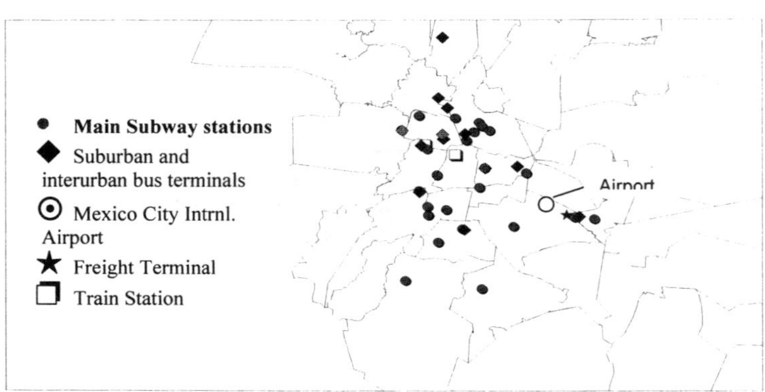

Source. Facilities and infrastructure. Internal document SETRAVI 1997

While main subway stations are regularly spread out across the city, we can see in **Map 2** that suburban and interurban bus terminals tend to concentrate in the north and east. This is so since most of the trips enter, through northern roads, from SM, Querétaro and Hidalgo states and eastern ones from SM as well as Puebla states. A freight terminal for non-industrialized food is located on the eastern boundary of the DF. Trucks heavier than 3.5 tons unload farm and agricultural products into light weight vehicles that distribute them throughout MCMA. The Mexico City International Airport is located right along the first orbital freeway loop, in the eastern border of the historical downtown district. In spite of its high traffic, this strategic facility is located practically in downtown and generates an important number of trips.

According to the 1994 Origin and Destination (O-D) Survey, 20.7% of trips in MCMA are made from home to school. Most of the trips take from 31 to 40 minutes. Higher education institutions and universities are concentrated in 5 boroughs that originate more trips and increase traffic at peak hours. **(Map 3)**

Map 3. Higher education institutions and universities in Mexico D.F. and surrounding areas of MCMA

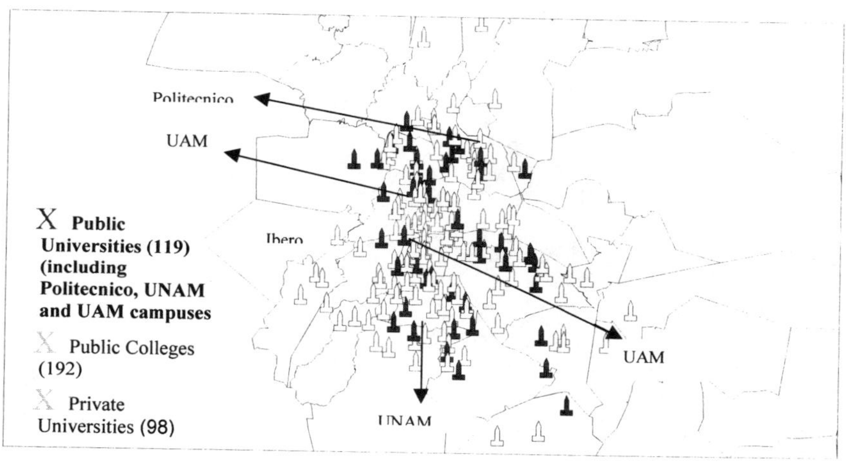

Source. Facilities and infrastructure. Internal document SETRAVI 1997

The percentage of trips from home to work is 48.1% of the total trips in the MCMA. Most of these trips last from 60 to 70 minutes. Map 4 shows that must of the government buildings are concentrated in Cuauhtemoc and Benito Juárez boroughs in the central area of the map.

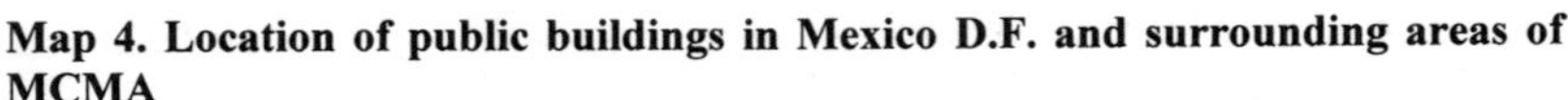

Map 4. Location of public buildings in Mexico D.F. and surrounding areas of MCMA

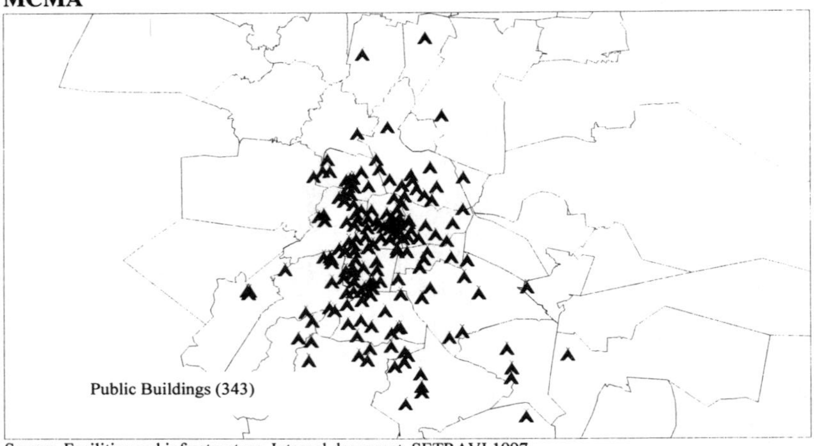

Source. Facilities and infrastructure. Internal document SETRAVI 1997

Chart 6 shows the labor force shares regarding each type of job, as reported for the 1990 Mexican Census. The MCMA is a tertiary urban economy having 70% of labor force in the services sector, 29% in the industrial sector and one percent remaining on the agricultural sector, as there still exist some agricultural land areas in the peripheral areas both in the Federal District and the surrounding municipalities.

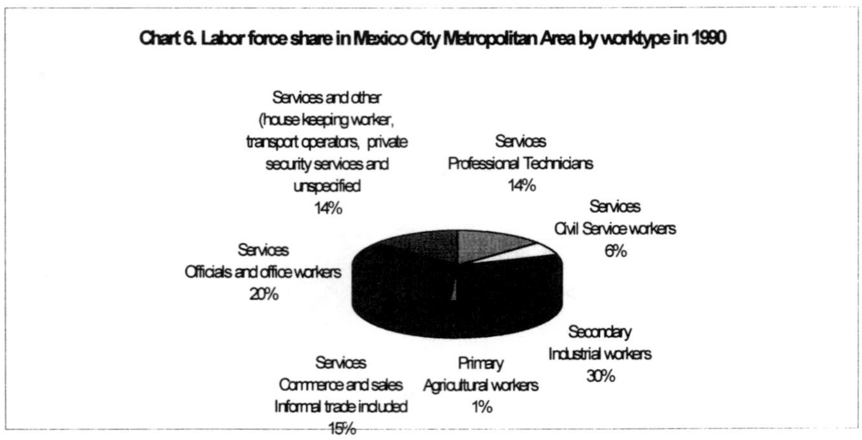

Source. Socio-economic Census in the Federal District, INEGI, 1995

Economic activity is more heavily concentrated to the west and south of the region. The proportion of population in the inner areas is falling as commercial activities and associated high land-values drive residents to outlying areas and force people with lower incomes to locate in unsuitable settlements far from services and employment.

Table 3 shows that predominant economic activities within Federal District are manufacturing, commercial, and services. By contrast, agriculture activities are almost non-existent.

Table 3. Gross Domestic Product GDP in Mexico D.F. according to economic activities by sector (Thousands of Pesos) 1980 to 1993

Type	1980	1985	1988	1993
Agriculture and fishing	2641	22 825	90 500	378 704
Mines	7472	26 345	225 940	736 980
Manufacturing	290140	2 735 533	24 642 480	58 517 064
Construction	59065	477 114	3 591 756	16 472 328
Electricity, gas and water	6253	56012	550 025	1 696 192
Commerce, restaurants and hotels	276125	2 433 023	20 681 207	57 937 983
Transportation, communications and warehouses	85606	818 035	8 232 109	29 398 595
Finance, insurance and real estate	90093	1 061 379	9 587 282	45 392 506
Social, personal services	279468	2 574 426	18 574 813	77 873 831
Banking	21238	242315	2 805 038	17 132 251
Total	1 075 626	9 935 376	83 371 073	271 271 933

Source. Socioeconomic Census. Mexico DF. 1995. INEGI

Travel Patterns

Differences in social and economic circumstances among the DF and SM are strongly reflected in MCMA's travel patterns. In 1990, most residents within VCT had low incomes with an average motorization rate of 12 persons per automobile. In the DF, most people are in the medium-income range and its motorization rate is 5.7 persons per automobile **(Table 4)**. While industry is fairly evenly distributed in relation to population in the DF and the SM, service employment is heavily concentrated in the DF where the density of this type of employment is 16 jobs per one hundred population compared with 3.4 in the VCT. This results in heavy volumes of commuting from the VCT to the DF.

Table 4. Index of motorization in the MCMA. Population and number of cars per person 1990-1995

Year	Population (Thousands)	Vehicles (Thousands)	Population/vehicles
1990	15047.7	2798	5.38
1991	15194.5	2763	5.5
1992	15343.4	3188	4.81
1993	15493.7	3468	4.47
1994	15645.6	3126	5
1995	15783.8	3195	4.94

Source: INEGI. National Census VII, VIII, IX, X and XI and 1995

Road Network

Now in 1998, there are 2,320 km of main roads and 8,400 km of secondary roads in the MCMA. The DF road network has 198 km of freeways, 310 km of Expressways with preferential busway lanes running north–south and east–west across the city and 552 km of main roads. The SM has 47.3 km of freeways, 618 km of primary roads, and 94 km of Expressways.

The DF is better provided with road space than the VCT. There are four types of road structures. The grid network of main roads is to be seen in the DF and parts of the VCT, particularly in the east. This provides a uniform distribution of capacity. In the northern and western parts of the MCMA is the "comb" network, which has a fairly regular layout that acts as a distributor from the main network and tends to saturate it.

There are also three incomplete ring roads centered from the historical downtown area along with several major radials. Finally, there are local street systems that have grown up as part of older settlements and do not form a consistent part of the general network.

Overall, the network shows great diversity and variety with few good connections between the DF and VCT. The two incomplete rings in VCT need to be finished to accommodate orbital traffic movements.

Table 5. Road Distribution

Road Classification	DF (km)	VCT (km)	TOTAL (km)
Freeways	198	47	245
Expressways	366	94	460
Primary Roads	552	618	1,170
Secondary Roads	8,000	N.A.	+ 8,000
Total	9,116	+ 759	+ 9,875

Source. Facilities and infrastructure. Internal document SETRAVI 1997

The low level of service detected at the metropolitan level reflects on the DF network's capacity analysis on 40 main corridors that were examined. Twenty roads, 74% out of total surveyed avenues, have service levels graded at "F", non-stable flow, with global speed below 20 kilometers per hour. 7.4% of the network, which is two roads, have service level "E", with average speed of 20 Km/h. One road, accounting for 3.7% of the surveyed network, has service level "D", poor stable flow, with speed between 20 to 30 km/h. Three more roads have service level "C", stable flow, with average speed above 30 and below 40 km/h; this represents 11.2% of the road network. Finally, only one road reaches service level "B", stable flow, with a speed above 40 km/h. Network's traffic average operation speed is 20 km/h and public transit operation speed is 17 km/h. The road surface characteristics also have an influence: in general, conditions are acceptable on most of the corridors; the service rating of the road surface averages out at 3 on scale from 0 to 5.

Chart 7 shows an increase in the volume of cars. However, the number of cars registered decreased from 1993 to 1995. The program called *Hoy No Circula* in which each car has to stop running once a week could be one of the reasons for this. The economic depression of 1994 also affected and discouraged purchasing new cars.

Chart 7. Traffic counts and number of vehicles in Mexico D.F. years 1986, 1992 and 1995. Source. Traffic Counts in Mexico City years 1986, 1992 and 1996. SETRAVI

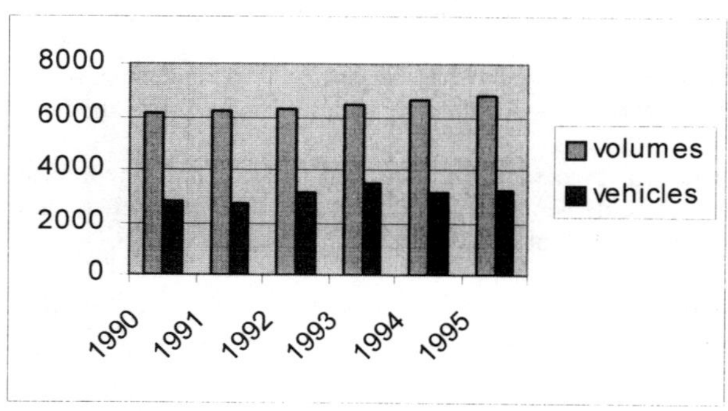

Modes of Travel

As shown in Table 7, the predominant mode of transportation is mainly composed of low capacity units called microbuses that meet almost 60% of the total demand of public transportation. The social, economic and environmental impact caused by this mode has been negative for the rest of the city. Metro is still the backbone of transportation in the Federal District that is complementary with the bus services. However, both modes ridership has been surpassed by microbuses in the last 10 years due to lack of control of the permits given to the owners of taxicabs and microbuses.

Table 7. Modes of Transportation, 1972-1994. Number of trips by thousands

Transport	**1972**		**1979**		**1983**		**1986**		**1994**	
Mode	Total	%	Total	%	Total	%	Total	%	Total	%
Metro	1.07	8.5	2.10	11.4	3.70	19.0	4.20	19.1	3.91	13.4
Bus	5.99	47.4	6.43	34.9	5.82	29.8	5.70	25.9	1.95	6.7
Trolley bus / Light Rail	0.58	4.6	0.61	3.3	0.34	1.7	0.70	3.2	0.17	0.6
Suburban bus	0.49	3.9	2.76	15.0	3.15	16.2	3.60	16.4	1.03	3.5
Privately operated microbuses / Taxicabs	1.59	12.6	2.39	13.0	1.99	10.2	2.30	10.5	16.85	57.6
Subtotal Public Transport	9.7	76.8	14.3	77.7	15.0	76.9	16.5	75.0	23.9	81.7
Automobiles / Others	2.93	23.2	4.11	22.3	4.50	23.1	5.50	25.0	5.34	18.3
Total Demand	**12.65**	**100**	**18.40**	**100**	**19.50**	**100**	**22.00**	**100**	**29.24**	**100**

Source: Data for 1972 - 1986: General Transport Coordination Office, Department of the Federal District. Comprehensive Transport and Road Plan 1988 - 2000. Data for 1994: Origin–Destination Survey of MCMA Residents' Travel Patterns, 1994. INEGI

Travel problems

Finally, the future growth prospects for the MCMA appear to be:

- The annual average growth rate has been reduced from 4.2%, between 1970-80, to 1.8%, between 1980-90, with less migration to the DF.
- There is a process of suburbanization with both medium and high-income planned settlements and unplanned low-income dwellers, at the expense of a light abandonment of the historical downtown district.
- Population for the year 2000 is estimated at 18 million people and, for the year 2010, at 20 million.

- Most growth will be in the SM as DF's spare capacity could be about 400 thousand, that could be handled through a higher density.

- There will be a process of activity restructuring with land-use changes as the population moves out from the central area to the periphery of the DF and to the VCT.

- Planning actions must aim to ease the transport problems of the urbanized area and transport policy should be used to shape its future growth.

- This extreme concentration of activity has generated excessive pollution and congestion, which has required the issuing of policies to decentralize and relocate some types of industry. Nevertheless, the MCMA remains the main focus of economic activity in Mexico.

Urban Public Transportation and Operations in Mexico City

Alejandro Villegas Lopez1

1. Urban Public Transportation System and Operations

Inhabitants of the Mexico City Metropolitan Area (MCMA) are served by several modes of transportation. The MCMA includes two jurisdictions. One is the Mexico City Federal District (DF by its Spanish acronym) which is the capital city of the Republic of Mexico. The second includes 27 municipalities of the surrounding State of Mexico (MS), called Valle Cuautitlan Texcoco (VCT by its Spanish acronym). Transit vehicles include short units known as minibuses, buses, trolleybuses, light rail trains, as well as light metro and metro cars. Taxicabs and private automobiles are also popular. A brief description of the MCMA public transportation system is as follows.

1.1 Private operated minibuses (*colectivos*) and paratransit (taxicabs)

Because of their adaptability and flexibility as well as the deficiency of scheduled public transport, this mode has grown rapidly in recent years. From 1.8 *colectivos* per thousand population in 1981, the provision increased to 5.3 in 1984, and since then the numbers slightly decreased to 3.2. From their illegal beginnings in the 1960s, the *colectivos* have graduated from automobiles to nine-seater vans and minibuses operating with regulated concessions. In 1990, average route length was about 14.5 km and operating speeds averaged 25 km/h. Ridership grew from about 6.3 million trips per day in 1988 to 18 million in 1997.

The frequent and widespread services make the *colectivos* attractive despite the fact that they are sometimes overcrowded. While fares have been much higher since 1987 (four times higher) than scheduled services, this difference was only 20% in 1990, which accounts, in part, for the recent upsurge in use. Fares in the DF are currently between 30% to 130% higher than those of metro, trolley buses and buses operated by the Mexico City government. This difference is larger in the VCT as fares are up to 430% higher. The services are uncontrolled except for the self-protection of concessionaires.

At the present time, the *colectivos* and taxi service in the MCMA operate 54,000 minibuses and 106,000 taxi cabs. In 1996, on average, the vehicles in this fleet were 5.8 years old. By contrast, in 1991, average obsolescence was 11 years old. The capacity of more than 70% of these units is less than 12 passengers. The *colectivos* serve basically as feders and sometimes as strong competitors to the metro network and operate on almost 2,000 routes in a network with a total length of more than 22,000 km.

1Transportation Planning Director, Mexico Capital City Government, Alvaro Obregon 269, 9o Piso, Col. Roma C.P. 06700, Delegacion Cuauthtemoc, Mexico, D.F.

Most of these service operators are grouped together into informal organisations; however, development of these services is based on the man-bus system, which is characterised by the individual operation and administration of units by their owners.

This structure is inefficient from both the transport and ecological points of view. It is estimated that *colectivos* operations generate about 100,000 direct jobs. Finally, a 1997 survey shows that each colectivo unit has a daily ridership of 688 passengers.

1.2 The DF Bus System

The state owned bus company (R1OO) was created in 1981 with the integration of nineteen private companies which were deemed to be providing inadequate services. Of the 6,345 buses inherited, only 25% were operational and most of the 48 garages had inadequate engineering facilities. The first four years saw heavy investment in vehicles, depots and equipment. Substantial revenue support, which came from taxpayers money, was used to keep fares low.

While traffic on R1OO services was growing at 4.6% a year up to 1987, in 1994 R100 ridership fell to less than two million trips per day (6.7% of all MCMA trips) due to lack of buses and ever increasing competition by private operators. While some garages had a staffing ratio of under four persons per bus, the company average was 7.2 which indicates a considerable waste and over-manning in some parts of the operation. Also, wages and fringe benefits of R100's 22,000 employees averaged over six times the national minimum wage.

Owing to R100's economic, administrative and operational inefficiency, the capital city government declared the company bankrupt in 1995. Consequently, public bidding is now taking place to grant new concessions for the bus transit routes that were previously covered by R100. To date, concessions have been granted to three companies that operate 790 new buses running on 61 routes that are strictly controlled in their operations and development **(Table 1).** The city government also entrusted the decentralised entity that operates the light rail train and the trolleybuses (Sistema de Transportes Eléctricos, STE) with the operation of 169 articulated buses. Additionally, there are 1,200 buses privately operated in the VCT by informal organizations similar to *colectivos* mode.

Table 1. Buses in the DF

Company	Number of Routes	Buses	Zone
Consejo de Incautación (state owned)	119	800	All the city
Transporte y Servicios Terrestres G	25	240	South
Servicios Metropolitanos y Transporte 17 de Marzo	17	240	North
Autotransportes Urbanos Siglo Nuevo	19	310	Southeast

1.3 Trolley buses and LRT - (Servicio de Transportes Eléctricos, STE)

The first electric streetcar line opened in 1900 and by 1919, the network had grown to 286 km and was providing 359,000 passenger trips per day. At its peak, between the two world wars, the system provided 85% of all passenger trips and by 1944 the network had grown to 330 km and was providing over a million passenger trips a day (55% of all trips). In 1947, the two private companies were combined into the publicly owned STE which by 1956 had 275 PCC cars running over 260 km of routes and 161 trolley buses running over 87 km of routes.

The Electric Transport System (STE) currently comprises a trolley bus network that covers 17 routes and 410 km with a fleet of 399 units **(Table 2).** It is expected that 200 new trolleybuses will begin to operate by 1999.

The light rail service currently has a single 12-km line with 18 stations between Xochimilco and Tasqueña station on Metro Line 2. It operates on an exclusive right-of-way along the *Calzada de Tlalpan* and the *Calzada Mexico-Xochimilco* in the south of the city. It operates with 10 trains and it is improving its yields. Between 1991 and 1995, passenger use per vehicle-km grew by 300% from 5.7 to 18.2 passengers per vehicle per km and from 1,437 to 7,067 passengers per train per day. As with the trolley buses, the light rail system has a low rate of operating cost recovery, thus requiring a subsidy, which, in 1994, was equal to 86% of costs.

1.4 The Metro - Sistema de Transporte Colectivo (STC)

Opened in 1969, the Mexico City metro now comprises ten lines with 149 stations over a 178-km network. It carries approximately 4 million passengers every day. Since construction began in 1967, the metro has grown at an average rate of 6 km per year. Since 1986, the network has been expanded by 43%.

Table 2. Transit Modes Operated by STE

	Trolley	Articulated Buses	Light Rail Train
Network Length (km)	410	415	31
Lines/routes	17	8	1
Stops	1,137	1,185	18
Cars	399	170	10 trains

In 1990, operating costs were $0.23 per car-km while revenues were only $0.07 per car-km. Fares would have to rise to $0.04 per trip to cover operating costs. In 1994, operating costs were $0.27 per passenger, while fare was only $0.12 . Investment per trip is between 10 and 15 times that required for a bus system. The system employs a staff of 10,633, which gives a productivity of 66.6 car-km per employee per day. Trip lengths range from an average of 2.9 km on line 4 to 8.1 km on line 3, averaging 7.1 km overall. The incomes of 92% of metro users are three times the minimum wage (approximately $270 per month in 1998) and 10% drive to the metro station. The Mexico City subway currently charges a flat fare of $0.15 per trip.

One feature of the system is the high level of ridership on the three central lines (1, 2 and 3), which carry 2.6 million passengers a day (65% of the total), with an average passenger density of 40,287 per km.**(Table 3).**

1.5 Individual transportation

Finally, of a total of 29 million trips generated per day in the MCMA in 1994, automobile users, driving 2.5 million cars with occupancy rates between 1.2 to 1.8 make persons per vehicle 5.34 million persons per vehicle. The remaining 24 million trips are made in public transport vehicles. It is estimated that private automobiles represent 95% out of total MCMA passenger transportation fleet.

2. Use by people

If we assume that high capacity transportation modes should move most of the trips and conversely that low capacity vehicles should move the least of trips, we conclude that use of MCMA transportation modes has been distorted in the last years. This situation has worsened since 1986 when the number of 27 seats minibuses and 9 seats vans has increased for transit services while the share of trips done by Metro, LRT, trolleybuses and buses has decreased.

Table 3. Supply and demand in the Mexico City's Metro

Line	Length (km)	Trains	Daily Passengers	%	Passengers per km
1.Observatorio - Pantitlan	18.828	42	881920	22.6	46840
2.Cuatro Caminos-Tasqueña	23.431	40	953461	24.4	40692
3. Indios Verdes – Universidad	23.609	44	786887	20.1	33330
4. Martin Carrera – Santa Anita	10.747	9	90142	2.3	8388
5. Politecnico – Pantitlan	15.675	17	227035	5.8	14484
6. El Rosario – Martin Carrera	13.947	12	124604	3.1	8934
7. El Rosario – Barranca del Muerto	18.894	18	215392	5.3	11400
8. Garibaldi – Constitucion de 1917	20.078	21	203676	5.1	10184
9. Pantitlan – Tacubaya	15.300	19	319862	8.1	20906
A. Pantitlan - La Paz (light rail train)	17.000	19	220134	5.5	12949
TOTAL	178.000	239	3908447	100.0	208107

Note: The number of passengers is 4,023,113 due to the fact Line 8 was studied from January to December 1994.
Source: Secretaría de Transportes y Vialidad, GDF., 1997.

In 1986, modal split share of medium and high capacity vehicles was 65% but, in 1998, it has fell below 20%. This situation brings important negative externalities such as congestion, inefficiency and air pollution.

The main conclusions about current modal split in the DF are as follows **(Chart 1)**

•14% of total daily trips were made using the metro

•Bus, trolley or light rail train made only 3% of total daily trips.

•59% of total daily trips were made in minibuses with a vehicle capacity less than 30 passengers per unit.

•20% of total daily trips were made on individual automobiles.

Chart 1, Evolution of the Modal Share in the DF

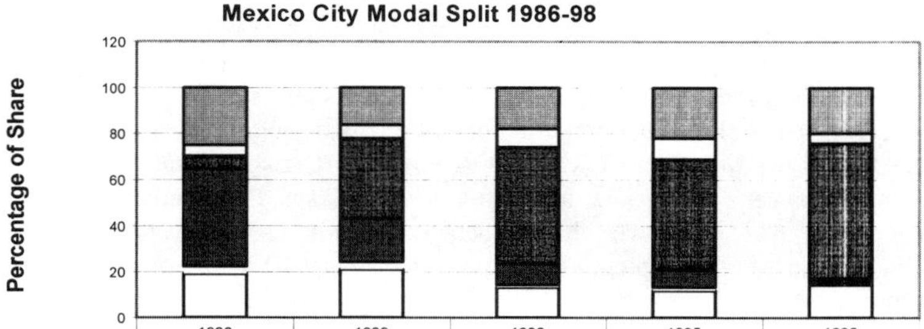

Mexico City Modal Split 1986-98

Percentage of Share

	1986	1989	1992	1995	1998
Individual automobiles	25	16.3	17.8	22	19.9
Taxicabs	5	5.9	8.2	9.3	4.4
Minibuses	5.5	34.6	50.7	47.8	58.6
Buses	42.2	19	9	7.6	1.9
LRT & Trolleybus	3.2	3.3	1.1	1.5	0.9
Metro rail	19.1	20.9	13.2	11.8	14.3

Year

Source: Programa Integral de Transporte, 1999. Secretaría de Transporte y Vialidad, GDF.

Organization and Financing of Transport

The Metropolitan Situation

While the DF and the VCT is a self-contained area, differences in transport resources and policy in addition to inadequate coordination mean that quite different service quality and fares are to be found in the two jurisdictions. The governmental structure is very complex with four levels ranging from the local municipalities state, Capital City and the federal government. In addition, there are a number of functional agencies with overlapping, and sometimes conflicting responsibilities.

Since 1978, both the DF and the MS have made major efforts towards achieving a global approach to metropolitan transport. These initiatives were limited, however, centering its efforts on the consolidation of an advisory body.

Later on, a series of studies emphasised the need for metropolitan-wide coordination from which in 1988 a metropolitan council (Consejo del Area Metropolitana, CAM) was formed to coordinate programs of the DF and the MS and subsequently a metropolitan transport authority (Consejo de Transporte del Area Metropolitana COTAM) evolved.

The legal framework has placed significant restraints on forming the metropolitan transport system. This situation has been improved by the recent amendment to the Mexican Constitution which provides the basis for establishing service-related commissions. Consequently, in 1995, the Mexico City Metropolitan Area Traffic and Transit Commission (COMETRAVI) was established with the participation of authorities responsible for the planning, operation and regulation of roads and transport in the DF and the MS. The COMETRAVI now also has a permanent Technical Staff comprised of experts in the area, in addition to facilities and equipment that enable it to carry out its tasks. Moreover, it has benefited from resources and credits for the implementation of a Comprehensive Road and Transport Plan for the Mexico City Metropolitan Area, which proposes several strategies aimed at interconnecting transport and air-quality needs.

Transport Financing

The DF spends about ten times more than the MS on transport as a percentage of its total budget. Much of the DF expenditure goes to public transport subsidies.

Automobiles are subsidized through the construction of roads and relatively cheap gasoline prices **(Chart 2)**.

Chart 2. Gasoline prices in selected countries

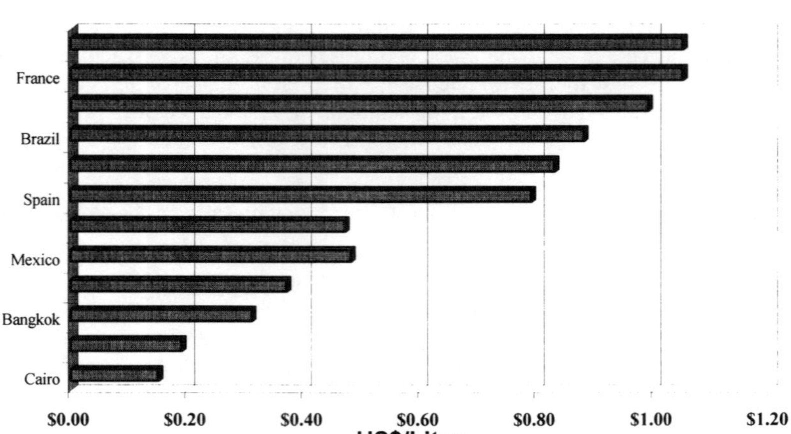

Furthermore, they do not completely pay for the high pollution costs they impose on the MCA's environment. State-owned transport operators do not cover their operating costs from fares, therefore subsidies are paid by the DF. Investment in infrastructure for transit facilities is funded by the DF fiscal budget, public debt and decreasingly by the federal government. Privately owned buses both in the DF and VCT do not receive operating subsidies and thus provide a poor level of service in terms of comfort and regularity. Public transport subsidies mean low fares, which makes the decisions regarding location, play little regard to transport costs, thereby generating unnecessary travel and creating additional demand. Furthermore, state-owned operators charge flat fares disregarding trip-length. Finally, the subsidies distort investment decisions, for example, in stimulating demand for new metro lines, when other modes might better meet transport needs.

Subsidies in the DF tend to take away funds from productive investment and show a significant bias towards the metro and light rail train (76 % of the total). Fares have risen within the DF from about $0.12 in 1994 to $0.15 in 1998. Fares charged by the modes operated by the Mexico City Government account for 4 % of the daily minimum wage per trip. Private *colectivos* charge 10% as fares in the VCT and reach up to 23% of the minimum wage per trip.

According to the 1994 origin and destination survey, which considered only trips made on one mode of transport, the average cost of a trip on private operated transport was almost three times higher than on services provided by the DF government. In turn, the cost of travel on suburban routes was 3.5 times greater, which is consistent with the longer trips. **Chart 3** shows average trip cost by transport mode in 1998.

Chart 3. Average Trip Cost by Mode of Transport

In 1998, the average cost on privately operated transport was up to four times higher than on the modes of transport operated by the DF Government

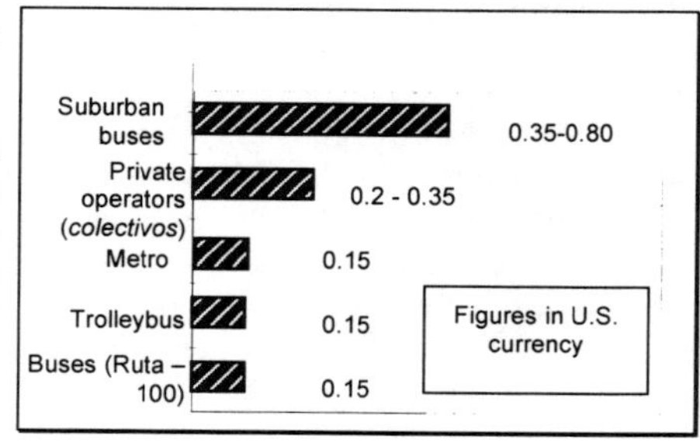

Fares on transport concessions vary in accordance with the trip distance. In 1996, fares in the DF were $0.15 for a trip of up to 5 km, $0.20 for between 5 and 12 km and $0.25 for more than 12 km, as **Table 4** shows. The minimum fare in the State of Mexico was 20% higher: the initial fare was $0.18 for the first 5 km, with an additional cent for each extra kilometer.

Table 4. Private Operators Fares

	Fare (US$)					
	1994		1996		1998	
Distance Travelled (km)	DF	MS	DF	MS	DF	MS
0 – 5	0.17		0.15	0.18	0.20	0.35
5 – 12	0.23		0.20	1 cent for each additional kilometre	0.30	0.40
More than 12	0.26		0.25		0.35	0.80

Source: "Main Results of an Evaluative Study of Collective Transport (Microbuses)," a study prepared for the Commission of Representatives of Public Transport Concessions. Consolidated average of the financial model. November 1994.

Among private operators (*colectivos*), deficient income management and organization constrain both savings and ability to contract debt in order to invest in new equipment and suitable maintenance facilities.

On the Threshold of the Year 2000

The rapid growth of the MCMA has led to a number of adverse economic, social and environmental consequences. The transport infrastructure and services differ between the DF and VCT, leading to considerable inefficiencies and inequality in the supply. The growth of the transport problems has been such that they have left behind the solutions proposed to deal with them. This has meant that planning has concentrated on dealing with immediate and pressing problems rather than looking forward to future needs.

Continued population growth will cause an increase in of the MCMA's transport problems, especially in the MS municipalities. Shortage of resources will require better use of existing transport facilities rather than massive investment in new technology and infrastructure. Effective coordination and cooperation between the DF and VCT authorities in relation to transportation will be essential. Without some sort of new organizational arrangements, no amount of investment will be adequate.

This reorganization should embrace all the institutions, authorities and companies concerned with transport within the entire MCMA. Operations, regulations and finance should all be included with a view to the provision of more equitable and efficient mechanism which will integrate the effort of both public and private companies. The main elements of this should be strengthened.

Control and Organization. There is an urgent need to consolidate the numerous organizations that have developed through the years of reactive planning. Informal organization of *colectivos* is an obstacle for efficiency and quality of transit services. An increased role for the private sector is essential but within a considered and co-ordinated framework covering both the public and private sectors. The development of substantial and well-managed operating companies competing for the market not within the market must be promoted.

Fares and Finance. Inequalities in fares and structures between the DF and VCT, resulting largely from different fiscal policies, need to be addressed. Imaginative solutions must take into account the increasing distances from the outlying and generally low-income areas to the centre. Fares charged by the modes operated by the Mexico City Government should cover operational costs as well as the travelled distance; however, the low income of passengers using these modes impose social restrictions to achieve these goals. Improved efficiency is needed to reduce operational costs, but the continuing need for resources to improve services must focus attention on new methods of financing for the MCMA as a whole, as well as ensuring that automobiles and colectivos make a greater contribution towards covering the actual social costs they are imposing.

Pollution. The greatest source of atmospheric pollution in the MCMA is from the large number of gasoline powered automobiles and microbuses. This is a matter of considerable public concern and both the DF and VCT authorities are acting to reduce emissions. This involves restricting the use of older cars one day in a week, and a progressive vehicle emissions control, as well as inspection and maintenance programs. Besides this, new natural compressed gas stations are scheduled to start operations during 1999 in order to fuel suitable transit vehicles that would alleviate pollution generated by transportation.

Physical and Operational Integration. Duplication and poor integration of public transport operations creates excessively long transfers and this needs to be improved. This is also a factor that negatively affects ridership in some metro lines. In particular, suburban bus operations should serve more metro terminals and penetrate further into the DF to complement bus services. The road network would improve with unified standards on both sides of the DF boundary. Integration of the urban transport system with the existing metro rail network should be studied in the context of planned urban development and the development of new services.

Metropolitan Coordination. The disparity in transport policy and resources between the DF and VCT cannot be dealt with solely through voluntary co-ordination. It will require changes in the power and responsibilities of the DF and the MS. Failure to resolve this issue has been the main obstacle to setting up an effective metropolitan authority. The unplanned expansion of the urban area without making provision for the necessary transport facilities must also be reviewed. Proper co-ordination between officials in charge of air quality, urban development and transportation is also required.

Information and Planning. The different agencies concerned with transport have different practices in data collection and processing, which means that several conflicting sources of transport information may exist. This makes sound analysis, planning, making public policies and evaluation extremely difficult. New metro lines should be carefully reviewed in order to overcome current underutilization of some existing lines. Better coordination of data collection and the training of technically competent managers for a strong non-political planning unit within the MCMA are necessary. Forecoming creation of the Mexico City Urban Transportation Institute, along with the Mexican National Autonomous University, will be an important step towards this end. Suitable planning tools are also needed, but those used in developed countries should be looked at critically as they are not necessarily oriented towards the real needs of the MCMA.

High Crime. In recent years high crime rates in the MCMA have reached the on-board safety of transit passengers, especially those riding taxicabs and *colectivos*. While specially tailored actions have to be implemented, the Mexico City Government has implemented a special program aimed at taxicabs, consisting of establishing safe spots where vehicles equipped with radios and passenger protection facilities are operated.

Alternative Fuels and Institutional Reform

Alejandro Villegas Lopez[1]

Abstract

This report addresses the actions that Mexico City Government is currently implementing. At present Mexico's Capital City Government is new since it is the first time that its mayor or chief executive is directly elected by voters. By the same token, investment as well as operational costs for public goods and services have to be locally financed; public debt has to be approved by the national deputies chamber; the federal government has reduced its financial support, and current government term is limited to three years (1997-2000).

Mexico City has also been facing some challenges during the last decade: overpopulation and high urban density, congestion, air pollution and high crime rates.

The altitude and topography of the Mexico City valley contribute to less efficient combustion, trapping toxic emissions generated from the tailpipes of more than three million cars, taxicabs, trucks, minibuses and buses running in the city.

However, in this context, there have been some opportunities to formulate new policies for urban transportation. This paper focuses on some specific actions such as implementation of alternative fuels, decentralization of duties from the Mexico City Central Government into local jurisdictions and creation of the Mexico City Urban Transportation Institute.

1. Alternative Fuels: Use of Compressed Natural Gas for urban transportation

For 15 years, the Mexico City Metropolitan Area (MCMA) has had a severe air pollution problem. The negative impact on public health from the high levels of air pollution can severely affect the quality of life, health care costs, economic productivity and the general well being of the MCMA population.

---

[1]Transportation Planning Director, Mexico Capital City Government, Alvaro Obregon 269, 9o Piso, Col. Roma C.P. 06700, Delegacion Cuauhtemoc, Mexico, D.F.

The Mexico City valley is a victim of its topography. Its 2,286 meters altitude reduces the level of oxygen by almost a third. As a result, motor vehicles produce twice the carbon monoxide and hydrocarbon emissions than they would at sea level. The surrounding volcanic range traps the noxious fumes and stifles the flow of cleansing air currents.

In general, it is estimated that 75% of all toxic emissions generated in the MCMA are produced by mobile sources such as motor fuel vehicles, including private cars, taxicabs, minibuses, buses, as well as light and heavy weight trucks. **(Chart 1).**

Fuel consumption in the MCMA includes gasoline, diesel, liquid petroleum gas and compressed natural gas, 56% of which is used for transportation and the other 44% for industry, services and electricity generation*. It is estimated that nearly three million motor vehicles run the MCMA roads and streets every day; private cars accounting for 85%. Among urban transportation vehicles, 45% of emissions come from private cars and 22% from taxicabs. Low capacity transit vehicles operated by private informal organizations like minibuses and vans originate 16% of pollution. Another 3% are generated by transit such as buses, trolley buses, light rail train and metro rail. Delivery and loading vehicles such as light and heavy weight trucks generate the remaining 14%. **(Chart 2)** These figures show that gasoline powered vehicles are the source of more than 83% of emissions.

Chart 1. MCMA Emisions Inventory 1994

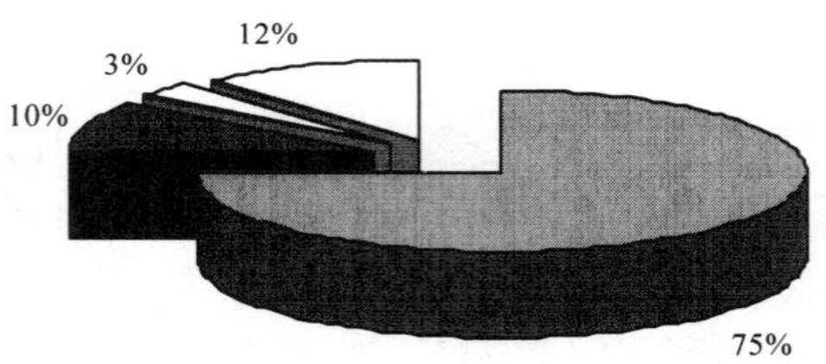

*Departamento del Distrito Federal, Gobierno del Estado de México, Secretaría del Medio Ambiente, Recursos Naturales y Pesca, Secretaría de Salud, Programa para Mejorar la Calidad del Aire en el Valle de México 1995-2000, FIDEICOMISO Ambiental, Mexicana Internacional de Impresiones, S.A. De C.V., México, D.F. 1997.

The air-pollution problem in the MCMA has two important factors that have to be taken into account. First are the specific geographical and weather conditions of the "Cuautitlan - Texcoco Valley" where the MCMA is settled, characterized by high solar radiation and low intensity wind flows. Both lead to the production of ozone and less favorable dispersion conditions. The second factor is a constraint due to the relevance of MCMA economic activities, which generate more than one third of the Mexican gross domestic product. Thus, the objective of reducing air pollution should be counterbalanced against the objective of encouraging the efficiency and development of the region's economy. Finally, a significant outcome from this constraint is that urban transportation supply has to meet the increasing urban passenger-trips-kilometers expansion. MCMA needs a policy to reduce toxic emissions generated by commuting, while maintaining economic development and urban transportation services supply.

Chart 2. MCMA Mobile Sources Emissions Inventory 1994

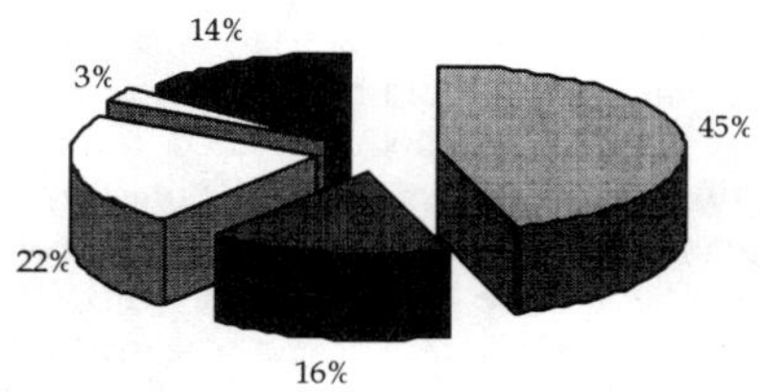

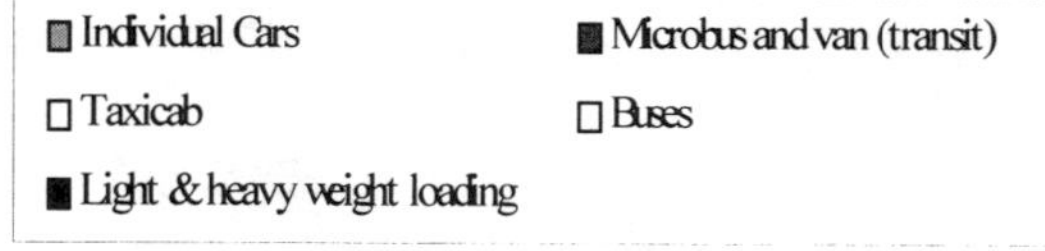

The main programs conducted by MCMA authorities, as responses to the air pollution problems are the following:

a) No driving day bans as a temporary measure while the others take effect. It consists of the restriction for a single car to be used once a weekday. The rationale for this measure assumes that 20% of the MCMA motor fleet stops running each day and average running speed is increased so the impact on pollution is reduced. The program is enforced through scheduling the ban based on the digits of the license's plates. High fines for those violating the restriction are enforced. Due to the fact that this program was continuously executed from 1989 to 97, the driving ban, as a permanent measure, turned into an incentive for motorists to pollute more rather than to control toxic emissions. It is so because many motorists started to buy a second car to be used during the day when the first car was banned to run. This second car usually was an older one and thus, producing higher amounts of pollution. Of course, this second car could be used the three remaining weekdays by a second family member.

Tolerable air quality standards were surpassed several days a year so, during this time, emergency actions were enforced including a second day restriction for each car to run instead of one. In the long run, the no driving ban policy will encourage motorists to invest in cleaner technologies.

As an alternative policy option it was considered that incentives for reducing pollution from private car commuting could be introduced by allowing newest car owners to run normally since their cars are fitted with three-way catalytic converters as well as fuel injection engines. Thus, the one-day without my car ban is still enforced for a second set of cars not fitted with fuel injection engines. A strongest ban is still applied for those cars not fitted with fuel injection engines neither with three-way catalytic converter. This set of cars, besides being enforced to stop running once a week, have to stop running during those days when the maximum air quality standards are surpassed as a part of the emergency actions taken.

b) Fuel Quality and Technology Improvement. In 1991, unleaded gasoline was introduced in Mexico as well as compulsory factory-installed use and proper operation and maintenance of three-ways catalytic converters. In 1992, LPG was introduced as a fuel for vehicles converted from gasoline powered to LPG, such as light and heavy trucks, taxicabs, and public passenger transportation vans and minibuses. Also, a pilot project was implemented to convert 50 police patrol cars into CNG; however, neighborhood opposition to operation of the supply station caused the plan to fail in 1994. In 1993, new models, factory-fitted with LPG engines, were produced. By 1997, some 23 thousand vehicles converted to use LPG as fuel were operating in the MCMA.

c) Compulsory Inspection and Maintenance Program along with gradually stricter standards so older vehicles without proper maintenance are pushed to stop running. For example, from 1975 to 93, tolerable emissions standards for new cars were reduced, in hydrocarbons (HC), from 2.5 to 0.25 grams per kilometer, and in carbon monoxide (CO), from 29.2 to 2.11.

These measures have had a certain impact; however, ozone levels still surpass acceptable air quality standards during 80% of the days of the year **(Chart 3).** At present, it has been considered to encourage another "clean" fuel such as CNG for fueling the most pollutant vehicle fleets such as those gasoline powered.

Chart 3. Ozone Trends 1988-95

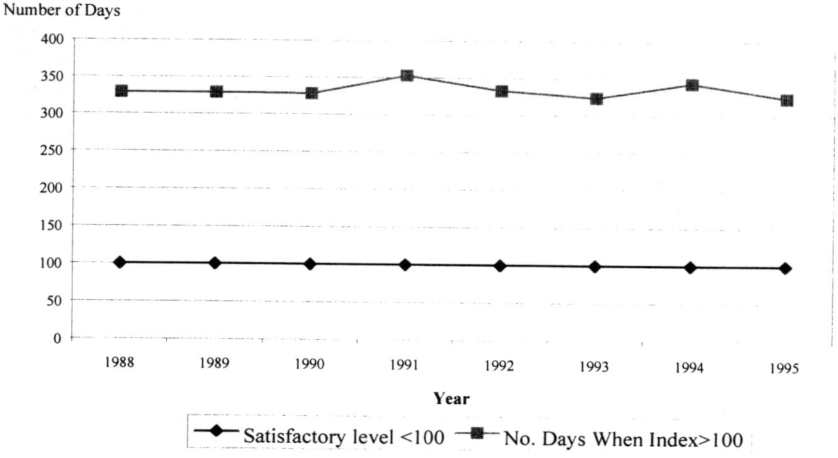

Engines fueled by CNG as compared with gasoline powered ones reduce carbon monoxide and hydrocarbon emissions by 90%, nitrogen oxides by 60% (only if fitted with a three way catalytic converter) and eliminate suspended particles, lead and sulfur dioxide. On the other hand, CNG engines as compared with diesel ones reduce those toxic emissions, which are ozone producer by 60%, as well as particles and sulfur dioxide emissions are almost eliminated. Briefly, it is considered that toxic emissions generated by CNG engines produce 90% less ozone and a 60% reduction in gases, which provoke earth overheating, as compared to those fueled by gasoline.

Sets of incentives have been established as a part of a policy to encourage the use of CNG for urban vehicles. First, the federal government enacted a Law by which the price of the non-compressed natural gas is fixed in order to be equal to 64% of gasoline price independent of the international price of gas. Second, an accelerated amortization of 100% is allowed for those vehicles converted to CNG. Finally, vehicles fueled by CNG will not be considered within the one-day without my car program.

The plan is focused, at its best, to those high mileage vehicles to be fueled by CNG. This fleet includes, for transit, 27 thousand microbuses and five thousand buses, as well as 89 thousand taxicabs. For freight and delivery, 338 thousand light truck vehicles. Among them, it is expected to convert or introduce 10 thousand vehicles in the short term and 45 thousand in the next five years. At present, 100 police patrol vehicles are fitted with dual gasoline/CNG engines so they are currently being fueled in the first CNG Station located in the close Naucalpan municipality, within MCMA.

The MCMA has a natural gas distribution pipe of 206 kilometers crossing, from west to east, the northern part of the city. This is part of the Mexico State plan to convert some five thousand gasoline powered microbuses to CNG. The company owning the station is financing the conversion and private operators are paying back the credit by being charged of an extra price for the fuel.

On the other hand, the Mexico City strategy has been to grant new bus transit routes concessions through bids that include preference for participants that include in their proposals buses by CNG.

2. Institutional Reform

A) Decentralization towards autonomous municipalities

Before 1997, Mexico's Capital City government used to be run by a chief executive appointed by the President and a centralized public administration including the division of its territory into 16 boroughs or jurisdictions called *delegaciones*. Present government is the first in our history directly elected by Mexico City voters as a part of important nation-wide reforms that are taking effect in Mexican political institutions. Another important implication is that we are witnessing a process of decentralization in which two steps can be distinguished: first, decentralization of power and duties from the federal level into the city level of government. Second, there is a movement towards further decentralization to convert current centralized boroughs into some sort of autonomous municipalities. In fact, recently enacted law considers direct election of local jurisdiction chiefs for the year 2000.

This reform is important in improving city government responsiveness since Mexican Federal District (DF by its spanish acronym) is a big capital city as to be centrally managed. For example, local boroughs' population is as big as Iztapalapa with 1.6 million or Gustavo A. Madero with 1.2 million, along with another four boroughs with a combined population of over half a million people each; the remaining 10 boroughs account for populations between 81 to 485 thousand people.

Regarding the urban transportation authorities, during the eighties a new type of traffic and transit committee was implemented within five of the 16 boroughs. However, the centralized type of government of the former Federal District Department, lack of economic and human resources, and indifference from public officials led to a failure in their implementation.

Currently, the scheme for meeting citizens demands for attention is shown in **Figure 1.** This process illustrates that, for those traffic and transit matters whose impact is local, it is useless to have them solved at the central level. On the other hand, their attention by the Transport and Traffic Secretary involves time and resources that the government at the city level requires for general and strategic planning, programming and policing functions.

Figure 1. Current process of response for citizen's transportation demands

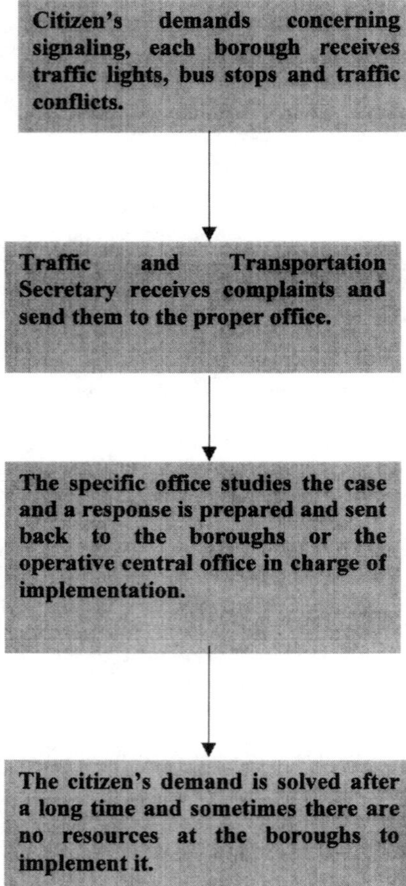

The new General Development Program issued by the Mexico City Government in 1998 establishes decentralization of some administrative duties as one of its main strategies. This strategy was taken into account by the local deputies' assembly, which issued, by late 1998, a new public administration law that confers greater power and duties to the local jurisdictions or *delegaciones*. This is so specifically in the fields of public parking lot regulation and fare enforcement, under the central supervision, policing, technical manuals and rules by the Mexico City Transit and Traffic Secretary.

In this context, the Mexico City Transit and Traffic Secretary is considering the implementation of local traffic offices within the 16 *delegaciones* or boroughs. There are several challenges such as limited budget for the real needs of each office, lack of general application rules, regulations and handbooks as well as high demand for meeting traffic and transit needs. For the near future, it is intended that those boroughs' offices will solve typical local impact traffic problems. A complete road inventory has been made so tertiary road operation and maintenance should be fully administered by local offices. In this local road network, borough offices will be in charge of changing traffic flow directions, minor improvements to road's geometry design, traffic lights operation, coordination and maintenance. Also, such offices will conduct traffic impact analysis for projected changes in urban land use.

With the decentralization of some functions into the *delegaciones,* better rates of response are expected for local demands as shown in **Table 1**:

Table 1. Expected DF government response rates.

	Before Implementation (Percentage)	**After Implementation (Percentage)**
Response of Federal District Transit and Traffic Secretary	62	95
Degree of Implementation	30	80
Degree of response	18.6	76

This decentralization is based on three points:

To create local offices for local traffic and transportation issues in each of the 16 boroughs in order to decentralize those activities which impact is only concerned with the tertiary road network. It is also an opportunity for the authority to come to a local level of response and performance.

The central Transit and Traffic Secretary continues to do strategic and long scale planning, coordination among cross-boroughs issues, programming, policing, regulation, operation and assessment of the overall sector.

Coordination and arbitrage among local offices will continue to be conducted by the Transit and Traffic Secretary.

B) Urban Transportation Institute

Problems in the urban transportation field have different perspectives: legal, administrative and economic, as well as technological and technical. It is important to note that the latter three aspects are very important. For example, suitable planning and decision making involves complex field surveys, comprehensive data collecting, database building both statistical and geographic, tax incidence analysis of policies and decisions and modeling tools that require a high degree of specialization regardless of political and administrative constraints. Applied research and training is required as well.

Institutional reform that is taking place within the Mexico City government also includes considerations regarding long term required decisions for the Capital City's urban transportation development. Proper continuity for implemented policies regardless of political changes in the local executive and legislative officials is necessary too. For this, institutional, human and technical resources are needed having characteristics of certain degree of specialization, expertise and autonomy from the political power and dynamics.

For this reason the Mexico City Government has signed an agreement with the Mexican National Autonomous University (UNAM by its Spanish acronym) in order to create the Mexico City Urban Transportation Institute. The main duties of this organization will be to develop research, training and high-tech consulting for the Mexico City Transit and Traffic Secretary as well as for other actors such as public and private operators and suppliers. A unique and broad information system, including a self-tailored urban transportation geographic information system, and permanent statistics database is considered to be built to help daily planning and making decisions.

The organization of the institute is thought to be self-sufficient and autonomous from the Mexico City government so it could become a necessary technical independent urban transportation authority.

As a part of the fiscal year 1998 budget an allocation of 800 thousand dollars was granted to the National University in order to set up the organizational and legal design of such an institution. Also, the transportation institute founding staff is already conducting some studies and surveys necessary for the planning process of the Mexico City Transit and Traffic Secretary. The urban transportation institute is settled in the graduate engineering school, however the idea is that the institute could be fed and linked by other schools such as architecture, economics and geography according to needs. Other important connections will be available to such academic institutions as the National Polytechnic Institute and the Metropolitan Autonomous University, which are also settled in Mexico City and have related graduate schools and programs.

SUMMARIES OF SPEECHES AT LUNCHEONS, DINNERS AND PANEL DISCUSSIONS

Ramakrishna Reddy Tadi[1]

Monday, March 22, 1999

Luncheon Speaker: George Wynne for Bill Millar of APTA

Wynne talked abmout the benefits of exchanging the experience in transit with other colleagues fro around the world. APTA, USDOT, and FTA have recognized this and TEA21 legislation provides such an opportunity. Wynne indicated that APTA and other federal agencies are promoting international workshops and sessions at their conferences that were bringing overseas practitioners. In Japan, for example, international study trips are an accepted part of the learning experience for government and industry officials and the cost of the trips are usually a line item in public and company budgets, whereas in the U.S., such trips are often viewed with considerable suspicion. Wynne feels that the innovations, adopted new products and processes that have been introduced by North American operators are direct results of their international travels and contacts. Some of the smart overseas models, which are being used to obtain private resources to implement public transit infrastructure, are gaining increasing acceptance in this country. For, example, FTA is currently promoting the concept of Bus Rapid Transit, a low cost alternative to commuter rail systems, is largely based on the successful 30-year experience of Curitiba in Brazil, according to Wynne.

Wynne also gave a few examples from Europe. MBTA's low floor cars are based largely on the European findings. Other examples are their management structure and automated fare collection. He concluded his presentation by referring to Shirley De Libero's (APTA's chair and CEO of the Houston System) study mission to Sweden, a couple of years ago, where most bus and rail lines in the Stockholm metropolitan area are operated under contract – a change that has produced savings of $145 million a year and garnered a cut in governmental subsidies.

Dinner Speaker: Professor Tony Ridley
Head, Department of Civil and Environmental Engineering
Imperial College, London, England

Prof. Ridley presented a historical perspective of Civil Engineering, UK Institution of Civil Engineers in particular over the past 150 years. He mentioned that engineers need to be political scientists in order to be successful. Only engineers can solve the world's environmental problems.

[1] California Department of Transportation, 464 W. 4th Street, San Bernardino, CA 92401, USA

According to Ridley, the U.K. Institution of Civil Engineers is the oldest professional institution in the world (founded in 1818). London Underground is the oldest, having started operation in 1863, followed by Budapest. In its 1995 publication, the Institution of Civil Engineers has indicated that civil engineers are responsible for much of the essentials of modern life: the muscles and senews which hold our society together (bridges, roads, railways, dams, airports, docks, tunnels); the provision and maintenance of its heart and lunge (clean water, natural resources in, waste out); transport for safe and efficient movement; and energy to make it all work (offshore gas and oil, nuclear, hydro, tidal, and wind power). Ridley indicated that for too long we, engineers have been in danger of taking too narrow a view of Civil Engineering. Simply to plan and design infrastructure is not good enough. Ultimately, implementation must be our aim – getting the right things done well. This involves according to Ridley, taking account of a whole range of issues including the political, financial and managerial. The world is increasingly global, and not just for the world's leading commercial business. Much of the wealth of developed and developing countries is created in major urban areas. Good transport system is necessary, though not sufficient, condition for wealth creation. Thus rail and other transit have a purpose greater than simply to improve mobility as per Ridley. He believes that we should recognize the 'pentagon' of issues involved in the development of transit projects: 1) hardware - conventional engineering, 2) software - personnel, marketing and other 'soft' issues, 3) ecoware - environmental matters, 4) finware - finance and funding, and 5) orgware - organizational and institutional matters. All of these are truly engineering today. In UK the government has offered a definition of 'Integrated Transport' a much used but little understood term. We want a transport system that is safe, efficient, clean and fair. A new approach, which brings together the public and private sectors in a partnership and benefits everyone, is needed. He gave the London Underground example and discussed other approaches using Singapore and Hong Kong examples.

There is nothing necessarily bad or good about a system, which requires operating subsidy on one hand or can be fully funded by private investors on the other. What tends to happen however according to Ridley is that great debates take place between these two extremes with little or no recourse to facts. As an example, during the course of the development of an urban rail project capital cost estimates will be made, though not often is any risk analysis carried out at the same time with proper estimation of contingency sums. Operating and maintenance costs will have been forecast, but the errors associated with these are frequently greater than those for capital costs. Rarely if ever, is any simple overall financial analysis carried out at an early stage. It can, and should be done while recognizing the uncertainty about figures at that stage. What is required is a 'ball-park' feel for the place of a project in the financial spectrum. This will, and should, influence design, procurement, political and many other decisions about the project, Ridley concluded.

8:30 pm – 10:00 pm
Panel Discussion on "the Future of Public Transportation"
Moderator: George Wynne, APTA

First Speaker: Dr. Murthy Bondada of Mehta & Associates, Inc spoke about Global Perspective on Mass Transportation Trends. Mass transportation use in a country depends on public affluence, urban population, size of the cities, and finally on urban transportation modes. As per the U.N. there are 45 developed and 125 developing countries in the world. The population of the developing countries is expected to increase by 2.07 billion (4.75 billion in 1998 to 6.82 billion in the year 2025) whereas in case of developed countries the growth is only .04 billion (1.18 billion in 1998 to 1.22 billion in 2025). The majority of the growth in developing countries will occur in urban areas. The U.N. sources also indicate that 20 cities with a population of more than 10 million will be concentrated in developing countries whereas only 6 cities in developed cities by the year 2015. Overall, out of 15 of the world's largest cities, 13 will be in developing countries and only two in developed countries by the year 2015. Bondada feels that due to limited operation and support, mass transportation in urban areas of large developed countries has no effect on global mass transportation trend. In developing countries, 60 to 90 percent of total trips are by transit. Infrastructure development is not in line with the rapid growth of automobiles in these countries. Mass transportation policies in most of the developing countries can be classified either as imitation policies or independent policies. Bondada concluded his presentation by discussing the pros and cons of each of these policies.

Second Speaker: Edward Thomas of the FTA gave an overview of current Public Transportation in the United States and future challenges/opportunities of the 21st Century. He indicated that safety and security, mobility and accessibility, economic growth and trade, human and national environment, and quality organization are the strategic goals of the U.S. Mass Transit. There was a slight increase in transit vehicle revenue miles by mode between 1993-1997. But, transit share of all trips has decreased eventhough transit ridership is relatively stable. TEA-21 allows flexibility and guaranteed funding for preventive maintenance, system integration, technological innovation, professional development for transit, and major capital investment for preliminary engineering of rail. Urban sprawl & congestion are some of the major challenges of the 21st century, according to Thomas. Between 1989-1993, AVR was down whereas VMT was going up rapidly. Average speeds for rail were down slightly and relatively constant for buses. He also talked about global warming. Integration with diesel, electric motor, battery, hybrid & inter-operability are the policies of FTA. Advanced automatic train control, new bus technologies (fuel cell, alternate fuel) and ITS are some of the examples of innovations in transit research and technology. Thomas concluded his presentation by talking about mobility management innovations such as coordination of services, regional traveler information, regional electronic payment, and vehicle-to-vehicle communication.

Third Speaker: Pierre Laconte of the International Association of Public Transport, Brussels, Belgium spoke about Worldwide Perspective on Urban Public Transportation. Approximately 45,000 in European Union countries and 80,000 world-wide die every year in traffic accidents. Laconte talked about CO_2 emissions from the transport sector and compared energy and emissions of different modes. He also discussed about congestion in Singapore, ozone and smog problems in Germany as well as effects of human actions and natural disasters on worldwide temperature increase. According to Laconte, U.S. produces maximum amount of CO_2 from fossil fuels. He further talked about urban space and gave examples of Zurich, Singapore, and Krakow where authorities have successfully implemented parking control schemes in parallel with public transport improvements. Laconte strongly believes that the use of valuable space by the car can and should be charged for. The U.K. Government's 1995 SACTRA Report concluded that most roads generate additional traffic. Due to this generated traffic, new roads rapidly become as congested as the existing ones. The new policies developed by the UK Government towards giving the car its place but not a dominant one, are in line with most recent best practices worldwide, according to Laconte. In Germany and Switzerland, the growths in GNP and in car ownership have not resulted in the decline of public transport but in its increase in parallel with the use of other transport modes. He also discussed about financing urban projects and the need to improve public transport through the use of information technology. New public transport infrastructure will usually provide the additional capacity at lower overall cost than equivalent new road capacity.

Final Speaker: Jean-Claude Ziv of Conservatoire National des Arts et Metiers, Paris, France spoke on Urban Mass Transportation Trends in Europe. More specifically, Ziv talked about transit system & travel patterns. He also briefly discussed about coordination and comprehensive planning for mobility enhancement in the European Countries.

Tuesday, March 23, 1999

8:30 pm – 10:00 pm
Panel Discussion on "Innovative Finance in Urban Transportation"

Moderator: Walter Kulyk
Director, Office of Mobility Innovation
Federal Transit Administration
Washington, D.C. (USA)

The first speaker of this evening was Richard Podolske of the World Bank, who spoke on Innovative Finance Strategies for Worldwide Public Transport Projects. He gave several examples of World Bank financing projects. Private investment flows to infrastructure projects average \$60 billion annually and more than 80 countries have nearly 600 active private infrastructure projects mostly in electric power and gas, telecommunications, sanitation and environmental services, water supply and sewerage, transport according to Podolske. In case of urban transport, financing includes for urban buses, urban metro and rail as well as urban road system. Podolske indicated that the role of government is shifting from infrastructure an services provision to facilitation and innovation. Diverse financing techniques are required to support this transition. In case of traditional government financing, government bears nearly all responsibility and risk whereas private sector allows for more efficient provision of infrastructure and services. Podolske further discussed various public/private options such as public ownership and public operation, public ownership and private operation, private ownership and private operation. He discussed different loan options available and gave examples of Buenos Aires Metro, Bangkok "Sky Train", and World Bank financing of bus services in Russia & Central Asia. Podolske concluded his presentation by indicating that engineers, public transport officials, and financiers will need to adapt as private sector provision of public transport infrastructure and services is growing rapidly.

The second speaker Mark Massman of Bechtel Enterprises, Inc., spoke on public/private partnerships in urban transportation. The focus of his presentation centered around projects of Bechtel. Bechtel is currently involved in Portland, Oregon. One such project is construction of the $125 million 5.5 mile extension of the existing Portland LRT System known as MAX to Portland International Airport (PDX). The expected completion date of this design/build project is December 2000. The second project, which Massman talked about, was 120-acre transit oriented, mixed use Portland Industrial Center (PIC) development project. The LRT extension demonstrates first of its kind in the U.S. to use cost sharing approach with private sector according to Massman.

The third speaker, Benjamin Redd of Raytheon Engineers and Constructors spoke on Design, Build, Operate, and Maintain (DBOM) concept in urban transportation. The United States, like many other countries, is just beginning to revitalize its infrastructure and transportation systems through public/private partnerships in order to sustain economic growth. Redd briefly discussed five different project delivery methods namely Design/Build (D/B), Design/Build/Operate/Maintain (DBOM), Build/Operate/Transfer (BOT), Build/Own/Operate (BOO), and Build/Transfer/Operate (BTO) with tables and charts. The inclusion of operation & maintenance in DBOM adds the combined benefits of innovation, quality, and life cycle costing according to Redd. All the above mentioned models offer significant advantages to clients in varying degrees (schedule savings, cost savings, etc.). For example, due to the elimination of multiple procurement cycles, savings in schedules can be realized as multiple tasks can be run concurrently instead of sequentially. Redd also gave examples of how D/B can save millions of dollars when compared to traditional way. Design/Build not only allows innovative financing options but also reduces construction completion risks to lenders thereby increasing financing options. Public/Private partnerships are becoming a necessary method of delivering transportation projects as traditional funding limits the number of projects a state can undertake. A critical component of successful partnerships is a reasonable allocation of risk. Project risk should be allocated to the partner best able to control and manage it. The bottom line for all parties, if used any of these models according to Redd, is that initial reduction in capital cost, reduction in life cycle cost/labor needs, improvement in operating efficiency and productivity, early utilization of system assets, and increased potential return on investment.

The final speaker of the evening was Thomas Bradshaw of Solomon Smith Barney, New York. The theme of his presentation was on private investments for public transportation in the United States. Bringing on-board the financial people early on in the process is most critical for the success of any project. Design/Build is the best option and solution resulting in under-cost and on-time for the highway projects according to Bradshaw. He explained revenue bond structure. He gave an example of the flow of Federal grant funds and indicated that nationally $41 billion are available under TEA21 for various transit related projects. Bradshaw feels that transportation is a good investment.

Questions & Answers:

In response to some of the questions from the audience, Podolske indicated private sector participation is more appropriate in developing countries due to weaker government funding. In response to another question, Tom Bradshaw gave the Canadian example where the government built and leased to private company to operate. Benjamin Redd feels that in spite of different tax structure in the U.S., Build/Operate/Transfer strategy can be equally applied in foreign countries.

Wednesday, March 24, 1999

Luncheon Speaker: Michael Bolton, Multisystems, Inc., Cambridge, MA (USA)

Bolton feels that systems need to be designed with people in mind. He gave an example of Ann Arbor, Michigan SMART Bus system. This system was initially started to meet the staggering ridership needs. Several surveys revealed that people do not use transit due to lack of accessibility and reliability. Bolton gave an interesting comparison of capacities between a bus and an automobile. He questioned why a bus with 30 persons is often equated with only 2 cars while performing the intersectional analysis. He strongly believes that a dialogue is necessary between the transit industry and traffic engineers and ITS could just do that. Bolton sited several examples from Nagoya (Japan), Curitiba (Brazil), and Puerto Rico of real time data transfer. He also briefly discussed about SMART CARD technology in his presentation and explained how Hong Kong public transit system is so critical to its population. Customer is more important and needs to be better served. Any amount of technology is a waste if customer needs are not addressed, according to Bolton.

1:15 pm – 2:45 pm
Panel Discussion on "Innovative Technologies and Techniques"
Moderator: Stephen Andrle, Manager, Transit Research Cooperative Research Program, Transportation Research Board, Washington, D.C. (USA)

First Speaker: James Larkins of Georgetown University spoke on Fuel Cell Development for Buses. He gave an overview of the world-wide trends in transportation fuel cells. Larkins presented some examples of fuel cell buses and explained measured vehicle emissions for various fuels (diesel, CNG, Methanol, etc.) as well as other features. Georgetown University's fuel cell powered transit bus program is supported by a grant from the FTA. It has introduced the first commercially viable, liquid-fueled, fuel cell powered transit bus at the APTA's Bus Operations Conference in May 1998. The fuel cell for this first commercial transit bus is derived from stationary electric utility power plants. Larkins presented a comparison of CO_2 emissions from a 30ft fuel cell bus to other types of buses. He concluded by saying that a proton exchange membrane fuel cell bus, which is currently being tested, will rollout in 1999 completing a stable of five fuel cell buses developed by Georgetown to foster the commercialization process.

Second speaker: John Wilson of Southern Coalition for Advanced Transportation in Atlanta discussed the recent developments in alternate fuel cells. Some of the main objectives of the consortium, which currently has $60 million funds from 5 states in 50/50 cost shared projects, are in the areas of research, design and development, technology development, and commercialization. TEA 21 authorizes $50 million per year in these areas. Wilson also talked about hybrid electric development and advanced vehicle technology programs. He presented several examples such as electric shuttle buses in Miami Beach and Chattanooga. Electric buses in Miami Beach, which resulted in huge ridership increase in addition to reduction in the auto traffic. Merchants were really thrilled.

Third Speaker: Alan Rumsey of Parsons Transportation Group spoke on Communication Based Train Control (CBTC), more specifically 1) characteristics and advantages, 2) implementation issues /standardization, and finally 3) New York City Transit Canarrie Line Pilot Project. Rumsey for example, indicated high resolution train location, wayside data communication, automatic train protection, reduced headways/ energy management, optimized schedule regulation are some of the advantages of CBTC. The goal of the Canarrie line project in the New York City was to implement the CBTC with minimal disruption. Rumsey concluded his presentation by reviewing the project implementation challenges.

Fourth Speaker: Hideo Hirasawa of HSST Development Corp., Tokyo, Japan discussed Low Speed Urban Maglev Development. The basic concept of the HSST is that it floats using an electromagnetic levitation system, propelled by the linear induction motor (LIM). Some of benefits of HSST include safety, environmental and aesthetic compatibility, ride comfort, low maintenance cost, and low guide-way construction cost. Hirasawa gave examples of HSST projects across the globe (east Nagoya line, Hiroshima Airport Line, Yakoma Dream Land Line in Japan, Foxwoods winter evaluation parking lot of the museum in Connecticut USA, Mexico City). He also briefly discussed about US/Japan cooperation and concluded his talk by suggesting that FTA can use Japan's low speed technology for its deployment and demonstration projects.

Fifth Speaker: Alan Danaher of Kittelson and Associates spoke on transit capacity and quality of service manual (TCQSM) they are developing under a grant from TCRP. According to Danaher, transportation profession currently lacks a consolidated and generally accepted set of transit capacity and quality of service definitions, principles, practices and procedures for planning, designing, and operating vehicles and facilities. Transit quality of service can be defined as measurements or predictions of how a transit route, facility, or system operates under specified demand, supply, and control conditions. After briefly speaking about a stake holders survey, Danaher discussed the various phases in the development of the manual. The first edition of the manual includes bus, streetcar, light rail, heavy rail, commuter rail, and automated guideway transit with potential future additions of ferries and minor modes (e.g., inclined plane, cable car, elevator). The manual mainly consists of Introduction, Concepts, Bus Transit Capacity, Rail Transit Capacity, Terminal Capacity, Quality of Service sections and finally Transit Glossary. Transit performance measures as well as transit system elements will also be discussed in the manual. TCQSM is expected to be published in mid 1999 and HCM2000 contains a subset of TCQSM material relating to on-street transit modes. Danaher's formal paper on this topic follows these summaries.

The final speaker of this final panel discussion of the conference was Michael Bolton of Multi-Systems, Inc. who spoke on ITS Applications in Mass Transit. Bolton basically discussed about kiosks and other new transit related technologies and concluded his presentation with a discussion on Hong Kong.

Development of Transit Capacity and Quality of Service Manual

Alan R. Danaher, P.E, AICP[1]

Abstract

The U.S. Transit Cooperative Research Program (TCRP) recently completed a two-year study to develop updated principles, practices, and procedures for transit capacity and quality of service. The study was conducted with the focus of developing a new *Transit Capacity and Quality of Service Manual,* which will be a comprehensive document summarizing transit capacity and quality of service concepts, procedures, and applications. Three other key components of the study included updating the transit chapters in the new year 2000 *Highway Capacity Manual* (HCM2000), identifying overall research needs in transit capacity and quality of service, and conducting research in identified areas.

The *Transit Capacity and Quality of Service Manual* (TCQSM) will be published in 1999, and will be in a format similar to the *Highway Capacity Manual.* There will be six sections of the document:

1. Introduction and Concepts,
2 Bus Transit Capacity,
3. Rail Transit Capacity,
4. Terminal Capacity,
5. Quality of Service, and
6. Glossary.

For each section of the TCQSM, procedures and applications related to different capacity/quality of service topics will be prepared. Sample problems are also presented.

----

[1]Principal Engineer, Kittelson & Associates, Inc., 610 SW Alder, Suite 700, Portland, OR 97205.

Introduction

The transportation profession in North America has lacked a consolidated and generally accepted set of transit capacity and quality of service definitions, principles, practices, and procedures for planning, designing, and operating vehicles and facilities. This is in contrast to the *Highway Capacity Manual* that defines quality of service and presents fundamental information and analytical procedures related to quality of service and capacity of highway facilities. In the absence of a comparable, authoritative document, the case for transit service and enhancements in a multi-modal environment is weakened. Therefore, there is a need for a *Transit Capacity and Quality of Service Manual.*

In 1995, the U.S. Federal Transit Administration (FTA) developed a research statement for a comprehensive research program on transit capacity and quality of service, which was submitted to TCRP for possible funding. The program included the development of a new *Transit Capacity and Quality of Service Manual,* and undertaking research on ten different transit capacity and quality of service topics. TCRP took this proposal and transformed it into the A-15 research project, *Development of Transit Capacity and Quality of Service Principles, Practices, and Procedures.* In addition to the preparation of a new TCQSM, the project included revisions to the transit chapters being prepared for the new year 2000 Highway Capacity Manual (HCM2000), as well as identifying a prioritized set of transit capacity and quality of service research needs, and undertaking the identified top priority research. Figure 1 illustrates the work tasks which were conducted in the A-15 project. The A-15 project was initiated in November 1996 and was completed in January 1999.

Input from Potential Users on the Content of the TCQSM

To help identify the desired content of a new *Transit Capacity and Quality of Service Manual* by potential users, a stakeholders survey was conducted at the outset of the A-15 project. Seven different groups were targeted, with a total of 411 questionnaires distributed:

- Large Urban Transit Systems (>200,000 population) - 181 surveyed,
- Small Urban Transit Systems (<200,000 population) - 105 surveyed,
- Paratransit Systems - 23 surveyed,
- Cities/Counties - 24 surveyed,
- Metropolitan Planning Organizations - 21 surveyed,
- State Departments of Transportation - 24 surveyed, and
- Professionals (Consultants, Academicians) - 23 surveyed.

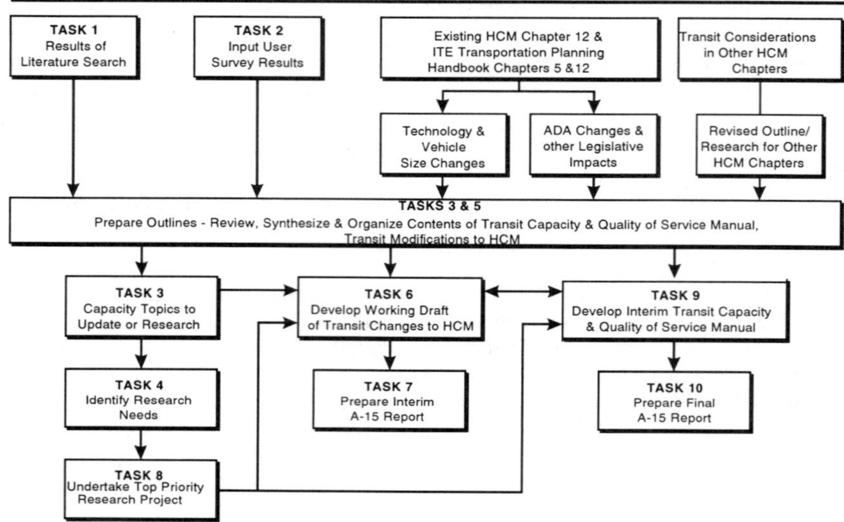

One of the questions asked was whether agencies or individuals used specific documents to analyze transit capacity and quality of service. The results are shown on Figure 2. The most frequently used document was the *Highway Capacity Manual* (23%). Many agencies not using published documents use in-house evaluation procedures. A third of the respondents indicated they do not have any procedures they use. Ninety-four (94%) of the survey respondents indicated they would use a new *Transit Capacity and Quality of Service Manual* if available.

A set of questions in the stakeholders survey related to the desired content of the new TCQSM. When asked if the TCQSM should have an "applications" theme, 79% answered "yes." Eighty-one percent (81%) agreed that the TCQSM should be structured similar to the *existing Highway Capacity Manual.* Ninety-one percent (91%) of the survey respondents agreed that examples illustrating different applications of transit capacity and quality of service analysis procedures should be included in the TCQSM.

In addition to the Stakeholders Survey, over 70 publications which addressed some element of transit capacity and/or quality of service were reviewed as part of the A-15 project, and summarized in an extensive annotated bibliography.

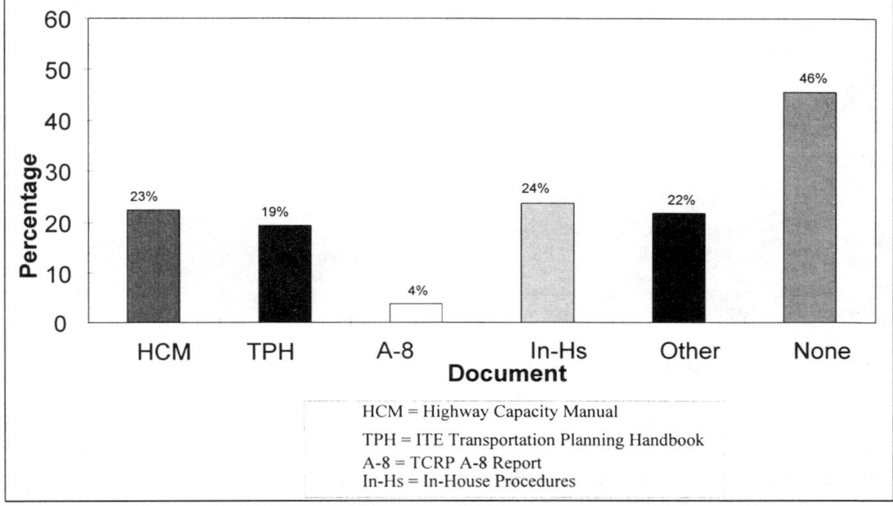

Figure 2 - Transit Capacity and Quality of Service Documents Used by Agencies

TCQSM Content Overview

The new *Transit Capacity and Quality of Service Manual* is intended to be a comprehensive guide for use by transit planners and engineers to find information on capacity characteristics of different types of transit systems, capacity analytical procedures, and quality and level of service concepts and measures. The manual will be a new TCRP Report publication, with the intent of being updated periodically under the guidance of the U.S. Transportation Research Board (TRB) Task Force on Transit Capacity and Quality of Service, which was established in 1998.

The TCQSM will address capacity and quality of service for the following transit modes:

Bus

- Fixed-Route
- Demand-Responsive

Rail

- Light Rail Transit/Streetcar
- Heavy Rapid Transit
- Commuter Rail
- Automated Guideway Transit.

Based on input from the A-15 stakeholders survey and discussion with TCRP and the research team, a detailed outline of the TCQSM was prepared. The TCQSM will be divided into six parts:

1. Introduction and Concepts,
2. Bus Transit Capacity,
3. Rail Transit Capacity,
4. Terminal Capacity,
5. Quality of Service, and
6. Glossary.

Over 200 exhibits will be included in the manual.

Relationship to Transit Provisions in the Highway Capacity Manual

In addition to the new *Transit Capacity and Quality of Service Manual,* transit provisions are proposed to remain in the *Highway Capacity Manual* (HCM), in particular the new HCM2000. Since 1985, there has been a chapter in the HCM devoted to transit (Chapter 12), which addresses transit vehicle and system characteristics, and associated capacity and level of service characteristics. In addition, the impact of transit on traffic operations of different highway facilities (freeway, arterials, signalized intersections) is addressed through adjustment factors in other chapters of the HCM. As part of the development of the new HCM2000, the Transit Subcommittee of the TRB Highway Capacity and Quality of Service Committee (which is responsible for updating the HCM), is coordinating transit modifications to the HCM2000.

Table 1 identifies the transit topics planned to be included in the HCM2000 vs. the *Transit Capacity and Quality of Service Manual.*

Part 1 - Concepts and Definitions

At the outset of the TCQSM, a general part on transit concepts and definitions will be provided. This part of the manual will provide the framework for the more detailed presentation of transit capacity and quality of service analytical procedures later in the manual. This part will define transit quality of service and level of service, and identify the different types of transit modes and service addressed in the manual. The concept of vehicle capacity vs. person capacity will also be addressed. The Table of Contents for the TCQSM is presented in Table 2.

Table 1 - Transit Topics Addressed in the *Highway Capacity Manual* vs. the *Transit Capacity and Quality of Service Manual*

Topics	**Highway Capacity Manual**	**Transit Capacity and Quality of Service Manual**
Glossary	X	X
General Concepts		
• Transit Capacity	X	X
• Transit Quality of Service	X	X
Vehicle Characteristics		
• Fixed-Route Bus		X
• Paratransit Bus		X
• LRT	X	X
• Rapid Transit	X	X
• Commuter Rail	X	X
• Advanced Guideway Transit		X
Transit System Data		
• Fixed-Route Bus	X	X
• Paratransit Bus	X	X
• Light Rail/Streetcar	X	X
• Rapid Transit		X
• Commuter Rail		X
• Advanced Guideway Transit		X
• Transit Preferential Treatments	X	X
Capacity/Quality of Service Scenarios		
• Station/Stop Waiting Area		X
• Vehicle Berthing		X
• On-Board Vehicle		X
• On-Line Operations	X	X
• System	X	X
• Simulation Modeling	X	X
Service Measures		
• Station/Stop Waiting Area		X
• Vehicle Berthing		X
• On-Board Operations	X	X
• System Coverage	X	X
• Effect of System Size	X	X
• Use of Transit Demand Models	X	X
Impact of Transit on Highway Operations		
• Freeway	X	
• Urban Arterial	X	
• Signalized Intersection	X	
• Unsignalized Intersection	X	
• Multi-Lane Highway	X	
• Rural 2-Lane Highway	X	
Sample Problems	X	X
References	X	X

Table 2 - *Transit Capacity and Quality of Service Manual* Table of Contents

Part 2 - Bus Transit Capacity

Analytical procedures to identify bus vehicle and person capacity will be addressed in this part of the manual. The procedures will address bus capacity for four different operating configurations:

1. segregated busway,
2. bus/HOV lane,
3. mixed traffic, and
4. demand-responsive.

Procedures for calculating passenger loading and bus dwell times at stops will be identified, as well as an assessment of the impacts of bus signal priority. The impact of different bus operating policies and the Americans with Disabilities Act on capacity will also be discussed.

Part 3 - Rail Transit Capacity

This part of the TCQSM will incorporate to a large part the recent TCRP A-8 Rail Transit Capacity research (Parkinson and Fisher, 1996). Rail capacity analytical procedures will be presented for rail rapid transit, light rail/streetcar, commuter rail, and automated guideway transit. Capacity factors such as train control and signaling, station dwells, passenger loading levels, and operating issues will be addressed.

Part 4 - Terminal Capacity

Capacity considerations related to bus stops and rail/bus stations will be presented in this part of the TCQSM. For bus stops, application of pedestrian level of service analysis procedures in identifying the required size of passenger waiting areas and adjoining sidewalks and crosswalks will be presented. Also procedures for determining the type and number of bus berths will be presented. At rail/bus stations, capacity characteristics and sizing of different terminal elements: platforms, escalators, elevators, stairways, walkways, fare gates, and ticket machines, will be addressed.

Part 5 - Quality of Service

The concept of transit quality of service will be addressed in detail in this part of the TCQSM. An overall framework of transit performance measures will be provided, along with selected service measures and how such measures should be applied to (1) policy and goal setting, (2) service assessment, and (3) facility planning and design.

Definitions

The definitions used for quality of service, level of service, performance measures, and service measures for the TCQSM are as follows:

Quality of Service. Quality of service is the overall measured or perceived quality of transit service form the user or passenger's point of view.
Transit Performance Measure. A quantitative or qualitative factor used to evaluate a particular aspect of transit service.
Transit Service Measure. A quantitative performance measure that best describes a particular aspect of transit service and represents the passenger's point of view.
Levels of Service. Six designated ranges of values for a particular service measure, graded from "A" (best) to "F" (worst), based on a transit passenger's perception of a particular aspect of transit service.

The proposed level of service definition is similar to the one applied in the *Highway Capacity Manual* (HCM). The TCRP A-15 stakeholder's survey also had transit professionals expressing a preference for this system.

Basic Framework

Four transit performance measure categories are proposed:

- *Availability.* Measures that address the spatial and temporal availability of transit service.
- *Quality.* Measures that address user comfort and convenience.
- *Capacity.* Measures that address the capacity of different components of a transit system.
- *Utilization.* Measures that address the usage of a transit system and the time required to serve passenger demand.

Measures in the first two categories, availability and quality, express the user's point-of-view; therefore, the TCQSM will present both service and performance measures for these categories. Measures in the last two categories, capacity and utilization, while important, generally do not directly express the user's point of view and, therefore, only performance measures will be provided.

It is proposed to use the term "availability" to refer to both the spatial and temporal availability of transit, to avoid confusion with the Americans with Disabilities Act usage of "accessibility." If transit service is located too far away from a potential user or it does not run at the times a user requires the service, that user would not consider transit service to be available. Assuming transit service is available to a potential user, the measures in the "quality" category can be used to evaluate a user's perception of the transit experience.

Different elements of a transit system require different performance measures. The following three categories have been developed:

- *Transit Stops.* Measures addressing transit capacity, utilization, and service quality at a single location. Depending on passenger volumes, scheduling and routing, and stop and station design, these measures will vary from one location to another.
- *Route Segments.* Measures addressing transit capacity, utilization, and service quality along a portion of a route (which can range from two stops to the entire length of a route). These measures should stay the same over the segment being analyzed regardless of conditions at an individual stop.
- *Systems.* Measures addressing transit capacity, utilization, and service quality for more than one route operating within a specified area (e.g., a district, city, or metropolitan area) or of a specified type (e.g., fixed-route vs. demand-responsive). System measures can also address door-to-door travel.

Combining the three transit system elements with the four performance measure categories produces the 3x4 matrix shown in *Table 3*. Proposed service measures are shown in capital letters, with other performance measures proposed to be discussed in the TCQSM shown in lower case.

Table 3 - Proposed Framework for Transit Performance Measures in the TCQSM

	Transit Stop	**Route Segment**	**System**
Availability	FREQUENCY passenger loading ped/bike access ADA accessibility	HOURS OF SERVICE auto access ADA accessibility	SERVICE COVERAGE % person-minutes of avail.
Quality	PASSENGER LOADING amenities reliability	RELIABILITY Transit/auto travel time Travel speed	TRANSIT/AUTO TRAVEL TIME travel time safety
Capacity	load factor d/c ratio failure rate (bus queuing)	seat miles/route mile d/c ratio	seat miles/capita
Utilization	passenger on/off volumes	Passenger throughput	passenger volume

ADA: Americans with Disabilities Act

Development of LOS Measures

The following four key considerations entered into the development of the proposed transit LOS system for the TCQSM:

1. *The transit LOS system should use an A-F scale*. This is based on the responses to the stakeholder survey during Phase 1 of the TCRP A-15 project. The benefits of this system are (1) decision-makers are already familiar with the A-F scale for highways, so less effort will need to be spent by transit agencies educating decision-makers, and (2) the public is already familiar with the A-F scale used for report cards.
2. *The measures should reflect a user's point-of-view.* LOS A, therefore, is not necessarily representative of optimum conditions from a transit operator's point-of-view.
3. *LOS F should represent an undesirable condition from a user's point-of-view.* A transit operator may choose to set higher standards based on their needs or policy goals.
4. *The thresholds for LOS A-E should represent points where a noticeable change in service quality occurs*. However, it is also desirable to have evenly-spaced ranges of values for each LOS grade, to the extent possible.

Part 6 - Glossary

The appendices will include a comprehensive glossary of transit terms described in the manual, as well as other common transit industry terms. Over 3,000 entries will be included.

Manual Style Guidelines

The new *Transit Capacity and Quality of Service Manual* will be in a format similar to the HCM2000. Basic style features for the manual include the following:

- metric and U.S. customary units (customary units in parentheses), and
- single-column text, with note column on the outside of the page.

In future updates of the TCQSM, it is likely that a CD-ROM component of the document will be prepared, which could be integrated with a software package including many of the analytical procedures identified in the manual.

Role of TRB Task Force on Transit Capacity and Quality of Service

In late 1996, the U.S. Transportation Research Board (TRB) established a new Task Force on Transit Capacity and Quality of Service. One of the missions of this task force in the future will be to shepherd the application and evolution of the *Transit Capacity and Quality of Service Manual,* and will be responsible for updates to the TCQSM. This task force will also promote research on transit capacity and quality of service topics. Strong liaison between the task force and other TRB committees will also be promoted.

References

Fitzpatrick, Kay; Hall, Kevin; Perkinson, Dennis; and Nowlin, Lewis (1996). Guidelines for the Location and Design of Bus Stops. *TCRP Report 19.* Transportation Research Board, Washington, D.C.

Garrity, Richard and Eads, Linda L. (1993). Bus Stop Accessibility: A Guide for Virginia Transit Systems and Public Entities for Complying with the Americans with Disabilities Act of 1990. *Transportation Research Record 1390.* Transportation Research Board, Washington, D.C.

Parksinson, Tom and Fisher, Ian (1996). Rail Transit Capacity. *TCRP Report 13.* Transportation Research Board, Washington, D.C.

Transportation Research Board (1994). Highway Capacity Manual. *TRB Special Report 209.* Washington, D.C.

Subject Index

Page number refers to the first page of paper

Author Index

Page number refers to the first page of paper